Active Electrical Distribution Network

A Smart Approach

Edited by

Baseem Khan
Hawassa University, Hawassa, Ethiopia

Josep M. Guerrero
Aalborg University, Denmark

Sanjeevikumar Padmanaban
Aarhus University, Denmark

Hassan Haes Alhelou
Tishreen University, Syria

Om Prakash Mahela
Power System Study Division, RVPN, Jaipur, India

Sudeep Tanwar
Nirma University, India

WILEY

This edition first published 2021

Registered Offices
John Wiley & Sons, Inc., 111 River Street, Hoboken, NJ 07030, USA
John Wiley & Sons Ltd, The Atrium, Southern Gate, Chichester, West Sussex, PO19 8SQ, UK

Editorial Office
The Atrium, Southern Gate, Chichester, West Sussex, PO19 8SQ, UK
For details of our global editorial offices, customer services, and more information about Wiley products visit us at www.wiley.com.

Wiley also publishes its books in a variety of electronic formats and by print-on-demand. Some content that appears in standard print versions of this book may not be available in other formats.

Library of Congress Cataloging-in-Publication Data
Names: Khan, Baseem, 1987- editor.
Title: Active electrical distribution network : a smart approach / edited by Baseem Khan, Hawassa University, Hawasa, Ethiopia, Josep M. Guerrero, Aalborg University, Denmark, Sanjeevikumar Padmanaban, Aalborg University, Denmark, Hassan Haes Alhelou, Tishreen University, Syria, Om Prakash Mahela, Power System Study Division, RVPN, Jaipur, India, Sudeep Tanwar, Nirma University, India.
Description: Hoboken, NJ, USA : John Wiley & Sons, 2021. | Includes bibliographical references and index.
Identifiers: LCCN 2020056560 (print) | LCCN 2020056561 (ebook) | ISBN 9781119599517 (hardback) | ISBN 9781119599456 (pdf) | ISBN 9781119599579 (epub) | ISBN 9781119599593 (ebook)
Subjects: LCSH: Smart power grids. | Electric power distribution.
Classification: LCC TK3105 .A35 2021 (print) | LCC TK3105 (ebook) | DDC 621.319--dc23
LC record available at https://lccn.loc.gov/2020056560
LC ebook record available at https://lccn.loc.gov/2020056561

Cover image: © petovarga/iStock/Getty Images
Cover design by Wiley

Set in 9.5/12.5pt STIX Two Text by Integra Software Services, Pondicherry, India
Printed and bound by CPI Group (UK) Ltd, Croydon, CR0 4YY

C9781119599517_280121

Table of Contents

Foreword

I am thrilled to address the foreword for an accomplished scholarly volume of the book entitled *Active Electrical Distribution Network: A Smart Approach*, as its purview and content provide economic and specialized activities with the fundamental factors for realizing and overseeing power and energy system approaches and for control strategies. The new upgrades in power and energy technology, notably in active distribution power system applications, control, optimization, renewables, energy management through interpretation, and simulation, provided a potentially massive market for energy operation and transformation. Advances in the latest handling of electrical engineering, including service and control, energy system security, energy efficiency, optimization, and renewable energy sources (wind, photovoltaic, fuel-cell, biogas/biomass, etc.), have contributed to the solution of new energy issues for any community and real problems. However, now the main claim is to retain high energy efficiency during application and to examine sustainable energy reserves. For trading with professional obstacles, numerous large spaces in power distribution technologies are being classified in this book, prominent ones being:

1. *Harmonics Mitigation in a Smart Distribution Network*
2. *Energy Control of the Active Distribution Network*
3. *Phasor Measurement Unit Placement*
4. *Smart Microgrid Integration and Optimization*
5. *Electric Vehicle Technology*
6. *Reconfiguration of the Smart Distribution Network*
7. *Demand Side Management Tools and the Smart Home Energy Management System*
8. *Smart Meter Technology*

This book gives answers to many challenging questions dealing with electrical distribution power and system optimization. It forwards an assortment of issues pertained to energy control and newly developed optimization routines. It is comprised of 22 chapters divided into eleven parts, which deal with material according to the raised sections. I recommend this book to investigators and professionals from the field of smart grid research and also for experts and engineers engaged in power system activity and management. I acknowledge the struggles of all the Editors to organize this book. The managing editors, Dr. Baseem Khan, Prof. Josep M. Guerrero, Prof. Sanjeevikumar Padmanaban, Dr. Hassan

Haes Alhelou, Dr. Om Prakash Mahela, and Dr. Sudeep Tanwer, have assiduously collected the chapters and evaluated and arranged them properly for further in-depth learning. I trust the readers will benefit from the power and energy system offered in this book as an explanation for contemporary power and energy transformation for Green Engineering for All.

Dr. Jens Bo Holm-Nielsen
Head of the Center for Bioenergy and Green Engineering
Aalborg University
Department of Energy Technology
Esbjerg, Denmark

Preface

In recent years, the electricity industry has been transformed from a regulated to a deregulated structure. This transformation introduced efficacy and market competition in the energy industry. Different sectors of the electricity industry are deregulated and developed into separate entities such as GENCO, TRANSCO, and DISCOMS. Further, smart grid technologies are going to be implemented in the electricity sector in various parts of the world. These technologies in the generation, transmission, and distribution sectors will supply the rapidly increasing energy demand. Such technologies are changing the modern energy systems very rapidly with increasingly multifaceted features. In the present energy sector, power generation, transmission, and distribution are not only the main functions of the energy utilities as system stability, reliability, efficacy, and security have become the major issues. These issues help energy utilities to manage, control, and operate the generation sources on a long-term basis, by reducing their effects on ecology and the environment. Planning and operation of the energy sector requires efficiency for maximizing the advantages for the utilities and also the consumers. Further, energy saving can be in a form of power generation that contributes to minimization of carbon emissions in the globe's climate. In order to solve these problems, various techniques are continuously emerging. Implementation of these smart grid technologies supports exploration of renewable energy sources for their optimum utilization. However, there is still a huge gap between existing power generation and load demand. To meet this energy gap utilities are having to depend on traditional reliable energy sources along with intermittent renewable energy sources such as solar, wind, fuel cells, etc. Thus, for reliable operation of the electricity industry traditional sources of energy production are still used in various parts of the world. The dependence on these traditional generation sources creates a problem for the environment. Thus, utilization of the renewable energy sources and techniques for supplying the demand is an important issue for enhancing energy security. In this regard, green energy sources have attracted the attention of researchers from all over the globe. These sources will enhance high integration of renewable energy sources in the utility grid. The main issue with these renewable energy sources is their variable nature of generation, which depends on environmental conditions. Due to this, energy generation from solar and wind sources fluctuate in nature and directly affect generation frequency, voltage, and waveform and therefore affect the quality and amount of supply of energy to the integrated grid system. It is expected that the energy demand will increase by 70% from the present demand in the year 2030. However, the traditional energy sources are continuing to deplete day by day, which is a grave concern of energy usage per capita. The energy consumption per capita describes the living standard of a country. Thus it is essential that more renewable energy sources must be found that are locally available with optimization methods for enhancing energy generation efficiency. The available issues of the active power distribution system clearly indicate that the available techniques of planning, control, and operation require more efficient solutions.

This book presents various issues of an active electrical distribution network with solutions provided. Further, the potential of an engineering-based approach for a renewable energy system will systemically quantify the various options at several levels of distribution network planning, control, and operation via modeling, simulation, control, and optimization-dependent frameworks. The successful implementation of these approaches in a number of real-life case studies further highlights the significance of this integrated system-wide approach. This book presents the importance of fundamental and applied research in an active electricity distribution system by developing mechanisms for the transfer of the new methodology, which is applicable to the real-time problem.

Description of the Book

This book aims to be an essential source, building on available literature in the field of active distribution network planning, operation, and control, providing further research opportunities in this field. This specific text is expected to provide the primary and major resources necessary for researchers, academics, students, faculties, and scientists across the globe to adopt and implement new inventions in vital areas of traditional and active smart distribution networks and their utilization and energy management. Therefore, the *Active Electrical Distribution Network: A Smart Approach* should provide a platform to share up-to-date scientific achievements in the field as well as related fields. The book contains 22 chapters of high-quality contributions from international leading researchers in the field of power and energy systems. These chapters are divided into 11 sections depending upon their scope. The following list gives the main objectives:

1. Introduction of conventional and smart electrical distribution network.
2. Identified and explored the existing issues in a traditional electrical distribution network.
3. Identified and explored the harmonic mitigation issues in a smart distribution network.
4. Identified and explored the various techniques required to move towards smart distribution of electrical energy.
5. Identified and explored the energy management in an active distribution network.
6. Identified and investigated the optimal placement problem of a phasor measurement unit.
7. Identified and investigated the smart microgrid integration and optimization in the active distribution network.

8. Identified and investigated the integration of the electric vehicle technology in the active distribution network.
9. Identified and explored the reconfiguration of the smart distribution network.
10. Identified and explored the demand side management mechanisms and smart home energy management system in the active distribution network.
11. Identified and explored the smart meter technology in the active distribution network.

This book aims to be an essential reference source, building on the available literature in the field of modern and future active electrical distribution network operation and control, providing further research opportunities in this field. This specific text is expected to provide the primary and major resources necessary for researchers, academicians, students, faculties, and scientists, across the globe, to adopt and implement new inventions in the thrust area of active electrical distribution networks and their utilization and energy management. According to the application domain and nature, the chapters in this book are categorized into the following parts:

Part I: Electrical Distribution Network: Conventional vs Smart
Part II: Existing Issues in the Electrical Distribution Network
Part III: Harmonics Mitigation in the Smart Distribution Network
Part IV: Towards Smart Distribution of Electrical Energy
Part V: Energy Management of an Active Distribution Network
Part VI: Phasor Measurement Unit Placement
Part VII: Smart Microgrid Integration and Optimization
Part VIII: Electric Vehicle Technology
Part IX: Reconfiguration of a Smart Distribution Network
Part X: Demand Side Management Mechanisms and a Smart Home Energy Management System
Part XI: Smart Meter Technology

From recent trends, it can be observed that research is mainly focused on an active electric distribution network, its structures, operation and control, renewable energy resource integration, power quality issues with renewable energy sources, and integration issues of electric vehicles and a smart metering system.

This book will be a source of motivation to the researcher and can provide an exchange of knowledge of energy efficiency, power quality, reconfiguration, demand side management, smart home energy management systems, electric vehicle technology, smart metering, and integration of renewable energy sources for better utilization of the active electric distribution network. In order to meet the load demand and to improve the efficiency of energy management, researchers are working in the area of integration of renewable energy sources, electric vehicle, smart meter, optimized operation of the microgrid system, and demand side management for the development of an active electrical distribution network.

A brief description of the chapters in this book is given as follows:

Chapter 1 discussed various electricity distribution structures and business models considering smart grid perspectives.
Chapter 2 discussed the distribution feeder segregation issue of the electrical distribution network.
Chapter 3 discussed the different technical, economical, and environmental issues associated with the traditional electrical distribution network.
Chapter 4 proposed a power quality mitigation technique in a distribution network using a battery energy storage system.
Chapter 5 performed grid power quality improvement using a bi-directional off-board electric vehicle battery charger in a smart city scenario.
Chapter 6 discussed the smart distribution of electrical energy.
Chapter 7 investigated the active distribution management system.
Chapter 8 discussed the role of Volt-VAr-W control in energy management.
Chapter 9 proposed active management of distribution networks.
Chapter 10 enhanced the performance of the state estimation algorithm through an optimally placed phasor measurement unit.
Chapter 11 discussed the integration and optimization of the smart microgrid system.
Chapter 12 proposed control algorithms for energy storage systems to reduce distribution power loss of microgrids.
Chapter 13 discussed various challenges and solution techniques for higher levels of wind energy penetration into the remote grid.
Chapter 14 investigated the forecasting aspects of the Internet of Things and machine learning for improving solar PV plant efficiency.
Chapter 15 proposed the modular design of nonlinear controllers for photovoltaic distributed generation systems.
Chapter 16 discussed vehicle-to-grid challenges and potential benefits for smart microgrids.
Chapter 17 discussed various test systems for the reconfiguration of radial distribution networks.
Chapter 18 investigated the different case studies for the distribution system reconfiguration.
Chapter 19 presented the genetic algorithm application in the distribution system reconfiguration.
Chapter 20 discussed demand response techniques and smart home energy management.
Chapter 21 discussed a sustainable building lightning solution for energy conservation in different geographical conditions.
Chapter 22 discussed the smart metering infrastructure for the smart grid system.

Acknowledgments

This book, *Active Electrical Distribution Network: A Smart Approach*, is the outcome of the inspiration and encouragement given by many individuals for whom these words of thanks are only a token of our gratitude and appreciation for them.

Our sincere gratitude goes to the peoples who contributed their time and expertise to this book. We highly appreciate their efforts in achieving this project. The editors would like to acknowledge the help of all the people involved in this project and, more specifically, the editors would like to thank each one of the authors for their contributions and the editorial board/reviewers regarding the improvement of quality, coherence, and the content presentation of this book.

Second, the editors would like to express their sincere thanks to Juliet Booker, Becky Cowan, and Sandra Grayson of Wiley Publishing Group for their continuous support and giving us an opportunity to edit this book.

The editors are thankful to our family members for their prayers, encouragement, and care shown towards us during the completion of this book. Thank you all.

We also express our gratitude to the GOD for all the blessings!!!

Baseem Khan
Hawassa University, Hawassa, Ethiopia
Josep M. Guerrero
Aalborg University, Aalborg East, Denmark
Sanjeevikumar Padmanaban
Aarhus University, Herning, Denmark
Hassan Haes Alhelou
Tishreen University, Latakia, Syria
Om Prakash Mahela
Power System Study Division, RVPN, Jaipur, Rajasthan, India
Sudeep Tanwar
Nirma University, Ahmedabad, India

List of Contributors

Hassan Haes Alhelou
Tishreen University, Syria

Sanjana Babu
Vellore Institute of Technology, Vellore, India

Baidyanath Bag
National Institute of Technology, Raipur, India

Laxmidhar Behera
Indian Institute of Technology Kanpur, Uttar Pradesh, India

Bharat Bhatia
Manipal University, Jaipur, India

Gajendra Singh Chawda
Indian Institute of Technology, Jodhpur, Rajasthan, India

Sajjad Fattaheian-Dehkordi
Sharif University of Technology, Iran

Hooman Firoozi
School of Technology and Innovations, University of Vaasa, Vaasa, Finland

Man Mohan Garg
Malaviya National Institute of Technology, Jaipur, Rajasthan, India

Mohammad Gholami
Sharif University of Technology, Iran

Esayas Gidey
Hawassa University, Ethiopia

Sunil Kumar Goyal
Manipal University, Jaipur, India

Josep M. Guerrero
Aalborg University, Denmark

Shu-Yuen Ron Hui
Imperial College London, UK

Sanjay Jain
R.K.D.F. University, Bhopal, M.P, India

Siddharth Jain
Vellore Institute of Technology, Vellore, India

Shilpa Kalambe
Dr. Babasaheb Ambedkar College of Engineering and Research, Nagpur, India

Bhojraj N. Kale
Dr. Babasaheb Ambedkar College of Engineering and Research, Nagpur, India

Hosna Khajeh
School of Technology and Innovations, University of Vaasa, Vaasa, Finland

Baseem Khan
Hawassa University, Ethiopia

Meher Preetam Korukonda
Indian Institute of Technology Kanpur, Uttar Pradesh, India

Pankaj Kumar
Department of Electrical Engineering, Rajasthan Technical University, Kota, Rajasthan, India

Hannu Laaksonen
School of Technology and Innovations, University of Vaasa, Vaasa, Finland

Rajesh Kumar Lenka
National Institute of Technology Roukela, Odisha, India

Meisam Mahdavi
São Paulo State University, Brazil

Om Prakash Mahela
Power System Planning Division, Rajya Vidyut Prasaran Nigam Ltd, Jaipur, Rajasthan, India

Ujwala Malkhandale
Priyadarshini Indira Gandhi College of Engineering, Nagpur, M.S., India

Hesam Mazaheri
Sharif University of Technology, Iran

Ashwini Ramachandran Nair
Vellore Institute of Technology, Vellore, India

Kinfe Negash
Hawassa University, Ethiopia

Mohamad Amin Rajabi Nezhad
Power Quality Research Lab, Iran University of Science and Technology, Tehran, Iran

Anup Kumar Panda
National Institute of Technology Roukela, Odisha, India

Bhuvnesh Rathore
MBM, Jodhpur, Rajasthan, India

Mohd Tayyab Saeed
Al-Mostajed Technologies Co. LLC, United Arab Emirates

P. Sanjeevikumar
Aarhus University, Denmark

Amit Saraswat
Manipal University, Jaipur, India

Yashwant Sawle
Vellore Institute of Technology, Vellore, India

Abdul Gafoor Shaik
Department of Electrical Engineering, Indian Institute of Technology, Jodhpur, Rajasthan, India

Pierluigi Siano
University of Salerno, Italy

Bhuwan Pratap Singh
Manipal University, Jaipur, India

Siew-Chong Tan
The University of Hong Kong, Hong Kong

M. Thirunavukkarasu
Vellore Institute of Technology, Vellore, India

Yun Yang
The Hong Kong Polytechnic University, Hong Kong

List of Abbreviations

AAC	Awareness and advertisement cell
ABB	ASEA Brown Boveri
ABR	Average billing rate
AC	Alternating current
ACO	Ant colony optimization
ADC	Analog to digital converter
ADE	Adaptive differential evolution
ADN	Active distribution network
ADSL	Asymmetric digital subscriber line
AFE	Analog front end
aFRR	Automatic frequency restoration reserve
AGC	Automatic generation control
AI	Artificial intelligence
AIS-ACO	Artificial immune systems and ant colony optimization
AMI	Advanced metering infrastructure
AMR	Automated meter reading
AMS	Advanced measurement station
ANFIS	Adaptive neural fuzzy inference system
ANM	Active network management
ANN	Artificial neural network
ANSI	American National Standards Institute
API	Application programming interface
ASCII	American Standard Code for Information Interchange
AST	Advanced systems technology
AT & C Losses	aggregate technical and commercial losses
AUTOCAD	Automatical computer aided design
BAN	Body area network
BBDC	Back-end DC–DC bidirectional converter
BCGA	Binary godification GA

BD	Big data
BEE	Bureau of energy efficiency
BES	Battery energy storage
BESS	Battery energy storage system
BEV	Battery electric vehicle
BOOT	Build–own–operate–transfer
BPL	Broadband over power lines
BPO	Business process outsourcing
CAES	Compressed air energy storage
CAESS	Compressed air energy storage system
CAIDI	Customer average interruption duration index
CAISO	California Independent System Operator
CAPEX	Capital expenditure
CA	Control action
CC	Constant current
CCM	Current control mode
CCOM	Communication Company
CCS	Combined Charging System
CEN	Comité Européen de Normalization
CENELEC	European Committee for Electrotechnical Standardization
CHAdeMO	CHArge de MOve
CHBDC	Cascaded H-bridge bi-directional converter
CHP	Combined heat and power
CIS	Customer information system
CL	Critical load
CMPC	Centralized model prediction control
CORMU	Control–Operation–Revenue Management Unit
COSEM	Companion specification for energy metering
CR	Condensation resistance
CTC	Communication technology consultant
CV	Constant voltage
CVR	Conservation voltage reduction
C-WET	Centre for wind energy technology
DA	Distribution automation
DAC	Digital-to-analog converter
DBS	Distributed battery systems
DC	Direct current
DCC	Distribution control center
DCES	Direct current energy storage
DCGA	Decimal Codification GA
DDS	Data Distribution Service
DDUGJY	Deen Dayal Upadhyay Gram Jyoti Yojana
DER	Distributed energy resource
DF	Deterministic forecast
DF	Distribution franchisees

DFACTS	Distributed FACTS
DFIG	Doubly fed induction generator
DFR	Distribution feeder reconfiguration
DG	Distributed generation
DiG	Diesel generator
DISCOM	Distribution company
DLMS	Device language message specification
DMMU	Distribution management and maintenance unit
DMS	Distribution management system
DNO	Distribution network operators
DOE	Department of Energy
DoS	Denial-of-service
DPTr	Distribution power trader
DR	Demand response
DRES	Distributed renewable energy source
DSA	Distribution system automation
DSG	Dispersed storage and generation
DSL	Digital subscriber line
DSM	Demand side management
DSO	Distribution system operator
DSP	Digital signal processor
DSR	Distribution system reconfiguration
DSTATCOM	Distribution static compensator
DTrF	Distribution transformer franchisee
DVD	Digital versatile disc
DVTr	Distribution VAR trader
EA	Evolutionary algorithm
EC	European Commission
EDCGA	Efficient DCGA
EISA	Energy Independence and Security Act
EMS	Energy management system
EPA	Environmental Protection Agency
ERCOT	Electric Reliability Council of Texas
ERP	Enterprise resource planning
ES	Electric spring
ESCO	Energy Service Company
ESR	Equivalent series resistance
ESS	Energy storage system
ETAP	Electrical Transient Analyzer Program
ETSI	European Telecommunications Standards Institute
EV	Electric vehicle
EVSE	Electric vehicle service equipment
FACTS	Flexible alternating current transmission system
FAN	Field area network
FC	Fuel cell

FCEV	Fuel cell electric vehicle
FCI	Faulted circuit indicator
FCR	Frequency containment reserve
FCR-D	Frequency containment reserve for disturbance condition
FCR-N	Frequency containment reserve for normal condition
FDLF	Fast decoupled load flow
FERs	Flexible energy resources
FESS	Flywheel energy storage system
FF	Feeder franchisee
FFR	Fast frequency reserve
FLC	Fuzzy logic control
FLISR	Fault location, isolation, and service restoration
FO	Fiber optics
FOR	Forum of regulators
FRR	Frequency restoration reserve
FSP	Feeder Segregation Plan
G2V	Grid to vehicle
GA	Genetic algorithm
GCHC	Grid current harmonic compensation
GDP	Gross domestic product
GENCO	Generation company
GEP	Generation expansion planning
GIS	Geographical information system
GNP	Gross national product
GPRS	General packet radio service
GPS	Global positioning system
GRIHA	Green rating for integrated habitat assessment
GSC	Grid side converter
GTEP	Generation–transmission expansion planning
GUI	Graphical user interface
GW	Gigawatts
HAN	Home area network
HEMS	Home energy management system
HSA	Harmony search algorithm
HT	High tension
HV	High voltage
HVAC	Heating, ventilation, and air conditioning
HVAC	High-voltage alternating current transmission
IAM	Integrated assessment model
IAN	Internet area network
IBDF	Input-based distribution franchisee
ICCT	Indirect current control theory
ICE	Internal combustion engine
ICOM	Insurance company
ICT	Information and communication technology

IDCGA	Improved DCGA
IEC	International Electrotechnical Commission
IED	Intelligent electronic device
IEEE	Institute of Electrical and Electronics Engineers
IEM	Intelligent energy management
IFR	Inertial frequency response
IGBC	Indian Green Building Council
ILP	Integer linear programming
IM	Induction Motor
IMPROGRES	Improvement of the social optimal outcome of market integration of DG/RES in European electricity markets
IoT	Internet of Things
IP	Interoperability
IR	Independent regulator
IRP	Integrated resource planning
ISM	Bands Industrial, scientific, and medical bands
ISO	International Organization for Standardization
ISO	Independent system operator
ISO-NE ISO	New England
ITC IT	consultant
ITCOM	Information technology company
ITU	International Telecommunication Union
KCL	Kirchhoff's current law
KVL	Kirchhoff's voltage law
LAN	Local area network
LCD	Liquid crystal display
LED	Light emitting diode
LEED	Leadership in energy and environment design
Li-Ion	Lithium-ion battery
LSGR	Light-to-solar gain ratio
LTE	Long-term evolution
LV	Low voltage
LVRT	Low-voltage ride-through
MAIFI	Momentary average interruption frequency index
MAN	Metropolitan area network
MAS	Multi-agent system
MATLAB	Matrix Laboratory
MCOM	Metering companies
MDMS	Meter data management system
MEDA	Maharashtra Energy Development Agency
mFRR	Manual FRR
MG	Microgrid
MGCC	Microgrid control center
MHBMO	Modified honey bee mating optimization
MHBO	Modified honey bee optimization

MISO	Midcontinent independent system operator
ML	Machine learning
MPC	Model predictive control
MPPT	Maximum power point tracking
MT	Micro-turbine
MTC	Metering technology consultants
MV	Medium voltage
MVA	Mega-volt amperes
NAN	Neighborhood area network
NE	North-East
NGO	Non-governmental organization
NISE	National Institute of Solar Energy
NIST	National Institute of Standards and Technology
NL	Non-critical loads
NR	Network reconfiguration
NR	Newton Raphson method
NT	Number of hours
NW	North-West
NYISO	New York independent system operator
O&M	Operation and maintenance
OCP	Optimal capacitor placement
OCPP	Open Charge Point Protocol
OLTC	On-load tap changer
OM	Operation and maintenance
OMS	Operating and management system
ONR	Optimal network reconfiguration
OpenDSS	Open distribution system simulator
Opex	Operational expenditure
OPP	Optimal PMU placement
PAFC	Phosphoric acid fuel pellets
PCC	Point of common coupling
PET	Power education toolbox
PEV	Plug-in electric vehicle
PFA	Project Finance Agency
PFR	Power flow router
PGA	Programmable gain amplifier
PGSA	Plant growth simulation algorithm
PHESS	Pumped hydroelectric energy storage system
PHEV	Plug-in battery electric vehicle
PHS	Pumped hydro storage
PI	Proportional–integral
PIU	Physical infrastructure unit
PJM	Pennsylvania–New Jersey–Maryland interconnection
PLC	Power line carrier
PLL	Phase-locked loop

PMU	Phasor measurement unit
PON	Passive optical network
PP&L	Pennsylvania power and light company
PPP	Public–private partnership
PQ	Power quality
P-RTU	Pole top units
PSAT	Power system analysis toolbox
PSO	Particle swarm optimization
PV	Photovoltaic
PVC	Polyvinyl chloride
QKD	Quantum key distribution
QoS	Quality of service
RE	Renewable energy
RER	Renewable energy resource
RESS	Renewable energies and systems
RME	Revenue management engine
RMS	Root-mean-square
RPC	Reactive power compensation
RSC	Rotor-side converter
RTDS	Real-time digital simulator
RTU	Remote terminal unit
SAIDI	System average interruption duration index
SAIFI	System average interruption frequency index
SAPSO	Self-adaptive particle swarm optimization
SARSA	State-action-reward-state action
SAS	Statistical analysis system
SC	Shading coefficient
SC	Supercapacitor
SCADA	Supervisory control and data acquisition system
SCC	Short circuit current
SCC	Social cost of carbon
SCIG	Squirrel-cage induction generator
SDH	Synchronous digital hierarchy
SE	South-East
SE	State estimation
SFLA	Shuffled frog-leaping algorithm
SG	Smart grid
SGAM	Smart grid architecture model
SGIP	Smart grid interoperability platform
SH	Sampled and holds
SHG	Self-help group
SHGC	Solar heat gain coefficient
SI	Swarm intelligence
SL	Smart loads
SMES	Superconducting magnetic energy storage system

SNE	Social networking engine
SOC	System-on-chip
SPP	South-west power pool
SPV	Solar photovoltaic
SPVDG	Solar photovoltaic distributed generation systems
SRRA	Solar radiation resource assessment
STATCOM	Static synchronous compensator
STLF	Short-term load forecasting
SW	South-west
T&D	Transmission and distribution
Taipower	Taiwan Power Company
TCP/IP	Transmission control protocol/Internet protocol
TCPD	Total cost per day
TEP	Transmission expansion planning
THD	Total harmonic distortion
TI	Training Institute
ToD	Time of day
Totex	Total expenditure
ToU	Time of usage
TPC	Taiwan Power Company
TRC	Trade and retail cell
TSA	Tabu search algorithm
TSO	Transmission system operator
UA	Uncertainty analysis
UPF	Unity power factor
UPS	Uninterruptible power supply
USA	United States of America
V2G	Vehicle to grid
VCM	Voltage-controlled mode
VDSL	Very high bit rate digital subscriber line
VEIC	Vermont Energy Investment Corporation
VGI	Vehicle grid integrated
VPP	Virtual power plant
VRB	Vanadium redox batteries
VSC	Voltage source Converter
VSI	Voltage stability index
VSI	Voltage source inverter
VT	Visible transmittance
VVO	Volt-VAR optimization
WAMS	Wide area monitoring system
WAN	Wide area network
WDM	Wavelength multiplexing
WE	Wind energy
WEG	Wind energy generating system
WiMAX	Worldwide interoperability for microwave access

WLAN	Wireless local area network
WLS	Weighted least squares
WMO	World Meteorological Organisation
WPAN	Wireless personal area network
WRIG	Wound rotor induction generators
WRR	World radiometric reference
WV	Wind velocity
ZBr	Zinc bromine
ZC	Zero crossings

Part I

Electrical Distribution Network: Conventional vs Smart

1

Electricity Distribution Structures and Business Models Considering Smart Grid Perspectives

Baidyanath Bag

Department of Electrical Engineering, National Institute of Technology Raipur, Chhattisgarh, India

1.1 Introduction

The deregulation process has faced many difficulties and various problematic sectors are still being explored and resolved [1–5]. A business model describes the strategy by which an organization creates, delivers, and captures values. A suitable business model articulates the value created to users by using the product or service. It also identifies the market segment, defines the value chain to estimate the cost structure and profit earning ability, judges the position of the firm with respect to other competitors, and provides the competitive strategy to the firm to build up its identity. Because after manufacture a product is sold in a market, after being generated electricity is also sold like a market product. Hence a business model is also necessary for electricity distribution in order to articulate its value position. The process of delivery of electrical energy from the point of its generation up to the point of consumption needs a business strategy in each and every step of its production. The major parts of the business in the electrical sector lie with generation, transmission, and distribution processes. Of these three major business procedures in the power sector, distribution is the most growing concern sector in the smart grid environment, where the motivation to hold a business model is to be found in each and every step of the job needed to carry out the task of power distribution among consumers. In the power distribution sector the major functions lies with:

- Infrastructure creation and its operation and maintenance.
- Creation of an organizational structure.
- Collection of big data and its analysis.
- Policy framing and framing different processes.
- Managing work, asset, and demand.
- Developing different models related to finance, business, system, and technology.
- Pricing, billing, collection, and awareness related activities.

An opportunity exists for a distinct business model for each of the above activities. Hence there is an essential need to find a suitable business model that will divide all the above actions into a distinct business. As the major concern of the changes under smart

Active Electrical Distribution Network: A Smart Approach, First Edition. Edited by Baseem Khan, Josep M. Guerrero, Sanjeevikumar Padmanaban, Hassan Haes Alhelou, Om Prakash Mahela, and Sudeep Tanwar.

grid initiatives exists in the power distribution sector and the varieties of these changes are large in number corresponding to their required expertise, a suitable business model for the power distribution sector is an utmost requirement to ensure smart grid implementation.

Various forms of business models adopted to carry out the task of power distribution across various parts of the globe have been summarized along with their merits and demerits. Considering the various functions and the various issues or challenges present in the power distribution sector under a smart grid environment, a new business model described by Jamasb, Thakur, and Bag [6] that has not been adopted by any utility is also discussed in this chapter. This new business model is expected to overcome all the challenges present while fulfilling all future desires of the power distribution sector.

1.2 Importance of the Business Model in the Smart Grid

The smart grid is more technology oriented as compared to the traditional grid. Value from new technology can be captured by a firm in two basic ways: through incorporating the technology in their current businesses or through launching new ventures that exploit the technology in new business arenas. Unlike other tangible products sold in the market that can be stored and sold at a later date, electricity cannot be stored and should be consumed as soon as it is generated. Hence, for other tangible products that are able to capture the value from new technology through storage, electrical businesses must use another way to capture value from new technologies that is more suitable for an electrical business. However, power distribution utilities function under the traditional grid operation mode and are following the trend of technology adoption through storage and hence are failing to capture the values from the newly developed technologies during their transition toward the smart grid. Thus, an alternative way is an essential need under the smart grid operation mode. Various technologies have been developed that could be useful for pursuing different tasks in the distribution business will have to be taken to the market through a venture shaped by a specific business model.

Unless and until it is commercialized, the inherent value of a technology remains latent. A correct business model helps in the commercialization of a new technology and may become the architecture of new revenue. If a firm fails to select a correct business model, technology will yield less value to the firm than might otherwise have been the case. If the firm fails consistently in selecting an appropriate business model, it fails its various commitments. Because of the non-availability of an appropriate business model for the power distribution sector, ideal technologies developed especially for the sector are not being utilized properly toward developing the sector. Various Information Communication Technology (ICT)-based technologies that have been developed especially for the sector are not being properly utilized because of the unavailability of an appropriate business model. Time of day (ToD) or time of usage (ToU) pricing, average billing rate (ABR) fixing, theft control, etc., are being designed through technology implementation, but because of the absence of a right business model, these technologies are inactive in fulfilling their objectives of demand side management.

Similarly, adequate technological developments have already been done for reactive power management and voltage stability control, but they are still failing to meet their objectives because of the absence of a correct business model. Again, the commercial losses (excluding theft) taking place in the sector are very high and cannot be addressed through technology development, although they can be addressed through implementation of a correct business model.

1.3 Electricity Distribution Structures and Business Models

In most countries, transmission systems have been expanded in a planned way based on detailed technical studies, but there is an unplanned and haphazard expansion in the distribution sector with the objective of meeting the growing urgent consumer demand. Due to this reason, the distribution system has become inefficient, leading to the following technical and governance challenges:

- Technical challenges: huge aggregate technical and commercial (AT&C) losses, poor power quality, low reliability, etc.
- Governance challenges: problems with things related to operation and maintenance, adequate metering, dealing with cross subsidy, energy billing, activity outsourcing, improving financial status of distribution utilities, mechanisms for attracting private investors, etc.

Hence, many distribution utilities across different countries are characterized as inefficient with a low productive system. To overcome technical- as well as governance-related difficulties, basic and radical restructuring of the power sector has become imperative [7]. Unbundling, corporatization, privatization, and outsourcing were the major initiatives under restructuring. As a result, four different models to carry out the function of power distribution have evolved across the globe, viz. government-owned Distribution Company, privately owned Distribution Company, Public–Private–Partnership Model, and Distribution Franchisee, as shown in Figure 1.1.

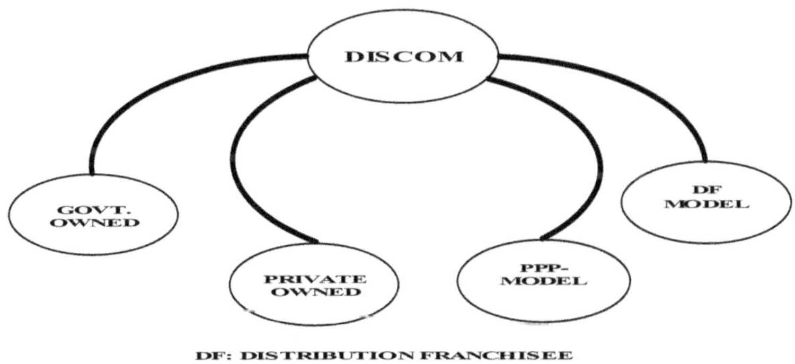

Figure 1.1 Different distribution models evolved.

1.3.1 Government-Owned DISCOM

In this case, ownership of the distribution company (DISCOM) lies with the government. Government-owned DISCOM can again be categorized into two categories, a single vertical structure and an unbundled (corporatized) structure. In the case of a single vertical structure, a single entity that is government owned performs all the tasks of power generation, transmission, and distribution. However, following the trends of reform, unbundling this single vertical structure led to segregation of generation, transmission, and distribution as distinct companies. Again this segregation was done either to single generation, single transmission, and single distribution companies or to multiple generation, multiple transmission, and multiple distribution companies. Some utilities unbundled their single vertical structure into two entities – one dedicated to transmission and another dedicated to both generation and distribution. Similarly, some utilities unbundled their single vertical structure into two entities – one dedicated to generation and another dedicated to both transmission and distribution. Figures 1.2 and 1.3 show various forms of government owned distribution companies.

From Figure 1.2, it is clear that the task of generation, transmission, and distribution is under the purview of a single entity, irrespective of whether there are single or multiple subsidiaries in generation or transmission or distribution.

Figure 1.3 consists of six different blocks where all the six blocks represent a distinct pattern of unbundled power distribution structure. In the first block, more than one subsidiary are being shown for both the generation and distribution segments whereas transmission has been shown as a single subsidiary because most of the countries kept a single transmission utility while undergoing unbundling. Although a vertical integrated structure and unbundled (corporatized) structures are different in structures no significant differences are found in their operations and functions. The relative concerns relating to this model are:

- High technical losses in distribution.
- No theft handling mechanism.
- No proper fault handling mechanism.
- Lack of sourcing and management.
- Overstaffing.

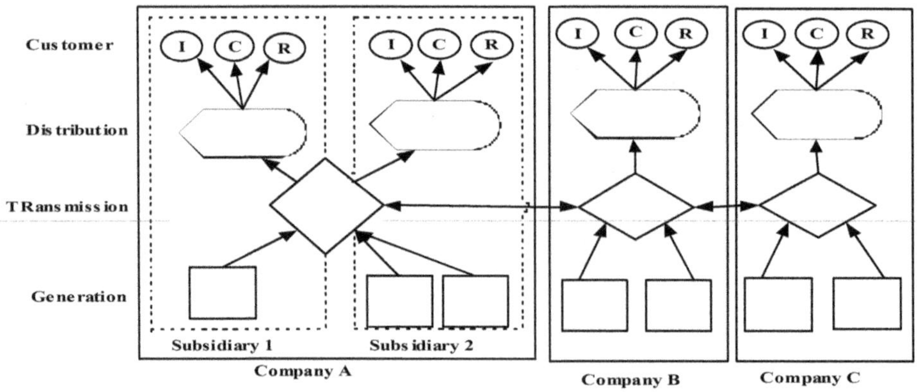

Figure 1.2 Single vertical structure.

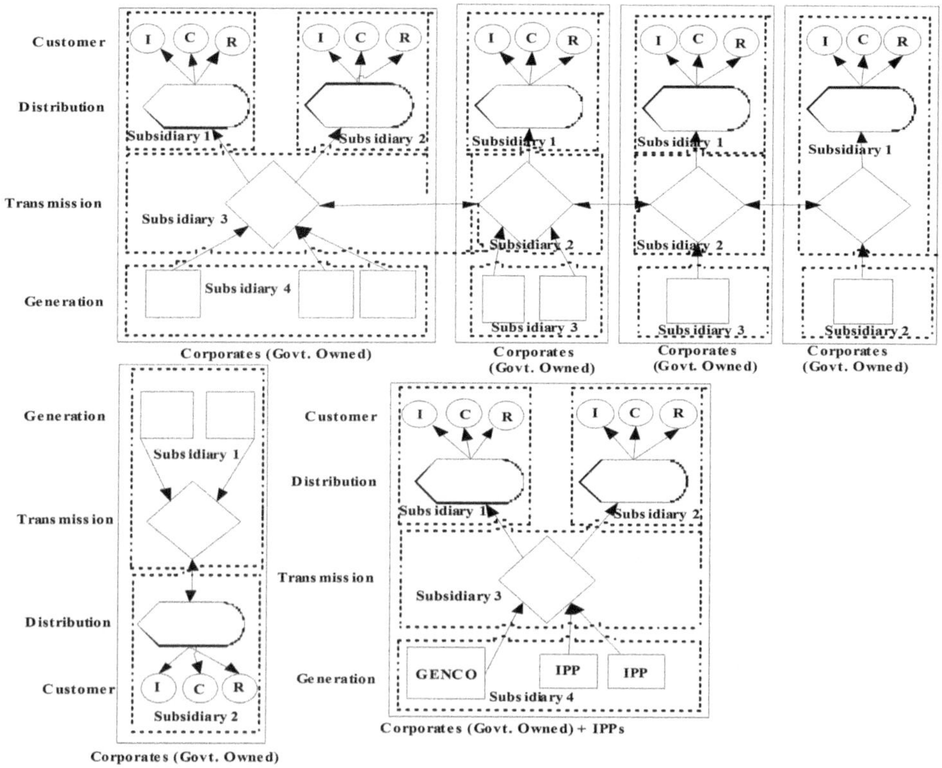

Figure 1.3 Unbundled structure.

- Poor collection efficiency.
- High rate of outages, blackouts, device failures, etc.
- Losses are not established.
- Commercial losses are not accountable.
- No outage handling and transactional mechanisms.
- Lack of enterprise resource planning (ERP) practice.
- Poor managerial efficiency.
- High fiscal deficit.
- No asset management and facility management.
- High political interferences.

1.3.2 Privately Owned DISCOMs

In this case, the ownership of the distribution lies with a private company. As a measure of private participation and introducing competition, the following steps were implemented:

- More than two licensees were allowed to sell power in the same area.
- Provision was made for different tariff rates based on different technical factors like load demand, voltage demand, power factor, etc.

- Licensees were allowed to do other business required for optimal utilization of assets but with prior permission.

The relative concerns with this model are:

- The private companies started running the power distribution business without bothering government norms and provisions.
- They failed to protect consumer interests.
- The tariff rate was high.
- They faced political resistance as privatization is not welcome in public services.
- The number of consumer protests increased.

1.3.3 Public–Private Partnership (PPP) Model

By leaving the distribution business totally in the hands of private players, government lost its control on the distribution sector and was not able to interfere when some norms or guidelines were being violated by the private companies. As a result, different governments kept some share in the distribution utility instead of giving the entire stake to the private player. This initiative led to the development of the concept of a Public–Private Partnership (PPP) model in the power distribution business. The PPP model encompasses various functions and obligations related to electricity supply in a licensed area. In the case of the PPP model, competitive bidding is the criteria used to select the concessionaire. Selected players would be responsible for operation, management, and maintenance related works. Apart from these tasks, the electricity supply, loss reduction, and looking after the interest of the consumers by maintaining quality of supply are also taken care by the concessionaire [8]. The ownership of the assets is not given to the concessionaire, rather the exclusive use of the distribution assets is allowed. Concession agreement and the applicable laws regulate the use of distribution assets.

The relative concerns with this model are:

- Tariff will be fixed separately for the PPP project area; this will bring multiple tariffs in the same state.
- May face political resistant as privatization is not welcome in public services.
- The PPP model needs attention for structuring, as it is complex in nature, so more conscious efforts are required from the government.
- The tariff may be different for urban and rural areas, so for uniform tariffs the government needs to provide subsidies.

1.3.4 Distribution Franchisee (DF) Model

This model is a hybrid arrangement between a concession and a management contract, which enables private players to carry out some power-related functions without letting the ownership of the distribution sector go to the private player. In the case of the DF model, a distribution licensee is legally allowed through the regulatory provisions to authorize a person/entity for electricity distribution in a specific area under its licensee area [9]. Table 1.1 shows the various forms of a DF model.

These four DF models all have their own popularity. The selection of a particular model is highly dependent upon the area of its implementation. For example, in rural areas, meter reading and bill distribution using vending machines, revenue collection, etc., compels the involvement of the franchisee. For this purpose the franchisee is given to individuals, panchayatiraj (the local government in the Indian subcontinent) institutions, self-help groups (SHGs), non-governmental organizations (NGOs), or groups of customers. Thus the first three models of DF models are suitable for operations in rural areas. However, in the case of the IBDF, the operational area is mostly preferred to be the urban area and the franchisee is assigned to undertake the functions of operation and maintenance along with the above tasks. Beside these activities, a franchisee is also allowed to undertake capital expenditure (CAPEX) for its allowed area. In the case of an IBDF franchisee scheme, the franchisee has to buy electricity from the distribution licensee at some predefined input rates, which are decided based on a competitive bidding process. Figure 1.4 shows the structure of an input-based distribution franchisee (IBDF) model.

Table 1.1 Various forms of DF model.

Franchisee model	Compensation	Area	Benefit to power sector
Outsourcing Outsourcing of some of the activities like billing and collection	Fixed fee	Rural	Improved operational efficiency
Collection-based revenue franchisee (CBRF) Allowed for billing and collection	Provisions for incentives and penalties based on collection while ensuring fixed percentage of collection toward compensation of labor used	Rural	Improved operational and collection efficiency
Revenue collection with O&M Operation and maintenance related responsibilities in addition to billing and collection	Fixed fee and incentive	Rural	Improved operational and collection efficiency Reduction in manpower
Input-based distribution franchisee (IBDF) Perform the same tasks as that of a distribution licensee except it must buy electricity from the licensee at some predecided input rates	Right on revenue	Urban	Improved operational and collection efficiency Reduction in manpower and distribution losses Avoided Investments

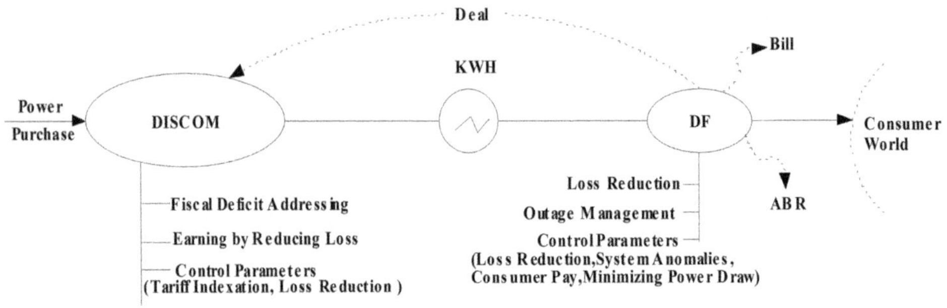

Figure 1.4 Input-based franchisee model.

Again, the input rate is decided by the licensee based on a loss reduction trajectory. As the tariff is the imposed control on a franchisee to reduce AT&C losses, the main objective of the franchisee is to reduce AT&C losses in its network. Thus the profit label of the franchisee totally depends upon its performance toward reducing losses as framed in a performance-based multiyear tariff framework. If the franchisee is not able to achieve the targeted loss reduction, it is also penalized and suffers from the loss. As a result, the primary focus of the DF model has shifted toward loss reduction. The DF model has tremendous potential toward reducing losses, improving collection efficiencies, and is synonymous with a privately owned DISCOM. Because of these features, this model has attracted lot of interest by various utilities. Involvement of private companies reduces the political interference and ensures adequate private capital investment. The relative concerns with this model involve:

- The takeover time is too long as the transition of asset and manpower requires an audit.
- Bid norms are such that only big players get a chance and therefore bidding is very restrictive: bidding talent becomes a problem.
- After a takeover, the focus of the DF shifts toward operations and asset management, thus neglecting the prime focus of loss reduction.
- Loss handling is weakly addressed as operation and maintenance supersede the loss handling system and method.
- Power management status is very poor as the DF is not allowed for market participation.
- Investment toward system up-gradation and augmentation related works are poor due to lack of availability of sufficient capitals.
- Investment needed for implementations, technologies, and innovations is missing because of the unavailability of adequate capital.
- Capital mobilization toward renovation and modernization works are not sufficient.
- DF is not able to ensure sufficient capital mobilization toward physical structure development and future investment because the tariff is not within its control.

1.4 Need for a Novel Business Model for Power Distribution under Smart Grid Environment

Poor innovation capability and adaptation of an inappropriate business model for the power distribution sector has acted as an obstacle toward incorporating the latest technologies, especially designed for the sector. For example, specific technologies have been designed for ToD/ToU metering, demand-side management (DSM) initiatives, loss reduction, distributed generation (DG) promotion, etc., but the adopted business model has failed in harnessing benefits from these specially designed technologies. Therefore, developing an appropriate business model that can capture values from these technologies is highly essential for the effective and efficient performance of the power distribution sector.

1.4.1 Description of the Novel Business Model

The power distribution function comprises two major activities, namely distribution network creation and retailing [10]. The development of an electricity distribution infrastructure capable of meeting the present as well as the future demand by the consumer is one of the most valuable assets but is very high capital intensive in almost every nation. After the infrastructure has been developed, increasing power quality, reliability, cost effectiveness, etc., create a demand for various ancillary services that impose various additional infrastructures in addition to the basic distribution infrastructure. For example, metering, communication, IT, controllers, protection infrastructure, etc., impose ancillary needs by the distribution sector. This again adds a huge capital requirement for the distribution infrastructure development. The huge capital required for this development is quite difficult for a government owned distribution company or a private distribution company to invest alone. Hence, outsourcing and partnership can be adopted as a possible solution toward capital mobilization for infrastructural development in the distribution sector.

The novel business model proposed [6] is shown in Figure 1.5. In this model of power distribution, various major tasks of power distribution have been proposed to be carried out through outsourcing. The PPP model, build–own–operate–transfer (BOOT) model, or Opex/Totex model has been adopted as the mode of outsourcing. The total structure of the power distribution business has been divided into three segments, viz. the physical infrastructure unit (PIU), the distribution management and maintenance unit (DMMU), and the control–operation–revenue management unit (CORMU). The following sections describes the various segments along with their functions.

1.4.2 Physical Infrastructure Unit (PIU)

Various physical entities such as DISCOM, metering company (MCOM), information technology company (ITCOM), and communication company (CCOM) are part of this unit, whose roles and functions are now given.

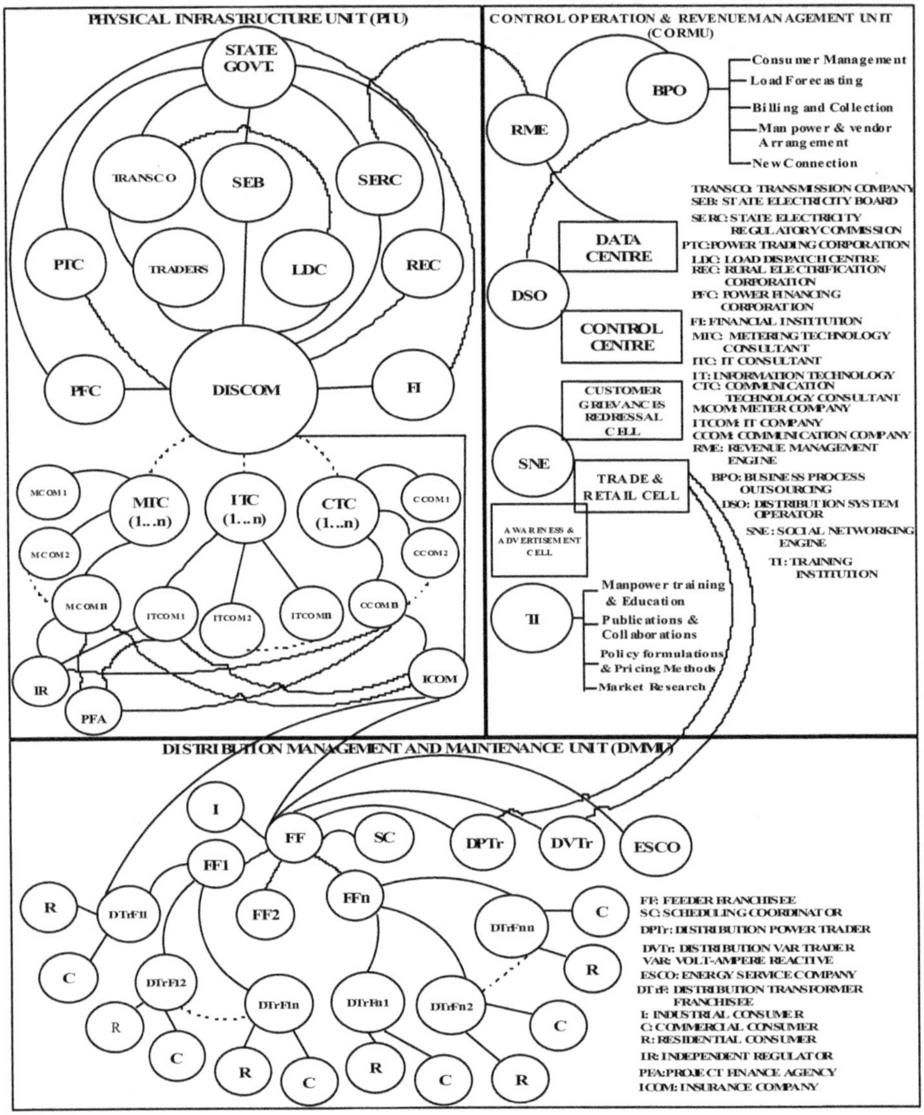

Figure 1.5 Nobel business model for power distribution.

1.4.2.1 DISCOM

DISCOM, which is either government owned or a corporatized government owned or a private company, is responsible for building all the physical infrastructures. Physical infrastructures include: towers and poles, overhead conductors, cables, transformers, etc. DISCOM is responsible for any expansion or upgradation work.

1.4.2.2 MCOM

Metering is an essential task to bring energy accountability, data collection, appropriate planning, device monitoring, risk reduction, etc. Meters are required at all different modes of power transactions, which is very huge in numbers. Therefore, metering activity is a big task in power distribution and hence the metering business can be carried out as a separate segment of power distribution business as MCOMs. An MCOM receives the metering contract from a DISCOM through metering technology consultants (MTC) hired by them.

1.4.2.3 ITCOM

Maintaining and managing a network database is an essential requirement in the power distribution sector, like many other sectors. It helps in planning and monitoring, decision making, creating transparency, keeping records, etc. In a power distribution business, it is required to deal with a very big data base as collecting and managing this large amount of data is not possible manually. Hence automation is highly essential where the information technology plays an important role. Automation should necessarily be preferred over manual activities and this distribution automation (DA) can provide power utilities with a long-term competitive advantage through better power reliability and an improved customer service [11]. As the data maintenance and management activity is a very big activity in a power distribution business, it can be carried out as a distinct business activity. This is the basic function of power distribution and hence there exists scope for involving different ITCOMs for data maintenance and management. To get the business contract from a DISCOM, an ITCOM has to approach it through an IT consultant (ITC).

1.4.2.4 CCOM

Very big data is required to be communicated across different nodes of the power distribution network to ensure the reliable, stable, and secure operation of the distribution networks for which a strong communication channel is essential. As the power distribution system is distributed across a huge area at which millions of nodes are present, the number of communication channels required is also huge. Hence, focusing on the development and management of this communication infrastructure with the focus on electrical infrastructure becomes difficult for the DISCOM. Furthermore, the demand response (DR) is also considered an integral part of the power system and market operational practices [12] and is considered a key focus area, which needs communication that is possible through smart grid initiatives [13]. Hence, the activities related to communication can be segregated as a distinct activity and a separate entity can be allotted to hold the communication related business as a CCOM. To get the business contract from a DISCOM, a CCOM must approach it through communication technology consultants (CTCs).

1.4.2.5 ICOM

There exists a risk of contingent and uncertain losses in every business. Hence to protect the business participants from such risk, involvement of an insurance company (ICOM) is essential, as in other sectors.

1.4.2.6 IR

It is highly essential for any organization to protect the interest of various stakeholders for which regulatory frameworks and standardizations are required. To carry out these tasks, roles of an independent regulator (IR) are necessary in the power distribution sector. An IR functions as a non-profit organization and will help all the stakeholders.

1.4.2.7 PFA

Like other businesses, there must be the provision for financing in the power distribution business as a very huge investment is required during the setting-up of different infrastructural requirements in the distribution sector. Hence there is a provision for a financing agency, called the Project Finance Agency (PFA), that has been made part of this model. Financial debt for current existing lending agencies is continuously increasing and there is a risk of financial contagion [14], so the newly proposed entity will solve the issue of financing in the power distribution business.

1.4.3 Distribution Management and Maintenance Unit (DMMU)

This unit is responsible for different activities related to supply management and network maintenance. It becomes difficult by a single entity to carry out all the various tasks related to maintenance and supply management activities. Therefore, outsourcing of various major distribution components such as a distribution feeder, a distribution transformer, etc., for their separate maintenance and management can lead to more efficient results. The financial capital required by the DISCOM for management and maintenance related works can be utilized in infrastructure development works by allowing private participants to carry out these tasks through outsourcing. To facilitate outsourcing, the concept of a feeder franchisee (FF) and distribution transformer franchisee (DTrF) has been proposed. Similarly, active power management, reactive power management, various other auxiliary services-based management are some other essential tasks that can be managed through the creation of a distribution power trader (DPTr), a distribution VAR trader (DVTr), and an energy service company (ESCO).

In this model FF will take part in power trading (both active and reactive) business through DPTr and DVTr. For the contract period, FF will be responsible for the management and maintenance of the allotted feeder. FF will do the business with all the DTrF connected to that particular feeder as its end-customers. Apart from DTrF, various industrial consumers will also buy power from the FF at HT (high tension) level. Similarly, DTrF will be responsible for all the management and maintenance-related works of the distribution transformer for the contract period. DTrF can do business directly with residential and commercial consumers. Required manpower to carry out different management and maintenance-related tasks can be hired through business process outsourcing (BPO).

1.4.4 Control, Operation, and Revenue Management Unit (CORMU)

To ensure efficient operation and sustainability in the power distribution business, control, operation, and revenue management-related tasks are highly essential. Hence making the provision for a separate entity that is not being allowed to carry out any commercial activity is highly essential. Different components of this unit each function as a non-profitable organization.

1.4.4.1 RME

Different revenue management-related activities of the DISCOM are being carried out through the RME (Revenue Management Engine). Energy pricing, billing and collection, consumer management, new connections, vendor management, etc., are being done by the RME. Except for energy pricing, all other activities of the RME are done through BPOs.

1.4.4.2 DSO

The Distribution System Operator (DSO) is responsible for all the activities such as control, monitoring, decision making, etc., required to maintain the stability of the distribution system. To do so, the DSO maintains a data center and a control center. The DSO also helps in promoting and monitoring renewable-based DG integration into the distribution network.

1.4.4.3 SNE

The SNE (Social Networking Engine) takes care of all the social activities and social responsiveness that are essential for the betterment of all the stakeholders. A unit name TRC (Trade and Retail Cell) under SNE takes care of all the tasks related to power trading and retailing. Similarly, awareness dissemination programs, media advertisements, publications, etc., are being done through an AAC (Awareness and Advertisement Cell) of the SNE. Other important tasks of the SNE include consumer rankings that ensure efficient use of electricity by creating tariff differences between efficient and inefficient consumers.

1.4.4.4 TI

The TI (Training Institute) has the essential responsibility of providing technical, financial, legal, and managerial training across various nodes of the distribution sector, which ensures an improved decision-making capability. TI may also help toward publications, collaborations, policy formulations, technical planning, plan execution monitoring, performance evaluation, and many other research-related activities.

1.5 Conclusion

The present study applies the business model concept to investigate how different models of power distribution fit themselves to be suitable for the requirements of a smart grid. Different business models have evolved across the globe to analyze the power distribution function. Based on the analysis, related concerns with a particular model have been sorted. A surprising result is that most of the adopted models are not suitable to accommodate the undergoing changes that are taking place as a result of smart grid implementations.

A clear result shows that it is necessary for the distribution utilities to strengthen their business model innovation capabilities to cope with the growing challenges. If a utility fails to adapt a suitable business model, it becomes difficult for the utility to transition from a traditional grid to a smart grid. Therefore, it becomes an inevitable task for the distribution utilities to innovate and adapt a suitable business model that ensures adequate infrastructural development in terms of basic electrical infrastructure, metering infrastructure, communication infrastructure, and IT infrastructure. Apart from this, an organization structure is also needed for management and maintenance works as well as control and operation-related tasks. Creating a mindset open to business model innovation could help utilities to overcome the various challenges under a smart grid environment. The research also has some implications for policy makers, regulators, and energy consultants.

The new business model is expected to overcome all the different challenges (as discussed in previous sections) faced by presently widely accepted and planned business models. Apart from this, separation of different activities into distinct businesses will lead to the following benefits:

- Create an opportunity for all different small players who are sound in technology related to the power business.
- Will ensure a competitive environment across various stages of power distribution.
- Will create huge job opportunities.
- Open a new business in the field of education and training.
- Tariff rates will be reduced.
- Ensure improved power quality.
- Ensure system stability and reliability.
- Not only big players but also other small players will get a chance to participate in business.
- Will attract many other expertise companies like IT, communication, insurance, trading, BPOs, etc., toward the power business.
- BPOs are paying huge salaries to employees and people because of their ideas. Thus, by attracting BPOs for participation in the distribution sector, intelligent people are being accommodated in the sector.
- It will help in promoting renewable-based distributed generations.
- Promote the needs and importance of the micro grid.
- Will bring down the level of AT&C loss within acceptable limits.
- Attract attention toward loss reduction apart from operation and maintenance.
- Create motivation for DSM-related activities.
- Automatic realization of ABR.
- This model suits both urban and rural areas in the context that some models are not suited for rural implementation.

References

[1] B. Khan, G. Agnihotri, and A.S. Mishra, "An approach for transmission loss and cost allocation by loss allocation index and co-operative game theory," *Journal of The Institution of Engineers (India): Series B*, vol. 97, pp. 41–46, 2016.

[2] B. Khan, G. Agnihotri, S.E. Mubeen, and G. Naidu, "A TCSC incorporated power flow model for embedded transmission usage and loss allocation," *AASRI Journal Elsevier*, vol. 7, pp. 45–50, 2014. ISSN: 2212-6716.

[3] B. Khan, G. Agnihotri, G. Gupta, and P. Rathore, “A power flow tracing based method for transmission usage, loss and reliability margin allocation,” *AASRI Journal Elsevier*, vol. 7, pp. 94–100, 2014.

[4] B. Khan, G. Agnihotri, P. Rathore, A. Mishra, and G. Naidu, “A cooperative game theory approach for usage and reliability margin cost allocation under contingent restructured market,” *International Review of Electrical Engineering*, vol. 9, no. 4, pp. 854–862, 2014.

[5] P. Rathore, G. Agnihotri, B. Khan, and G. Gupta, “Transmission pricing scheme under contingency conditions in open access market,” *2014 Eighteenth National Power Systems Conference (NPSC)*, Guwahati, pp. 1–6, 2014.

[6] T. Jamasb, T. Thakur, and B. Bag, “Smart electricity distribution networks, business models and application for developing countries,” *Energy Policy*, vol. 114, March 2018, pp. 22–29, 2017, Elsevier. doi: 10.1016/j.enpol.2017.11.068.

[7] D.P. Sharma, P.S.C. Nair, and R. Balasubramanian, “Performance of Indian power sector during a decade under restructuring: A critique,” *Energy Policy*, vol. 33, pp. 563–576, 2005.

[8] Task Force. “Report on private participation in power distribution, planning commission,” Government of India, 2012, [Online]. Available at: www.infrastructure.gov.in.

[9] T. Thakur, B. Bag, and S. Prakash, “A critical review of franchisee model in electricity distribution sector in India,” *The Electricity Journal*, vol. 30, no. 5, pp. 15–21, 2017, doi: 10.1016/j.tej.2017.05.001.

[10] J. Roman, T. Gomez, A. Munoz, and J. Peco, “Regulation of distribution network business,” *IEEE Transactions on Power Delivery*, vol. 14, no. 2, pp. 662–669, April 1999.

[11] D. Gutschow and M.O. Kachieng’a, “Making business sense of distribution automation systems: The case of Eskom, South Africa,” *IEEE Transactions on Power Systems*, vol. 20, no. 1, pp. 272–278, February 2005.

[12] J. Medina, N. Muller, and I. Roytelman, “Demand response and distribution grid operations: Opportunities and challenges,” *IEEE Transactions on Smart Grid*, vol. 1, no. 2, pp. 193–198, September 2010.

[13] Deloite. “Sector specific inventory and institutional strengthening for PPP mainstreaming”, Pre-feasibility Report on Distribution Franchisee in ESCOMs, 2012.

[14] S. Pargal and S.G. Banerjee, “More power to India: The challenge of electricity distribution” Washington, DC: World Bank. © World Bank, 2014. [Online]. Available at: https://openknowledge.worldbank.org/handle/10986/18726 License: CC BY 3.0 IGO.

Part II

Existing Issues in the Electrical Distribution Network

2

Existing Problems Related to Electrical Distribution Network, Part 1

Distribution Feeder Segregation

Baidyanath Bag

Department of Electrical Engineering, National Institute of Technology Raipur, Chhattisgarh, India

2.1 Introduction

Traditionally the electricity supply industry was monopolistic in nature with generation, transmission, and distribution owned by the Government. This system has also brought inefficiency and corruption due to policies and regulations. Many state-owned utilities are financially not solvent due to theft and lack of recovery of revenue. Further, in this model it was technically difficult to identify various issues in the distribution sector. Therefore, new mechanisms are required to make the system more efficient [1–5].

The Indian agriculture sector is one of the key sectors of the Indian economy and accounts for 18% of the country's GDP and employs 50% of its labor force [6]. However, it is considered as an inefficient sector as far as the use of electricity is concerned. The major reasons for the continued inefficiency of the agricultural sectors are due to heavy subsidies or free electricity supplies and the absence of any provision of incentives for efficient use. The inefficient use of ground water for agricultural purposes and the use of inefficient motors instead of electric pumps further reduces the overall efficiency. More than 80% of the country's water resources is consumed by the agricultural sector [6] and more than 60% of water used for irrigation is ground water [7]. Hence the necessity to segregate the agriculture feeder from the domestic feeder was acknowledged by Indian utilities.

To promote the implementation of feeder segregation in India, the Government of India has been trying since 1996 but it has only been effective since 2013 by projecting the process of feeder segregation as a national level scheme. The main objectives of this scheme are to guide, monitor, finance, and promote the process of feeder segregation in different agricultural-intensive Indian states. This scheme is included under Deen Dayal Upadhyay Gram Jyoti Yojana (DDUGJY) and a total cost of Rs. 4303.3 million was approved in December 2014 to separate the agricultural and non-agricultural feeders. The other achievable benefits through feeder segregation are the availability of 24×7 three-phase supplies to rural consumers, better regulation of supplies to agricultural customers that ensures improved load management, reduction in line losses, and an improvement in ground water husbandry. As the electricity is made available to the farmers either free of cost or at a highly subsidized rate, farmers misuse this facility by

Active Electrical Distribution Network: A Smart Approach, First Edition. Edited by Baseem Khan, Josep M. Guerrero, Sanjeevikumar Padmanaban, Hassan Haes Alhelou, Om Prakash Mahela, and Sudeep Tanwar.

running the agricultural pumps for more than the required number of hours actually required to irrigate the crops. Hence there are wastages in electricity as well as ground water. Thus, to avoid the misuse of electricity by the farmers, power is being supplied to the agricultural feeders for a restricted number of hours by segregating the feeder catering to the agricultural loads from the domestic loads. However, the possibility of theft of line conductors, transformers, and other electrical elements fitted to the agriculture feeder increases during the off hours in these feeders. Such practical incidences of theft have been experienced by many Indian utilities who have implemented feeder segregation. Hence this is one of the biggest hurdles and challenges to the agriculture feeder segregation scheme.

Electric motors are mainly used to lift ground water. Data published by India Infraline in 2014 shows that more than 15 million grid-connected pump sets and around 8 million diesel-powered pump sets are being used for irrigation purposes. As the electricity is supplied at highly subsidized rates, farmers use inefficient, low-cost pump sets and pump ground water excessively – more than is required. Today, almost all agriculture-intensive states subsidize electricity to agricultural customers, either fully or partially [8]. The state government is required to meet the subsidy but any delayed or partial release of the subsidy by the state government places an unplanned and unbudgeted burden on the utilities. This constraint often encourages utilities to restrict power to non-remunerative agricultural pump sets to about 6–8 hours a day.

This restriction in supply hours has led farmers to use an automatic starter. The use of an automatic starter results in a heavy initial load while power goes and comes, causing a heavy burden on the infrastructure. Again, many farmers do not use capacitor banks, which causes voltage fluctuations and a poor power factor. The above factors lead to frequent motor burnouts and hence farmers have moved to using low-cost pump sets, which are less efficient but fluctuation-resistant. Other reasons that contribute to the vicious cycle of inefficiency are the reluctance of state utilities and political expediencies. Different studies show that the average efficiency of the agricultural pumps used in India ranges between 20 and 30% as opposed to that of the Bureau of Energy Efficiency (BEE) star-rated pumps of more than 55% efficiency.

Hence, to replace these inefficient pump sets with efficient ones, an agriculture demand side management (DSM) program has been initiated by the Indian Government, but this program did not succeed due to many technical and non-technical reasons. Hence the process of feeder segregation has been adopted as the best solution for improving efficiency through efficient use of electricity and better husbanding of ground water. However, the possibility of theft of line conductors, transformers, and other electrical elements fitted in to the agriculture feeder increases during the off hours in the feeders. Hence, to deal with the problem of theft, a network reconfiguration can be used as a theft handling method. Network reconfiguration done by changing the normal topological structure of the distribution feeder to a new objective-based structure through closing and opening of the sectionalizing and tie switches. If the agricultural feeder is included as part of the feeder reconfiguration, electricity will be made available during off-feed hours and hence the thieves will not take the risk.

2.2 Review of Status of Research and Development in the Subject of Feeder Segregation

As far as the work related to the agriculture feeder segregation is concerned, no research has been initiated throughout the world in this area. Agriculture feeder segregation is a unique trick adopted in India that is based on its contemporary situation. However, ample research has been carried out worldwide in the area of network reconfiguration (NR), which is being considered as a very important task for real-time operation of the power distribution system. The essential role and performance of NR as well as the availability of a number of techniques for its implementation have made NR very useful for the distribution sector. NR has shown its suitability to be useful for the proposed objectives of this research proposal. The basic objectives of network reconfiguration include technical loss reductions, improving the voltage profile, minimizing outages, reducing device failures, minimizing overloading, smoothing peak demand, and increasing the network reliability. Some of the research work done worldwide in the area of network reconfiguration will be considered. From available data, Merlin and Back [9] proposed the concept of network reconfiguration for the first time in 1975. To determine the minimal-loss condition in a spanning tree structured distribution network, they made use of heuristic techniques. Similarly in 1997, it was found that the voltage instability problem in the distribution system, which spreads the corresponding transmission system, created major blackouts in the S/SE Brazilian system. This made the engineers think about an effective way of preventing this loss reduction, which included a network reconfiguration. Network reconfiguration is a non-linear optimization problem. Different researchers and industry persons throughout the world have made use of different optimization techniques to solve network reconfiguration problems. Various optimization techniques used to solve network reconfiguration problems include particle swarm optimization (PSO) [10], genetic algorithm (GA) [11], ant colony optimization [12], modified honey bee mating optimization (MHBMO) [13], branch exchange technique [14], hybrid heuristic technique [15], and fuzzy logic approach [16]. Most of the effective researches carried out in the area of network reconfiguration have made use of heuristic search techniques. Olamaei et al. [13] presented an efficient algorithm for multiobjective distribution feeder reconfigurations based on the modified honey bee mating optimization (MHBMO) approach. The main objectives of the distribution feeder reconfiguration (DFR) problem solved by this group are to minimize the real power loss and the deviation of the nodes' voltage. Hemmatpour et al. [17] presented a novel reconfiguration mixed with distributed generation planning via considering the voltage stability margin. This describes a novel approach for optimum reconfiguration and optimal location of DGs in distribution networks based on a hieratical two-stage optimization problem to improve the power system voltage stability margin and reduce active power losses. Savier et al. [18] presented a method for the allocation of power losses to consumers connected to radial distribution networks before and after network reconfiguration in a deregulated environment. Loss allocation is made in a quadratic way, based on identifying the real and imaginary parts of the current in each branch. The network reconfiguration algorithm is based on the fuzzy multiobjective approach and the max–min principle is adopted for multiobjective optimization in a fuzzy framework.

Wiselin et al. [15] proposed a hybrid technology to solve the distribution system reconfiguration problem. The mixture of the plant growth simulation algorithm (PGSA) and greedy and heuristic-based fuzzy operation has been used. The results obtained from the proposed optimization approach provide a detailed description of the state of different switches. A new distribution network reconfiguration approach using a tree model has been proposed by Dolatdar et al. [19]. From their research, it was shown that optimal reconfiguration involves the selection of the best set of branches to be opened, one each from each loop, for reducing resistive line losses and relieving overloads on feeders by shifting the load to adjacent feeders. However, since there are many candidate switching combinations in the system, the feeder reconfiguration becomes a complicated problem. A network reconfiguration technique that uses the distance measurement technique algorithm based on a two-stage methodology has been explained by Kashem et al [14]. Shirmohanunadi et al. [20] presented a reconfiguration of electric distribution networks for resistive line loss reduction. They described an efficient and robust heuristic method for the reconfiguration of distribution networks in order to reduce their resistive line losses under normal operating conditions. They claimed that their approach can be used in both the planning and operation environments because of its computational efficiency. Chiang et al. [21, 22] proposed an optimal network reconfiguration for distribution systems based on a two-stage solution methodology and a modified simulated annealing technique. They developed a solution algorithm to the network reconfiguration problem, which is a constrained, multiobjective, non-differentiable optimization problem. Their solution algorithm allows the designer to obtain a desirable, global, non-inferior point in a reasonable computation time. Nara et al. [11] presented implementation of a network reconfiguration using a genetic algorithm for a distribution system loss minimization. They concluded that NR is a complex mixed integer programming problem that is difficult to solve using a mathematical programming approach. Hence, they used a genetic algorithm as a search or optimization algorithm based on the mechanics of natural selection and natural genetics. Civanlar et al. [23] investigated the distribution feeder reconfiguration for loss reduction. They proposed a feeder reconfiguration as a planning and/or real-time control tool in order to restructure the primary feeders for loss reduction. Since electric utilities have become increasingly interested in the problems associated with interfacing of dispersed storage and generation (DSG), Rizy et al. [24] presented operational and design considerations for electric distribution systems with DSG. Three studies have examined the operational and planning problems associated with connecting small power sources to an electric distribution system. An efficient algorithm for load balancing of transformers and feeders by switching operation in large-scale distribution systems has been explained by Aoki et al. [25]. This work presents a systematic and practical algorithm for load balancing of transformers and feeders by automatic sectionalizing switch operation in large-scale distribution systems of a radial type. The algorithm is developed by extending an approximation algorithm for load transfer desired by two different transformers. Lee et al. [26] described and summarized the straightforward method developed and applied by Advanced Systems Technology (AST), a division of Westinghouse Electric Corporation, for The Pennsylvania Power and Light Company (PP&L). Liu [27] presented an expert system operational aid for restoration and loss reduction of the distribution sector. This work presents a new application of expert system techniques to the restoration of distribution systems. Primary feeders are typically radial in structure. To restore a maximal number of zones, several steps are followed: group restoration, zone restoration, and, if necessary, load transfer. Based on the methodology, an expert system was developed that can be utilized as an on-line aid to system operators in a distribution supervisory control and data processing system (SCADA) environment. Implementation of heuristic search strategies for distribution feeder reconfiguration is explained by Taylor [28] where a method is described for feeder reconfiguration with the potential for handling realistic operating constraints. The approach taken is to set up a decision tree to represent the various switching operations available. A best-first tree searching strategy, based on heuristics, is used to evaluate the various alternatives. Chan et al. [29] presented a methodology for determining the critical switches in the distribution system. The daily load curves of various types of load customers are used in a three-phase load flow program. The program calculates the hourly current loading of distribution feeders and current flows through all line switches. A distribution system comprising nine distribution feeders in the Taiwan Power Company (Taipower) is selected for computer simulation. Kennedy et al. [30] developed particle swarm optimization in which the evolution of several paradigms is outlined and implementation of one of the paradigms is discussed. Benchmark testing of the paradigm is described and applications, including nonlinear function optimization and neural network training, are proposed. The relationships between particle swarm optimization and both artificial life and genetic algorithms are described.

2.3 Indian Experiences with Agriculture Feeder Segregation

India is the world pioneer for the process of feeder segregation, especially in the area of creating a dedicated feeder for the agriculture sector. The process of feeder segregation in India has been initiated since 1996 but has been effective since 2013 by projecting the process of feeder segregation as a national-level scheme. The main objectives of this scheme are to guide, monitor, finance, and promote the process of feeder segregation in different agricultural-intensive Indian states. The scheme is included under Deen Dayal Upadhyay Gram Jyoti Yojana (DDUGJY) and a total cost of Rs. 4303.3 million has been approved to use in December 2014 to separate the agricultural and non-agricultural feeders. The process of feeder segregation was implemented by the state of Punjab for the first time in India in 1996–1997 and was executed on a turnkey basis in 2003–2004 [6]. In line with the state of Punjab, Gujarat opted for feeder segregation through Jyoti Gram Yojna. As an initiation towards feeder segregation, different pilot projects were being implemented by Andhra Pradesh in 2010. Growing trends and the benefits of feeder segregation has motivated many other Indian states, such as Karnataka, Uttar Pradesh, Bihar, Haryana, Rajasthan, Maharashtra, and Madhya Pradesh.

The concept of feeder segregation in India is familiar only among the utilities population, especially those working on the feeder segregation projects. Other technical experts as well as academicians have not taken any initiative to understand the concept of feeder segregation and hence to carry out the research and development activities in this domain. Hence, the research status in this area is negligible. Although some agencies like World Bank, Power Line (ABB), and some government agencies have been involved in publishing some studies upon feeder segregation, their studies have been limited to the status of implementation of this plan, the budget involved, and the associated challenges with feeder segregation. However, many Indian researchers have become involved in the research area of network reconfiguration, which has been considered to be one of the important techniques to deal with the issues lying with feeder segregation. Details of some of the contributions of Indian researchers in this area are now given. Goswami et al. [31] proposed the feeder reconfiguration of distribution system for loss minimization. This work reports a power flow based heuristic algorithm for determining the minimum loss configuration of radial distribution networks. The algorithm is based on the concept of the optimum flow pattern, which is determined by solving the KVL and KCL (Kirchhoff's voltage and current laws) equations of the network. Niknam et al. [32] developed a hybrid evolutionary algorithm for distribution feeder reconfiguration. This algorithm presents a new method to reduce the distribution system loss by feeder reconfigurations. This new method combines the self-adaptive particle swarm optimization (SAPSO) with the shuffled frog-leaping algorithm (SFLA) in an attempt to find global optimal solutions for the distribution feeder reconfiguration (DFR).

Increasing requirements of urbanization, industrialization, and modernization demands further expansion and development of the national power grid and, nonetheless, with a better efficiency and an enhanced voltage stability. To satisfy the above objective, Nath et al. [33] have verified a network reconfiguration as an effective and suitable method. A fuzzy multiobjective approach for network reconfiguration of distribution systems has been proposed by Das [16]. This work presents an algorithm for network reconfiguration based on the heuristic rules and fuzzy multiobjective approach. Multiple objectives are considered for load balancing among the feeders and also to minimize the real power loss, deviation of node voltage, and branch current constraint violation, while being subject to a radial network structure in which all loads must be energized. Ahuja et al. [12] proposed an AIS-ACO hybrid approach for a multiobjective distribution system reconfiguration. This proposes a hybrid algorithm based on artificial immune systems and an ant colony optimization trick for distribution system reconfiguration, which is formulated as a multi-objective optimization problem. Kalambe et al. [34] presented loss minimization techniques in a distribution network, a bibliographical survey that describes the three main techniques of loss minimization that includes network reconfiguration. Karthikeyan et al. [35] has proposed a novel criterion for segregating the feeder with a proposal based on the slope of the curve between the feeder losses versus the receiving end voltage. Application of this criterion results in the most minimal loss configuration for any given loading condition. An existing switching indices criterion and a switching algorithm criterion have been considered and the results are compared with that of the proposed slope criterion, which results in the most minimal loss configuration.

2.4 Importance of Feeder Segregation in the Context of Current Status

The Indian agricultural sector is the third largest consumer of electricity after the industrial and domestic sectors but the revenue realization from this sector is very low [36]. The share of agriculture in total electricity consumption is significant, i.e. 23% in 2011–2012,

18.08% in 2017–2018, and 7.69% in 2018–2019, but the share of the agriculture sector in total revenue generation is very poor; it was only 7% in 2011–2012. This reason and the reasons explained in Section 2.1 shift the focus of the Indian Government to give the utmost urgent attention to the agriculture sector. As a result, the Indian Government has initiated the concept of agriculture feeder segregation.

The Government of India thought to bring up a national level scheme that will guide, monitor, finance, and promote the process of feeder segregation in different agricultural intensive states. Thus, the Ministry of Power requested the forum of regulators (FOR) to work upon a framework to draw up a scheme at national level for feeder segregation of rural and agricultural consumers and suggest measures for effective metering. The report recommended that a national level program for feeder segregation be implemented in a calibrated manner while allowing states to have the flexibility to design the project to suit their specific requirements, subject to adequate power being available at the national level.

Although various Indian states, as explained, have implemented the process of feeder segregation, their rate of implementation is very slow. Many Indian states have even not thought of the initiation of the segregation process to date. However, Indian utilities that have adopted the process of segregation are not able to harness all such benefits that they had envisaged through feeder segregation. The actual theft or the possibility of theft of line conductors, transformers, and other electrical elements fitted in to the agriculture feeder is one of the main hurdles behind all the above issues. None of the utilities have claimed that they have found the trick to handle this issue of theft, and no researcher has yet published any article proposing a solution to deal with the issue of theft. These facts are demotivating and weaken the confidence of Indian utilities to take part in the feeder segregation plan.

Hence the process of network reconfiguration can be used to address one of the biggest challenges in promoting a national level scheme. Apart from its basic objective, the inclusion of network reconfiguration will also help in reducing technical losses, improving voltage profile, minimizing outages, reducing device failures, minimizing overloading, smoothing peak demand, and increasing the network reliability. It will further contribute toward the national target of achieving AT&C loss minimization. The outcomes of this proposed chapter will give thrust to the implementation of the Feeder Segregation Plan (FSP) across different Indian states. It will enhance the confidence of various Indian utilities and provide motivation in implementing the FSP.

2.5 Threats with Feeder Segregation and Possible Solutions

The basic objective behind this chapter is to discuss a theft handling mechanism for protecting various power equipment/appliances fitted in agricultural feeders during off-feed hours. This theft handling mechanism is based on a network reconfiguration in which the normal topological structure of the distribution feeder is changed to a new objective-based structure through closing and opening of the sectionalizing and tie switches. If the agricultural feeder is included as part of the feeder reconfiguration, electricity will be made available during off-feed hours, and hence thieves will not take any risk. The following broad steps are required to be taken when developing the theft handling mechanism:

- Step 1: To develop a method to decide the locations of different sectionalizing and tie switches in such a way that the inclusion of some of the section of the agricultural feeder in the network reconfiguration will not energize the connected pumps.
- Step 2: To develop a method for the selection of tapping points for the agricultural pumps in such a way that the pumps do not draw power during off-feed hours.
- Step 3: To deduce the best suitable method for network reconfiguration based on the contemporary situation and the demand for the Feeder Segregation Plan.
- Step 4: To draw up a plan for distribution system automation (DSA) to manage the network reconfiguration.
- Step 5: To determine the optimal switching frequency for the network reconfiguration while considering all its dynamics and impacts.
- Step 6: To validate the above proposed methodologies using high-end tools like ETAP/MATLAB.

Wherever the distribution system in not automated and switching is done manually, the positions of the tie switches are provided based on the ease of reachability and in India where the same practice is being adopted. However, wherever the distribution system is automated and there is scope for implementing network reconfiguration, placing the switches randomly will not serve the purpose. Hence technical studies are being carried out in deciding their numbers and positions. Many research publications are available in this context. As far as steps 1 and 2 are concerned, they are totally new concepts and hence no research is available in this area.

Developing the automation features for the distribution system operation involving an agriculture feeder in a network reconfiguration will totally depend upon the methodologies used for achieving steps 1 and 2. Hence, unless and until some methodology has been developed for steps 1 and 2, it is worthless to design a method for distribution system automation (DSA) considering a feeder reconfiguration. A number of automation models are available for the distribution system. Adequate technologies have been developed in the area of distribution system automation. The main components used for distribution system automation (DSA) are remote terminal units (RTUs), pole top units (P-RTUs), distribution control center (DCC), distribution automation software, communication network, IT network, etc. Simply by changing the operational strategy in an existing distribution system automation model, the task of network reconfiguration based on the set objectives can be achieved. Hence, designing the methodology for achieving step 4 can easily be done.

Switching is one of the important phenomenon in network reconfiguration. Network switching can be done as an hourly, daily, weekly, monthly, or seasonal reconfiguration plan, based on the load curves. Wherever the distribution system in not automated and switching is done manually, frequent switching operations are not possible. Under such a situation, constant load models (peak or mean value) are used for network reconfiguration related studies. However, wherever the distribution system is automated, dynamic (real-time) network reconfiguration is possible with the use of automatic (or remote controlled) switches. The flow chart presented in Figure 2.1 will explain the methodology for deducing the optimal switching frequency [37].

The most important task required for the proposed mechanism is network reconfiguration. The next section describes the various tasks required, their methodology, and the various processes for implementation of the mechanism.

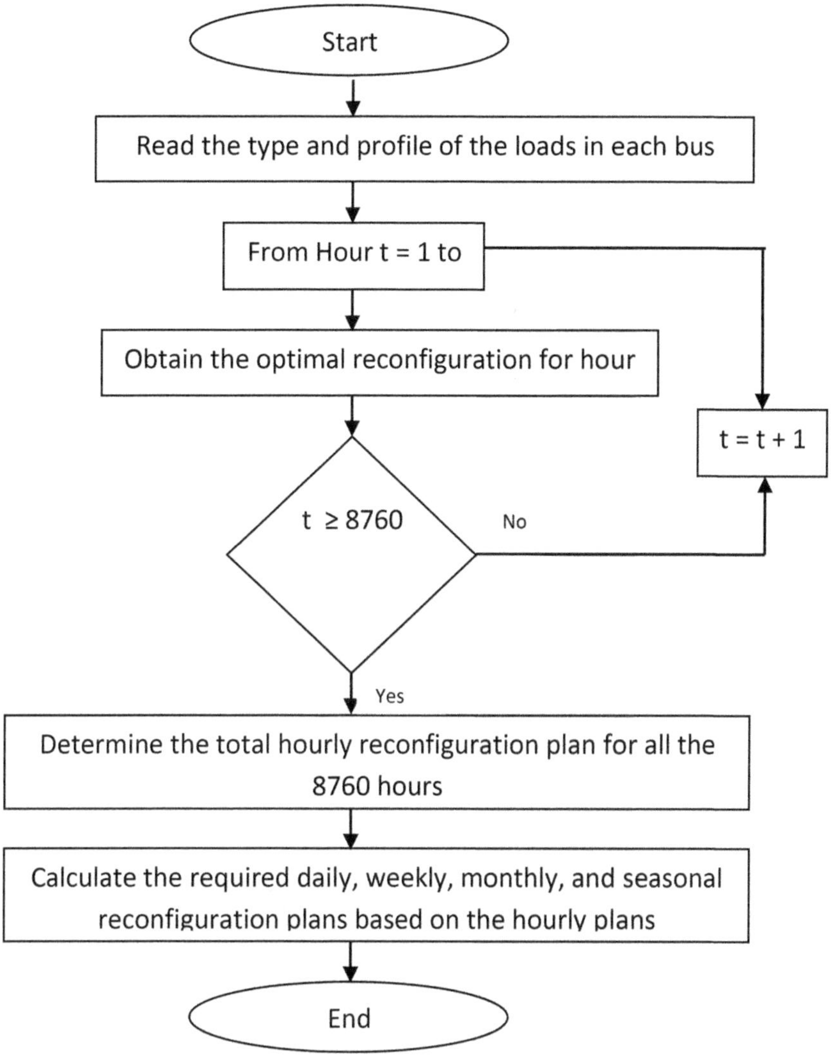

Figure 2.1 Methodology for deducing the optimal switching frequency.

2.6 Feeder Reconfiguration

The distribution system provides power to the end users from a set of distribution feeders, which are normally configured radially due to a protection point of view. Two types of switches, i.e. sectionalizing switches and tie switches are placed in every distribution network. During normal operations, sectionalize switches are kept closed and tie switches are kept open. In case of any requirements related to protection coordination or reconfigurations, these switches play the dominant roles.

Since many sectionalizing switches and tie switches may be present in a distribution network, the task of reconfiguration becomes complex due to huge possible combinations of these switches. Hence the task of reconfiguration becomes a complicated combinatorial, non-differentiable, constrained optimization problem. A huge amount of time is required for calculating network losses during all the different possible combinations. Thus to reduce the time and the number of steps involved, heuristic methods are mostly being used by different researchers across the globe for reconfiguration of the radial distribution system.

Let us try to understand the methodology through an example. Assume that the methodology is being applied to an IEEE 33-bus radial distribution test system, as shown in Figure 2.2. In this figure the dotted lines represent the tie switches and the solid lines represent the sectionalizing switches.

There are 33 buses, 32 sectionalizing switches, and 5 tie switches. Suppose the total loads connected to the network are (3715 + j2300) kVAh. The tie switches (33, 34, 35, 36, and 37) are normally kept open and the sectionalizing switches (1–32) are normally kept closed. Using this normal configuration, load flow will be carried out to calculate the total technical loss (I^2R), individual bus voltages, and different line flows. By using the load flow data, the voltage across all tie switches will be computed, which will be equal to the voltage difference between the two buses across which that tie switch is connected. Suppose that the total technical loss obtained through the normal case load flow is 150 kW. Similarly, voltages across different tie switches can be calculated using load flow data, as shown in Table 2.1.

Now, let us consider that the loss obtained, i.e. 150 kW, is too high and there is scope for a loss reduction for which network reconfiguration is selected. The basic task required in network reconfiguration is to alter topological structures of the network by changing the status of the tie and sectionalizing switches. At the start any one tie switch will be closed. However, whenever the tie switch is closed, one loop will be formed in the network, which will violate the network operational constraint, i.e. radiality. Hence, to eradicate this

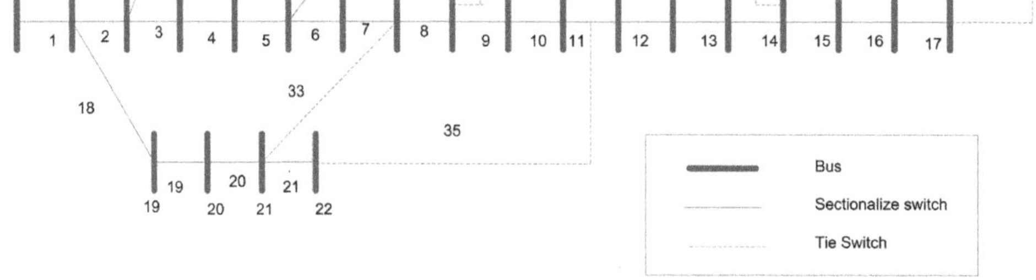

Figure 2.2 IEEE 33-bus radial distribution system.

Table 2.1 Voltage difference across all open tie switches after first switching.

S. No.	Tie switch number	Voltage difference across tie switch
1	33	0.071
2	34	0.027
3	**35**	**0.085**
4	36	0.023
5	37	0.058

violation, any sectionalizing switch of the network that is essentially part of the loop formed must be opened. Further, while opening the sectionalizing switch, care must be taken that every node is getting a supply. After opening the sectionalizing switch, the load flow will be run for the restructured case. The total technical loss (I^2R), individual bus voltages, and different line flows for this restructured case will be observed. From the observed data, it must be verified that the current flow through none of the line is violating the thermal limit of the line.

As per the proposed methodology, any one of the tie switches is closed and one of the sectionalizing switches constituting the loop is made open in the first step. Generally, the tie switch across which the voltage difference is highest among all the tie switches and is also more than a specified value is preferred to be closed first. For example, in the network shown in Table 2.1, tie switch 36 will be closed first as the voltage difference across it is highest among all the ties switches. Similarly, the sectionalizing switch that is opened first is the one that is connected to any one bus between the two supporting buses of the tie switch; generally the bus with the lower voltage is preferred. The reconfigured network obtained with the closing of the tie switch and opening of the sectionalizing switch will be considered as the base case for running the load flow. Suppose x_1 is the total technical loss (I^2R) taking place in the network for the first base case, as obtained from the load flow. In the next step, the presently open sectionalizing switch will be closed and its adjacent sectionalizing switch in the same loop will be open. For this second reconfiguration, again the load flow will be run and the above parameters will be observed while checking for all different constraints explained above. Suppose the total technical loss (I^2R) obtained in this case is x_2. If x_2 is greater than x_1, then the first reconfiguration will be the best reconfiguration corresponding to tie switch 1. However, if x_2 is less than x_1, then similar steps will be repeated by closing the presently opened sectionalizing switch while simultaneously opening the next adjacent sectionalizing switch within the same loop. This process will be repeated until a reconfiguration is obtained for which the value of the total technical loss (I^2R) is more than its current previous value. Suppose the process continued for nth number of sectionalizing switches in the loop and x_n is the corresponding technical loss (I^2R). Let this x_n obtained from nth reconfigurations is found to be more than its previous value, i.e. x_{n-1}. Then the process will be stopped and $(n - 1)$th reconfigurations will be considered as the best reconfiguration corresponding to tie switch 1.

Now in the next step, considering the best reconfiguration corresponding to tie switch 1 as the next base case, a similar process will be adopted for tie switch 2 and the respective optimal reconfiguration, and a corresponding total technical loss (I^2R) will be obtained. Tie switch 2 will be selected by using the same procedure as adopted for selecting tie switch 1 while considering the remaining tie switches other than tie switch 1. In the above selection of tie switch 2, the load flow data obtained from the latest reconfiguration, i.e. the best reconfiguration corresponding to tie switch 1, will be used. The final reconfiguration obtained by practicing all the procedures adopted in the case of tie switch-1 will be considered as the best reconfiguration corresponding to tie switch 2.

In the next step, the best reconfiguration corresponding to tie switch 2 will be considered as the base network and a similar process will be carried out for other remaining tie switches and their respective optimal configuration, and a corresponding total technical loss (I^2R) will be obtained. The reconfiguration that will be obtained from the last step by practicing the above explained procedure will be considered as the optimal network reconfiguration.

From the above explanation, it can be concluded that a large number of steps are required for obtaining each best reconfiguration corresponding to a particular tie switch. Hence the process of network reconfiguration becomes a complicated, combinatorial, non-differentiable, constrained optimization problem. There are even networks where millions of nodes are used. For such networks, investigating all possible options is impractical. The time required to calculate the total technical loss (I^2R) is also too huge. Therefore, it is essential to develop some trick for reducing the number of steps involved and the time required for calculating technical loss. Hence, heuristic techniques are thought to be the most suitable techniques for the network reconfiguration process.

The main objective while implementing network reconfiguration is to minimize the technical loss. Hence the objective function for the network reconfiguration problem is a loss function and is required to be minimized. While optimizing this objective function, different technical and non-technical constraints are required to be addressed, which will now be discussed.

2.6.1 Objective Function

The objective is to minimize the system total technical loss (I^2R) described by:

$$\min f(x) = \sum_{l=1}^{N_l} r_l i_l^2 \qquad (2.1)$$

where N_l = number of lines in the network

r_l = resistance of the line l

i_l = current flowing through line l

2.6.2 Constraints

1. Bus voltage limit:

$$V_{\min} \le V_k \le V_{\max} \qquad (2.2)$$

2. Line limits:

The current flowing through any line (i_l) should not exceed the current limit of that line ($i_{l\max}$).

$$i_l \le i_{l\max} \tag{2.3}$$

3. Radiality:

To ensure the radiality, the following assessment parameters are being used

- The number of loops while closing all different switches must be:

$$N_{lps} = \left(N_b - N_{bs}\right) + 1 \tag{2.4}$$

N_b = total number of branches
N_{bs} = total number of buses

- Number of branches should be:

$$N_b = \left(N_{bs} - N_s\right) \tag{2.5}$$

N_{bs} = number of buses
N_s = number of sources

- All the loads should get a supply every time of reconfiguration.

The use of heuristic approaches can reduce the complexity of the network reconfiguration problem. In this chapter, a heuristic approach has been used to solve network reconfiguration problems. The following points show the detail outlines of the used heuristic algorithm.

2.6.3 Algorithm

1. Read the system input data required for load flow.
2. Run the load flow program for the distribution network and compute the power loss and voltages at various nodes.
3. Calculate the differences in voltages $\Delta V_{tie}(x)$ between the two buses across which different tie switches are placed to be connected; $x = 1, 2, 3, \ldots S_{tie}$, where S_{tie} = number of tie switches.
4. Form a table for the above calculated voltage differences for all of the tie switches.
5. From the above table, select the tie switch for which the value of $\Delta V_{tie}(x)$ is maximum. Suppose it is the tie switch '*n*' such that $\left(\Delta V_{tie,\max} = \Delta V_{tie}(n)\right)$.
6. Compare the value of the $\left(\Delta V_{tie,\max}\right)$ with a set threshold value (ϴ). If $\Delta V_{tie,\max}$ is greater than the threshold value, go to the next step, else stop.
7. Compare the node voltages of the two nodes across which the tie switch '*n*' is connected. Select the node with the smallest value. Let it be the node '*m*' with node voltage V_m.
8. Close the tie switch '*n*' and simultaneously open the sectionalizing switch connected to node '*m*'.

9. Run the load flow for the above reconfigured distribution network and compute the power loss. Let it be P_{loss}.
10. In the next step, close the previously opened sectionalizing switch and open its adjacent sectionalizing switch within the same loop.
11. Run the load flow for the above reconfigured distribution network and compute the power loss. Let it be P'_{loss}.
12. If $P'_{loss} \geq P_{loss}$, declare the present reconfiguration as the best reconfiguration corresponding to tie switch n, else replace P_{loss} by P'_{loss} and go to step 10.
13. i = number of iterations. If $i < S_{tie}$ then replace i by $i + 1$ and go to step 2 to repeat the program to find out the best reconfiguration corresponding to other tie switches, else go to next step.
14. Consider the current reconfiguration as the optimal reconfiguration of the distribution network and run the load flow to calculate the total technical loss (I^2R), individual bus voltages, and different line flows.
15. Stop.

At last, the total technical loss (I^2R) obtained during the best reconfiguration corresponding to tie switch n ($n = 1, 2, \ldots, n$) and normal case load flow will be compared. Among all these $n + 1$ cases, the case for which the total technical loss (I^2R) obtained will be minimum will be adopted as the operational network.

2.7 Conclusion

The basic objective behind this chapter is to discuss a theft-handling mechanism for protecting various power equipment/appliances fitted in agricultural feeders during off-feed hours. The discussed method is based on network reconfiguration in which the normal topological structure of the distribution feeder is changed to a new objective-based structure through closing and opening of the sectionalizing and ties switches. If the agricultural feeder is to be included as the part of the feeder reconfiguration, electricity will be made available during off-feed hours, and hence the thieves will not take any risk. The main tasks involved with the discussed methodology are as follows:

- To develop a method to decide the locations of different sectionalizing and tie switches in such a way that the inclusion of some of the section of the agricultural feeder in network reconfiguration will not energize the connected pumps.
- To develop a method for the selection of tapping points for the agricultural pumps in such a way that the pumps do not draw power during off-feed hours.
- To deduce a best suitable method for network reconfiguration based on the contemporary situation and as per the demand of the Feeder Segregation Plan.
- To draw up a plan for distribution system automation (DSA) to manage the network reconfiguration.
- To determine the optimal switching frequency for network reconfiguration while considering all its dynamics and impacts.
- To validate the above proposed methodologies using high-end tools like ETAP/MATLAB.

The potential threat in the theft of line conductors, transformers, and other electrical elements while segregating the agriculture feeder can be reduced through the use of the discussed scheme. Apart from its basic objective, the inclusion of network reconfiguration will also help in reducing technical losses, improving voltage profile, minimizing outages, reducing device failures, minimizing overloading, smoothing peak demand, and increasing the network reliability. It will further contribute towards the national target of achieving AT&C loss minimization. The outcomes from the discussed methodology will give thrust to the implementation of the Feeder Segregation Plan (FSP) across different Indian states. It will enhance the confidence of the various Indian utilities and provide motivation in implementing the FSP.

As the network reconfiguration has been proposed to be implemented through distribution system automation in the discussed work, this methodology can be initiated as a demand-side management (DSM) program. The nature of the discussed scheme is such that the utilities can implement the plan without depending upon consumer mind sets to control their consumption. The discussed scheme will also help toward improved management of environmental resources through husbanding ground water.

References

[1] P. Rathore, G. Agnihotri, B. Khan, and G. Gupta, "Transmission pricing scheme under contingency conditions in open access market," *18th IEEE National Power Systems Conference (NPSC)*, pp. 1, 6, 18–20 December 2014.

[2] P. Rathore, G. Agnihotri, B. Khan, and G. Gupta, "Transmission pricing based on power flow tracing methodology with consideration of system reliability," *IEEE Students Conference on Engineering and Systems (SCES)*, pp. 1–6, May 2014.

[3] A.S. Mishra, B. Khan, and G. Agnihotri, "Embedded cost allocation with Shapley approach using tracing and sensitivity factors to identify coalition cost: A comparative study," *Proceedings of The EE Centenary Conference, IISc, Bangalore*, 15–17 December 2011.

[4] B. Khan, G. Agnihotri, G. Gupta, and P. Rathore, "A power flow tracing based method for transmission usage, loss & reliability margin allocation," *AASRI Journal Elsevier*, vol. 7, pp. 94–100, 2014.

[5] B. Khan, G. Agnihotri, P. Rathore, A. Mishra, and G. Naidu, "A cooperative game theory approach for usage and reliability margin cost allocation under contingent restructured market," *International Review of Electrical Engineering*, vol. 9, no. 4, pp. 854–862, 2014.

[6] Forum of Regulators, Framework to draw up a scheme at national level for feeder segregation of rural and agricultural consumers and suggest measures on effective metering, Ministry of Power, Government of India, December 2014.

[7] World Bank, Lighting rural India: Load segregation experiences in selected states, February 2014.

[8] B. Bag and T. Thakur, "Energy management using volt/var control of end-use appliances without affecting their performances and lives," *Journal of Renewable and Sustainable Energy*, vol. 9, no. 13, pp. 035504: 1-035504, 2017.

[9] A. Merlin and H. Back, "Search for a minimal-loss operating spanning tree configuration in an urban power distribution system," *Proceedings of 5th Power System Computation Conference (PSCC)*, Cambridge, UK, pp. 1–18, 1975.

[10] X.-M. Yu, X.-Y. Xiong, and Y.-W. Wu, "A PSO-based approach to optimal capacitor placement with harmonic distortion consideration," *Electric Power Systems Research*, vol. 71, pp. 27–33, 2004.

[11] K. Nara, A. Shiose, M. Kitagawoa, and T. Ishihara, "Implementation of genetic algorithm for distribution systems loss minimum reconfiguration," *IEEE Transactions on Power Systems*, vol. 7, no. 3, pp. 1044–1051, 1992.

[12] A. Ahuja, S. Das, and A. Pahwa, "An AIS-ACO hybrid approach for multi-objective distribution system reconfiguration," *IEEE Transactions on Power Systems*, vol. 22, no. 3, pp. 1101–1111, 2007.

[13] J. Olamaei, T. Niknam, and A. Sirous Badali, "Distribution feeder reconfiguration for loss minimization based on modified honey bee mating optimization algorithm," *Energy Procedia*, vol. 14, pp. 304–311, 2012.

[14] M.A. Kashem, V. Ganapathy, and G.G. Jasmon, "Network reconfiguration for load balancing in distribution networks," *IEEE Proceedings-Generation, Transmission and Distribution*, vol. 146, no. 6, pp. 563–567, 1999.

[15] J. Wiselin and P. Shankar, "Distribution system loss reduction through hybrid heuristic technique," *International Journal of Computer Applications*, vol. 41, no. 6, pp. 0975–8887, 2012.

[16] D. Das, "A fuzzy multi-objective approach for network reconfiguration of distribution systems," *IEEE Transactions on Power Delivery*, vol. 21, no. 1, pp. 202–209, 2006.

[17] M.H. Hemmatpour, M. Mohammadian, and M. Rashidinejad, "A novel reconfiguration mixed with distributed generation planning via considering voltage stability margin," *Amirkabir International Journal of Electrical and Electronics Engineering*, vol. 43, no. 1, pp. 23–34, 2011.

[18] J.S. Savier and D. Das, "Impact of network reconfiguration on loss allocation of radial distribution systems," *IEEE Transactions on Power Delivery*, vol. 22, no. 4, 2007.

[19] E. Dolatdar, S. Soleymani and B. Mozafari, "A new distribution network reconfiguration approach using a tree model," *International Journal of Computer, Electrical, Automation, Control and Information Engineering*, vol. 3, no. 10, 2009.

[20] D. Shirmohammadi and H.W. Hong, "Reconfiguration of electric distribution networks for resistive line loss reduction," *IEEE Transactions on Power Systems*, vol. 4, no. 1, pp. 1492–1498, 1989.

[21] H.D. Chiang and J.J. Rene, "Optimal network reconfiguration in distribution systems: Part 1: A new formulation and a solution methodology," *IEEE Transactions on Power Delivery*, vol. 5, no. 4, pp. 1902–1908, 1990.

[22] H.D. Chiang and J.J. Rene, "Optimal network reconfiguration in distribution systems: Part 2: Solution algorithms and numerical results," *IEEE Transactions on Power Delivery*, vol. 5, no. 3, pp. 1568–1574, 1990.

[23] S. Civanlar, J.J. Grainger, H. Yin, and S.S.H. Lee, "Distribution feeder reconfiguration for loss reduction," *IEEE Transactions on Power Delivery*, vol. 3, no. 3, pp. 1217–1223, 1988.

[24] D.T. Rizy, W.T. Jewell, and J.P. Stovall, "Operational and design considerations for electric distribution system with dispersed storage and generation (DSG)," *IEEE Transactions on Power Aparatus System*, vol. 104, pp. 2864–2871, 1985.

[25] K. Aoki, H. Kawabara, T. Satoh, and M. Kanezashi, "An efficient algorithm for load balancing of transformers and feeders," *IEEE Transactions on Power Delivery*, vol. 3, no. 4, pp. 1865–1872, 1998.

[26] R.E. Lee and C.L. Brooks, "A method and its application to evaluate automated distribution control," *IEEE Transactions on Power Delivery*, vol. 3, pp. 1232–1240, 1988.

[27] C.C. Liu, S.J. Lee, and S.S. Venkata, "An expert system operational aid for restoration and loss reduction of distribution system," *IEEE Transactions on Power Delivery*, vol. 3, pp. 619–625, 1988.

[28] T. Taylor and D. Lubkeman, "Implementation of heuristic strategies for distribution feeder reconfiguration," *IEEE Transactions on Power Delivery*, vol. 5, pp. 239–245, 1990.

[29] C.S. Chen and M.Y. Cho, "Determination of critical switches in distribution systems," *IEEE Transactions on Power Delivery*, vol. 7, no. 3, pp. 1443–1448, 1992.

[30] J. Kennedy and R. Eberhart, "Particle swarm optimization," *Proceedings of IEEE International Conference on Neural Networks*, pp. 1942–1948, 1995.

[31] S.K. Goswami and S.K. Basu, "A new algorithm for the reconfiguration of distribution feeders for loss minimization," *IEEE Transactions on Power Delivery*, vol. 7, no. 3, pp. 1484–1491, 1992.

[32] T. Niknam and E. Azad Farsani, "A hybrid evolutionary algorithm for distribution feeder reconfiguration," *Science China Technology Science*, vol. 53, pp. 950–959, 2010.

[33] N. Suman and S. Rana, "Network reconfiguration for electrical loss minimization," *International Journal of Instrumentation, Control and Automation (IJICA)*, vol. 1, pp. 2231–1890, 2011.

[34] S. Kalambe and G. Agnihotri, "Loss minimization techniques used in distribution network: Bibliographical survey," *Renewable and Sustainable Energy Reviews*, vol. 29, pp. 184–214, 2014.

[35] S.P. Kartikeyan, J. Raglend, P. Prathyusha, and D.P. Kothari, "A new methodology for distribution system feeder reconfiguration," *International Journal of Energy*, vol. 2, no. 1, pp. 9–15, 2008.

[36] Energy Statistics, 2019 (Twenty Sixth Issue), Central Statistics Office, Ministry of Statistics and Programme Implementation, Government of India, New Delhi, 2019.

[37] L. Zhechao, S. Jazebi, and F. de Le′ On, "Determination of the optimal switching frequency for distribution system reconfiguration," *IEEE Transactions on Power Delivery*, vol. 32, no. 4, pp. 2060–2069, 2017.

3

Existing Problems Related to Electrical Distribution Network, Part 2

Technical, Economical, and Environmental

Shilpa Kalambe[1], *Sanjay Jain*[2], *Bhojraj N. Kale*[3], *and Ujwala B. Malkhandale*[4]

[1] *Electrical Engineering Department, Dr. Babasaheb Ambedkar College of Engineering and Research, Nagpur, M.S., India*
[2] *Electronics and Electrical Engineering Department, R.K.D.F. University, Bhopal, M.P., India*
[3] *Mechanical Engineering Department, Dr. Babasaheb Ambedkar College of Engineering and Research, Nagpur, M.S., India*
[4] *Electrical Engineering Department, Priyadarshini Indira Gandhi College of Engineering, Nagpur, M.S., India*

3.1 Introduction

The rigorously growing advancements in the technologies and theories discovered for upgrading and updating the performance of a power system have affected the traditionally trailing operating scenario. This ultimately has unveiled many fields for researchers to study and show their skills in bringing back the normal state of the system after incorporating all those theories efficaciously. This era of updating and upgrading has witnessed the increasingly mature distribution system that is systematized as active distribution networks cope with complicated conventional power grids [1–5]. The issues of deleting natural resources, tremendously rising fossil fuel prices, grid security, and economic and environmental concerns have convinced governments around the globe to showcase favorable approaches toward the development of evolving microgrids. Impending microgrids facilitate a high rate of renewable penetration and proved to be significant building blocks of smart grids. Anyone can predict the well-advanced distribution system and will be well acquainted with all the pros and cons of existing issues it is facing. The attention-seeking issues of the distribution system, which certainly require to be addressed for efficient planning and restructuring of it, are as follows:

1. Losses
2. Inadequacy of the traditional distribution system structure to cope with the day-to-day enhancement of the load requirement
3. Deterioration of the grid over the course of time
4. Impact of worsened climatic conditions

Active Electrical Distribution Network: A Smart Approach, First Edition. Edited by Baseem Khan, Josep M. Guerrero, Sanjeevikumar Padmanaban, Hassan Haes Alhelou, Om Prakash Mahela, and Sudeep Tanwar.

5. Reliability of the system
6. Contingency analysis
7. Reverse power flow due to inappropriate allocation of distributed generators
8. Reactive power management
9. Voltage profile management
10. Network restructuring
11. Impacts of distributed generator insertion
12. Grid security
13. Economic operation of the system
14. Stability of the system

All these issues are strongly endorsed by the network operators, economists, and researchers working in the investigations of various parameters that hamper the performance of the power system, outcomes of newly authorized structure insertion, evaluation of load patterns, and their survival through preplanned as well as sudden strategies implemented for improving the system efficiency. This chapter presents a thorough comparative analysis of the state-of-the-art of addressing all these technical as well as non-technical problems relevant to distribution system utility.

3.2 What Is the Distribution System?

The distribution system is an integral part of a power system that distributes electric power to an end user at a low or medium voltage level. It is the part of the power network that connects the generation and transmission network to the consumer's meter [6]. It is a major part of the power system and reaches every nook and corner of the consumer's property. Its main objective is to transfer electricity to consumers using the defined standards of all performance parameters such as efficiency, quality, and reliability. The main concern of researchers in this area is minimizing energy losses and refining performance indices such as stability, contingency, and voltage profile [7]. Thus, while designing or upgrading an existing structure of the distribution system all its features and problems should be thoroughly studied and analyzed to exhibit optimum utilization of the resources. The substantial concerns of the distribution system are losses, power quality, grid security, economy, and unsuitable existing structures to cope with ever increasing demands [8].

3.3 Existing Problems Related to the Electrical Distribution Network

A comprehensive survey of the literature available in this field reveals various technical, economical, and environmental problems in the distribution systems, which are listed in Figure 3.1.

All the problems mentioned in Figure 3.1 are handled by network operators with the available tools and knowledge of the severity of a situation. The categorized description of these issues is given in this section.

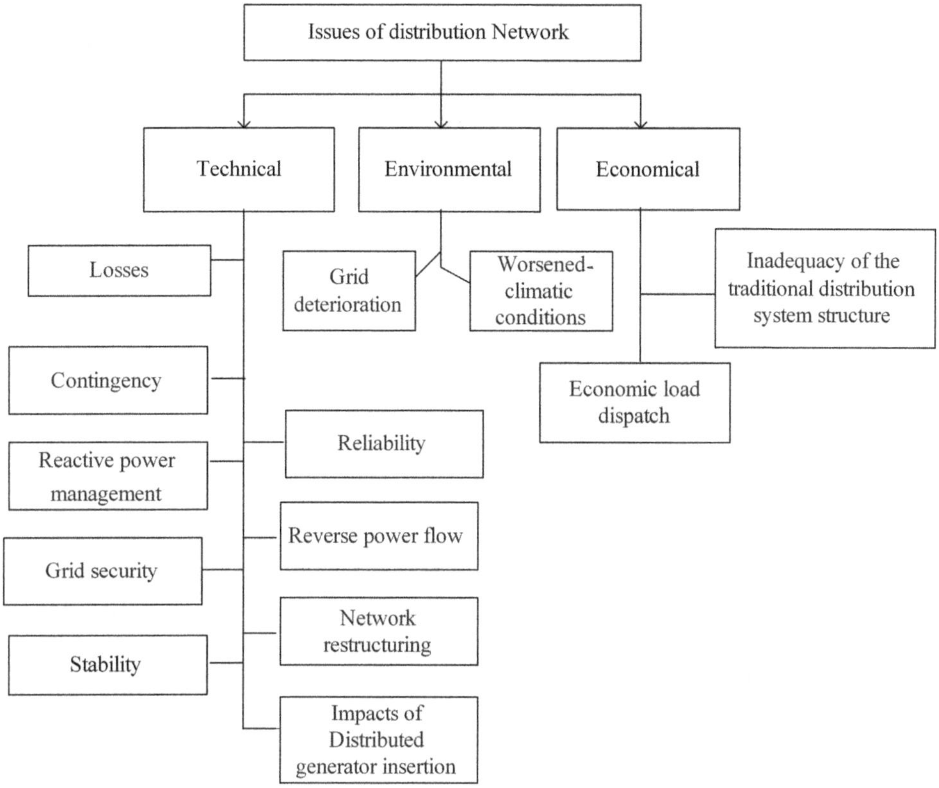

Figure 3.1 Problems in the distribution systems.

3.3.1 Technical Problems

Various technical problems often found in the distribution system are as follows:

1. Losses
2. Inadequacy of the traditional distribution system structure to cope with the day-by-day enhancement of the load requirement
3. Worsened climatic conditions
4. Reliability of the system
5. Contingency analysis
6. Reverse power flow due to inappropriate allocation of distributed generators
7. Reactive power management
8. Voltage profile management
9. Network restructuring
10. Impacts of distributed generator insertion
11. Grid security
12. Stability of the system

These problems persist from a long way back in the system and continued effort towards reducing their impact on the performance of the system are showing effective results, but much more is still impeding implementation.

3.3.1.1 Distribution Losses

A distribution system offers a linkage between a high-voltage transmission system and low-voltage consumers where heat loss in an overall network is high due to the high current. Distribution system operators have a financial incentive to diminish losses in these networks. This incentive is evaluated as the difference of the evaluated unit price between real and standard losses. Thus, if the real losses are higher as compared to standard losses then the distribution system faces the economic loss and in the opposite situation it gains the profit. Ultimately, this is revealed as a well-researched area as all available methodologies differ from each other in terms of selection of a tool for loss minimization, problem formulation, or problem solution methods executed for obtaining the best solution. Plenty of published formulations and methodologies of loss reduction, like network reconfiguration, capacitor allocation, increased voltage in the distribution system, conduct or grading, distributed generator (DG) insertion, etc., are showcasing their efficacy in improving the performance of the system.

Among all the predefined problems, the most important issue in the distribution system is the unavoidable unintended performance of the distribution sector due to improper design criteria, which leads to high transmission and distribution (T&D) losses, including electricity theft and unrealistic tariffs to consumers. This aspect highlights the financial status of the distribution companies, which is extremely stressed and leads to inadequate investment by the electricity sector, both in generation and in distribution. It also hampers the contribution of private investors toward generation due to the problems projected about the capacity of the unviable distribution companies to pay for power [9]. The power and energy losses in distribution grids are mostly related to energy conversion into heat subjected to the circulation of proportional currents in the line conductors, which is described as the Joule effect [10]. The transmission of electric energy over long lengths are subjected to power losses. The major part of these energy losses derives from the Joule effect in transformers, power lines, and other components in the power system. The energy is lost as heat in the transmission lines. The investigations have revealed that the overall losses lie in the range between 8 and 15%; thus its evaluation and analysis play an important role in calculating the efficiency of the distribution system. Pricing and appropriate cost allocation defining the precise contribution of operators in a power network in the total costs are counted as the most important problem identified along with all other important issues of the restructuring of the power industry. Fair allocation of price among all the contributors in the system is a complicated but still essential part of operating criteria in the electric power market [11].

3.3.1.2 Reliability of the System

The distribution system links the complicated structure of the transmission system to the individual customers. The existing power system is smart, twofold, active, interconnected, interdependent, load-sharing, and phased-mission systems [12]. Subsequently, to showcase all these features efficiently it has undergone many modifications in its structure, such as:

- Incorporation of new protection devices
- Inclusion of distributed generators in the systems
- Reactive power compensating devices
- Capacitor bank requirements

Inclusion of all such foreign structures may affect the reliability of the system. Reliability of a power system is commonly defined as a measure of the ability of the system to provide the product to the customers with a passable stream. It is considered to be one of the most important performance criteria of the power sector, which includes the planning, performance, and maintenance of electric power systems [13]. Reliability engineering is a subdiscipline of systems engineering that focuses on the ability of system components to function without failure. It enhances the system output by reducing the failure rate of the system. It describes the ability of a utility to function under defined criteria for a quantified period of time. Reliability and quality are the two faces of the coin so need to be paid due attention. It is a well-known fact that is commonly discussed that the probable cause of almost 90% of failures that occur in the distribution systems are due to reliability issues. Thus the first work was oriented to reliability indices, defined as objectives to attempt to improve performance measures in the electricity market. Overviewing all these concerns of updating the qualitative and quantitative output of the distribution system by considering the importance of new technologies, its effective and fruitful implementation, revealed that reliability aspects and energy sustainability are two principal issues of evolvement in any society [14].

3.3.1.3 Contingency Analysis

Contingency analysis plays a crucial role for understanding the impacts of sudden line outages due to failures of equipment, transformers, transmission lines, etc. This is mainly reflected in the system due to overburdened parts of the network caused by an irrational power flow in the lines. Its analysis is essential to take the necessary action for maintaining the security and reliable functioning of the system. The offline as well as online execution to envisage the effect of a distinct possibility of getting contingency in the system lines is a monotonous task due to the huge sizes and complication of the network. It has been found that only selected contingencies may result in severe situations in the power system and these can be tackled through precise contingency analysis. The procedure implemented for investigating and evaluating these conditions of overcrowding of power lines is referred to as contingency analysis, which can be done by calculating performance indices for each contingency criterion [15]. The analysis of such situations provides the eminent output of digital computing technologies of managing the various performance criteria, such as Energy Management Systems, SCADA systems, etc. The focus of these systems is primarily to give the network operator information regarding statistics improvement, grid security status, reliability, contingency, etc. Technically this is achieved by conducting a power flow analysis as there is the possibility of single or many contingency scenarios in the network, including lines, transformers, and generators. A mathematical execution of a contingency analysis is a critical function because of its prodigious extent of an information-gathering process in the execution of a problem [16]. It is considered as the most significant process in distribution system planning and operation. It has mainly been found that a line outage may lead to overburdened branches, sudden system voltage variations resulting in huge losses to the system in terms of economy, efficiency, and execution of all-important performance parameters, so the contingency analysis is often used to calculate violations in the predefined norms of the distribution system [17].

3.3.1.4 Reverse Power Flow Due to Inappropriate Allocation of Distributed Generators

Distributed generation (DG) is considered as an essential part in future distribution power systems. Grids are eventually being converted into smart grids with annexation of advanced and improved, technologically sound distributed generation (DG) sets, along with their advanced control systems, adaptive structures, effective sensors, information and communication technologies [18]. The DG insertion has changed the complete mode of operation of the distribution systems. The traditional one directional flow of power in the lines has changed to a two-way mode, which leads to an urgent need to redesign the protection systems. In smart grids the complicacy of the transformed structures and technologies of the power system is monitored and meticulously controlled by sensors, smart meters, digital controllers, and adaptive technologies [19].

3.3.1.5 Reactive Power Management

The renewable generation is connected to the distribution network at the low-voltage up to the high-voltage (HV) level. The insertion of this power can cause an objectionable voltage rise. Reactive power is also required to sustain and transfer this power in the transmission network. If distribution network operators (DNOs) regulate the reactive power locally, the distribution generators can contribute to reduce the many technical problems of the system. Reactive power production as well as its consumption, as required in the system by generators, urge the network operators to control operation of systems like voltages, stability, load angle, frequency, and power flow throughout the contributing system. Reactive power is one of the most important factors encountered for ensuring the smooth flow of the apparent power in the power lines. Generators are considered to be the prime source for reactive power and need to comply with the reactive power demand of the system and load. Due to many undesirable factors and inappropriate functioning of a few components or parts of the system, there may be evidence of an unrealistic flow of excessive reactive power in the transmission periphery, which may cause voltage instability and uneconomic network performance. Thus the reactive power control in a minor as well as a major power network is important, particularly for a steady load. However, the control of reactive power is difficult for network operators in large systems, where the system consists of long-length power lines and highly variable load because of widely and frequently varying meteorological conditions [20].

3.3.1.6 Voltage Profile Management

Increasing deployment of distributed generation units in the distribution network brings new challenges, such as voltage fluctuations, power quality hampering, and reverse power flows. The sudden changes in electric power generation from the DGs due to climate conditions or their internal glitch might cause high-voltage variations. In such critical scenarios both low load requirements and high power contributions from renewable sources possibly results in voltage, power, and load angle limit violations. To ride through this state of a highly approved approach is the disconnection of such units or curtailment of power generation, which may lead to several disadvantages, such as a sudden reactive power imbalance, harmonic distortion, voltage instability, and malfunction of protection devices [21]. Thus the DG insertion has proved to be a boon accompanied with many technical challenges that are definitely required to be analyzed in order to enjoy all the benefits of DG insertion in the distribution networks.

3.3.1.7 Network Restructuring

Due to the low-voltage–high-current operating scenarios of the distribution system, I^2R losses are the major concern of DNOs and need to be taken care off to improve the performance of the system. Network restructuring is one of the methods to deal with this problem, which is generally used in low-voltage distribution systems. It is a very effective and conclusive method to save the electrical energy. Distribution systems consist of a large number of interconnected radial and mesh networks. The network configuration and structures of distribution systems may be varied through switching operations to reallocate loads in the feeders with exhaustive planning. This can be executed through two types of switches, which are often used in primary as well as secondary distribution systems:

- Sectionalizing switches (normally closed switches) or
- Tie switches (normally open switches)

These types of switches are particularly structured for protection and configuration supervision of complicated distribution systems. Network reconfiguration is the commonly used process of altering the distribution systems topology by changing the open/closed position of the switches. Reconfiguration is primarily applied for:

1. Service restoration during faulty conditions
2. Load balancing for
   a. release of overburdened power lines and
   b. to improve the overall voltage profile
3. Planning the line outages for maintenance conditions and
4. Loss reduction

Loss minimization executed by the switching operation is found to be the elementary control action in network restructuring. The prime activity is the closing of the switches in an opened branch and opening the switch in a closed one by maintaining the network state radial. However, due to the possibility of several candidates switching combinations in the system at the same time, it is a complex combinatorial problem. This distinct feature of the availability of numerous switch combinations makes it a discrete optimization.

Recurring changes in formation can cause outages or transient problems. However, network restructuring has proved to be a complex combinatorial decision-making execution for contributors to trace, which generally requires extensive numerical designs and scheming, and affects the management of the control strategies of protective devices. In feeder reconfiguration, it is usually assumed that the protecting elements are precisely harmonized and correlate when the feeder configurations are altered by acute switch operations, which are achieved through essential protective device planning and coordination, usually carried out for secure configuration [22].

3.3.1.8 Impacts of Distributed Generator Insertion

Continuously growing energy requirements resulting due to industrialization, environmental concerns with diminishing fossil fuel resources, network congestion as a result of overburdened lines, and declining technical performance are the key motivating factors for the day-to-day increase in integration of both renewable as well as non-renewable distributed generation (DG) units and transition of the prevailing power systems into restructured and advanced systems. Improving the technological and economic outcomes as a result of DG allocation is the biggest challenge for distribution network operators (DNOs) for both types of distributed generators. Distribution system engineers showed strong interest in these tiny energy sources due to their technical, economic, and environmental enticements. Its impending assistance includes improved system reliability, greenhouse gas emissions, and energy loss reductions [23]. However, the increased rate of DG deployment provides a prime challenge to distribution network operators because traditional distribution networks are not designed to connect power generation facilities on their doorsteps. With the integration of DGs in these passive networks, wide ranges of alternations and sometimes parameter violations result in variability and uncertainty in the operation of renewable energy resources (RERs). It has been found that renewable DGs reveal many associated power quality (PQ), characteristics hampering, and performance violation challenges to the operators. Power quality is one of the most significant characteristics of renewable DG systems because today's loads are excessively sensitive to PQ disturbances. Therefore, the need for innovative power quality improvement techniques becomes a foreseeable feature of the restructured power system [24].

3.3.1.9 Grid Security

The changed scenario of the distribution system with many technological changes, such as network restructuring, distributed generation insertion, increasing complicity to relieve the network contingency, creates many problems in its normal operation. One of the significant problems is the grid security in terms of:

- Stability
- Altering the status of protection components, such as setting of relays, circuit breakers, etc.
- Types of relays that were installed at the time of installation by keeping the one-directional state of distribution systems required to be changed due to the bidirectional operating scenario
- Increased MVA capacity of the conductors

All these issues strongly recommend urgent alteration in the present structure to cope with newly installed technologies.

3.3.1.10 Stability of the System

Stability is the ability of a power system to maintain steady and acceptable limits and standards of performance parameters such as voltage and load angle at all buses under both normal and abnormal (faulty) network conditions [25]. The load on the distribution network shows conscious behavior due to variations in consumer demands. In certain industrial as well as domestic areas, it is often observed that under certain critical loading conditions and faulty situations the distribution system experiences a voltage collapse, which is also called an instability or unstable operation of the system. The main factor causing voltage instability is the inability of power systems to comply with the demand of reactive power and maintenance of required parameter, demand load correlation [26].

3.3.2 Economic Problems

3.3.2.1 Inadequacy of the Traditional Distribution System Structure to Cope with the Day-by-Day Enhancement of the Load Requirement

The available structure was designed to supply a load that was much less than the ever-increasing load the power system is bearing nowadays. Structure analysts and power system engineers work hard to investigate the means to make the available system compatible to sustain this continuously growing demand. Network modification, renewal at certain sectors, or insertion of distributed generation to make it more flexible are some of the remedies that have been incorporated in the system to make it compatible to sustain the enhanced burden of the ever increasing load.

3.3.2.2 Economic Operation of the System

Evaluation of distribution system losses that occurred through generating units loading plays an important role in efficiency evaluation of distribution systems. All the enhancements made for reactive power compensation required in distribution systems require an investment and proportional maintenance cost, which must be examined along with the outputs of generating units and energy loss reduction achieved by the compensating devices, along with the benefits of quality and reliability advances, which have proved to be the qualitative goals targeted with the use of these devices. The aspects that are responsible for load dispatching have operated with a precise evaluation of generation costs of each discrete unit, but its assessment certainly necessitates an adequate knowledge of the impacts on system losses occurring from load shifting among various available generation units. In order to function at maximum efficiency it is necessary to understand and control appropriate generator scheduling, the allocation of an optimal cost of system losses, and thereafter evaluating the correlation of these costs with station operational costs. Such a methodology is considered to be an important tool in the maintenance of the operation of existing and working units and in the planning and allocation of proposed generating stations [27].

3.3.3 Environmental Problems

3.3.3.1 Deterioration of the Grid with the Course of Time

This factor is completely encompassed by civil engineering but still affects the electrical engineering segment. Like all other plants, power sector structures do not remain untouched by the adverse effects of the progressing course of time. Due to the advance of time and its degrading effects, the efficacy of the system is hampered, which in turn is accompanied by the economic as well as efficiency losses. Restructuring and load up-gradation needs vast investments and through planning to incorporate the compactible structures to balance the load requirements and sustain the overburdened demands. The adverse effects of deterioration of the system can be changed by implementing the following means:

- Preventive maintenance
- Replacement of degrading units
- Annual renewal planning
- Implementation of the latest technologies to improve the performance
- Annual performance audit

Apart from these, many other strategies have been found to be implemented by network operators to improve the system performance and enhance the revenue.

3.3.3.2 Impact of Worsened Climatic Conditions

Like all other sectors the distribution system is also greatly influenced by impacts of climatic conditions. The performance degradation due to variations in renewable energy generation, which is often subjective to climate variations, is uncontrolled by human interferences and obligations and is the major concern of evaluation of distribution system operation. The often found mismatch of generation and the electricity consumption results in load shading, load curtailment, and deficiency of electricity supply during an electricity consumption peak [28].

3.4 Conclusion

The effective operation of an active distribution network requires regulation and control, load forecast, its analysis and execution of adaptive technologies for resilient system operation, grid security monitoring and parameter optimization monitoring, etc. All these operations require advanced data management methods like energy management systems, SCADA, etc. However, to obtain the maximum output from those technologies, all the technical, economic, and environmental issues faced by distribution systems will be well researched, discussed, analyzed, and evaluated. It is also essential to improve performance and quality of power delivery by a distribution network to consumers. The state-of- the-art method of addressing these issues has been presented in this chapter, highlighting the major concerns for a distribution system that offers the best way to adopt installation of new technologies and resources in its territory.

References

[1] B. Khan, H.H. Alhelou, and F. Mebrahtu, "A holistic analysis of distribution system reliability assessment methods with conventional and renewable energy sources," *AIMS Energy*, vol. 7, no. 4, pp. 413–429, 2019.

[2] Y. Kifle, B. Khan, and P. Singh, "Assessment and enhancement of distribution system reliability by renewable energy sources and energy storage," *Journal of Green Engineering*, vol. 8, no. 3 (2), pp. 219–262, July 2018.

[3] B. Khan and P. Singh, "Optimal power flow techniques under characterization of conventional and renewable energy sources: A comprehensive analysis," *Journal of Engineering*, vol. 2017, Article ID 9539506, 16 pages, 2017.

[4] T.F. Agajie, B. Khan, H.H. Alhelou, and O.P. Mahela, "Optimal expansion planning of distribution system using grid-based multi-objective harmony search algorithm," *Computers & Electrical Engineering*, vol. 87, 106823, 2020.

[5] Y. Kifle, B. Khan, and J. Singh, "Designing and modeling grid connected photovoltaic system: (Case study: EEU building at Hawassa city)," *International Journal of Convergence Computing*, vol. 3, no. 1, pp. 20–34, 2018.

[6] Electric Power Distribution Engineering, TuranGonen, CRC Press, United States, 3rd Edition, 2014.

[7] T.M. Krishna, N.V. Ramana, and S. Kamakshaiah, "A novel algorithm for the loss estimation and minimization of radial distribution system with distributed generation," *2013 International Conference on Energy Efficient Technologies for Sustainability*, Nagercoil, 2013, pp. 1289–1293.

[8] S. Kalambe and G. Agnihotri, "Loss minimization techniques used in distribution network: Bibliographical survey," *Renewable and Sustainable Energy Reviews*, vol. 29, pp. 184–200. 2014, R. Sangepu and V. Muni, Effect of power quality issues in power system and its mitigation by power electronics devices, *Discovery*, vol. 28, no. 105, pp. 72–79, 2015.

[9] R.M. Holmukhe, "Electricity distribution sector in India: Key challenges for service to customer & way out," *International Journal of Application or Innovation in Engineering & Management*, vol. 5, no. 7, pp. 116–119, July 2016.

[10] E.M. García, A. Águila, I. Isaac, J.W. González, and G. López, "Analysis of voltage profile to determine energy demand using Monte Carlo algorithms and Markov Chains (MCMC)," *Power Engineering Conference (UPEC) 2016*, 51st International Universities, pp. 1–6, 2016.

[11] M.H. Salar and M.R. Haghifam, "Transmission loss allocation in power systems using artificial neural network," *2010 IEEE International Conference on Power and Energy*, Kuala Lumpur, 2010, pp. 729–734.

[12] S.S. Sarma, V. Madhusudhan, and V. Ganesh, "Evaluation and enhancement of reliability of electrical distribution system in the presence of dispersed generation," *2016 International Conference on Signal Processing, Communication, Power and Embedded System (SCOPES)*, Paralakhemundi, 2016, pp. 357–362.

[13] P.K. Goswami, S. Chowdhury, S.P. Chowdhury, Y.H. Song, and J.K. Das, "Reliability evaluation of distribution system," *2007 42nd International Universities Power Engineering Conference*, Brighton, 2007, pp. 158–166.

[14] R. Billinton and J. Billinton, "Distribution system reliability indices," *IEEE Transactions on Power Delivery*, vol. 4, no. 1, pp. 561–568, January 1989.

[15] A. Roy and S. Jain, "Contingency analysis in power system," Thesis of Master of Engineering in Power Systems & Electric Drives, Thapar University, Patiala, 2011.

[16] R. Bacher, "Graphical interaction and visualization for the analysis and interpretation of contingency analysis results," *Proceedings of Power Industry Computer Applications Conference*, Salt Lake City, UT, USA, 1995, pp. 128–134.

[17] V.J. Mishra and M.D. Khardenvis, "Contingency analysis of power system," *2012 IEEE Students' Conference on Electrical, Electronics and Computer Science*, Bhopal, 2012, pp. 1–4.

[18] U.S. Department of Energy, Smart Grid, Available at: http://www.oe.energy.gov/smartgrid.htm. [Accessed 13 August 2020].

[19] B. Jiang and Y. Fei, "Dynamic residential demand response and distributed generation management in smart microgrid with hierarchical agents," *Energy Procedia*, vol. 12, no. 2011, pp. 76–90, 2011.

[20] M. Singh, T.V. Vardhan, J. Pradhan, and K.S. Meera, "Reactive power management in transmission networks," *2017 7th International Conference on Power Systems (ICPS)*, Pune, 2017, pp. 568–572.

[21] T.U. Okeke and R.G. Zaher, "Reactive power management for distributed generation: Motivation and solutions," *2013 International Conference on New Concepts in Smart Cities: Fostering Public and Private Alliances (SmartMILE)*, Gijon, 2013, pp. 1–4

[22] H.W. Hong, C.T. Sun, V.M. Mesa, and S. Ng, "Protective device coordination expert system," *IEEE Transaction on Power Delivery*, vol. 6, no. 1, pp. 359–365, 1991.

[23] S. Ruiz-Romero, A. Colmenar-Santos, R. Gil-Ortego, and A. Molina-Bonilla, "Distributed generation: The definitive boost for renewable energy in Spain," *Renewable Energy*, vol. 53, no. 2013, pp. 354–364, 2013.

[24] M. Bajaj and A. Singh, "Grid integrated renewable DG systems: A review of power quality challenges and state-of-the-art mitigation techniques," *International Journal of Energy Research*, vol. 44, pp. 26–69, 2019.

[25] P. Kundur, Power System Stability and Control. New York: McGraw-Hill, 1994.

[26] N.K. Roy A, H.R. Pota, and M.J. Hossain, "Reactive power management of distribution networks with wind generation for improving voltage stability," *Renewable Energy*, vol. 58, pp. 85–94, 2013.

[27] J.B. Ward, J.R. Eaton, and H.W. Hale, "Losses in power transmission networks," *Electrical Engineering*, vol. 69, no. 5, pp. 451–451, May 1950.

[28] S. Ruifeng, Z. Penghui, Z. Jie, L. Niu, and H. Xiaoting, "Multidispatch for microgrid including renewable energy and electric vehicles with robust optimization algorithm," *Energies*, vol. 13, no. 2813, pp. 1–15, 2020.

Part III

Harmonics Mitigation in the Smart Distribution Network

4

Power Quality Mitigation in a Distribution Network Using a Battery Energy Storage System

Gajendra Singh Chawda[1], Om Prakash Mahela[2], and Bhuvnesh Rathore[3]

[1] *Department of Electrical Engineering, Indian Institute of Technology, Jodhpur, Rajasthan, India*
[2] *Power System Planning Division, Rajya Vidyut Prasaran Nigam Ltd., Jaipur, Rajasthan, India*
[3] *Department of Electrical Engineering, MBM, Jodhpur, Rajasthan, India*

4.1 Introduction

With increased industrialization and power demand, power engineers are looking for new and efficient ways for power generation and their sources. For this, renewable energy (RE) technologies are the best and most promising option/answer, along with an advantage of reduced carbon emissions [1]. Battery energy storage systems (BESSs) have been proved to be an efficient emerging technology for short-term power storage for power utilization for mitigation of various power quality (PQ) disturbances, stability, reliability, and protection [2–5].

In the current scenario, the demand for good quality of power in the distribution grid enhances significantly due to frequent use of smart equipment. These types of equipment are the primary cause of harmonics. This distribution equipment draws distorted sinusoidal current and behaves as nonlinear loads, which deteriorates the quality of supply in the network of the distribution grid [6]. Thus, PQ is defined by any deviation in frequency, voltage quality, and current quality of the electrical signal [7]. The major PQ issues include a low power factor, unbalanced voltage, reactive power requirement, and harmonics level in three-wire distribution systems. These issues generally originate due to the proliferation of nonlinear loads. It causes mal operation, failure, or damages of smart electrical equipment of the distribution network [8]. Hence, the nature of the distribution load at the point of common coupling (PCC) is of significant concern when analyzing the impact of the utility network performance given PQ in compliance of the IEEE 519-2014 standard [9].

Literature details the various smart solutions for the aforementioned problems in the distribution system. For this purpose, research on different design architectures of BESS for power quality application is under development [10]. The battery integrated with FACTS devices is one of the smart solutions for the distribution network [11, 12]. Also, there can be various topologies of the converters like a four-quadrant converter topology with which

Active Electrical Distribution Network: A Smart Approach, First Edition. Edited by Baseem Khan, Josep M. Guerrero, Sanjeevikumar Padmanaban, Hassan Haes Alhelou, Om Prakash Mahela, and Sudeep Tanwar.

bidirectional energy flow can be achieved. The voltage source converter (VSC) has been used for modulating the energy flow in mostly BESS controllers for controlling active and reactive power simultaneously and independently [13]. An adequate charge controller is needed for appropriate charging of the battery, for when a battery is fully charged (constant voltage), its charge acceptance capability and charging current are both reduced [14, 15]. Thus, it takes a long time for full charging at a constant voltage. Hence, various control algorithms have been proposed in the literature, such as an instantaneous current control theory for PQ conditioning with BESS [16], a synchronous reference frame control theory for PQ mitigation in the presence of motor load [17], an adaptive linear element-based control for PQ improvement in the presence of multifarious loads [18], a least mean control algorithm for PQ mitigation in a weak grid environment with inductive nature loads [19], and application of a control algorithm for RE power generation [20, 21] for power quality mitigation and maximum power extraction. Moreover, the performance comparisons of these algorithms have been presented in [22] for the power quality application.

Apart from all the control algorithms, as mentioned above, indirect current control theory (ICCT) has the additional ability for fast regulation of the battery charging level and distributed system stability for transient conditions. Also, the proposed control algorithm provides a constant DC link voltage with fewer fluctuations.

This chapter aims to design an ICCT controlled DSTATCOM-BESS configuration for power conditioning in a distribution system with a balanced and unbalanced linear, nonlinear, and induction motor load. This detailed analysis helps beginners in selecting the type of load for the specific BESS-based configuration. The basic configuration of ICCT controlled DSTATCOM-BESS system is illustrated in Figure 4.1.

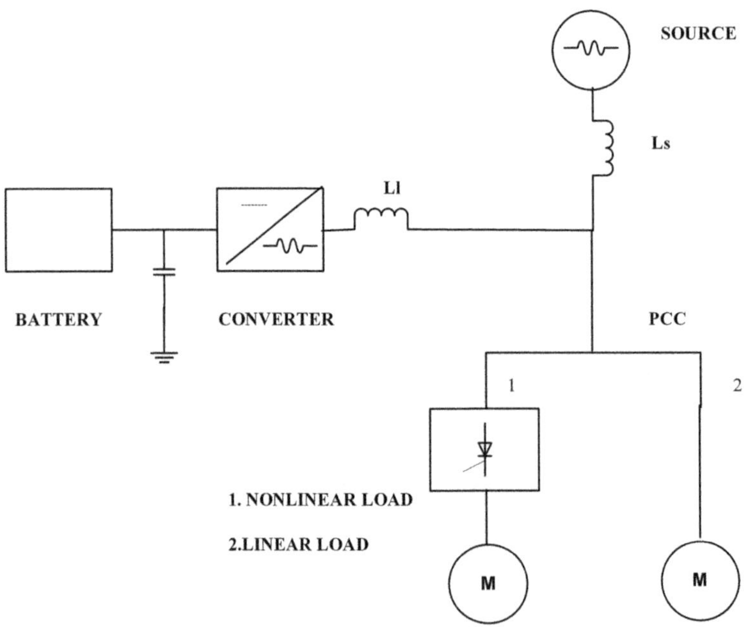

Figure 4.1 ICCT controlled DSTATCOM-BESS system connected at PCC.

4.2 System Configuration of D-STATCOM with BESS

The proposed ICCT-controlled DSTATCOM-BESS configuration is illustrated in Figure 4.2. This system is composed of various components such as a three-phase voltage source converter, energy storage device, ripple filter, and specific controller. The application of the proposed system is best suited for rural areas, where loads are unpredictable [23].

- **Three-phase VSC:** The VSC is the core component of the D-STATCOM. During an abnormal condition, the VSC generates compensating current to support the voltage at the PCC under load unbalancing and provides reactive power compensation during a nonlinear load connection.
- **Energy storage:** A Conventional STATCOM has only a DC-link capacitor for supporting the reactive power. However, the requirement for active and reactive power at a coupling point is frequent, while an unbalanced nonlinear load is connected. Hence, the DSTATCOM-BESS configuration is introduced for maintaining the DC-link voltage.
- **Filter:** A small rating of the capacitive filter is introduced at the PCC for mitigating the ripples from the signals.
- **Controller:** The proposed ICCT control algorithm estimates the gating signals for the accurate output voltage of the DSTATCOM-BESS. Also, the proposed algorithm has proved to be the best-suited control algorithm for power quality mitigation applications.

4.3 BESS Operation and Its Control

The proposed ICCT-controlled DSTATCOM-BESS system under current controlled operation is presented in Figure 4.3.

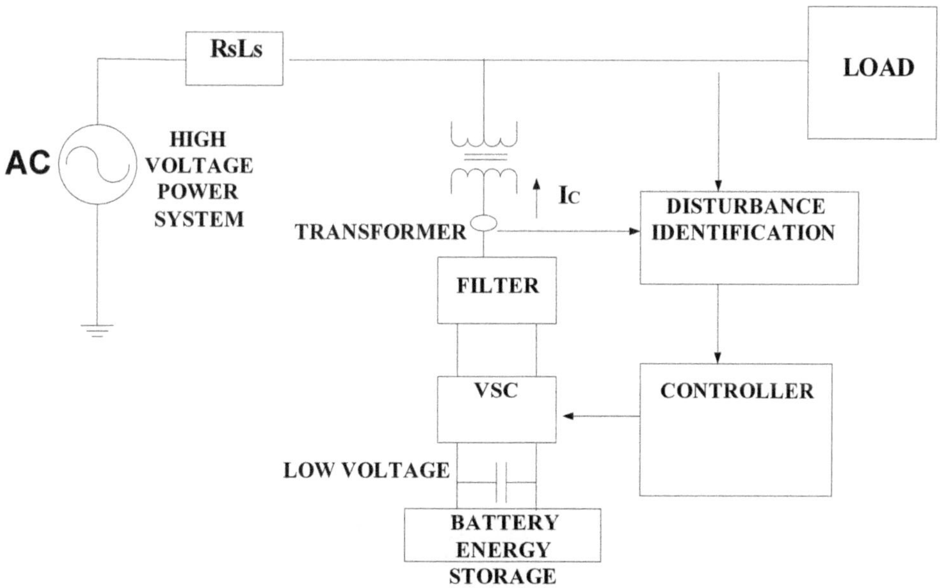

Figure 4.2 System scheme of D-STATCOM with BESS.

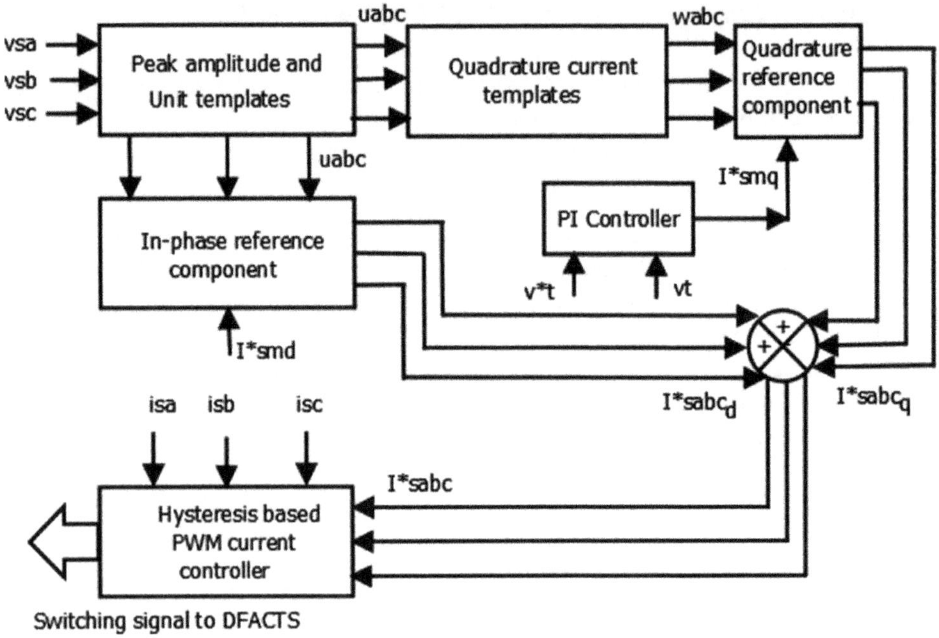

Figure 4.3 ICCT-based control algorithm for DSTATCOM-BESS.

The ICCT-based control algorithm consists of two components of three-phase reference currents with respect to phase-voltages, i.e. the in-phase component and the quadrature component. I^*_{sad}, I^*_{sbd}, and I^*_{scd} are the in-phase components of the reference source currents. These are required for charging the battery of the BESS system which supports active power to the connected load. However, the other techniques for the generation of reference source currents for maintaining power quality are well discussed in [24].

The single point operation of load levelling can be illustrated by assigning a constant value of I^*_{smd}. However, the multiplication of the obtained value of in-phase voltage unit vectors (U_a and U_c) are calculated by dividing the AC voltages (V_a, V_b, and V_c) using the peak amplitude (V_m), which provides the in-phase component of the reference currents (I^*_{sad}, I^*_{sbd}, and I^*_{scd}). Also, estimation of the quadrature component of the source reference currents (I^*_{saq}, I^*_{sbq}, and I^*_{scq}) is required for accurate control of the supply voltage at the PCC. Moreover, the quadrature unit vectors (W_a, W_b, and W_c) are perceived from the in-phase voltage unit vectors (U_a, U_b, and U_c). The peak amplitude of the voltage (V_m) is sensed at a coupling point and compared with the calculated reference voltage. This generates the voltage error. The obtained error is further passed through the proportional-integral voltage controller. The output of this controller (I^*_{smq}) ensures that the AC voltage control and provides the amplitude of reactive current component required for the system. The multiplication of W_a, W_b, and W_c with I^*_{smq} helps to estimate the quadrature component of the reference source signals $\left(I^*_{saq}, I^*_{sbq}, \text{and } I^*_{scq}\right)$. The addition of the quadrature component and in-phase component of the above-estimated reference signals provides the reference source currents (I^*_{sa}, I^*_{sb}, and I^*_{sc}). However, the hysteresis-based current controller is implemented for the generation of gating pulses for the DSTATCOM-BESS system [25–27].

4.4 Discussion of the Simulation Results

The proposed system has been examined under balanced and unbalanced loading conditions for load levelling and balancing. Power quality disturbances are also mitigated using the proposed approach. The indirect current control (ICCT) theory is implemented to generate switching signals for the shunt connected DSTATCOM-based BESS, as illustrated in Figure 4.4.

4.5 Results Analysis under a Balanced System

The proposed system is analyzed under a balanced system condition that considers linear, nonlinear, and induction motor loads for demonstrating the performance on harmonic eliminations. The system is analyzed for source voltage (V_s), terminal voltage (V_t), source current (I_s), compensating current (I_c), load currents (I_l), battery voltage ($V_b b$), and battery current ($I_b b$) based signals. The parameters under the system consideration are as follows:

Source: $V_{peak} = 415$ V, 3-Phase, 50 Hz
Linear load: $\text{Resistive}_{\text{load}} = 1000\ \Omega$, $\text{Inductive}_{\text{load}} = 1\text{e}^{-3}\text{H}$
Nonlinear load: Rectifier-based inductive load with $R_{\text{load}} = 1000\ \Omega$, $L_{\text{load}} = 1\text{e}^{-3}\text{H}$
Induction motor load: Nominal power = 3730VA, 415 V, 50 Hz and Tm = 10, hysteresis controller band 0.1
Battery and controller parameter: $V_{oc} = 720$ V, $R_{b1} = 0.1\ \Omega$, $R_{b2} = 1000\ \Omega$, $C_{b2} = 1688$ F, $C_{dc} = 5000$ F, $I^*_{smd} = 35$
PI controller: Proportional constant = 043, Integrator constant = 0.15

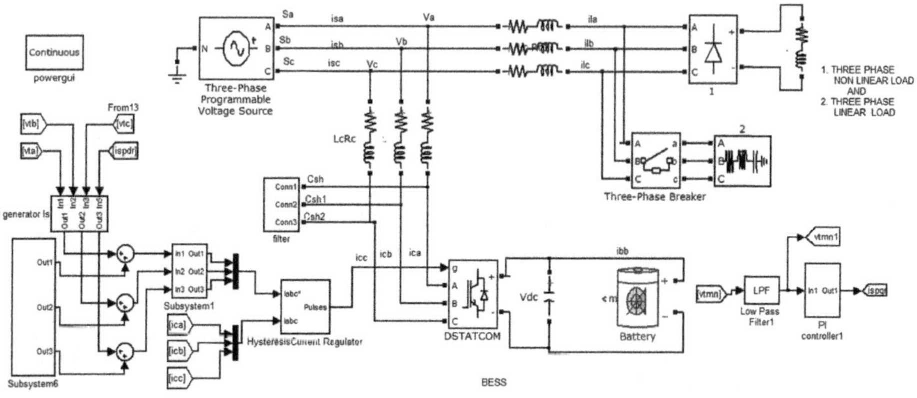

Figure 4.4 MATLAB simulation of indirect current control theory controlled for the DSTATCOM-BESS system.

4.5.1 Response of DSTATCOM-BESS under a Balanced Linear Load

The response of the DSTATCOM-BESS system with a balanced linear load is as depicted in Figure 4.5. The linear load varies from 1000 Ω to 1 Ω to test the battery charging operation. It has been perceived that the source voltage (V_s) and currents (I_{sa}) are balanced and in-phase under the battery charging condition ($I_b b$). Load currents (I_l) are observed to be balanced when keeping the balanced terminal at its reference voltage. Also during the operation the compensating current (I_c) is in balance and continuously provides the required current.

4.5.2 Response of DSTATCOM-BESS under a Nonlinear Load

The response of the DSTATCOM-BESS system with the nonlinear load is depicted in Figure 4.6. It has been visualized that the ICCT is initially fixed in the constant current control mode operation. Hence, the system's other load is removed in order to demonstrate the off-peak condition and under this condition, batteries are now starting to charge. Also, due to efficient control, the reference voltage is regulated in order to maintain a constant charging current. It has also been seen that the DSTATCOM provides a compensating current to maintain the source current and voltage sinusoidal and is balanced by keeping the terminal voltage at its calculated reference voltage. Moreover, the source current harmonics and load current harmonics are illustrated in Figures 4.7 and 4.8 and are visualized at 0.44% and 30.31%, respectively, which shows the proposed system's superiority.

4.5.3 Response of DSTATCOM-BESS under an Induction Motor (IM) Load

The response of the DSTATCOM-BESS system with an IM load is as depicted in Figure 4.9. It is well known that if T_m is positive, the machine operates as a motor, and if it is negative,

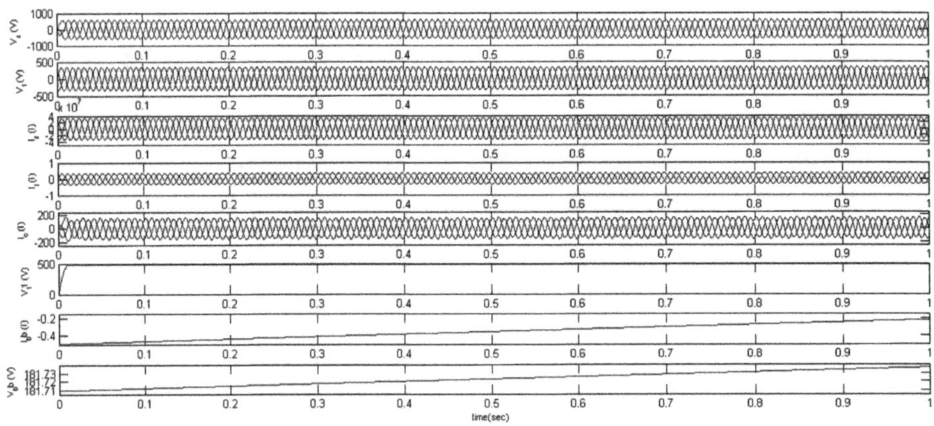

Figure 4.5 Response of DSTATCOM-BESS under a balanced linear load.

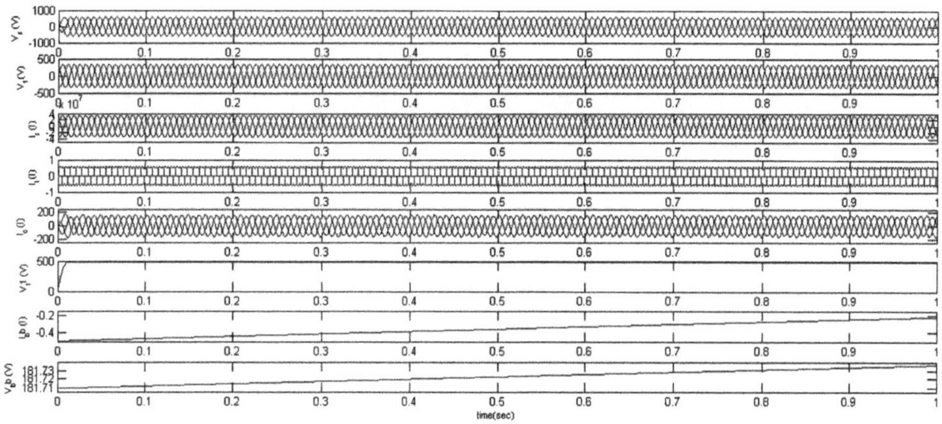

Figure 4.6 Response of DSTATCOM-BESS under a nonlinear load.

the machine operates as a generator. Hence, the sign of the mechanical torque directs the mode of operation. It has been visualized that the motor takes a large starting current at the time of starting. Accordingly, DSTATCOM injects the compensating current to maintain

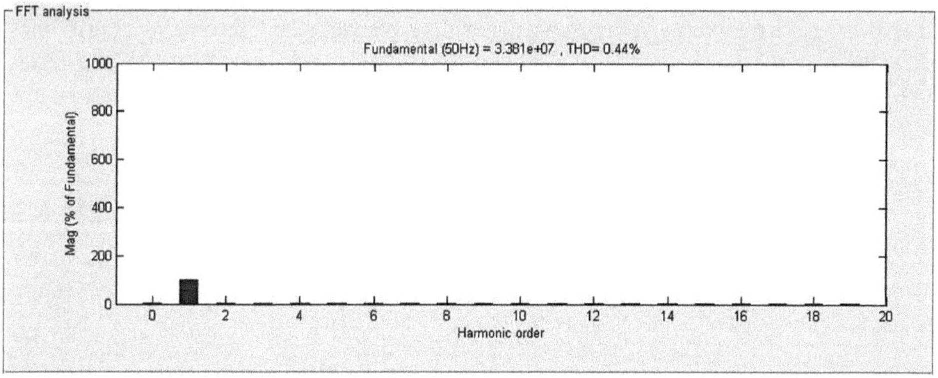

Figure 4.7 Source side frequency spectrum.

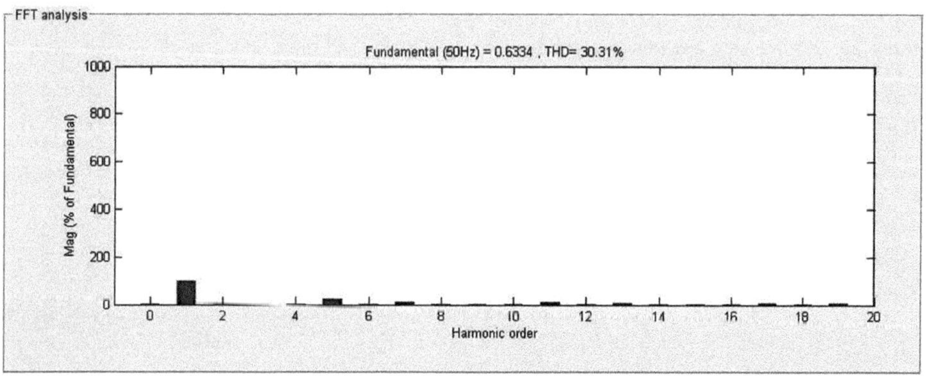

Figure 4.8 Load side frequency spectrum.

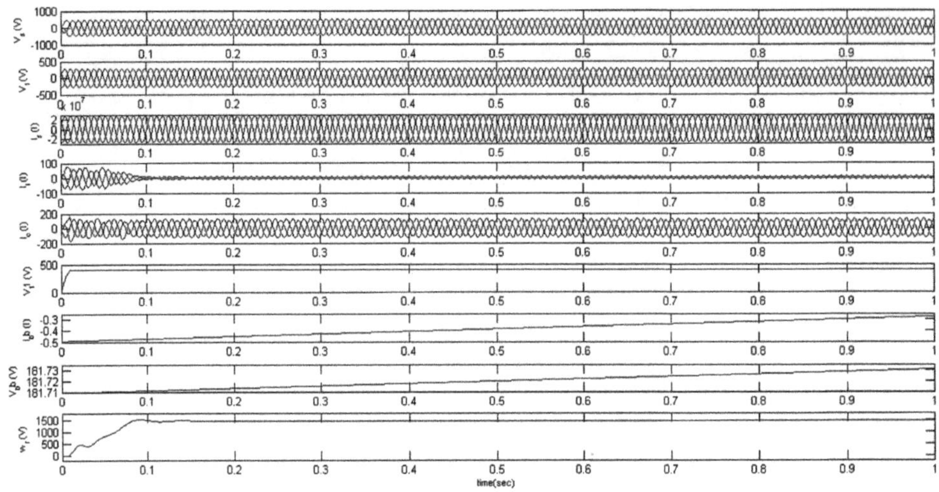

Figure 4.9 Response of DSTATCOM-BESS under an induction motor load.

the voltage at PCC and current sinusoidal. During this condition, the terminal voltage is also kept at its reference voltage and the battery starts charging. Also, the grid side and load side harmonics are depicted in Figures 4.10 and 4.11, respectively. The obtained harmonics are well below the international standard IEEE519-2014. Moreover, Table 4.1 illustrates the harmonics-based comparative analysis of all considered loads. The perceived results establish the performance of the DSTATCOM-BESS system.

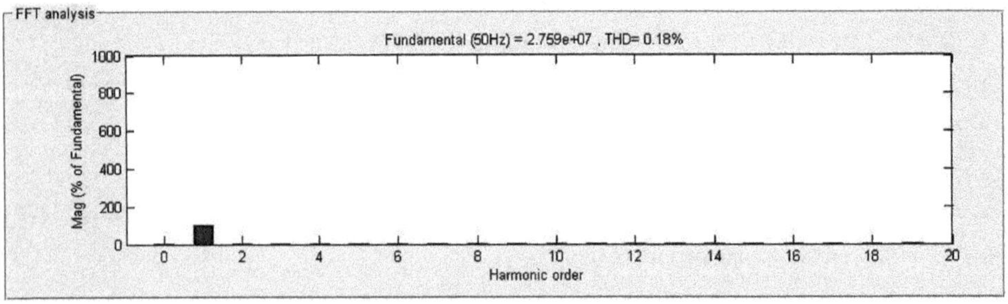

Figure 4.10 Source side frequency spectrum.

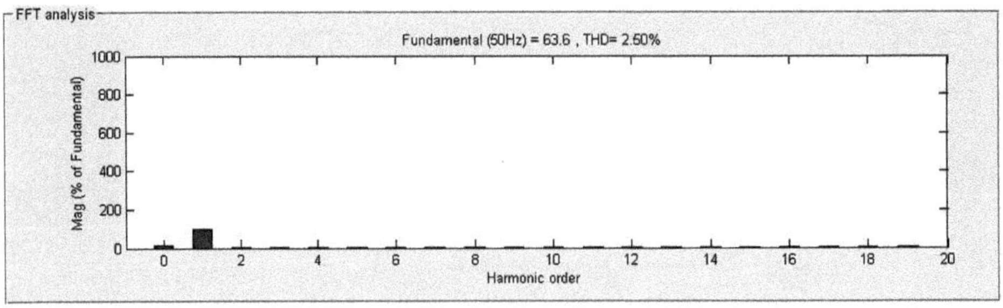

Figure 4.11 Load side frequency spectrum.

Table 4.1 Harmonics analysis under balanced loads.

		Indirect current control theory controlled DSTATCOM-BESS	
S. No.	**Type of load**	**Source side THD**	**Load side THD**
1	Linear load	0.57%	0.57%
2	Nonlinear load	0.44%	30.31%
3	Motor load	0.18%	2.50%

4.6 Results Analysis under an Unbalanced System

4.6.1 Response of DSTATCOM-BESS under an Unbalanced Linear Load

The response of the DSTATCOM-BESS system with an unbalanced linear load is as depicted in Figure 4.12. The unbalancing in the linear load starts from 0.25 seconds to 0.5 seconds. The intention of load unbalancing (I_l) from three-phase to two-phase and then to one-phase and again reverse illustrates the proposed BESS-based system's performance. It has been perceived that the source voltage (V_s) and current (I_s) are balanced and in-phase during and after the unbalancing. The battery charging is slightly disturbed during unbalancing but maintains a continuous charge (I_bb). The whole operation compensates the current (I_c) from the DSTATCOM continuously injected while keeping the balanced terminal voltage (V_t) at its calculated reference voltage. Also the source and load side current are visualized in Figures 4.13 and 4.14.

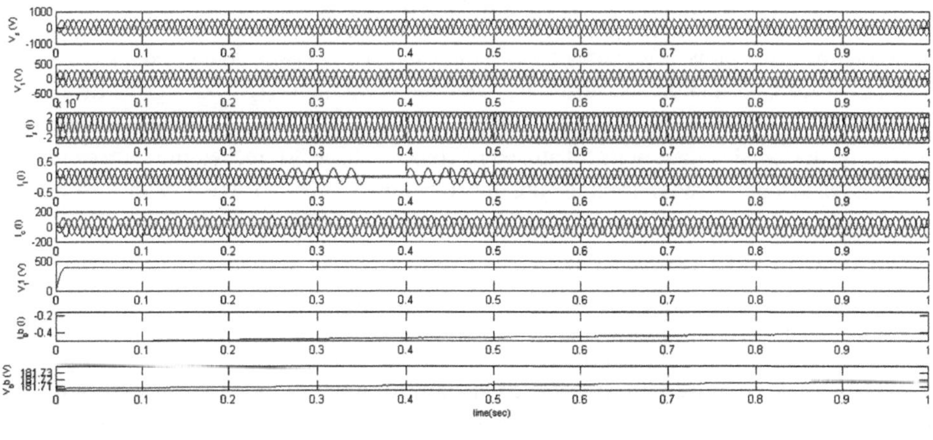

Figure 4.12 Response of DSTATCOM-BESS under an unbalanced linear load.

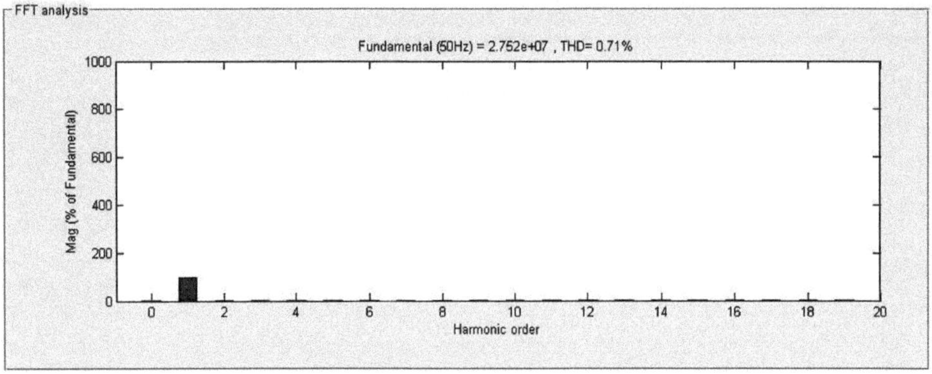

Figure 4.13 Source side frequency spectrum.

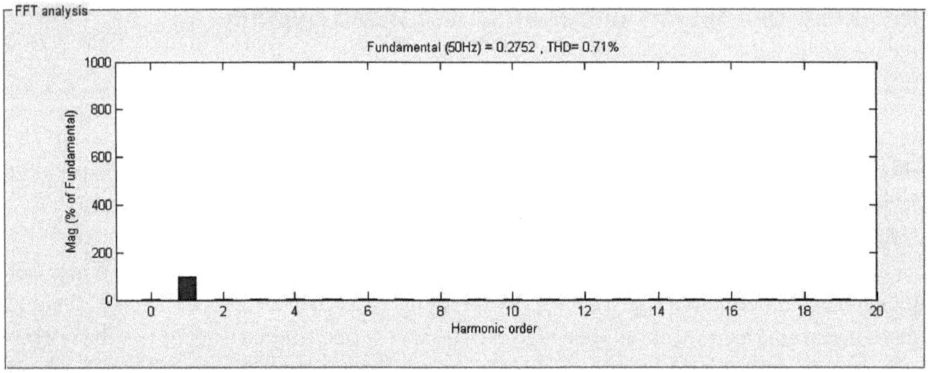

Figure 4.14 Load side frequency spectrum.

4.6.2 Response of DSTATCOM-BESS under an Unbalanced Nonlinear Load

The response of the DSTATCOM-BESS system with an unbalanced nonlinear load is as depicted in Figure 4.15. At 0.25 seconds, the three-phase load is transformed to two-phase, and at 0.3 seconds, one-phase and loads are disconnected at 0.35 seconds. The *d* component of currents is fixed at 35 A. The proposed BESS system supports load levelling and balancing during these loading conditions. The connected nonlinear load is again connected in the same sequence at 0.4, 0.45, and 0.5 seconds. It has been seen that the DSTATCOM provides a compensating current to maintain the source current and voltage sinusoidal and is balanced when keeping the terminal voltage at its calculated reference voltage. Moreover, the source current harmonics and load current harmonics are illustrated in Figures 4.16 and 4.17 and are visualized at 0.19% and 30.31%, respectively, which shows the proposed system's superiority.

4.6.3 Response of DSTATCOM-BESS under an Unbalanced IM Load

The response of the DSTATCOM-BESS system with an unbalanced induction motor (IM) load is depicted in Figure 4.18. It has been visualized that it draws a heavy inrush of current, leading to voltage disturbances. These disturbances were effectively minimized using the proposed DSTATCOM-BESS system by injecting the dynamic reactive power at

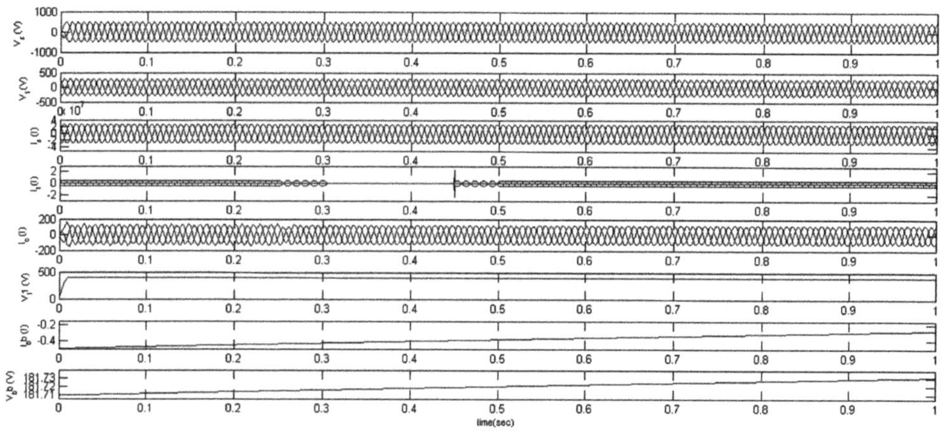

Figure 4.15 Response of DSTATCOM-BESS under an unbalanced nonlinear load.

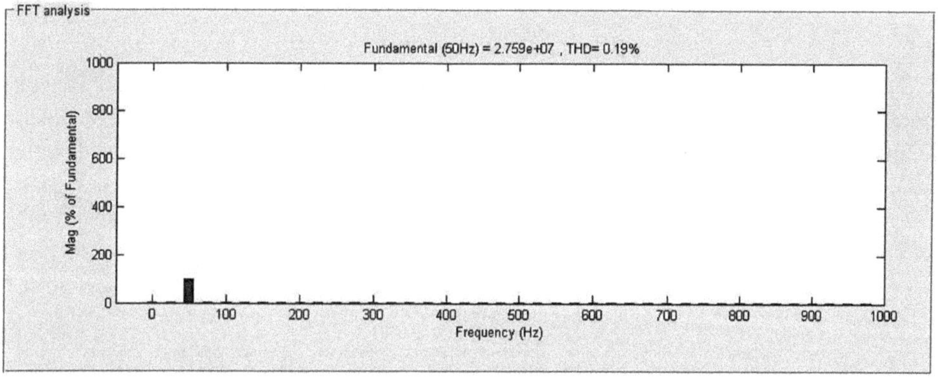

Figure 4.16 Source side frequency spectrum.

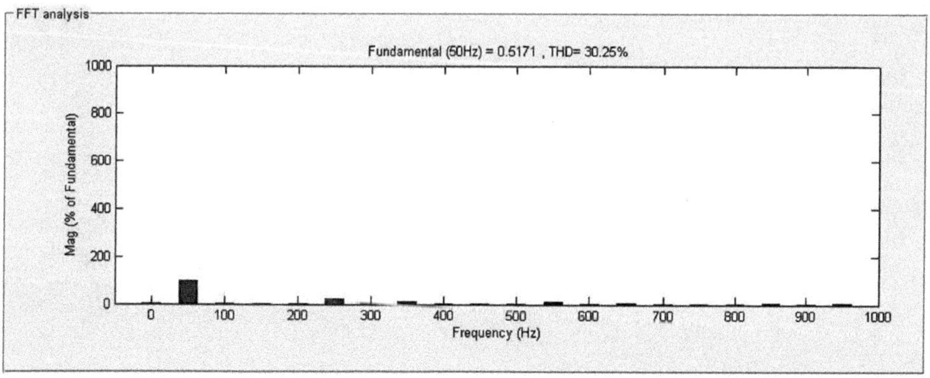

Figure 4.17 Load side frequency spectrum.

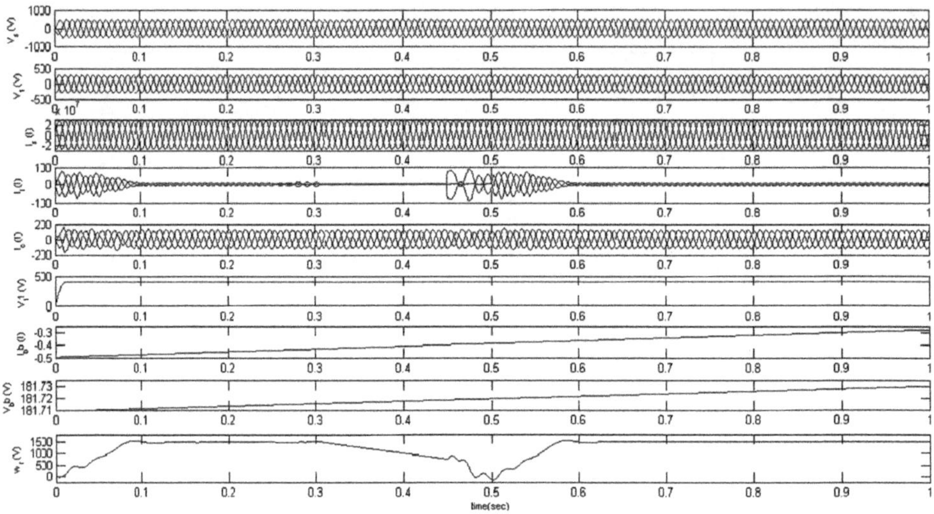

Figure 4.18 Response of DSTATCOM-BESS under an unbalanced motor load.

PCC to regulate the voltage. Similar to previous loading conditions, load unbalancing is created, and it was found that the PCC voltage remains constant during an IM load. During this condition, the terminal voltage is also kept at its reference voltage and the battery starts charging. The grid side and load side harmonics are depicted in Figures 4.19 and 4.20, respectively. The obtained harmonics are well under the international

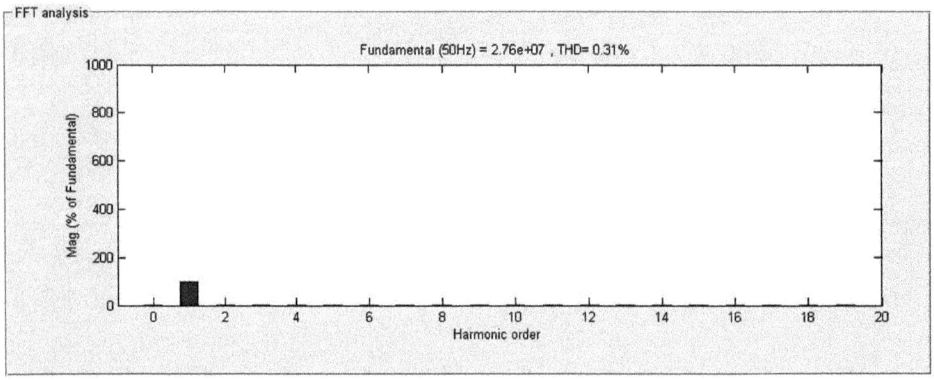

Figure 4.19 Source side frequency spectrum.

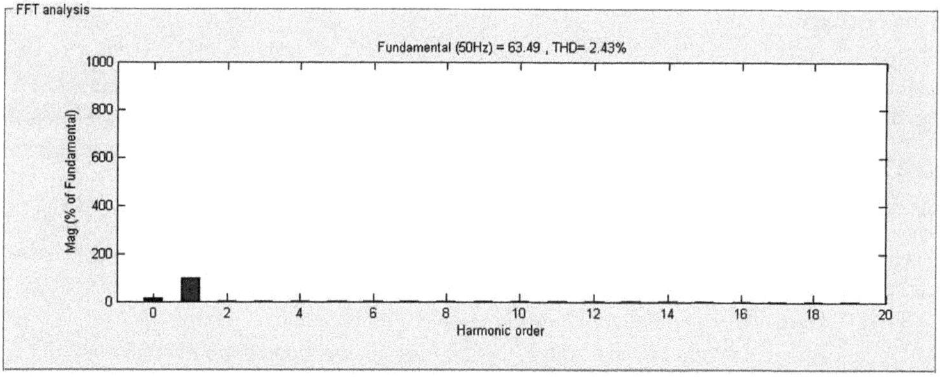

Figure 4.20 Load side frequency spectrum.

standard IEEE519-2014. Moreover, Table 4.2 illustrates the harmonics-based comparative analysis of all considered loads. The perceived results reveal the performance of the DSTATCOM-BESS system.

4.7 Conclusion

An indirect current control theory-based control algorithm has been strongly performed for DSTATCOM-BESS to improve power quality in the presence of balanced/unbalanced linear, nonlinear, and induction motor loads. The sudden changes associated with the active power and reactive power, the dc-link voltage, have been considered in designing the ICCT controlled DSTATCOM-BESS system. The particularized working boundaries of the dc-link voltage in the control algorithm ensure the DSTATCOM reliable operation, which protects the inverter during large load fluctuations. The proposed system has been used to solve the challenges associated with the loading conditions of the distribution network. It has been verified that the system's power quality is mitigated under the IEEE-519-2014.

Table 4.2 Harmonics analysis under load unbalancing.

		Indirect current control theory controlled DSTATCOM-BESS	
S. No.	**Type of load**	**Source side THD**	**Load side THD**
1	Linear load	0.71%	0.71%
2	Nonlinear load	0.19%	30.25%
3	Motor load	0.31%	2.43%

References

[1] M.T. Lawder, B. Suthar, P.W.C. Northrop, Sumitava De, C.M. Hoff, O. Leitermann, M.L. Crow, S. Santhanagopalan, and V.R. Subramanian, "Battery energy storage system (BESS) and battery management system (BMS) for grid-scale applications," *Proceedings of the IEEE*, vol. 102, no. 6, pp. 1014–1030, 2014.

[2] B. Khan, S. Degarege, F. Mebrahtu, and H. Alhelou, "Energy storage system and its power electronic interface," *Handbook of Research on New Solutions and Technologies in Electrical Distribution Networks*, IGI Global, Pennsylvania, United States, pp. 309–321, 2020.

[3] P.K.D. Pramanik, N. Sinhababu, B. Mukherjee, S. Padmanaban, A. Maity, B.K. Upadhyaya, J.B. Holm-Nielsen, and P. Choudhury, "Power consumption analysis, measurement, management, and issues: A state-of-the-art review of smartphone battery and energy usage," *IEEE Access*, vol. 7, pp. 182113–182172, 2019.

[4] R.S. Bhatia, B. Singh, D.K. Jain, and S.P. Jain, "Battery energy storage system based power conditioner for improved performance of hybrid power generation," In *2008 Joint International Conference on Power System Technology and IEEE Power India Conference*, IEEE, pp. 1–6, 2008.

[5] B. Singh, R. Niwas and S.K. Dube, "Load leveling and voltage control of permanent magnet synchronous generator-based DG set for standalone supply system," *IEEE Transactions on Industrial Informatics*, vol. 10, no. 4, pp. 2034–2043, 2014.

[6] F. Mebrahtu, B. Khan, P. Sanjeevikumar, P.K. Maroti, Z. Leonowicz, O.P. Mahela and H.H. Alhelou, "Harmonics mitigation in industrial sector by using space vector PWM and shunt active power filter," *2020 IEEE International Conference on Environment and Electrical Engineering and 2020 IEEE Industrial and Commercial Power Systems Europe (EEEIC/ I&CPS Europe)*, IEEE, pp. 1–6, 2020.

[7] O.P. Mahela and A.G. Shaik, "Topological aspects of power quality improvement techniques: A comprehensive overview," *Renewable and Sustainable Energy Reviews*, vol. 58, pp. 1129–1142, 2016.

[8] G.S. Chawda, A.G. Shaik, M. Shaik, S. Padmanaban, J.B. Holm-Nielsen, O.P. Mahela, and P. Kaliannan,"Comprehensive review on detection and classification of power quality disturbances in utility grid with renewable energy penetration," *IEEE Access*, vol. 8, pp. 146807–146830, 2020, doi: 10.1109/ACCESS.2020.3014732.

[9] "IEEE recommended practice and requirements for harmonic control in electric power systems," In: *IEEE Std 519-2014 (Revision of IEEE Std 519-1992)*, pp. 1–29, 11 June 2014, New York: IEEE, 2014.

[10] O.P. Mahela and A.G. Shaik, "Power quality improvement in distribution network using DSTATCOM with battery energy storage system," *International Journal of Electrical Power & Energy Systems*, vol. 83, pp. 229–240, 2016.

[11] E. Hossain, M.R. Tür, S. Padmanaban, S. Ay, and I. Khan, "Analysis and mitigation of power quality issues in distributed generation systems using custom power devices," *IEEE Access*, vol. 6, pp. 16816–16833, 2018.

[12] G.S. Chawda, A.G. Shaik, O.P. Mahela, S. Padmanaban, and J.B. Holm-Nielsen, "Comprehensive review of distributed FACTS control algorithms for power quality

enhancement in utility grid with renewable energy penetration," *IEEE Access*, vol. 8, pp. 107614–107634, 2020, doi: 10.1109/ACCESS.2020.3000931.

[13] B. Khan and P. Singh, "Optimal power flow techniques under characterization of conventional and renewable energy sources: A comprehensive analysis," *Journal of Engineering*, 2017, pp. 1–16, 2017.

[14] R.S. Bhatia, S.P. Jain, B. Singh, and D.K. Jain, "Power conditioning of an uncontrolled micro-hydroturbine-driven induction generator for distributed generation using a battery energy storage system," *International Journal of Energy Technology and Policy*, vol. 5, no. 5, pp. 604–618, 2007.

[15] V. Virulkar, M. Aware and M. Kolhe, "Integrated battery controller for distributed energy system," *Energy*, vol. 36, no. 5, pp. 2392–2398, 2011.

[16] G.S. Chawda. "Battery energy storage system for power conditioning of distributed energy system," M.Tech. Thesis, National Institute of Technology, Kurukshetra, India, June 2013.

[17] G.S. Chawda and A.G. Shaik, "Fuzzy logic based control algorithm for DSTATCOM connected to weak AC grid," *2018 2nd International Conference on Power, Energy and Environment: Towards Smart Technology (ICEPE)*, Shillong, India, pp. 1–6, 2018. doi: 10.1109/EPETSG.2018.8659093.

[18] G.S. Chawda and A.G. Shaik, "Performance evaluation of ADALINE controlled DSTATCOM for multifarious load in weak AC grid," *2019 IEEE PES GTD Grand International Conference and Exposition Asia (GTD Asia)*, Bangkok, Thailand, IEEE, pp. 356–361, 2019. doi: 10.1109/GTDAsia.2019.8715994.

[19] G.S. Chawda and A. Gafoor Shaik, "Adaptive reactive power control of DSTATCOM in weak AC grid with high wind energy penetration," *2019 IEEE 16th India Council International Conference (INDICON)*, Rajkot, India, IEEE, pp. 1–4, 2019. doi: 10.1109/INDICON47234.2019.9029010.

[20] G.S. Chawda and A.G. Shaik, "Smooth grid synchronization in weak AC grid with high wind energy penetration using distribution static compensator," *2019 2nd International Conference on Smart Grid and Renewable Energy (SGRE)*, Doha, Qatar, pp. 1–6, 2019. doi: 10.1109/SGRE46976.2019.9020671.

[21] G.S. Chawda, O.P. Mahela, N. Gupta, M. Khosravy, and T. Senjyu, "Incremental conductance based particle swarm optimization algorithm for global maximum power tracking of solar-PV under nonuniform operating conditions," *Applied Sciences*, vol. 10, no. 13, p. 4575, 2020.

[22] B. Singh and J. Solanki, "A comparative study of control algorithms for DSTATCOM for load compensation," *2006 IEEE International Conference on Industrial Technology*, IEEE, pp. 1492–1497, 2006.

[23] S. Kiros, B. Khan, S. Padmanaban, H.H. Alhelou, Z. Leonowicz, O.P. Mahela, and J.B. Holm-Nielsen, "Development of stand-alone green hybrid system for rural areas," *Sustainability*, vol. 12, no. 9, p. 3808, 2020.

[24] B. Singh, A. Chandra, and K. Al-Haddad, Power Quality: Problems and Mitigation Techniques. Hoboken, NJ: John Wiley & Sons, 2014.

[25] B. Singh and V. Verma, "An indirect current control of hybrid power filter for varying loads," *IEEE Transactions on Power Delivery*, vol. 21, no. 1, pp. 178–184, 2005.

[26] V. Khadkikar, A. Chandra, and B.N. Singh, "Generalised single-phase PQ theory for active power filtering: Simulation and DSP-based experimental investigation," *IET Power Electronics*, vol. 2, no. 1, pp. 67–78, 2009.

[27] H. Kim, Y. Taesik, and S. Choi, "Indirect current control algorithm for utility interactive inverters in distributed generation systems," *IEEE Transactions on Power Electronics*, vol. 23, no. 3, pp. 1342–1347, 2008.

5

Grid Power Quality Improvement Using a Bidirectional Off-Board EV Battery Charger in Smart City Scenario

Rajesh Kumar Lenka[1], Anup Kumar Panda[1], and Man Mohan Garg[2]

[1] *Department of Electrical Engineering, National Institute of Technology, Rourkela, India*
[2] *Department of Electrical Engineering, Malaviya National Institute of Technology, Jaipur, India*

5.1 Introduction

Nowadays, electric vehicles (EVs) have attracted great attention in developed countries because of reduced gasoline usage and greenhouse gas emissions [1]. In a study presented in [2, 3], it is predicted that the usage of on-road EVs will reach 64% in the United States (US) and 45% in Australia of the total on-road vehicles by the end of 2030. One of the key factors for the growing penetration of EVs is the increasing installation of off-board fast-charging stations. Off-board charging stations extend the traveling distance miles/time of EVs with battery power [3]. Further, recent advancements in battery technologies hastened the practice of transportation electrification at higher power levels. Hence, installation of off-board fast-charging stations at underutilized places like shopping malls and parking areas has been encouraged to expedite the charging of EVs [3, 4]. Currently, off-board charger coupling standards like CHAdeMO and CCS, support DC charging up to 50 kW, 100–600 V, to meet different battery levels.

The off-board charger power electronic circuitry can be unidirectional or bidirectional operated. The unidirectional battery chargers are cheap, simple, and draw charging power from the utility grid at an almost unity power factor (UPF) [5, 6]. The bidirectional operation stands for the transfer of active power in two ways. The first is to draw the active power from the grid to the vehicle (G2V) for battery charging. The second is to deliver the battery-stored energy from the vehicle to the grid (V2G). Hence, the bidirectional operation can integrate more than one function to an EV charger. The V2G operation of the EV charger is fascinating in the smart grid due to the stored energy of EV batteries [7, 8]. Even though the stored energy in the EV batteries potentially can achieve the storage requirement of the

Active Electrical Distribution Network: A Smart Approach, First Edition. Edited by Baseem Khan, Josep M. Guerrero, Sanjeevikumar Padmanaban, Hassan Haes Alhelou, Om Prakash Mahela, and Sudeep Tanwar.

grid, deprivation of the EV batteries during V2G active power exchange is still a concern [9, 10]. Conversely, the use of chargers to supply power quality ancillary services like voltage regulation, RPC, GCHC, and power factor correction without using EV batteries preserves battery life [11]. The use of an on-board charger for ancillary function has some inherent disadvantages, such as flexibility of the charger and a limited range. Therefore, off-board chargers are favorable candidates to provide ancillary services as compared to on-board chargers as they are more flexible to power levels. Furthermore, it allows the charger owner to generate some revenue by supplying these ancillary services to the grid. Hence, the off-board charging stations with embedded grid ancillary functions at underutilized public places like shopping malls, parking areas, office complexes, residential complexes, and restaurants encourages the use of EVs.

Focusing on grid ancillary services, this chapter briefly discusses RPC and GCHC by using a bidirectional off-board EV charger. In a conventional power system, the reactive current and harmonic current are supplied by the energy source. This causes additional losses in the reactance of the long transmission and distribution network, hence reducing the system efficiency. A further harmonic voltage drop at line reactance reduces the voltage quality of the system. Therefore, it is preferred to generate these load demands locally. Moreover, the residential load, such as a compressor, refrigerator, washing machines, microwave, and smart devices, etc., draws reactive power and harmonic current from the system for which the customers do not pay enough. Conversely, a bidirectional EV charger can supply reactive power and harmonic current locally without using additional VAR compensators and active filters. These two operations of EV charger in service to the grid is the focus of this chapter.

In [12] the proposed EV charger is designed to provide RPC to the grid. However, to achieve uninterruptible RPC, EV batteries regulate the DC link voltage. Hence, the battery is exposed to unwanted DC-link ripples that affect its life. Moreover, it goes under more charging and discharging cycles that reduce battery life. Similarly, in [13, 14], the charger is used to serve the grid with reactive power support. However, using EV batteries to regulate the DC link voltage reduces its performance and life. Furthermore, the RPC operation of the EV charger simultaneously with G2V or V2G operation is less explored in the literature. Whereas the charger control strategy proposed in this chapter regulates the DC link by using the grid instead of EV batteries and enables the simultaneous operation of the charger in more than one operating mode. The application of the EV charger to mitigate grid current harmonics by operating as an active power filter is introduced in [15]. Further, in [15–20], the study is extended by developing new control strategies for achieving harmonic compensation by using an EV charger. During harmonic compensation, the EV batteries are used to sustain the DC link voltage [15, 17–19]. In [20], the grid regulates the DC link voltage, but this chapter does not discuss the simultaneous operation of the charger in more than one operating mode. The EV charger control function was developed only to serve the grid by providing harmonic compensation but was not able to charge the EV along with providing this service. A multifunctional EV battery charger is proposed in [19] with both GCHC and RPC functions. Still, it does not explore the simultaneous operation of the charger in more than one operating mode at a single time. In [21], a power quality assessment and event detection in distribution network with wind energy penetration are performed by using stockwell transform and fuzzy clustering. In [22], assessment of power quality in the utility grid integrated with wind energy generation is performed. In [23], recognition of power quality issues associated with grid integrated solar photovoltaic plant in an experimental framework is performed. In [24], recognition of complex power quality disturbances using an s-transform-based ruled decision tree is performed. In [25], harmonics mitigation in the industrial sector by using a space vector PWM and a shunt active power filter is presented.

In this chapter, an efficient control strategy is developed for a bidirectional off-board EV battery charger, which is used to achieve grid ancillary services when requested from the grid. The charger controller also enables the ancillary services simultaneously with the EV battery charging or delivering the EV battery stored energy. Further, the proposed charger topology provides galvanic isolation to make the EV charger highly reliable in practical applications. The EV charger controller ensures the unity power factor (UPF) during G2V and V2G operation by generating the reference current by using the active component of load current and synchronizing with the generated synchronizing voltage template. For GCHC, the control algorithm estimates the active component of the nonlinear load current by using a resonant filter. The reactive power command is synchronized with 90° shifted synchronizing voltage template to generate a reactive reference current. The presented EV charger controller ensures effective operation without using a PLL, additional power calculations, and low-pass filters. Hence, this reduces the controller design complexity and computational burden.

5.2 EV Battery Charging Model

The off-board EV battery charging model configuration is depicted in Figure 5.1. The grid facing front-end AC–DC cascaded H-bridge bidirectional converter (CHBDC) is a single excited voltage source converter. The CHBDC has three H-bridge modules per phase, as shown in Figure 5.2. Each H-bridge output is connected to the primary side of a single-phase toroidal core transformer. The series-connected secondary windings added the secondary voltage of each transformer to give a per phase output voltage. As presented in Figure 5.2, each H-bridge per phase contributes equally, i.e. 33.33%, of the output phase

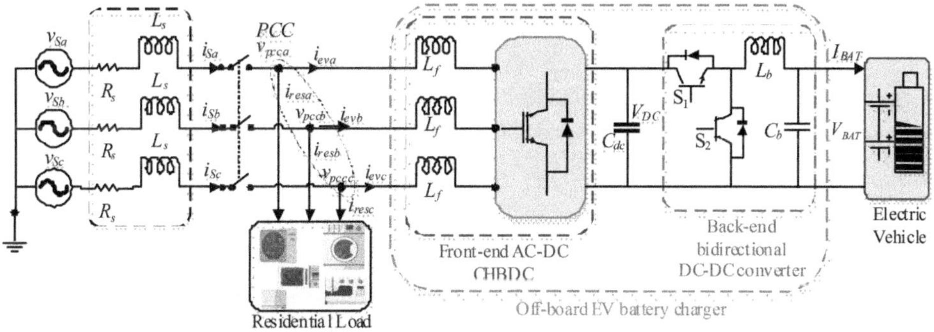

Figure 5.1 EV battery charger model.

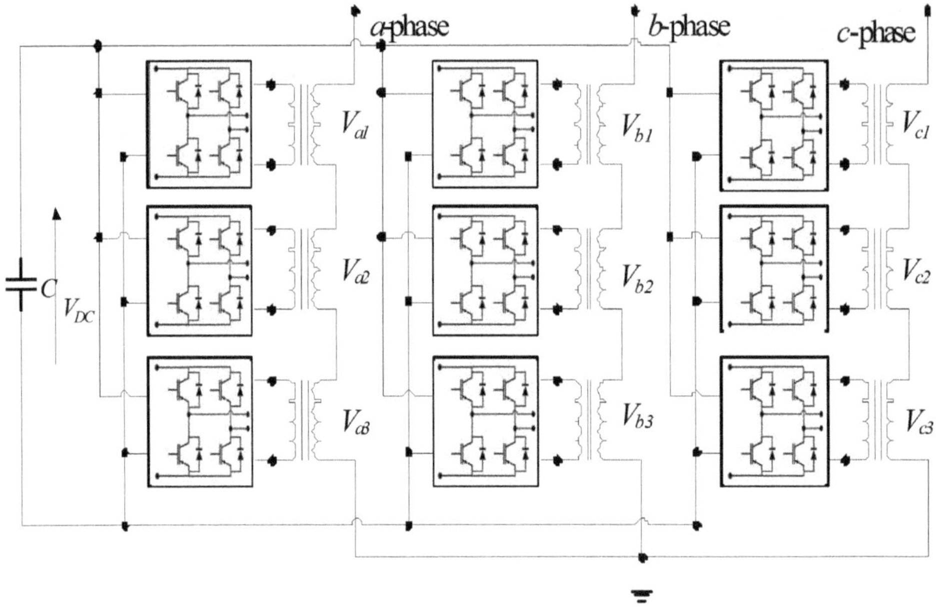

Figure 5.2 Circuit configuration of grid facing converter.

voltage. The application of toroidal transformers is seen in a front-end converter as a high-frequency link, but its application elevates the converter performance compared to conventional transformer-based prototypes [26, 27].

The use of the toroidal core transformer facilitates the converter to work with a single DC excitation voltage. Furthermore, it eliminates the additional use of voltage matching sensors required to maintain equal power distribution among the modules [26]. The added advantages with the presented H-bridge configuration are very low charging current ripples, both current and voltage control capabilities, and provision of galvanic isolation [27]. The grid facing CHBDC is coupled to the power grid through the L-filter to improve the converter output voltage quality. Due to the multilevel structure, the L-filter is sufficient to eliminate higher-order switching harmonics. The BBDC is used to charge/discharge the EV batteries. The detailed circuit configuration of the BBDC is shown in Figure 5.1. The presented configuration can operate in two modes (buck and boost) by controlling the two switches (S_1, S_2). The control of switch S_1 operates the BBDC as a buck converter and the boost mode can be achieved by controlling S_2. Hence, the charging and discharging of the EV batteries can be achieved by operating the BBDC in the buck and boost mode.

5.3 Control Strategy

The charger control algorithm is designed to achieve GCHC and RPC by using the EV battery charger and simultaneously use it as a battery charger or power generator without changing the charger hardware configuration. The charger controller serves two main objectives. First, it exchanges active power with the grid by operating either in the G2V mode to charge the EV batteries or in the V2G mode to supply battery stored energy to the grid. Second, it offers GCHC and RPC when requested from the grid, in a timely manner. The charger controller extracts the active load current component and synchronizing voltage template from the nonlinear load current and PCC voltage, respectively. The control strategy to extract the synchronizing voltage template is depicted in Figure 5.3. First, the measured PCC voltage (v_{PCC}) is transferred to the $\alpha\beta$ frame using the Clarke transformation. Then the $v_{pcc\alpha}$ and $v_{pcc\beta}$ voltages are processed through the resonant filter as given in Eq. (5.1) to extract the fundamental PCC voltage in the $\alpha\beta$ frame. The performance of the resonant filter completely depends on the selection of w_{bp}. The bode plot of the resonant filter for different values of w_{bp} is shown in Figure 5.4. The magnitude and phase plot shows that the filter imposes a zero phase shift and unit magnitude at the fundamental grid frequency $f_{grid} = 50\,\text{Hz}$. Furthermore, it can be observed that the smaller values of w_{bp} ensures better selectivity of the filter. On the other hand, the larger values of w_{bp} ensures

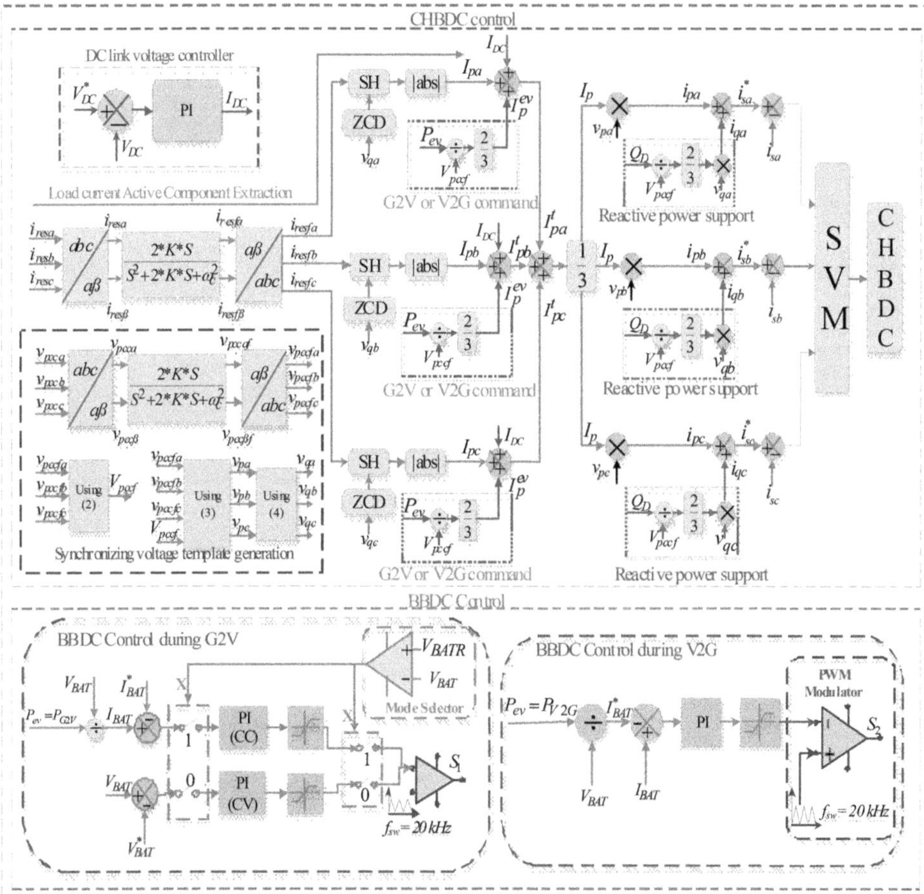

Figure 5.3 EV charger controller block diagram.

improved dynamic response of the filter. Therefore, the selection of the value of w_{bp} is a tradeoff between the dynamic response and selectivity of the filter. Here, the value of w_{bp} is selected to be 40.

$$v_{pcc\alpha f} = \frac{2w_{bp}\, s}{s^2 + \left(2w_{bp}\, s\right) + \left(2\pi\, f_{grid}\right)^2}\, v_{pcc\alpha} \tag{5.1}$$

The inverse Clarke transformation of $v_{pcc\alpha f}$ and $v_{pcc\beta f}$ gives the fundamental PCC voltage in the *abc* frame. After getting the fundamental PCC voltage, the peak magnitude (V_{pccf}) is calculated as below:

$$V_{pccf} = \sqrt{\frac{\left(v_{pccfa}\right)^2 + \left(v_{pccfb}\right)^2 + \left(v_{pccfc}\right)^2}{(3/2)}} \tag{5.2}$$

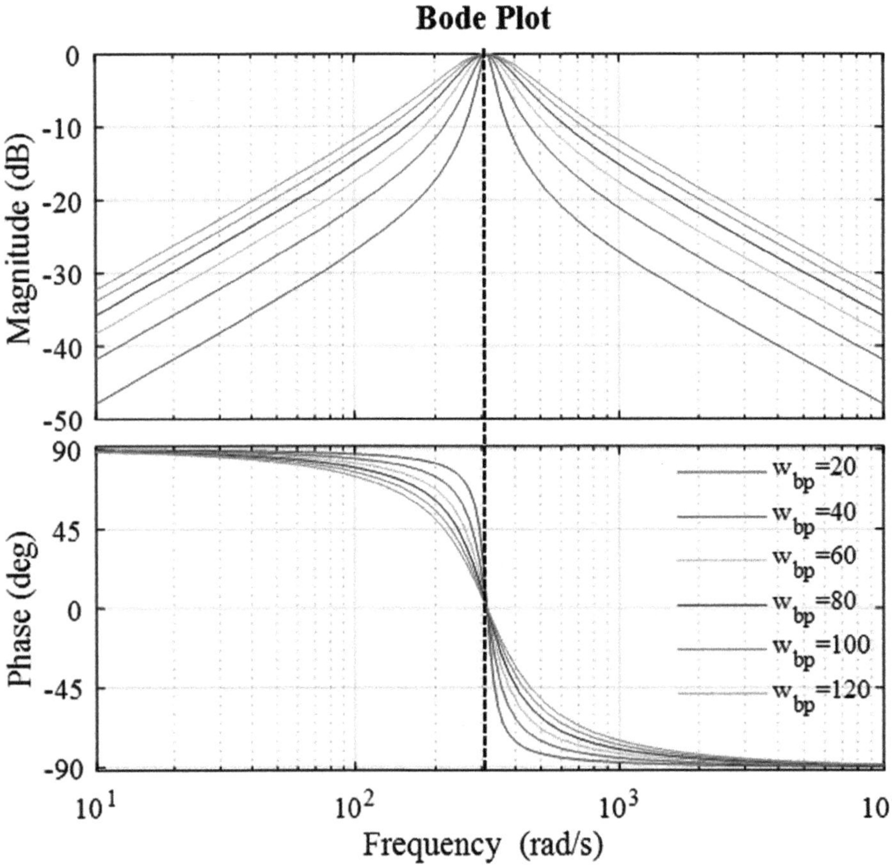

Figure 5.4 Frequency domain response of the resonant filter.

With this the in-phase synchronizing voltage template v_{pa}, v_{pb}, and v_{pc} and the 90° shifted synchronizing voltage template v_{qa}, v_{qb}, and v_{qc} are estimated as follows:

$$v_{pa} = \frac{v_{pcca}}{V_{pccf}}, v_{pb} = \frac{v_{pccb}}{V_{pccf}}, v_{pc} = \frac{v_{pccc}}{V_{pccf}} \tag{5.3}$$

$$\begin{bmatrix} v_{qa} \\ v_{qb} \\ v_{qc} \end{bmatrix} = \begin{bmatrix} 0 & -1/\sqrt{3} & 1/\sqrt{3} \\ \sqrt{3}/2 & 1/2\sqrt{3} & -1/2\sqrt{3} \\ -\sqrt{3}/2 & 1/2\sqrt{3} & -1/2\sqrt{3} \end{bmatrix} \begin{bmatrix} v_{pa} \\ v_{pb} \\ v_{pc} \end{bmatrix} \tag{5.4}$$

The active component of the load current is estimated by sample and hold (SH), the fundamental frequency current component at zero crossings of the 90° shifted synchronizing voltage template v_{qa}, v_{qb}, and v_{qc}. The required fundamental frequency current component is extracted by using a resonant filter, as shown in Figure 5.3. The nonlinear residential load current in the $\alpha\beta$ frame, i.e. $i_{res\alpha}$ and $i_{res\beta}$, can be represented as the sum of two components, as given by

$$i_{res\alpha} = i_{res\alpha f} + i_{res\alpha h}, i_{res\beta} = i_{res\beta f} + i_{res\beta h} \tag{5.5}$$

The terms $i_{res\alpha f}$ and $i_{res\alpha h}$ represent the fundamental frequency component and harmonic distortions term of $i_{res\alpha}$, respectively. To extract the fundamental frequency component $i_{res\alpha f}$ from $i_{res\alpha}$, $i_{res\alpha h}$ needs to be removed, which is achieved by using a resonant filter. After finding the $i_{res\alpha f}$ and $i_{res\beta f}$, an inverse Clarke transformation gives a fundamental frequency component in the *abc* frame. With this, the active components of the residential load current I_{pa}, I_{pb}, and I_{pc} are estimated by sample and hold i_{resfa}, i_{resfb}, and i_{resfc} at zero crossings of v_{qa}, v_{qb}, and v_{qc}, depicted in Figure 5.3. Furthermore, the charger exchanges active power with the utility grid during the G2V and V2G operating modes. During the G2V operating mode, the front-end CHBDC draws active power from the grid to charge the batteries. The PCC voltage $v_{pcc}(t)$ and EV charger current $i_{ev}(t)$ are expressed as in Eqs (5.6) and (5.7) and the mean active power is given by Eq. (5.8):

$$v_{pcca}(t) = V_{pccf} \sin(\omega t) \tag{5.6}$$

$$i_{eva}(t) = I_{ev} \sin(\omega t - \varphi) \tag{5.7}$$

$$P_{ev} = 3 \frac{V_{pccf}}{\sqrt{2}} \frac{I_{ev}}{\sqrt{2}} \cos\varphi \tag{5.8}$$

Considering only the active power exchange between the grid and charger, the PCC voltage and charger current are in phase, i.e. $\cos\varphi = 1$. With this, the active component of the current can be expressed as

$$I_p^{ev} = \frac{2}{3} \frac{P_{ev}}{V_{pccf}} \tag{5.9}$$

Ignoring the converter losses, the active power P_{ev} represents the battery charging power in the G2V mode and the battery power delivered to the grid in the V2G mode. Hence, the active component of the EV current can be expressed as in Eq. (5.10), where positive P_{ev} stands for battery charging, i.e. G2V operation of the EV charger, and negative P_{ev} represents the battery power delivery in V2G operation.

$$I_p^{ev} = \frac{2}{3} \frac{\pm P_{ev}}{V_{pccf}} \tag{5.10}$$

In a grid-connected operation, the grid compensates the CHBDC losses. Furthermore, the grid also regulates the DC link voltage of the CHBDC. Hence, the total active current supplied by the grid is the sum of the active current demand by the load, charger, and loss component I_{DC}, as given in Eq. (5.11). The loss component I_{DC} is obtained as given in Figure 5.3.

$$I_{pa}^t = I_{pa} + I_p^{ev} + I_{DC}, I_{pb}^t = I_{pb} + I_p^{ev} + I_{DC}, I_{pc}^t = I_{pc} + I_p^{ev} + I_{DC} \tag{5.11}$$

The average active current component per phase is obtained as

$$I_p = \frac{I_{pa}^t + I_{pb}^t + I_{pc}^t}{3} \tag{5.12}$$

Now, the sinusoidal current corresponds to the active component is obtained as follows:

$$i_{pa} = I_p\ v_{pa}, i_{pb} = I_p\ v_{pb}, i_{pc} = I_p\ v_{pc} \tag{5.13}$$

The charger also provides RPC when requested from the utility grid in a timely manner. By using Eqs (5.6) and (5.7), the reactive power at the AC side of CHBDC is given as

$$Q_{ev} = 3 \frac{V_{pccf}}{\sqrt{2}} \frac{I_{ev}}{\sqrt{2}} \sin \varphi \tag{5.14}$$

The charger current corresponding to the RPC is $I_{ev} \sin \varphi$. Therefore, Eq. (5.14) can be represented as

$$I_{ev} \sin \varphi = \frac{2}{3} \frac{Q_{ev}}{V_{pccf}} \tag{5.15}$$

When the grid demands a reactive power Q_D, then Q_{ev} becomes Q_D and the reactive current component $I_{ev}\sin\varphi$ corresponds to I_q. Hence, the current component at the grid side that corresponds to the reactive power demand Q_D is given as

$$I_q = \frac{2}{3}\frac{\pm Q_D}{V_{pccf}} \tag{5.16}$$

Where positive, Q_D means that the charger draws reactive power from the grid and negative Q_D represents the charger delivering reactive power to the grid. Now, the sinusoidal current corresponds to the reactive component is obtained as follows:

$$i_{qa} = I_q\ v_{qa}, i_{qb} = I_q\ v_{qb}, i_{qc} = I_q\ v_{qc} \tag{5.17}$$

With the sinusoidal active current component (i_{pa}, i_{pb}, i_{pc}) and the reactive current component (i_{pa}, i_{pb}, i_{pc}) the grid reference current is estimated as

$$i_{sa}^* = i_{pa} + i_{qa}, i_{sb}^* = i_{pb} + i_{qb}, i_{sc}^* = i_{pc} + i_{qc} \tag{5.18}$$

After getting the reference current, the switching pulses for CHBDC are evaluated by using the space vector modulation technique, as shown in Figure 5.3.

By considering the EV charger to charge lithium-ion batteries, the battery charging process runs through two stages. In the first stage, the reference battery charging current is set to a proper power level under a constant current (CC) until the battery voltage touches the rated permitted voltage level (in this chapter 90% of battery capacity). In the second stage, the battery charging is carried out at a constant voltage (CV) with a decreasing current until the current reaches the rated cutoff level. During charging, the BBDC operates as a buck converter by switching S_1 to control the battery charging current (I_{BAT}) and voltage (V_{BAT}). The controller structure of BBDC during the G2V operation is depicted in Figure 5.3. The variables are the same as in Figure 5.1. In V2G operation, the EV charger delivers the battery-stored energy to the utility grid. Whenever EV is coupled to the grid, G2V operation is the primary function of the charger. However, with the integration of a BBDC, power flow in both directions can be possible for some period. The power flows from the batteries to the grid as required by the utility grid with the convenience of EV owners. During this mode, the CHBDC is controlled to maintain a 180° phase shift between the EV current and the grid voltage. The EV charger control algorithm receives the required energy (reference power $P_{ev} = P_{V2G}$ and the time interval) command to generate the reference current. Disregarding the EV charger power losses, the reference battery current (I_{BAT}^*) can be estimated by using the following equation:

$$I_{BAT}^* = \frac{P_{V2G}}{V_{BAT}} \tag{5.19}$$

5.4 Simulation Results and Discussion

This chapter developed a MATLAB simulation model of the proposed EV battery charger to verify its steady state and transient performance with the proposed control technique. The parameter specification used to design the charger simulation model is listed in Table 5.1. The control input P_{ev} decides the charger-operating mode. With $P_{ev} = P_{G2V}$, the charger draws P_{G2V} amount of active power from the grid to charge the EV batteries and operates in the G2V mode. Similarly, by setting $P_{ev} = P_{V2G}$ the charger delivers P_{V2G} amount of active power to the grid and operates in the V2G mode. The reactive power command Q_D decides the reactive power requirement of the grid. Positive P_{ev} and Q_{ev} represents the power supplied from the grid to the charger and vice versa. The charger operation first starts by charging the battery (G2V) with an agreement that it will supply the reactive power and harmonic compensation whenever requested from the utility grid. It allows the controller to operate with a different charging power if the grid requests reactive power. The detailed working operating modes of the EV charger during reactive power compensation implemented in 6 seconds is listed in Table 5.2. For easy understanding, a linear load is considered to show the RPC capability of the charger. The grid requests both the inductive and capacitive reactive power from the charger. Further, the study was extended with the nonlinear load by connecting and bridge rectifier with the R-C load and L-filter to show the harmonic compensation operation of the charger.

5.4.1 Linear Loading

The performance of the charger during the G2V operating mode with RPC service to the grid is shown in Figure 5.5. The reference charging power and reactive power demand in a timely manner are listed in Table 5.2. The recorded results show three-phase PCC voltage $\left(v_{pcc(abc)}\right)$, three-phase EV current $\left(i_{ev(abc)}\right)$, charger active power $\left(P_{ev}\right)$, charger reactive

Table 5.1 Simulation parameter specifications

Parameters	Specifications
Charger apparent power	11 kVA
CHBDC filter	$L_f = 2.5$ mH (25 A)
BBDC elements	$L_b = 3.7$ mH, $C_b = 660$ μF
Grid impedance (Z_s)	$R_g = 0.1$ Ω, $L_g = 1.6$ mH
DC link capacitor (C_{DC})	2200 μF/500 V
Transformer (CHBDC)	1 kVA, 1 - ϕ, toroidal core
Supply system	230 V rms, 50 Hz
EV battery	Nominal voltage = 192 V $V_{BATR} = 0.9*192$V

Table 5.2 EV charger simulation operating states

States	Time (s)	Active power P_{ev} (kW)	Reactive power Q_{ev} (kVAR)	Harmonic compensation
G2V	Start - 1.5	$P_{G2V} = 11$	$Q_D = 0$	Inactive
(EV charging)	1.5–3.5	$P_{G2V} = 6.3$	$Q_D = -9$	Inactive
	3.5–5	$P_{G2V} = 8$	$Q_D = 7.5$	Inactive
V2G	Start - 1.5	$P_{V2G} = -11$	$Q_D = 0$	Inactive
(Battery power delivery)	1.5–3.5	$P_{V2G} = -6$	$Q_D = -9.2$	Inactive
	3.5–5	$P_{V2G} = -9$	$Q_D = 7.5$	Inactive

power (Q_{ev}), and DC link voltage $\left(V_{DC}\right)$. Positive P_{ev} and Q_{ev} represent the power supplied from the grid to the charger. Similarly, negative P_{ev} and Q_{ev} represent the power delivered to the grid from the charger. The simulation results of the controller performance for all scenarios are presented in the zoomed version. Figure 5.5b shows the charger operation shifted from the G2V operating mode to G2V with a capacitive reactive power support command with $Q_D = -9\,\text{kVAr}$. The output EV power follows the power command. The transition from the capacitive to inductive RPC is shown in Figure 5.5c. During the transition, the amplitude of the EV current does not change, as the charger power is limited to its rated value. Similarly, Figure 5.6a shows the performance of the charger during the V2G operating mode with RPC service to the grid. Figure 5.6b and c shows the zoomed version of the simulation results during the V2G mode with both the capacitive and inductive reactive power commands in a timely manner, as given in Table 5.2. The charger responds to the power command changes from inductive to capacitive in less than two grid cycles. This proves the quick response of the charger control function to the utility grid command.

5.4.2 Nonlinear Loading

The performance of the charger during the G2V operating mode with the harmonic compensation operation is shown in Figure 5.7. The results show "a-phase" PCC voltage $\left(v_{pcca}\right)$, source current (i_{sa}), residential load current (i_{resa}), EV current (i_{eva}), and DC link voltage (V_{DC}). The controller performance is presented in per phase manner with harmonic compensation requests from the grid at 0.3 seconds. Before initiation of the harmonic compensation, the grid current is distorted with THD of 23.43%. It supplies the required load demand and EV battery charging power. After the initiation of the harmonic compensation operation, the source current becomes nearly sinusoidal with THD of 3.21%. In this case, the reactive powers demand $Q_D = 0\,\text{kVAR}$. The charger operation starts with charging the battery. Hence, EV current is in phase with PCC voltage. At 0.3 seconds, the grid requests GCHC service from the charger. As a result, the EV charger delivers the harmonic current to the grid and draws the charging current from the grid.

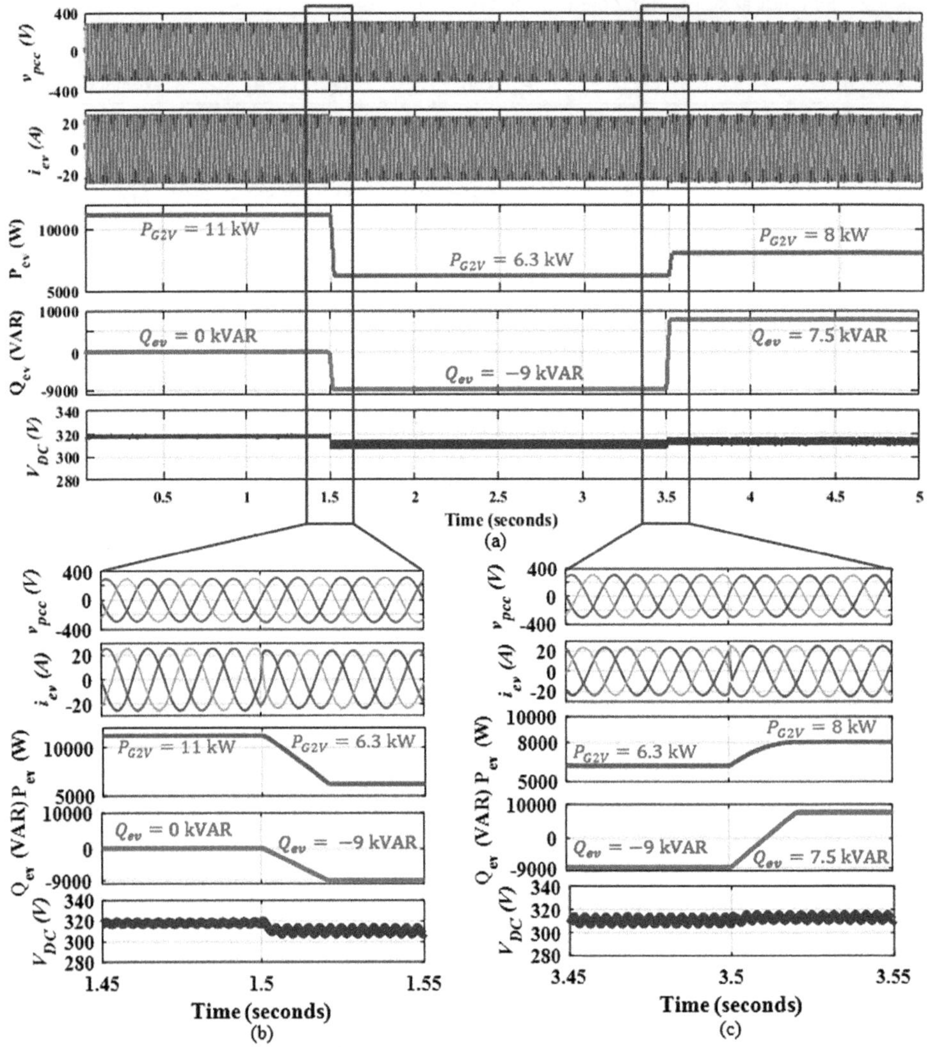

Figure 5.5 (a) Performance of the charger with battery charging and RPC operation, (b) transition from $Q_D = 0$ to capacitive reactive power demand, (c) transition from capacitive reactive power to inductive reactive power demand.

Similarly, Figure 5.8 shows the charger performance during V2G operation with harmonic compensation. The grid requests harmonic compensation at 0.3 seconds. Before initiation, the source current is harmonically distorted with THD of 28.26%. The EV supplies battery stored energy to the grid with $P_{V2G} = 6\,\text{kW}$. The EV current is out of phase with the PCC voltage and supplies the reference active power to the grid. Furthermore, the charger supplies the harmonic compensation current to the grid at 0.3 seconds. The grid current after compensation becomes nearly sinusoidal with THD of 2.56%. The charger

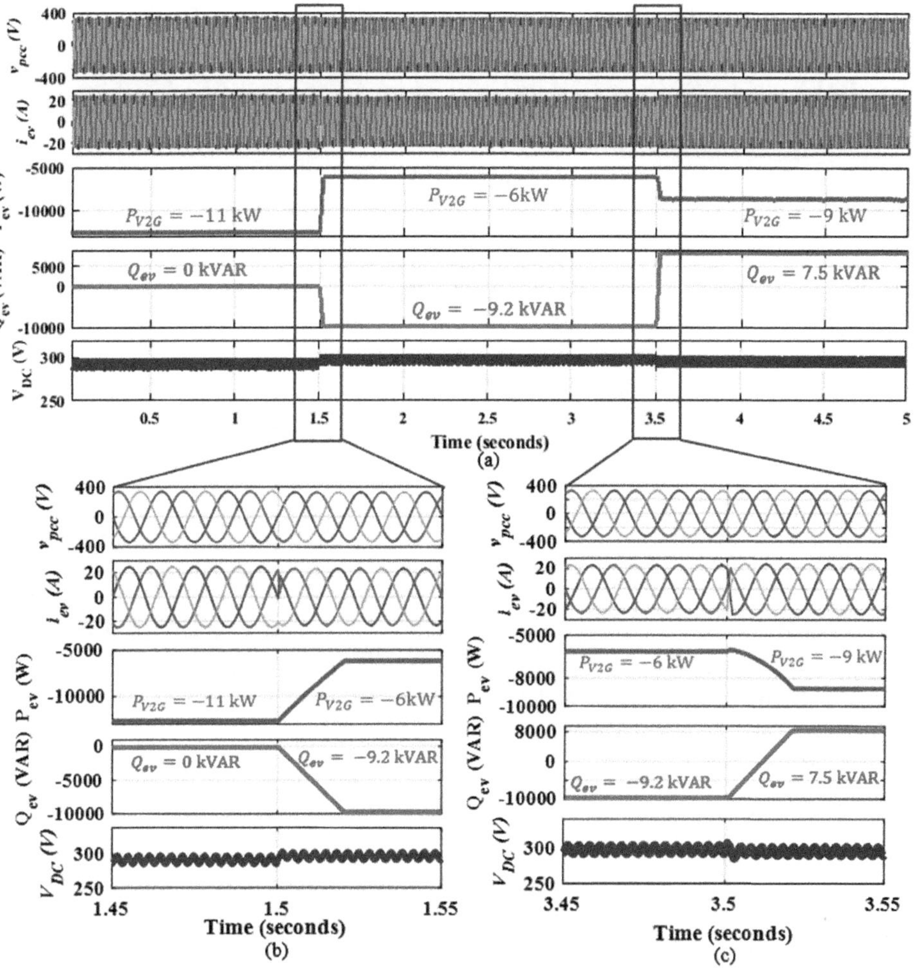

Figure 5.6 (a) Performance of the charger with battery power delivery and RPC operation, (b) transition from $Q_D = 0$ to capacitive reactive power demand, (c) transition from capacitive reactive power to inductive reactive power demand.

performance with both harmonic compensation and capacitive reactive power demand from the grid during battery charging is depicted in Figure 5.9. The charger operation starts with charging the EV batteries. At 0.3 seconds, the grid requests harmonic compensation, and at 0.4 seconds, the grid requests capacitive reactive power for 0.1 sec, by setting $Q_D = -7.5\text{kVAR}$. The grid current after harmonic compensation lags the PCC voltage, as the residential load is mostly inductive. To improve grid power quality, the grid requests capacitive reactive power of $Q_D = -7.5\text{kVAR}$ to compensate for the load reactive power demand. After reactive power compensation, the grid current becomes nearly in-phase with the PCC voltage, as shown in Figure 5.9. This proves the efficacy of the EV charger control function in providing GCHC and RPC service to the utility grid when requested.

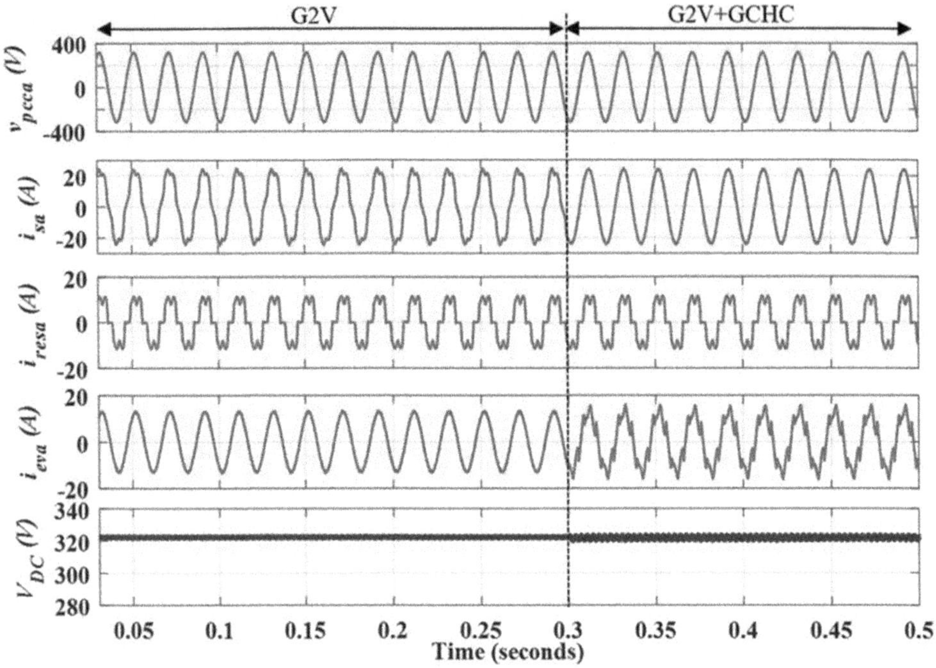

Figure 5.7 Performance of EV charger during battery charging with harmonic compensation.

5.5 Conclusion

In this chapter, an efficient control strategy is proposed to operate the EV charger as a grid ancillary service provider and simultaneously use it as a battery charger or power generator. This is achieved by incorporating the EVs as an active element that can store, consume, and provide energy. The proposed charger configuration offers galvanic isolation to improve its reliability and safety in practical applications. The developed control algorithm operates satisfactorily in different operating states and the modes of operation are well executed following the power command. The charger provides both capacitive and inductive RPC to the grid when requested, and the transaction from capacitive to inductive reactive power support takes less than two grid cycles. Therefore, the charger has improved the dynamic response. During the harmonic compensation operation, the EV charger limits the grid current THD to a level set by IEEE standard 519, i.e. less than 5%. The proposed control strategy can effectively provide both the ancillary grid services to the grid seamlessly. During operation, the charger draws the required active power from the grid to sustain the DC link voltage. Therefore, the EV batteries are not affected during RPC and GCHC operation of the charger. Moreover, it reduces the additional charging and discharging cycle, thus improving the battery life. The results obtained with the proposed controller validate its performance effectively during different power command scenarios. The proposed charger topology can provide an effective solution for RPC and GCHC in smart city applications.

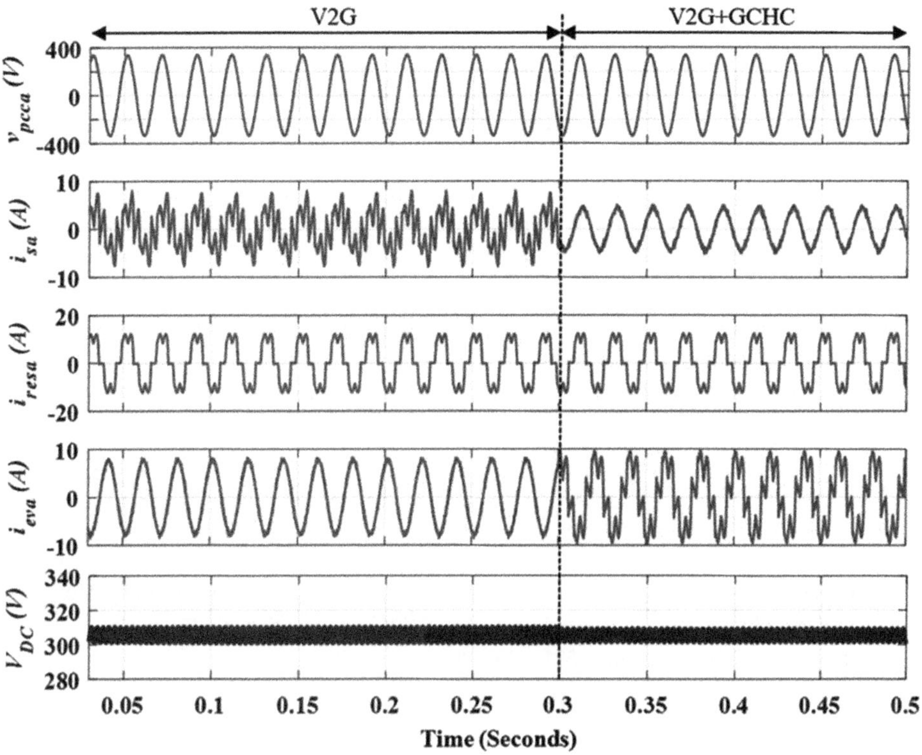

Figure 5.8 Performance of EV charger during battery power delivery with GCHC operation.

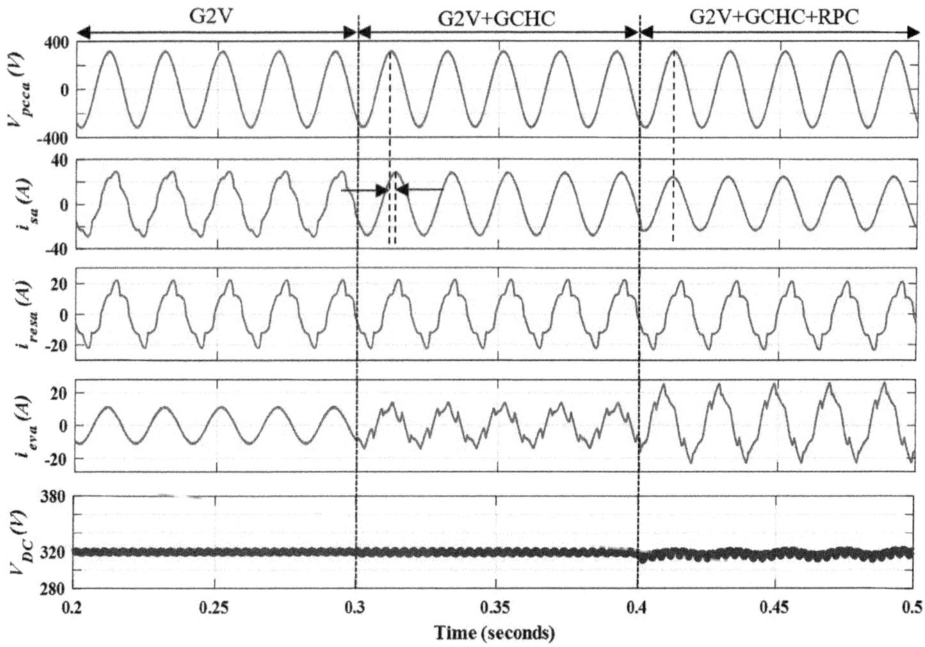

Figure 5.9 Performance of EV charger during battery charging with GCHC and RPC operation.

References

[1] S.S. Williamson, A.K. Rathore, and F. Musavi, "Industrial electronics for electric transportation: Current state-of-the-art and future challenges," *IEEE Transactions on Industrial Electronics*, vol. 62, no. 5, pp. 3021–3032, May 2015.
[2] J.G. Pinto, V. Monteiro, H. Gonçalves, and J.L. Afonso, "Onboard reconfigurable battery charger for electric vehicles with traction-to-auxiliary mode," *IEEE Transactions on Vehicular Technology*, vol. 63, no. 3, pp. 1104–1116, March 2014.
[3] M. Yilmaz and P.T. Krein, "Review of battery charger topologies, charging power levels and infrastructure for plug-in electric and hybrid vehicles," *IEEE Transactions on Power Electronics*, vol. 28, no. 5, pp. 2151–2169, 2013.
[4] A. Khaligh and S. Dusmez, "Comprehensive topological analysis of conductive and inductive charging solutions for plug-in electric vehicles," *IEEE Transactions on Vehicular Technology*, vol. 61, no. 8, pp. 3475–3489, October 2012.
[5] M. Su, H. Wang, Y. Sun, J. Yang, W. Xiong, and Y. Liu, "AC/DC matrix converter with an optimized modulation strategy for V2G applications," *IEEE Transactions on Power Electronics*, vol. 28, no. 12, pp. 5736–5745, 2013.
[6] A. Kuperman, U. Levy, J. Goren, A. Zafransky, and A. Savernin, "Battery charger for electric vehicle traction battery switch station," *IEEE Transactions on Industrial Electronics*, vol. 60, no. 12, pp. 5391–5399, 2013.
[7] S.E. Letendre and W. Kempton, "The V2G concept: A new model for power?" *Public Utilities Fortnightly*, vol. 140, no. 4, pp. 16–26, February 2002.
[8] W. Kempton and S.E. Letendre, "Electric vehicles as a new power source for electric utilities," *Transportation Research*, vol. 2, no. 3, pp. 157–175, September 1997.
[9] M. Yilmaz and P.T. Krein, "Review of the impact of vehicle-to-grid technologies on distribution systems and utility interfaces," *IEEE Transactions on Power Electronics*, vol. 28, no. 12, pp. 5573–5689, December 2013.
[10] C. Zhou, K. Qian, and W. Zhou, "Modeling of the cost of EV battery wear due to V2G application in power systems," *IEEE Transactions on Energy Conversion*, vol. 26, no. 4, pp. 1041–1050, December 2011.
[11] G. Buja, M. Bertoluzzo, and C. Fontana, "Reactive power compensation capabilities of V2G-enabled electric vehicles," *IEEE Transactions on Power Electronics*, vol. 32, no. 12, pp. 9447–9459, December 2017.
[12] D.B. Wickramasinghe Abeywardana, P. Acuna, B. Hredzak, R.P. Aguilera, and V.G. Agelidis, "Single-phase boost inverter-based electric vehicle charger with integrated vehicle to grid reactive power compensation," *IEEE Transactions on Power Electronics*, vol. 33, no. 4, pp. 3462–3471, April 2018.
[13] M. Nikkhah Mojdehi and P. Ghosh, "An on-demand compensation function for an EV as a reactive power service provider," *IEEE Transactions on Vehicular Technology*, vol. 65, no. 6, pp. 4572–4583, June 2016.
[14] M. Restrepo, J. Morris, M. Kazerani, and C.A. Cañizares, "Modeling and testing of a bidirectional smart charger for distribution system EV integration," *IEEE Transactions on Smart Grid*, vol. 9, no. 1, pp. 152–162, 2018.

[15] M.C.B.P. Rodrigues, H.J. Schettino, A.A. Ferreira, P.G. Barbosa, and H.A.C. Braga, "Active power filter operation of an electric vehicle applied to single-phase networks," *Proceedings of IEEE/IAS International Conference on Industry Applications*, Fortaleza, pp. 1–8, 2012.

[16] M. Falahi, H.-M. Chou, M. Ehsani, L. Xie, and K.L. Butler-Purry, "Potential power quality benefits of electric vehicles," *IEEE Transactions on Sustainable Energy*, vol. 4, no. 4, pp. 1016–1023, October 2013.

[17] A. Javadi and K. Al-Haddad, "A single-phase active device for power quality improvement of electrified transportation," *IEEE Transactions on Industrial Electronics*, vol. 62, no. 5, pp. 3033–3041, May 2015.

[18] V. Monteiro, J.G. Pinto, and J.L. Afonso, "Operation modes for the electric vehicle in smart grids and smart homes: present and proposed modes," *IEEE Transactions on Vehicular Technology*, vol. 65, no. 3, pp. 1007–1020, 2016.

[19] T. Na, Q. Zhang, J. Tang, and J. Wang, "Active power filter for single-phase Quasi-Z-source integrated on-board charger," *CPSS Transactions on Power Electronics and Applications*, vol. 3, no. 3, pp. 197–201, September 2018.

[20] A. Verma and B. Singh, "Multi-objective reconfigurable three-phase off-board charger for EV," *IEEE Transactions on Industry Applications*, vol. 55, no. 4, pp. 4192–4203, July–August 2019.

[21] O.P. Mahela, B. Khan, H.H. Alhelou, and P. Siano, "Power quality assessment and event detection in distribution network with wind energy penetration using Stockwell transform and fuzzy clustering," *IEEE Transactions on Industrial Informatics*, vol. 16, no. 11, pp. 6922–6932, November 2020, doi: 10.1109/TII.2020.2971709.

[22] O.P. Mahela, B. Khan, H. Haes Alhelou, and S. Tanwar, "Assessment of power quality in the utility grid integrated with wind energy generation," *IET Power on Electronics*, vol. 13, no. 13, pp. 2917–2925, 14 October 2020, doi: 10.1049/iet-pel.2019.1351.

[23] O.P. Mahela, A.G. Shaik, N. Gupta, M. Khosravy, B. Khan, H.H. Alhelou, and S. Padmanaban, "Recognition of power quality issues associated with grid integrated solar photovoltaic plant in experimental framework," *IEEE Systems Journal*, 2020. doi: 10.1109/JSYST.2020.3027203.

[24] O.P. Mahela, A.G. Shaik, B. Khan, R. Mahla, and H.H. Alhelou, "Recognition of complex power quality disturbances using S-transform based ruled decision tree," *IEEE Access*, vol. 8, pp. 173530–173547, 2020, doi: 10.1109/ACCESS.2020.3025190.

[25] F. Mebrahtu, B. Khan, P. Sanjeevikumar, K.M. Pandav, Z. Leonowicz, O.P. Mahela, and H.H. Alhelou, "Harmonics mitigation in industrial sector by using space vector PWM and shunt active power filter," *20th IEEE International Conference on Environment and Electrical Engineering, IEEE I&CPS'20*, Madrid, Spain.

[26] F. de León, S. Purushothaman, and L. Qaseer, "Leakage inductance design of toroidal transformers by sector winding," *IEEE Transactions on Power Electronics*, vol. 29, no. 1, pp. 473–480, January 2014.

[27] A.R. Dash, A.K. Panda, R.K. Lenka, and R. Patel, "Performance analysis of a multilevel inverter based shunt active filter with RT-EMD control technique under ideal and non-ideal supply voltage conditions," *IET Generation, Transmission & Distribution*, vol. 13, no. 18, pp. 4037–4048, September 2019.

Part IV

Toward Smart Distribution of Electrical Energy

6

Smart Distribution of Electrical Energy

Yashwnat Sawle[1], *Siddharth Jain*[2], *Sanjana Babu*[3], *and Ashwini Ramachandran Nair*[4]

[1,2,3,4] *School of Electrical Engineering, Vellore Institute of Technology, Vellore, India*

6.1 Introduction

In the last few years, a few countries have strongly increased access to electricity. It was achieved by reinforced political cooperation and government grants facilitating measures and strategies, grid and off-grid balance, and good infrastructure. However, still millions of people still live their lives without electricity. It affects various areas of development indicators, such as education, food security, health, poverty, and is an obstacle to the progress of a massive percentage of the world's population. An electrical grid is a complex system designed to supply power to the consumers from its generation station. The system has evolved from a small local network to a network that runs for kilometers and serves millions of consumers today [1].

6.2 Generation, Transmission, and Distribution

The grid has innumerable complicated interconnections. It has three major sections – Generation, the Transmission, and the Distribution of electricity. The different phases in which power is transmitted from generating stations to households or transmitting stations over the line are Generation and Distribution [2]. The voltage level at which the current is flowing at each phase is the primary difference between them. Generation depicts the stage of power supply that transmits bulk current to substations that are nearest to the electricity demand from generating stations over long distances. Electrical distributors must minimize the current passing through the transmission grid to lower voltage levels before it is transmitted to a household or company. Distribution is the final phase of power supply to the consumers. Power is delivered directly at the rated voltage via the distribution system. The power plant unit of electricity generated does not correspond with the units distributed to consumers. The distribution loses a small proportion of the units. This is called transmission and distribution losses in the system [3].

Active Electrical Distribution Network: A Smart Approach, First Edition. Edited by Baseem Khan, Josep M. Guerrero, Sanjeevikumar Padmanaban, Hassan Haes Alhelou, Om Prakash Mahela, and Sudeep Tanwar.

6.3 Smart Energy

Smart energy is the method of installing energy-efficient devices. It reduces cost and also concentrates on renewable sources that foster eco-friendliness. Though existing energy grid and networks have major investments, the system cannot handle emerging energy sources. Smart energy has increasingly become a preferred technology for the future [4]. A more diversified supply of power reduces air population and pollutants triggering global warming, so independence from the conventional sources that damage our environment are some of the benefits of smart energy. In today's world, smart energy is becoming increasingly relevant and is a top priority. This has led to expanded investment, which has a great impact on users, energy suppliers, and the environment.

6.4 Energy Independence and Security Act of 2007

On 19 December 2007, EISA was signed by former US President George Bush. It supports energy efficiency targets for Federal agencies outlined in Executive Order 13423. EISA aims to improve energy independence and security, provide more clean, renewable energy, give consumer protection, enhance the commodity, enhance energy protection, boost vehicle fuel economy, increase the production of renewable energy, enhance buildings and vehicle efficiency, and improve the Federal Government's Energy Act [5].

6.5 Challenges in the Existing Grid

The power grid is the most convoluted and quite important system. The conventional power grid constitutes various operations like generation, distribution, and transmission. It transmits the generated power to a variety of facilities and distributes it to the required customers. It can be defined as the power web that covers the generating stations, transformers, and the interconnection of transmission lines, distribution lines, substations, and consumers. The power stations are located near the sources used for the production of electricity [6]. The power generated is increased to a voltage that is higher so that it links with the grid. On arrival at the substations, the power will be decreased from the transmission voltage to the level of the distribution voltage. Figure 6.1 shows the traditional architecture of power flow in an electrical network.

Here are some of the issues and concerns that both developed as well as developing countries need to address when implementing the smart grid.

6.5.1 Technology Development

During the last few decades, there had been considerable advances in ICT. However, to make the grid more intelligent it is essential to build a new integrated communication system, one that can withstand attacks and is very secure from the current World Wide Web. Advanced sensor systems will be built to incorporate phase measured information and automated circuit breakers in smart premises and the grid to enable minimal

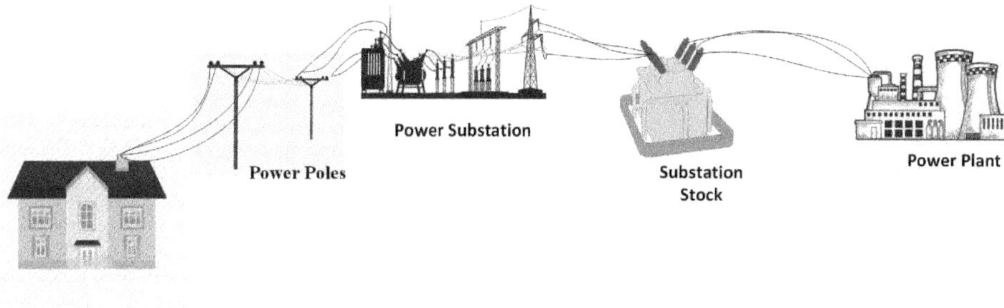

Figure 6.1 Traditional architecture of power flow in an electrical network.

disruption and provide for high-speed electrical devices. The smart grid should be able to use advanced control methods and make decisions on its own.

6.5.2 Quality Power

Electricity usage per capita is just around one-fourth of the world average in emerging countries like India. Therefore, for the next few years, a considerable increase in electricity system networks is required to supply all households with energy in a reliable way. Smart grid vision must also be assured by the standard of delivery. To ensure the quality of supply for all households, the current grid must be restored and enlarged.

6.5.3 Reducing the Losses of Transmission and Distribution

The loss of transmission and distribution (T&D) must be minimized to international levels. Operational losses due to a weak grid and economic losses such as fraud or a decline in output are the key factors causing T&D loss. These problems should be taken seriously by the utilities.

6.5.4 Renewable Energy Integration

Renewable energy sources are fluctuating and unpredictable and require complex systems for integrating with existing grids. There are also numerous issues with the modulations generated by the complex electronic circuits used to integrate the systems. The need for large areas of land is another drawback to renewable energy generation. The dependability of protective circuits to separate them from the present grid, where necessary, is a major problem for integrating renewable energy sources. The lack of skilled people and the lack of an optimum location for execution will also influence the integration of renewable energy.

6.5.5 Customer Support

The lack of consumer awareness of energy issues is one of the significant challenges in the execution of intelligent grid systems. Consumer support is a must when implementing the intelligent grid to reduce high consumption and encourage the production of distributed renewable energy. The implementation of the intelligent grid improves energy supply quality and reliability. However, consumers should know about the advancements and benefit from the grid.

6.6 Smart Grid (SG)

The existing power grid is a technological achievement, but its load is expanded beyond its efficiency. To accommodate the digital architecture and systems, we need to step ahead and design a modern model of a power grid that can control the ever-growing challenges and demands of power supply in the twenty-first century. A smart grid (SG) is a digital electricity infrastructure used to distribute electricity via two-way digital communication to consumers. It should increase efficiency, maximize the flexibility and robustness of the energy distribution chain, decrease energy intake and expenses, regulation, and connectivity within the distribution chain, and also facilitate surveillance.

6.6.1 Need for Smart Grid

To monitor the operation of the equipment, this technology was previously used in two ways, which can be described as an expanded version of analog technology. However, the SG has been more feasible in enforcing the availability of Internet connectivity in most households. To adapt quickly to developments in smart grid systems, smart grid devices relay data that helps operators, consumers, and automated equipment. Multiple government agencies around the globe have promoted the use of smart grids to monitor and resolve global warming, give immediate sustainability, and provide energy independence [7]. In order to monitor energy usage, the whole smart grid network is automated. An individual company's energy usage and its expenses are measured with grid architecture integrated with energy efficiency tools. There is a rise in the cost of energy and demand. Smart grid electricity utilities help to reduce usage at cost-efficient, fast demand periods by giving customers data on actual energy use and power prices. It is an incredible opportunity to stabilize ecological and financial well-being by shifting to a new age of sustainability, stability, and performance.

6.6.2 Traditional Grid vs Smart Grid Comparison

In Table 6.1, we can find a comparison between the functions of traditional grid vs smart grid.

6.6.3 Smart Grid Implementation

This promotes the involvement of customers, helps industries to expand and to create a profit, guarantees high-quality and reliable power that minimizes losses, can be run more effectively, and can restore faster after a power interruption. It guarantees more efficient power transmission, reduced maintenance, decreased peak demand, and costs for infrastructure that eventually lowers electricity prices for customers. It has enhanced security and improves the application of electrical generation services that incorporate renewable energy

Table 6.1 Traditional grid and smart grid comparison.

Function	Conventional grid	Smart grid
Flow of information	The flow of data or information in one way	The flow of data and information in either direction
Generation	Centralized generation	Distributed generation
Consumer participation	No consumer participation	Active consumer participation
Resilient to attacks	Slow recovery	Rapid recovery
New equipment	Limited opportunity	Growth of new market
Metering	Manual	Automatic
Power flow	Limited	Comprehensive, automated
Reliability	Vulnerable to failures and cascading outages	Outage prevention

projects. The smart grid provides consumers with the control to handle the resources and to select their energy use and allows a higher level of user involvement. This would help customers to save money by integrating real-time prices. Environmental stability and economic gains are generally addressed by the smart grid by aiding in monitoring electrical needs [8].

The power grid capacity is not constant and can change rapidly. It can aid customers to reduce energy consumption momentarily in the case of massive loads. Its decentralized infrastructure allows users to produce energy using any suitable choice on-site. Clarifying the advantages in terms that people understand by truthful public awareness strategies is the way to promote consumer adoption. Two long-term problems will conflict with smart grid adoption by consumers. First, though there is an assurance return on investment, volatility might manifest market differences. Second, customers should not expect to face an unfair cost to pay for the transformation. There are major aspects of implementing a smart grid. Some of the key influences are market models, customer education, security, atmospheric conditions, and funding. Policy constraints and technological developments play a part in a smart grid drive. A smart grid adds utility security to the nation by leveraging more existing and sustainable suppliers and offers maximum value to boost efficiency and stability. Advancement is a matter of integrating state and federal interest in business conditions [9]. Depending on various deployment considerations, such as improved reliability of transmission, the potential of suppliers and consumers to leverage innovations in information technology are a few of the advantages of the SG.

6.6.4 Smart Grid Architecture

Smart grid architecture consists of several functions.

6.6.4.1 Power Grid Functions

It performs functions to distribute energy effectively and intelligently, and to combine the generation and distribution of distributed renewable energy. It interferes with device and power management through the network to end-user communication.

6.6.4.2 Network Functions

It interferes by providing flexibility, secure transmission of data, data transfer metering, data aggregation, real-time data transmission, and other functions.

6.6.4.3 Smart Metering Functions

It involves interaction with the end-user, network, management, and security groups. It executes functions to track and manage metering appliances and interpret meter data. It interferes with the device to set up a meter database and pricing details and interfaces with the network to compile and transfer meter data; it can communicate with the end-user through portals and edge routers. This list of functions often facilitates real-time monitoring and security by accurate reporting and analysis of an incident or warning.

6.6.4.4 Energy Control Functions

It carries out responsibilities for monitoring and managing distributed sources and supporting facilities like electric vehicle component charging of a plug in an electric vehicle (PEV).

6.6.4.5 End-User Functions

It comprises reacting to energy demand, constructing power efficiency and mechanization, local electricity generation and storage, and charging for PEV. It cooperates with the implementation of the demand response (DR) for variable marketing details and manages the energy usage of household appliances and construction equipment.

6.6.4.6 Application Functions

It contains functions for device system information management (e.g. code structure, semantics, and storage), customer information management (e.g. payment and client membership), energy demand, and reasonable prices, as well as energy DR monitoring and regulation. This category of functions deals with the end-user feature, smart metering, power control, administration, and protection. This position will interact with the end-user position, the energy distribution system, and the grid feature to handle environmental factors.

6.6.4.7 Management Functions

It comprises functions for managing systems in all the functional blocks. This group interferes with other functional groups and includes various system management, application management, device management, and network management.

6.6.4.8 Security Functions

It interferes with the rest of the functional groups concerning the physical, system, and operational groups.

6.6.5 Summary of the Smart Grid Introductory Section

In this topic, we have discussed what a smart grid is and what challenges we are facing in the existing grid. We have also discussed the implementation of the smart grid and its architecture.

6.7 Role of Communication and Network Technology in a Smart Grid

6.7.1 Introduction

The basic structure electrical grid outline has not been modified for about a hundred years. Experience has shown us that the twentieth-century electric grid is not very well qualified to address twenty-first century needs/challenges. The usage of fossil fuels, especially coal, in the power grid causes environmental risks. Unlike fossil fuels, renewable energy (RE) sources offer alternative energy sources that are technologically effective and environment friendly [10]. The new smart grid concept evolved to tackle the challenges of the existing electric grid. The SG can be seen as a new electric power system for improved performance and dependability by means of automated control systems and advanced communication technologies [11]. Efficient and real-time knowledge has become the main factor for reliable power distribution within the smart grid. There are no communication capabilities for the current power system, while a smart grid infrastructure has sensing and advanced communication technology. Figure 6.2 shows a brief contrast between the conventional grid and the SG.

6.7.2 Comparability of Communication Technologies between the Traditional Grid and the Smart Grid

The flow of power takes place from main power plants to customers with little productivity and data sharing. The conventional grid communication systems are used to collect data from a few sensors at the transmission and distribution points, to control signals transmitted, and to detect errors. Data acquisition is carried out by supervisory control and data processing systems (SCADAs). The SG features need substantially more sensors and actuators than the conventional grid. The SG is utilized at all grid component levels: power plants and substation facilities, turbines, transformers, and end-users. The sensors collect data and share information between devices and data centers. The actuators regulate all components in the system. To handle such a massive data flow, the smart grid needs an improved, reliable, and efficient communication system capable of delivering safe communications in real-time. The communication system needs to provide a large bandwidth to ensure high information transfer rates.

6.7.3 Communication Standards of Smart Grid

Studies involving SG communication networks and protocols are performed concurrently by a variety of companies and countries. Several officially accredited organizations work on standardization of SGs like the Institute of Electrical and Electronics Engineers (IEEE), the European Standardization Committee, and the American National Standards Institute's International Telecommunication Union (ITU), and more [12]. The IEEE has set up a set of local area network (LAN) standards to make sure network interfaces and cabling are secure. Project 802 concentrates on norms that depict a network's physical elements. IEEE Standards 802 predate the design. Table 6.2 offers a list of IEEE Standards 802.

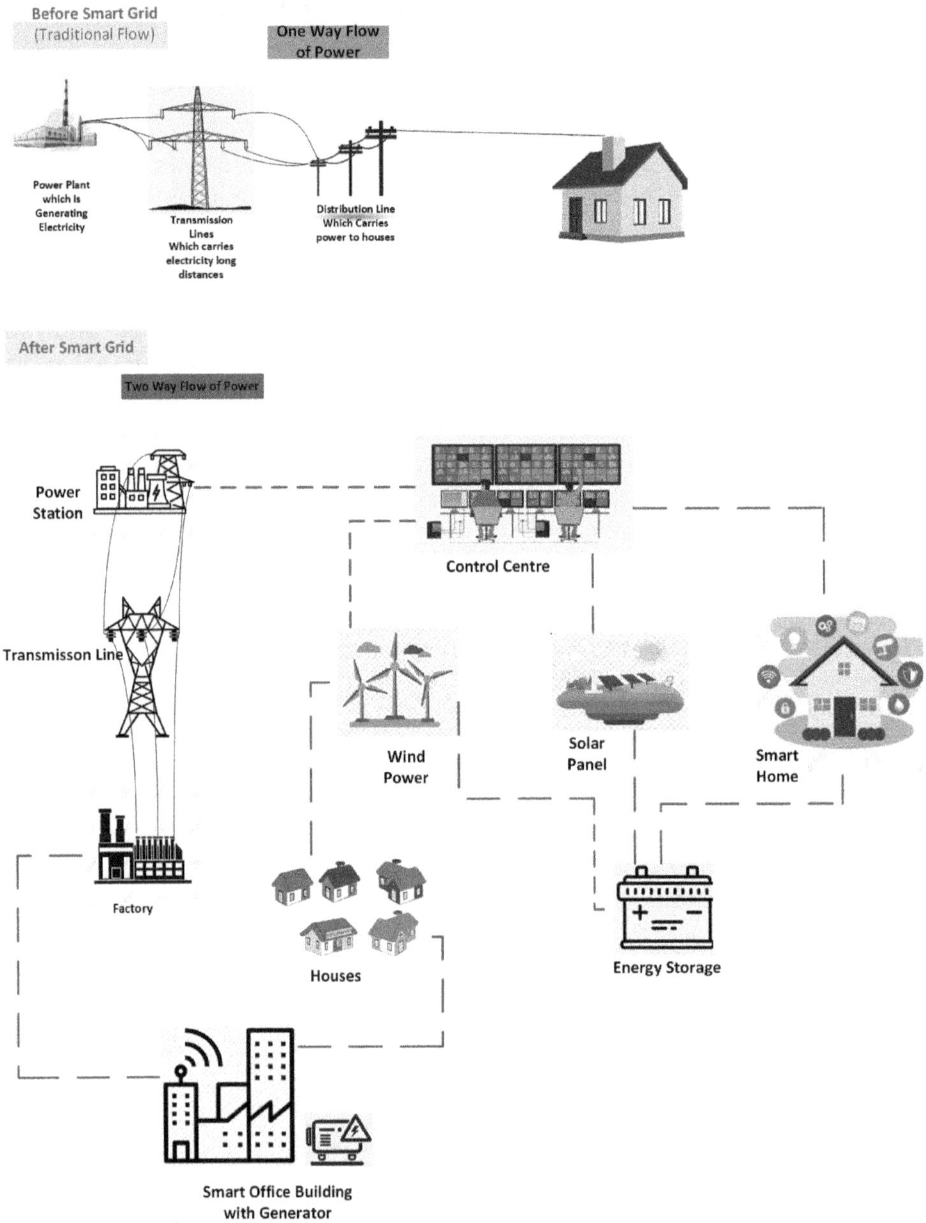

Figure 6.2 Comparison of communication architecture between a traditional grid and a smart grid.

6.7.4 Characteristics of the Smart Grid

The power grid is strongly influenced by the diverse environment, consumers as well as the industry, and emerging infrastructure challenges. These needs are much more important than earlier and will lead the power grid to extend and improve its functions to more intelligent functionality, thus exploiting the fast transmission system [13]. A smart grid is supposed to solve some of the challenges. Those challenges are now explored.

Table 6.2 IEE 802 Standards.

Standard	Name
802.1	Internet Working
802.2	Logical Link Control
802.3	Ethernet LAN
802.4	Token Bus LAN
802.5	Token Ring LAN
802.6	Metropolitan Area Network
802.7	Broadband Advisory Technical Group
802.8	Fibre-Optic Technical Advisory Group
802.9	Integrated Voice/Data Network

6.7.4.1 Digitalization

A particular digital system for quick and accurate sensing, measuring, communicating, calculating, regulating, security, simulation, and maintenance of the whole system of transmission is used for the smart transmission grid. This is now the core feature that makes other smart features easier to achieve. This framework offers easy-to-use visualization for critical circumstances and a high tolerance for human errors.

6.7.4.2 Flexibility

Flexibility includes four areas for the smart grid transmission system.

1. Extends to the future through the implementation of innovative technology using a different kind of generation of electric power.
2. Ability to adapt to different geographical sites and climate conditions.
3. Multiple management measures to harmonize distributed control systems.
4. Streamlined control strategies.

6.7.4.3 Intelligence

The smart distribution grid uses innovative technologies and human experience. Through online time-domain analysis like system stability and safety assessment, self-awareness of the system operation status is available. Self-healing is worked out by organized security and control systems to improve the safety of the transmission grid.

6.7.4.4 Resiliency

The smart transmission grid can quickly and safely distribute electricity to consumers in the event of external or internal disruptions or risks. The system can automatically reconfigure after threats, natural disasters, blackouts, or part failures with fast self-curing. Using online computing and research, fast and versatile network operation and controls like deliberate isolation, even in emergencies, are possible.

6.7.4.5 Sustainability

The reliability of the SG is adequate, effective, and environmentally sustainable. Increasing electricity demand must be fulfilled by cost-effective energy alternatives, enhanced energy efficiency through electrical and grid distribution technologies, as well as a reduction in network congestion. Lower emissions and carbon reduction, taking into account climate and environmental changes, should use creative innovations.

To achieve features of the smart grid and its characteristics, the upcoming technologies should include various methods.

6.7.4.6 New Materials and Alternative Clean Energy Resources

Advanced technologies and devices will increase power supply quality by improving energy transfer, minimizing energy losses, and reducing infrastructure costs. The high infiltration of renewable sources of energy reduces problems between both the development of human society and the environment.

6.7.4.7 Advanced Power Electronics and Devices

The latest technology will greatly enhance electricity distribution quality and power flow control stability.

6.7.4.8 Sensing and Measurement

Smart sensing, measurement, and innovative instrumentation techniques are used for basics for communication and sensing.

6.7.4.9 Advanced Computing and Control Methodologies

Modelling and simulation for complex power grids can be achieved in real-time via high-performance computing, parallel, and distributed technologies. The situational awareness accuracy will be enhanced for more effective operations as well as control techniques. The entire customer-centered electrical distribution system will be integrated via advanced control methodologies and modern distributed management concepts.

6.7.4.10 Intelligent Technologies

Smart technologies will make it feasible to create a smart transmission grid with logical thinking, knowledge exploration, and self-awareness, which are important components integrated into the development of smart technologies.

6.7.5 Communication Network Architecture of the Smart Grid

The SG's communication infrastructure is shown in a multilayer in Figure 6.3. This architecture includes

- Home area network (HAN)
- Neighborhood area network (NAN)
- Wide area network (WAN)

1.Backbone/ Network Core 2.Nerwork Substation.	Substation Distribution Network	Customer Premises
WAN	NAN/FAN	HAN/NAN
Bulk Generation and Transmission	Distribution	Customer

Figure 6.3 Three-layered infrastructure of a smart grid.

6.7.5.1 Home Area Network (HAN)

A limited area (10 meters), usually an apartment or a specific office, is operated by the HAN. Compared with two other networks, the HAN has a comparatively low data transmission rate of hundreds of bits/second. The HAN is comprised of a wired or wireless modem that provides a broadband Internet connection among multiple users. It allows communication and resource sharing over a network connection between various devices like computers, mobiles, etc. When implementing smart grids, all smart home appliances that use electricity and smart meters will be connected to the HAN.

6.7.5.2 Neighborhood Area Network (NAN)

Within a range of 100 meters, limited to a few urban buildings, the NAN is used and controlled. The transmission rate of the NAN is up to 2 kbps. The power line carrier (PLC), Wi-Fi, and cellular technology can be used to incorporate the NAN.

6.7.5.3 Wide Area Network (WAN)

Within a large space of 10 kilometers, the WAN is used and controlled with multiple NAN and LDCs. All elements of smart grids, including the operator control center, power generation, and the delivery of renewable energy, are WAN-based. The Ethernet, WiMAX, 3G/LTE, and microwave transmission are possible to introduce the WAN. The WAN transmission rate is exceptionally high, with up to a few Gbps.

6.7.6 Various Communication Technologies for the Smart Grid

The communications system is an essential part of the SG infrastructure. The SG networks use a combination of wired, wireless, and hybrid technology communications technologies.

6.7.6.1 Wired Communication

A. Fiber-Optic Communication

One of WAN's critical technologies for communication is its noise tolerance. The different ways of networking, including the passive optical network (PON), are wavelength-division multiplexing (WDM) and synchronous digital hierarchy (SDH). Depending on applications that are focused on the response time and service quality that can be supplied by the network, the communication method is chosen. The fiber communication method to be chosen depends on applications that are focused on the response time and service quality that can be supplied by the network. The durable solution for the SG is optical fiber supported communications. Because of the high initial investment and maintenance costs, they can still be expensive.

B. Digital Subscriber Line (DSL)
DSL is the suit that transfers data over telephone lines for wireless subscriber lines. The main advantages are that it is simple to use in the SG environment as it is utilized immediately by electric utilities at no additional cost for additional usage. DSL choices like ADSL for asymmetric DSL support 8 Mbps downstream and 640 Kbps upstream, ADSL 2 + up to 24 Mbps, and 1 Mbps downstream and upstream, respectively. ADSL is also accessible as an upstream version and VDSL provides up to 52 Mbps upstream and 16 Mbps upstream, but only for short distances (for very high bit DSL).

C. Power Line Communication (PLC)
PLC is a wired technology for both the transmission of data across existing cables. This procedure feeds high-frequency transmission electric lines and changes the transmission with data. PLC has several successful applications such as management of energy, smart metering, automation for home/house, and more. It is also appropriate for rural smart grid applications without any other existing communication networks. Because of the use of existing grid infrastructure, PLC is a cost-efficient technology.

6.7.6.2 Wireless Communication

A. ZigBee
ZigBee is also an IEEE 802.15.4 standard and is referenced as the low-rate personal area networks (WPANs). ZIGBEE is indeed an ideal technology for smart lighting, power management, smart home, and automatic meter reading, etc. It is the ideal technology for the physical and median access layer. The unlicensed ISM bands are 868 MHz in Europe, 915 MHz in the USA, and 2.4 GHz throughout the world. The data range between 20 kbps and 250 kbps, i.e. 250 kbps with 2.4 GHz, 40 kbps with 915 MHz, and 20 kbps with 868 MHz. ZigBee offers up to 100 meters of coverage and ZigBee Pro offers approximately 1600 meters of coverage.

B. Wireless Local Area Network (WLAN)
It is a fast, wireless technology, referred to as Wi-Fi. It works on the basis of the IEEE 802.11 standards, which includes 802.11, 802.11a, 802.11b, 802.11g, and 802.11n. WLAN works in unlicensed ISM frequency bands of 2.4 GHz, 3.6 GHz, and 5 GHz. Standards IEEE 802.11x require data speeds from 2 Mbps to 600 Mbps and can cover ranges up to 100 meters. For HANs and NANs, WLAN contact is yet another exciting option. WLAN offers high-speed, convenient, and efficient communication and does allow communications up to 100 meters in range. WLAN devices have higher costs and power usage than other wireless short-range solutions like ZigBee and Z-Wave.

C. Wireless Mesh
It comprises many nodes, which include mesh clients and routers. Every node will act as just a signal repeater in a wireless network and automatically transfer a message from node to node. As it can execute multihop routing, a wireless mesh network could offer broad

coverage. Mesh networks with various wireless technologies such as 802.11, 802.15, and 802.16 can be implemented. Even so, they could not transmit high-speed data but are subject to interference from many other wireless devices.

D. WiMAX

It is a wireless 4 G system for metropolitan area networks (MANs), according to the standards sequence of IEEE 802.16 (IEEE 802.16–2004, 802.16e). In addition to an unlicensed 5.8 GHz band, WiMAX works in 2.3, 2.5, 3.3, and 3.5 GHz frequency bands. It can cover distances up to 50 km and offers data rates of up to 75 Mbps and also has low latency (10–50 m). WiMAX networks may be used for applications such as processes for transmission and distribution control, and smart metering. WiMAX is costly and demands high strength.

6.7.7 Monitoring, PMU, Smart Meters, and Measurement Technology

The SG system includes the upgrading of sensing, measuring, and measuring equipment at all grid levels [14]. The data required for grid and power market monitoring will be generated by these components.

6.7.7.1 Wide Area Monitoring Systems (WAMSs)

For maximum capability of the grid and the prevention of disruptions, WAMSs are developed by the utilities. The WAMS gives warning regarding systems disturbance for both the prevention and mitigation of a system-large blackout by offering real-time data on durability. For the precise stamping of observations in the transmission system, WAMS utilizes sensors across the network along with GPS satellites.

6.7.7.2 Phasor Measurement Units (PMUs)

Phasor measurement units or synchro phasors offer service providers a time-stamped picture of the power system. They give information such as positions and other network parameters, the phasor bus voltage, and the phasor branch current. At the same time, phase measurements are performed with high precision, which allows the operator to determine the correct angular difference between the different positions. PMUs provide a high voltage and high precision current at a speed of 2.88 kHz.

6.7.7.3 Smart Meters

Smart meters are divided into two distinct features: sending energy consumption information to customers (end-users) for cost management and sending information to companies that control the load factor, high load needs, and create usage information-based pricing strategies. The development of smart meters is intended to utilize water, gas, and electricity. Smart meters provide consumers with information about how much they will pay per unit of electricity used, and how and when they will use electricity. This will give rise to better price details and more.

6.7.8 Future Scope

Future work in this field will include the consideration of smart grid routing protocols between smart devices in homes, smart meter, AMI networks, and the operator's controller. It includes examining another class of protocol families in all smart grid segments for every network type.

6.7.9 Summary of Communication and Network Technology Section

We have examined similar networks and networking technologies that can be applied for SG communication infrastructure on consumers and distribution. We have described its advantages and disadvantages for several applications to make the best option for communication technologies in SG. We believe the option of future power grid communication technologies is a crucial problem for developing and optimizing SG communications.

6.8 Load Flow Studies on the Smart Grid

6.8.1 Introduction

In electrical power systems, the load capacity is variable because the consumption patterns change every day, every month, and each year. Many triggers for load variation are as follows: during summer and winter, residential loads vary, particularly in northern areas [15]. Because of the time difference between day and night the load rises because it travels northwards. Depending on the economic conditions, industrial loads vary. Whenever a country's economic situation worsens, primary plants are affected and are slowed down [16]. Industrial loads are dependent on residential loads and are less affected by climate changes. The agriculture load is seasonal. The changes in climate conditions, therefore, affect them. Load flow studies are crucial for both systems planning and functioning of the system. The smart grid communication network consists of numerous media and technologies. Analysis of efficiency is among the essential issues for smart grid communication.

6.8.2 Need for Low Flow Studies

- Power system architecture
- Power system planning
- Power system development
- Provision of instructions for optimal power system operation
- Instructions for different energy system studies

6.8.3 Smart Grid Load Flow Enhancement Capabilities

Though many could provide real-time data, current power flows that can assess voltage, angle, and streams, MW/MVAr, and scheduling dispatch are generally offline. The smart grid load phase comprises the following steps to enhance load flow capabilities:

- Radial or mesh network for data acquisition.
- Data communication life to ensure the viability of networks.
- Formulation and determination of initial conditions of the Y-bus for described device interconnection under analysis.
- Real and reactive power mismatch and mismatch regulation by changing the initial conditions are usually known as the hot state.
- Identify a practical voltage and angle to portray the power demand gap.

6.8.4 Load Flow Problems in the Smart Grid

- Transmission and distribution adaptability criteria to handle renewable energy load flows.
- Self-adaptation for effective communication.
- High impedance topology for random and unknown distribution networks requiring intelligence analysis tools.
- Using FACTs devices for power electronics building blocks is required since reverse power flow technology is possible.

6.8.5 Methods for Performing Load Flow Studies

- For load flow analysis, we build a network of buses attached to a voltage-specified source bus through lines or switches.
- A similar load type (consisting of an inductor, shunt condenser, or combination) may be available for each bus.

The most effective methods for load flow are

- Gauss-Seidel
- Newton-Raphson
- Fast decouple

6.8.5.1 Gauss-Siedel Method

This method uses the nodal equations of Kirchhoff's current law:

$$P_i = \sum_{j=1}^{n} |V_i||V_j||Y_{ij}||\cos(\theta_{ij} - \delta_i + \delta_j)| \tag{6.1}$$

$$Q_i = -\sum_{j=1}^{n} |V_i||V_j||Y_{ij}||\sin(\theta_{ij} - \delta_i + \delta_j)| \tag{6.2}$$

Equations (6.1) and (6.2) are called static load flow equations.

$$I_i = \frac{P_i - jQ_i}{V_i} \tag{6.3}$$

$$\mathrm{V}_i = \frac{1}{Y_{ii}}\left(I_i - \sum_{\substack{j=1 \\ j \neq 1}}^{n} Y_{ij} V_j\right) \tag{6.4}$$

For the (k + 1)th iteration, the voltage equation becomes

$$V_i^{(k+1)} = \frac{1}{Y_{ii}}\left|\frac{P_i - jQ_i}{(V_i^k)} - \sum_{j=1}^{i-1}(Y_{ij}\ V_j^{k+1}) - \sum_{j=i+1}^{n}(Y_{ij}\ V_j^k)\right| \tag{6.5}$$

Advantages:

- Simple to use.
- The need for computer memory is low.
- Less iteration processing time.

Disadvantages:

- The convergence rate is slow.
- This are so many iterations.
- As the bus numbers rise, there are more iterations.

6.8.5.2 Newton-Raphson Method (NR)

The Newton-Raphson approach is mathematically superior to the Gauss-Seidel process because of quadratic convergence. This method is efficient for an extensive power system. The Jacobian matrix provides a straightforward relationship of small changes in voltage angles as a whole to the voltage level as a whole with small changes in active and reactive power as a consequence of both ΔP and ΔQ:

$$\begin{bmatrix} \Delta P \\ \Delta Q \end{bmatrix} = \begin{bmatrix} J \end{bmatrix} \begin{bmatrix} \Delta \delta \\ \Delta V \end{bmatrix}$$

Here

$$\begin{bmatrix} J \end{bmatrix} = \begin{bmatrix} J1 & J2 \\ J3 & J4 \end{bmatrix}$$

The partial differential matrix is called the Jacobian matrix [J]. The active power and reactive power in Eqs (6.1) and (6.2) are determined by the elements of the Jacobian unit.

Advantages:

- Greater convergence.
- Less is expected for the number of iterations.
- Number of iterations regardless of the size of the method.
- Large-power systems load flow solution.

Disadvantages:

- Solving technique is hard.
- Each iteration has more to calculate.
- More screen time.
- Specifications for computer memory are high.

6.8.5.3 Fast Decoupled Method

The fast decoupling procedure simplifies the Jacobian matrix with small angles to remove comparatively small elements of the Jacobian matrix. The method is among the important approaches employed in the study of power systems. However, with a high resistance to the impedance ratio method, there is low convergence. The interaction between V and δ with the flow of active and reactive power often leads to a poor intersection. A modification overcomes current power injection equations instead of models.

Advantages:

- Fast processing time for large power systems, accomplished by the small Jacobian matrix size.

Disadvantages:

- Convergence characteristics for delivery systems with high-resistance to reactance ratio lines compared with the Newton-Raphson traditional approach.

6.8.6 Analytical Comparison Between Different Methods of Load Flow Studies

The Gauss-Seidel approach is the oldest load flow solution technique. The low voltage and reacting power conditions are the most uncomplicated, accurate, and usually tolerant. There is a stable and quadratic rate of convergence for the Newton-Raphson (N-R) process. The relationship between the time of convergence and device size is only linear. The problem occurs when the initial voltages vary considerably from the actual values [17]. This approach is, therefore, appropriate to start with the float voltage.

The N-R method has a quadratic form of convergence, whereas it is linear for the Gauss-Seidel method. Researchers conclude that the best N-R method has a quadratic form of convergence, whereas it is linear for the Gauss-Seidel method approach. Quadratic convergence uses correctly selected starting values of *V* and delta. A common method is currently used in which voltage levels are set to specify values where the values are explicitly indicated; otherwise, the slack bus size is fixed to the voltage scale. Although the N-R method does not consist of less iterations, it is not quicker than the Gauss-Seidel method.

The other downside for the N-R approach is the storage constraint. The Newton-Raphson method requires a large amount of computing space in contrast to the Gauss-Seidel method. Within several iterations, the quadratic convergence results in an accurate precision similar to the exact solution. The Jacobian matrix consists of the voltages and the power angle variables and adjusts all these values. However, the Newton-Raphson process coefficients lead to constant values just after a few iterations.

The methods of the fast decoupling load flow (FDLF) are taken from the NR estimate. In the fast decoupled load flow (FDLF) method, certain assumptions and estimations are made, increasing the number of iterations. The convergence rate of the FDLF methods is linear in comparison to the quadratic convergence rate of the NR process. The initial voltage and reactive power specifications have minor impacts on FDLF methods compared to the NR method.

It can be assumed that easily decoupling the load flow increases the voltage profile and the line loss. However, wide angles across rows and control devices that have a strong effect on active and reactive power flow complete the NR expression and may be needed. In contrast to the methods of Newton-Raphson and quick decoupling, the explanation for Gauss-Seidel's lengthy investment and short calculation time is explained in Gauss-Seidel technology to increase the numerical juggling movement, while the Newton-Raphson method increases quadratically and the quick decoupling method increases the geometrical movement. The Gauss-Seidel method is useful for smaller systems with less computational complexity because of its useful computational features, while the Newton-Raphson method is best for rapid convergence and accuracy. Gauss-Seidel, Newton-Raphson, and quick decoupling methods for power flow studies, symmetry error estimates, position, and types of defects can be very helpful.

6.8.7 Enhanced Load Flow Capabilities in a Smart Grid

- Datalink is available to ensure the viability of the network.
- Radial or mesh network data acquisition.
- Y-bus formulation to demonstrate the interconnection of the device being examined and the initial conditions defined.
- Solving the inconsistency and regulation of the inconvenience by modifying the initial conditions usually referred to as the hot situation.
- Determining a suitable statical voltage and angle to mitigate misaligned power requirements (actual power and reactive power demand).

6.8.8 Summary of Load Flow Studies Section

In this chapter, several essential tools for smart grid architecture, operation, and performance analysis have been addressed – these methods include load flow, the optimum flow of power, and contingency assessment. Classical method reviews and information on each solution's integration into the intelligent grid have been presented.

6.9 Stability Analysis for a Smart Grid

6.9.1 Introduction

The electric power grid is a complex, nonlinear, non-stationary system that includes thousands of modules like generators, transformers, transmission lines, and highly developed electronics-based control devices, as well as consumer costs [18]. The grid's complexity has been further enhanced with the implementation of the distribution of the SG adds advanced transmission, data processing, and control methods to conventional power grids, and growing use of renewable energy like wind and solar power and rapid expansion of plug-in electric vehicles, among others. The smart grid has led to a much more distributed production, and two-way power flows between the customers and the grid, with semi-autonomous subsystem control [19]. Since the smart grid brings a new aspect and variability to the power system, stability is an issue with the smart power supply based on topography.

6.9.2 Stability Assessment

This is defined as the power system's ability to sustain acceptable voltage levels in any system bus during normal operation and to return after a disturbance to an acceptable operating point. However, power system stability is affected by new SG technologies introduced in conventional power grids [20]. With the development of the phasor measurement unit (PMU), however, the voltage range of smart grid systems is straightforward to evaluate. The nature of renewable sources can cause the system's voltage fluctuations and power oscillations. SG systems provide new ways to track and manage the grid's transmission network. PMUs can take measured voltage and current samples, which are synchronized into the universal time via a global positioning system.

With the phasor measurement unit (PMU), the voltage stability indices can be established. With generation of the waveform every 30 seconds, PMU provides more data about the voltage and the phase angle. Suppose any optimization technique can be used to produce missing information by using meta-heuristic algorithms if PMUs are not implemented. The voltage stability index (VSI) calculation is used after producing the data [21, 22]. The VSI generated is dynamic and gives the voltage stability data. Its value ranges between 0 and 1. The platform is more reliable if the value is closer.

6.9.3 Reliability Assessment

Security, reliability, as well as susceptibility are several concepts for discussion. All those are handled by reliability, which is an average or estimated system level of service for some moment [23–25]. A smart grid assessment includes consideration of cyber network performance since it is not entirely vulnerable or error-free. Some aspects of reliability are as follows:

- Design, planning, and operational cybersecurity.
- Capacity to maintain control of frequency and voltage.
- Standards of device interconnection.
- Increasing dependence on assets at the distribution level.
- Meet the requirements for the reliability of bulk systems.
- Stability of the system for inertia maintaining.
- Control and protection systems coordination.
- Ride through disturbance.

To find the reliability of the SG, the reliability concept application is used. Reliability approaches have become an essential method for analyzing and comparing the efficiency and development of the SG. Depending on the situation or even on the component, reliability parameters vary. For a reliability analysis, the transition rates of a system should not be continuous for some time. They will vary depending on the weather conditions from time to time since the failure of the power grid, as well as its reparation time, is linked to the environmental condition [26]. The distribution system's reliability depends on many factors, such as the capacity of the load, the customer base, and maintenance, as well as the type and age of the equipment. Depending on several factors like load capability, customer base, maintenance, type, and size of machinery, the distribution system reliability seems to be dependent. The reliability tool for the distribution system has been developed to ensure accurate comparisons between many competitive design alternatives. In terms of absolute and consistent reliability indices, many other utility companies, as well as regulatory agencies, define the reliability of distribution as well as transfer circuits.

6.9.3.1 System Average Interruption Duration Index (SAIDI)

This is the sum of all customer failure times. SAIDI is the average length of prolonged outages for all consumers under a certain period (usually one year), for the system as a whole and its average duration of errors for the every consumer. After even a continued power failure, it analyses the performance of the services, in hours and minutes.

6.9.3.2 System Average Interruption Frequency Index (SAIFI)

This is the amount of all extended consumer outages. SAIFI is the average interruption frequency, generally for one year, for consumers and is represented as a rational number.

6.9.3.3 Customer Average Interruption Duration Index (CAIDI)

This is the ratio of the system average interruption duration index to the system average interruption frequency index. For all those consumers who witnessed interruptions, the average interruption duration is CAIDI, which is stated in hours and minutes.

6.9.3.4 Momentary Average Interruption Frequency Index (MAIFI)

This is the count of momentary outages per number of consumers. MAIFI is an average frequency of momentary outages, which is represented as a rational number.

6.9.4 Resilience Assessment

When deformed elastically and then after the unloading process, resilience can be defined as the property absorbing energy by the material. It involves making sure the resiliency and control of smart grids are sufficient to achieve the deployment of material and, at the same time, are challenging. Grid resilience is indeed a crucial feature for the grid's future. The system's self-healing uses information, control, sensing, as well as modern communications to recognize unexpected circumstances and also to reduce their bad impacts. Resilience refers to a system's ability to regenerate after a failure occurs. Resilience is the ability to regenerate after a failure occurs in a system. Resilience is closely linked to the restoration of a power grid after a blackout.

6.9.5 Smart Grid Protection Issues

There are significant drawbacks in existing grid structures, for example, manual error identification, independent voltage control, non-optimized power flow, as well as minimal management of power. Also, such traditional grid infrastructures are highly mutually dependent. A fault or an unwanted event that can happen at any point in the grid can, therefore, quickly result in a significant portion of the system, and further damage can occur to the overall grid. The grid disadvantages have to be removed and improved grid solutions implemented to solve this problem.

Smart grids over certain technological developments aimed at improving the Grid, like in-time defect identification, real-time management of power, remote switching, and dynamic loss reduction simulation. A reliable communications system that is available in the SG concept to help in the acquisition of data, protection, and automatic control of the SG is needed to identify possible weaknesses and find rapid solutions. The effective selectivity issue, which is important during system protection failure events, is an important advantage of SG. The extension of defects with short response time could be avoided. This can be done by rapid security and remote control systems. Automatic detection and isolation of faults can be executed via advanced SG structure measurement, security, and communication features. The selectivity principle provides many options for ensuring customer's continuous energy supply once a failure is observed.

6.9.6 Summary of the Stability Analysis Section

In this section we have discussed the focus on resilience, reliability, stability indices for performance measurement of the SG. We have also discussed smart grid protection issues.

6.10 Renewable Energy in the Smart Grid

6.10.1 Introduction

It is obvious that both smart digital technology and energy storage work together to increase grid flexibility and stability. Some of the advantages associated with combining energy storage with smart grids are

- Transmission of electricity would be more efficient and efficient.
- Systems return to the operational state more easily after power outages and related problems.
- The cost of maintenance and production is minimized.

In many countries, green energies are being developed to reduce emissions of carbon dioxide and to distribute electricity. The architecture and installation of a smart grid include the simulation, study, and risk assessment of sources of renewable energy such as solar, wind, biomass, and fuel cells [27]. The distinctive and dynamic characteristics of renewable energy sources cause an imbalanced supply with increasing demand and results in fluctuations in the power system frequency or voltage. An energy storage system (ESS) is a solution to enhance the power fluctuations in the generation of renewable energy. This system helps in smoothing the output and matching contract positions.

There is no direct correlation between electricity demand and renewable energy output. When the supply is greater than the demand, store storage will store resources and support the pressure when demand exceeds supply. This is extremely beneficial if the low demand factors correlate with rising wind power [28].

When assessing and evaluating an energy storage device for a particular project, multiple parameters should be evaluated. These parameters include life, life cycle, electricity, self-discharge rates, effects on the environment, capital costs, and so on.

6.10.2 Smart Grid Renewable Energy System

The power grid will need vast quantities of traditional power backup and massive energy storage to accommodate higher renewable energies. These are required in order, depending on the day, season, and natural variations with the power created. As the electricity grid cannot accommodate this switch today, renewable energy sources are very expensive [29].

6.10.3 Advantages and Challenges of a Smart Grid Renewable System

The advantages are as follows:

- Allowing for more cost-effective integration of renewable resources while enhancing power quality and reliability.
- Integration of customers into the network as active participants in the power sector, improving energy efficiency, and reducing greenhouse emissions by decreasing demand peaks.
- Voltage and load controls, depending on the marginal cost of production, can also be applied to minimize operational costs

The obstacles faced for smart grid implementation justify the proposition of value between service providers and consumers accompanied by regulatory restrictions and quality requirements that inhibit the smart grid [30].

6.10.4 Renewable Energy and Distributed Generation in the Smart Grid

A shift in the production of electricity is required around the world to tackle climate change and enhance energy security. As a result, funding and share in power production are increased for renewables and distributed generation. The development of renewables in an immutable environment is the greatest obstacle for smart grid systems developers. Distributed generation neither receives market signals nor does the management of the networks engage for two reasons [31–33].

First, spread supply often comes from renewable sources and is therefore given priority for fixed feed-in rates and taken-off market prices. Second, distribution network generators are small for systems management purposes and are not fitted with technology nor with functionality in balancing markets. Also, one of the concerns that has been shown is that increased renewable shares will cause distributor network congestion. The intermittent production from renewable sources and the absence of dispatch capacity can cause other problems [34, 35].

6.10.5 Different Types of Renewable Energy Sources

6.10.5.1 Solar Energy

This is the energy from the sun and heat produced using solar power, solar thermal photovoltaics, etc. Solar power is usually known as passive or active. Active solar power involves the application of photovoltaic devices for leveraging power, concentrating solar energy, etc., whereas passive solar power includes the orientation of the building according to the sun. It is a vital power source due to the huge volume of solar radiation. Using photovoltaic (PV) cells, solar energy can be produced. This technology can be a single PV panel, a PV array, or a set of multiple PV panel strings. It does not have a negative environmental effect and needs less servicing.

Solar energy improves output at the PV to concentrate large distributed amounts of sunlight to a confined space, allowing a small ray of a beam by using lenses, mirrors, and tracking systems.

The PV output will vary due to solar exposure and surface temperature. We may determine the forecast information accessible at the location, a nearby location with comparable irradiance, or the official solar atlas or database for these statistical indicators. PV system variance experiments with different functions such as Gaussian (normal) and beta probability density functions were mathematically modeled to solar irradiance. To include PV into the intelligent system, volatility and transformation technologies must be examined. Beta and Rayleigh density functions are part of the mathematical models and probability distributions used for PV analysis. The fuzzy/GA application monitoring system can acquire appropriate power supply points. The setting and scale of PV use traditional and artificial intelligence approaches requiring real-time data assessments.

6.10.5.2 Wind Energy

Wind energy or wind power is a mechanical power for the operation of electric generators by wind turbines and other work, including milling/pumping. Wind power is renewable and has a slightly lower environmental footprint than fossil fuel emissions. Without extra funding, wind turbines generate electricity at reduced costs: a pump, engine, blades, and a connecting assembly composed of a wind turbine. Compared to PV, this is less costly. No harmful contaminants are created by the turbines. However, it has considerable drawbacks such as unregulated performance, inadequate wind, etc.

6.10.5.3 Biomass-Bio Energy

Organic matter, including maize, wheat, wood, and soybean, and other waste that may contain chemicals and materials, is used to acquire bioenergy. Bioelectricity stems from gasification processes, which process gas into gas turbines for electricity production. Biomass may be transferred directly to fuels such as biodiesel produced from oil, alcohol, or maize petrol. Historically, biomass is used for cooking and heating in many emerging countries. It produces electricity only if there are large-scale bioproducts, and the transformation process is carried out. CO_2 and other pollutants are produced by biomass. The optimal planning approach and distribution of biomass need the opportunity to incorporate simulation uncertainty into modern system theory principles.

6.10.5.4 Hydropower

Hydro or water power is the power that can be utilized for productive uses by falling or quick-running water energy. The biggest renewable electricity resource today is hydropower. Owing to water variations, hydropower generators operate at a varying rate. The combustion motor is like a turbo generator. The hydraulic turbine turns water strength into the mechanical energy of rotation.

6.10.5.5 Fuel Cells

The power supply of fuel cells in smart grids has improved. It is composed of natural gas, methanol, and gasoline. This is up to 65% effective. These are environmentally sustainable assets. Fuel cells are relatively low in pollution and have minimal environmental effects. Because of their efficiency, they are well suited for green power and luxury fuel. Fuel cells produce a medium-high thermal performance. The phosphoric acid fuel pellets (PAFCs), which can be used in very high temperatures and thus require an additional cooling device to cool the tank, are usually used to run low-temperature membrane exchange proton pellets that do not have any chemicals including acids.

6.10.5.6 Geothermal Heat Pumps

A core heating or cooling system, which changes heat to and from the surface, is a geothermal heat pump. It is based on access from ponds sunk into the soil to deep steam or hot water. Conversion is made by the pumping of hot water to run traditional steam turbines that generate electricity from the electric motor. The water is recycled to drain the tank back into the soil. Geothermal plants, including a dry drain, burst steam, and a binary loop, are possible in significant amounts. These are used in diverse uses, including heat pumps, irrigation, fisheries, livestock, food manufacturing, etc.

6.10.6 Electric Energy Storage Applications

6.10.6.1 Pumped Hydro Storage

Pumped hydro-electric (PHS) storage retains potential energy from altitude variations in water depths. PHS is a typical type of energy storage that offers comparatively higher performance, high power capability, high storage volume, with high cooling space, long life, and short cycle costs. In the past, a turbine and a pump were divided for operation, but now a spinning engine is being used as a motor and a generator for cost savings due to the advancement of technologies.

When using PHS for varying levels of renewable energy supply variability, we need to verify that ramp rates are sufficient for the volume of output. This is going to expand the accessibility of such resources [30].

6.10.6.2 Compressed to Air Energy Storage

This machine uses an actual basement location, such as a rock shaft, a deserted mine, or a salt container, for high-pressure gas storage. CAES is a practical mechanism of reduction of wind instability for the leveling of wind and energy conservation purposes. They have good productivity and financial viability. Due to its high storage capacity, the system is supposed to reduce wind dumping when demand for electricity is low and to help move wind energy generation to times of high grid output strength [30].

6.10.6.3 Batteries

A. Lead–Acid Batteries

The ones most often used are the rechargeable electrochemical devices for use in a low-range capacity. Their main application is in solar photovoltaic devices. These batteries are used in communications, vehicles, and other applications. They have high dependability, good surge capability, high performance, continuous power supply, voltage stability, etc., but have poor performance for energy management due to their short life, regular maintenance, etc.

B. Nickel–Cadmium Batteries

They are denser and have a longer lifetime than lead–acid batteries. They are great for a continuous power supply and start-up applications. However, they are the cause for an improvement in regulatory restrictions on hazardous chemicals. NiCd batteries have a specialized purpose for compatibility with VRES that makes them undesirable: the memory effect. Since solar and wind power are non-dispatchable and contain forecast errors, these batteries cannot operate effectively without creating memory impact issues.

C. Zinc-Bromine Batteries
Zinc-bromine batteries (ZBr) are also known as flow batteries, where one of their liquid electrolytes flows through the reactor in an external storage tank to create electricity. The energy storage is, therefore, not connected to the power capacity; the depth of discharge is ignored and the self-discharge is very limited. This makes it useful for a whole range of applications. They have high efficiency, peaking capacity, and a high charge life cycle, etc.

D. Vanadium Redox Batteries
These batteries suit small- and medium-scale applications. They are used for load balancing, remote area power systems, renewable energy stabilization, a continuous power supply, etc. VRB batteries have a very low environmental impact. They can be used to reduce variations in solar and wind power. VRB batteries have efficiency up to 75%, a high life cycle with proper maintenance, but as they have a low energy density, they need a large amount of space [36].

6.10.7 Renewable Energy Pricing

The price of renewable energies is an important success variable. The renewable energy pricing model is based on that of fossil fuels but contains special renewable source characteristics. The use of smart grid technology will bring benefits to consumers through the use of electric system assets to safely meet consumer energy requirements at lower monetary and environmental costs. In determining the electricity rate structure, marginal costs, the load pattern, social criteria, and the income requirement of power utilities have been taken into account.

6.10.7.1 Pricing
Marginal costs mean incremental costs resulting from the most suitable adaptation to the increasingly rising demand per unit of the power generation/distribution systems. The estimation of electricity tariff based on these costs reflects the actual energy production and distribution costs. Marginal costs can be divided into four tiers in the electricity sector: production, transmission, distribution, and retailing.

6.10.7.2 Load Pattern
In 1997 the rate of time of use (TOU) was implemented as a result of a shift in load pattern. The TOU rate for the current time of the day (TOD) customer and new power users was provided as the alternative rate. Under the TOU rate, the rates will be costly for the peak and less costly for the off-peak season. The new charging trend has shifted from what it used to be; that is, there is a lower demand for the charge curve on Saturday, Sunday, and the official holidays on Monday to Friday.

6.10.7.3 Revenue Requirement of the Service Providers

The electricity service providers estimate their financial status and the average electricity rates that would offer the financial status according to the parameters laid down. Every year, income is referred to as the tax requirement. Explicit assumptions, including assumptions on the price of the fuel, inflation rates, and customer price indices, improved efficiency of the transmission system, distribution system, and retail sector, are key to estimating the financial status.

6.10.7.4 Social Criteria

For deciding electricity tariffs, the main political and social conditions are as follows [37–44]:

- For an individual consumer group, standardized national tariffs should be applied.
- Subsidies should remain for residential consumers, particularly small residential consumers who have low intakes.
- The structure of energy tariffs should be structured to better represent the other market categories.

6.10.8 Summary of the Renewable Energy Section

This chapter describes the renewable sources in the SG. Due to their less negative impact on the environment, they have been widely used with smart grids. However, their irregular trends pose a threat to the SG and, as a result, methods to store renewable energy sources have been found. We have discussed various renewable energy sources, various storage methods, and renewable energy pricing in this section.

6.11 Interoperability in the Smart Grid

6.11.1 Introduction

Interoperability (IP) addresses open technology architecture and the software systems so they can interact with other systems and technologies. Technology layout links many types of equipment like hardware and software to realize SG capabilities. Interoperability is the capacity of networks, frameworks, gadgets, technology, and features to securely and effectively interchange and utilize data with small or no inconvenience to the client [30]. The ability to communicate with other standards, frameworks, and protocols can also be specified.

6.11.2 Smart Grid Interoperability Platform (SGIP)

The SGIP is made from a semantic model that requires operations based on the data. In addition to data-driven interoperability, communication-driven interoperability is supported by communication middleware that supports communication with both client–server and publish–subscribe.

6.11.3 Functional and Non-functional Requirements

The SGIP's functional needs are profoundly affected by the types of communication protocols. Owing to data incoherence, conventions used in the existing grid are often dependent on users and service providers, based on the restricted transmission of data. As existing grids have made progress to increased transmission of the data heading to SG, measurable, versatile, and stable conventions are required that allow data-centered communication [36]. There was increasing concern in the implementation of publish–subscribe protocols like DDS in SG in efforts to counter this. DDS is node-based, where data can be published and subscribed by each node. DDS is secure as it does not have a broker, which may be a single failure point. DDS offers 22 quality of service (QoS) attributes that are very efficient in protecting and managing SG. However, DDS may be large for some devices with restricted computing capabilities that are tackled by DDS configurations.

Publish–subscribe communication is heterogeneous to the communication between client and server that is used extensively in the conventional grid. We suggest supporting both transmissions; therefore, applications participating in both transmissions can develop and communicate efficiently.

The functional needs for the SGIP are as follows:

1. Participants should be able to register and recognize resources in the client–server contact. They must also be capable of broadcasting and supporting and recognizing other participants. This often involves the management of connections (e.g. building sessions).
2. The SGIP should help the reconstruction of obtained data items. This includes allocating data memory according to its form and scale. It can be handled by QoS policies in publish–subscribe communications (e.g. received message frequency), and even an object registry to store and access data objects should be considered.
3. The QoS attributes should be configurable to ensure efficient communication. In the correspondence between client and server, this helps a client to question the reliability level of a server before requesting a service. Contact can be restricted to a publish–subscribe contact to only those users that accept the integration of QoS.
4. In publish–subscribe communication, a content-based subscription should be accepted by the SGIP. This is accomplished by screening the quality of the data collected following the requirement set up by the recipient. (e.g. unique values in subjects).

Reliability in SG is just another significant QoS. Consider wireless sensors tracking the health devices. They communicate to locate a fault within the equipment. Sensors can be affected by adverse ecological factors such as wind and rain and thus affect dependability. Smart meters and other sensors (mainly wireless) connect with the power suppliers, e.g. for monitoring electricity use and collecting price information. This includes frequent, continuous, and reliable updates. The SGIP should be configured to resolve the specifications of such reliability. They also have to remember other consistency factors, such as access control.

6.11.4 Smart Grid Interoperability Panel (SGIP) Architecture

The SGIP aims to help SG systems with data-driven and communication-driven interoperability panels (IPs). The architecture consists of two levels, i.e. data processing and communication logic. Data management deals with the management of transmitting data [45]. It comprises the modules for the API, regulating the use of resources, and reconstruction of the object. Applications are supported by the API module with SGIP facilities, including sending and receiving of data, management of resources, QoS setup, and discovery mechanisms. It is responsible for allocating data storage memory to the information management module. With regard to device computing power, various techniques can be used for memory allocation. The reconstruction module of objects is related to rebuilding the data to make them usable for implementation. The primary data objects are retrieved from serialized files and stored in a repository of global data. They are saved in local memory when needed and provided for use to applications. The above common semantic model supports the reconstruction process because both the original and the reconstructed objects in the common semantic model are variants of the same semantic entities that allow data-driven interoperability. Object reconstruction is accountable for the representation of the data objects to the topics through a publish–subscribe contact.

The communication logic layer consists of the client and server communication, publish and subscribe communication, the discovery mechanism, and the contact-setting modules. Encouraging two different communication paradigms on the same network makes the program that runs on the machines involved in both interactions easier to construct and communicate.

6.11.5 Validation Requirements

SGIP validation may be conducted with functional requirements, non-functional requirements, data-driven and communication-driven IPs.

6.11.5.1 Validation of Functional Requirements

Everyone may verify functional specifications by running several test cases, depending on user scenarios. It can be designed to encourage gradual testing (e.g. from the case of sub-use to the case of main use). Events of usage can also be granted preference based on criticality. For example, cases of use relating to control and security should be provided with a higher preference, while cases of use relating to maintenance can be given a lower preference.

6.11.5.2 Validation of Non-Functional Requirements

One can validate non-functional specifications via QoS testing. A lot of QoS policies in SG are about time, which needs synchronization of time. For example, control data that are transmitted from a SCADA device to an IED must be receipted within 20–30 milliseconds per IEC 61850, a policy of QoS latency in a publish–subscribe communication.

6.11.5.3 Validation of Data-Driven Interoperability

The data-driven interoperability panel could be verified via communication between various areas in the SG. Three instances are:

1. Between substation and management – examples are SAS and SCADA/EMS.
2. Between the management and distribution – examples are EMS and DMS communication.
3. Between distribution substations – examples are DMS and SAS communication.

6.11.5.4 Validation of Communication-Driven Interoperability

Interoperability that is powered by connectivity could be tested using diverse sorts of equipment. The SGIP can be optimized for a system based on their facilities and operation, the part they play in communication (e.g. publisher, server). On the other hand, it is designed to provide sending and releasing features for a bigger system.

6.11.6 Summary of the Interoperability Section

In this section, we found that the SG consists of numerous systems with various functional capabilities and needs. Successful attainment of an SG needs seamless interoperability among involved components.

6.12 Smart Grid Standards

6.12.1 Introduction

Although elements of the "intelligent grid" are currently being implemented and pilot projects in cities have been incorporated, more problems need to be answered before the intelligent grid is understood. Standards are not the least of these, although current standards used, e.g. distributed development, standards of metering, communications standards, new ones for the different interfaces, must be established. These involve interfaces from the source to appliances in homes and companies that need to connect to in-house electrical transmission, storage, accounting, charge balance, etc. in real-time via the grid [10]. Standards are the compilation of parameters that specify the quality or the efficiency of the commodity for a specific interface.

6.12.2 Standards for Smart Metering

There are different means of using smart meters, and this contributes to various metering communication device specifications. Automated meter reading (AMR) often demands intermittent energy transmission, whereas advanced metering infrastructure (AMI) often demands bidirectional communication. The standards that define the open communication system are IEC 62056 and ANSIC12.22 [46]. IEC 62056 describes layers for system measurement and smart transmission according to the Companion Specification for Energy Metering (COSEM). ANSI C12.22 defines the transmission and meter reading recovery from and to outer networks and is utilized in any grid.

6.12.2.1 Iec 62056

IEC 62056 is a collection of standards in sharing data for the measurement of electricity. It is the standard DLMS/COSEM international model and is planned for media applications. A meter transfers ASCII data via a serial port to a local portable device. The digital data are

Table 6.3 List of classification standards.

Standards	Functions
IEC 62056-1-0:2014	An optimization model for smart meter
IEC 62056-3-1:2013	Usage of LAN with the provider on twisted pair
IEC 62056-4-7:2014	DLMS/COSEM transport layer for IP networks
IEC 62056-5-3:2017	DLMS/COSEM application layer
IEC 62056-6-1:2017	Object identification system (OBIS)
IEC 62056-6-2:2017	COSEM interface classes
IEC 62056-6-9:2016	Mapping data models and protocols between the Generic Information Interface Messages Profiles (IEC 61968-9) and DLMS/COSEM (IEC 62056)
IEC 62056-7-3:2017	Profiles of wired and wireless M-bus connectivity for local and adjacent networks
IEC 62056-7-5:2016	Local data transmission profiles for Local Networks (LN)
IEC 62056-7-6:2013	The three-layer, connection-oriented HDLC-based communication profile
IEC 62056-8-3:2013	Communication profile for PLC S-FSK neighborhood networks
IEC 62056-8-5:2017	Narrow-band OFDM G3-PLC communication profile for neighborhood networks

typically obtained either by a signal from an LED or a pair of wires. This is a half-duplex protocol.

The general protocol is a sequence of "sign on," in which a portable device analyzes the measuring unit. The portable device approaches the given meter by number during sign-on. The meter and the device interact with different specifications during transmission and reception, such as the maximum frame length, if numerous frames could be discharged without recognizing the frames. Table 6.3 provides a list of classification standards.

6.12.2.2 ANSI C12.22

ANSI C12.22 is the American National Standard for Protocol Specification linking the networks. ANSI C12.22/IEEE Std 1703 sets out communication networks that are an advanced metering infrastructure (AMI) feature of smart networks. The facilities and scope of ANSI C12.22/IEEE Std 1703 include ANSI C12.22/IEEE Std 1703 and ANSI C12.22/IEEE Std 1703 Nodes, which are operated by clustered centers, network circuits, and portals. The AMI networks could be developed to support the field area networks (FAN) and the home/premise area networks of the utility as a smart grid. ANSI C12,22/IEEE 1703 domains may be developed as a field network. The nodes functioning within these fields must be defined with unique ApTitles, which should be enrolled (ECMX).

6.12.3 Smart Grid Communication Standards

6.12.3.1 Ieee P2030

This standard presents a smart grid integration guide for electricity and the service for this power system and end-use applications and loads. The framework offers instructions on the interpretation and concept of SG connectivity with end-use applications and loads in an electric power system.

6.12.3.2 Ieee P1901

This is the standard for fast networking devices with a speed of 500 Mbit/s in the physical layer through power lines, generally known as broadband over power lines (BPL).

6.12.3.3 Iec 62351

The above standard tackles safety challenges in the smart grid. Data transmission protection via data encryption, protected connectivity, playback avoidance, and intruder monitoring are various security priorities.

6.12.3.4 Plc G3

PLC G3 enables fast and incredibly efficient Internet protocol-based communication through existing electric lines that allow information and regulation signals to be transmitted through local smart grid generation, transmission, and distribution networks. These are built to establish comprehensive links among intelligent grid components to enable automated pricing and demand regulation strategies for consumer loads to be implemented and traditional and renewable distributed renewables like solar and wind farms to be easily connected.

6.12.4 Summary of Smart Grid Standards

This section specifies the various standards in smart grids. It describes IEC 52056 and ANSI C12.22 that are used in smart metering.

6.13 Cyber Security in Smart Grids

6.13.1 Introduction

Smart grid security is vital if any important electric grid element is to establish a safe and secure power plant service in an emergency. A stable, smart grid eliminates the risk for a system outage or malfunctioning equipment. A large shutdown, which could escalate to a rippling outage, may occur due to the absence of adequate security protocols. Smart grid protection concerns must also be resolved with supreme importance to secure this vital electricity grid infrastructure and guarantee a stable electricity supply to end-users. To enhance the stability, safety, and performance of the smart grid, cybersecurity is the phrase that gained significance from the advancement of the smart grid infrastructure by the expanded use of digital communication and management technologies [47].

Electric power infrastructure is modernized in a smart grid environment by incorporating the present and future demands and specialized features for its customers. A cyber system is incorporated into the power system to make the smart grid happen [48]. It can damage critical components of the power system, which can disturb the power supply and cause an unstable system, which results in high economic penalties.

In the SG, the power grid and the technologies like communication and data of the cyber system are eminently linked, which creates new safety issues. Problems with SG have to tackle emerging problems for secure, effective, and consistent activity. An SG is a mix of components, including software devices, hardware devices, and communication devices. Some of the threats to the cyber-physical SG are:

1. **Complexity increase:** The latest smart grid interconnections generate higher complications.
2. **Threats in cascading faults:** Since the physical system and cyber system are closely related, breakdown due to an abrupt strike or a deliberate attack on one of them will impact the other, which can result in possible cascading faults.
3. **Rising potential opponents:** The rise of network nodes provides access points for system hackers, thus raising the threats.
4. **Data privacy concerns:** The development of smart electronic systems increases information storage and data transfer, creating privacy challenges and consumer data breaches.

The factors for the cyber attack include financial interests, injuries, outraged staff, corporate sabotage, terrorist attacks, etc. Accessibility, honesty, privacy are the three top-level priorities for cybersecurity. Cybersecurity not only defends against internal and domestic cyber threats, but it also prohibits illegal access and manipulation of data within a network. It also requires cyber security.

6.13.2 Cyber Security Standards

Cybersecurity standards applied to substation protection are commonly used and the risk evaluation for a device is defined. ISO 28000 explicitly defines safety protection for a supply chain [49].

6.13.2.1 Ieee 1686

This is the IEEE standard for cybersecurity capabilities for intelligent electronic devices (IEDs) substations. This applies to any IED where the consumer needs confidentiality, transparency, integration, and service auditability of IED. The functions of IED include:

- Security by special configurations of user ID and password.
- No way to interrupt or override the user's ID/password.
- Supporting different levels of utilization of IED functions and features.
- Having a time-out function that signs out the user automatically.
- Recording in a sequential circular buffer trail with an audit list with cases according to the order of the occurrence.
- Monitoring security linked operation and tracking deliver information via the SCADA communications system in real-time.

6.13.3 Cyber Security Risks Through Mitigation

The process of mitigation of minimizing insecurity errors includes:

- Defining detailed security requirements.
- Identification and evaluation of details to be dealt with.

- Reviewing already existing security architectures.
- Implementation of a vulnerability risk reduction strategy.
- Monitoring of communication channels.
- The system verifies connection with the network during the specified period.
- Installation of the device with centralized operations, functional and security measures.

6.13.4 Summary of Cyber Security Section

This section deals with the cyber attacks and security issues associated with smart grids and methods to mitigate the cybersecurity risks.

References

[1] Shuwen Niu, Yanqin Jia, Wendie Wang, Renfei He, Lili Hu, and Yan Liu. Electricity consumption and human development level: A comparative analysis based on panel data for 50 countries, *International Journal of Electrical Power & Energy Systems*, vol. 53, pp. 338–347, 2013.

[2] Henrik Lund, Poul Alberg Østergaard, David Connolly, and B.V. Mathiesen, Smart energy and smart energy systems, *Energy*, vol. 137, pp. 556–565, 2017.

[3] T. Ackermann and V. Knyazkin. "Interaction between distributed generation and the distribution network: Operation aspects," *IEEE/PES Transmission and Distribution Conference and Exhibition*. Vol. 2, IEEE, 2002.

[4] B.V. Mathiesen, H. Lund, D. Connolly, H. Wenzel, P.A. Østergaard, B. Möller, S. Nielsen, I. Ridjan, P. Karnøe, K. Sperling, and F.K. Hvelplund, Smart energy systems for coherent 100% renewable energy and transport solutions, *Applied Energy*, vol. 145, pp. 139–154, 2015.

[5] G. Ravikumar, S.A. Khaparde, and R.K. Joshi, "Towards CIM implementation challenges in control centers worldwide and the Indian power grid," *2014 Annual IEEE India Conference (INDICON)*, IEEE, 2014.

[6] S. Mei, X. Zhang, and M. Cao, *Power Grid Complexity*. Springer-Verlag Berlin Heidelberg, Heidelberg, Germany,2011.

[7] G. Dileep, "A survey on smart grid technologies and applications," *Renewable Energy*, vol. 146, pp. 25892625, 2020.

[8] T.M. Chen, "Smart grids, smart cities need better networks [Editor's Note]," *IEEE Network*, vol. 24, no. 2, pp. 2–3, 2010.

[9] Y. Yu and W. Luan, "Smart grid and its implementations," *Proceedings of the CSEE*, vol. 29, no. 34, pp. 1–8, 2009.

[10] V.C. Gungor, D. Sahin, T. Kocak, S. Ergut, C. Buccella, C. Cecati, and G.P. Hancke, "Smart grid technologies: Communication technologies and standards," *IEEE Transactions on Industrial Informatics*, vol. 7, no. 4, pp. 529–539, 2011.

[11] T. Ma, K. Li, S. Zhu, X. Zheng, C. Guo, and L. Li "Discussion about information and communication technology of smart grid [J]," *Electric Power Automation Equipment*, vol. 5, 2010.

[12] Z. Zhu, S. Lambotharan, W.H. Chin, and Z. Fan, "Overview of demand management in smart grid and enabling wireless communication technologies," *IEEE Wireless Communications*, vol. 19, no. 3, pp. 48–56, 2012.
[13] Y. Lei, J. Li, and B. Hou, "Power distribution and utilization communication network for the smart grid," *Power System Technology*, vol. 35, no. 12, pp. 14–19, 2011.
[14] M. Kuzlu, M. Pipattanasomporn, and S. Rahman, "Communication network requirements for major smart grid applications in HAN, NAN, and WAN," *Computer Networks*, vol. 67, pp. 74–88, 2014.
[15] A. Cataliotti, V. Cosentino, D. Di Cara, P. Russotto, E. Telaretti, and G. Tinè, "An innovative measurement approach for load flow analysis in MV smart grids," *IEEE Transactions on Smart Grid*, vol. 7, no. 2, pp. 889–896, 2015.
[16] A.B. Eltantawy and M.M.A. Salama, "A novel zooming algorithm for distribution load flow analysis for smart grid," *IEEE Transactions on Smart Grid*, vol. 5, no. 4, pp. 1704–1711, July 2014.
[17] M. Sedghi and G.M. Aliakbar. "Analysis and comparison of load flow methods for distribution networks considering distributed generation," pp. 27–32, 2012.
[18] A. Farraj, E. Hammad, and D. Kundur, "A cyber-physical control framework for transient stability in smart grids," *IEEE Transactions on Smart Grid*, vol. 9, no. 2, pp. 1205–1215, 2016.
[19] A.K. Singh, R. Singh, and B.C. Pal, "Stability analysis of networked control in smart grids," *IEEE Transactions on Smart Grid*, vol. 6, no. 1, pp. 381–390, 2014.
[20] R.G. Wandhare and V. Agarwal, "Novel stability-enhancing control strategy for centralized PV-grid systems for smart grid applications," *IEEE Transactions on Smart Grid*, vol. 5, no. 3, pp. 1389–1396, 2014.
[21] K. Negash, B. Khan, and E. Yohannes, "Artificial intelligence versus conventional mathematical techniques: A review for optimal placement of phasor measurement units," *Journal of Technology and Economics of Smart Grids and Sustainable Energy*, vol. 1, pp. 10, 2016.
[22] K. Negash, B. Khan, D. Tesfaye, and M. Gayathri, "Optimal placement of phasor measurement unit for system observability," *Proceedings of IEEE International Conference on Computing Methodologies and Communication (ICCMC 2017)*, Erode, India, pp. 73–77, 18–19 July 2017.
[23] Y. Kifle, B. Khan, and P. Singh, "Assessment and enhancement of distribution system reliability by renewable energy sources and energy storage," *Journal of Green Engineering*, vol. 8, no. 3 (2), pp. 219–262, July 2018.
[24] D. Anteneh and B. Khan, "Reliability enhancement of distribution substation by using network reconfiguration a case study at Debre Berhan distribution substation," *International Journal of Economy Energy and Environment*, vol. 4, no. 2, pp. 33–40, 2019.
[25] B. Khan, H.H. Alhelou, and F. Mebrahtu, "A holistic analysis of distribution system reliability assessment methods with conventional and renewable energy sources," *AIMS Energy*, vol. 7, no. 4, pp. 413–429, 2019.
[26] P. Gopakumar, M. Jaya Bharata Reddy, and D.K. Mohanta, "Letter to the editor: Stability concerns in the smart grid with emerging renewable energy technologies," *Electric Power Components and Systems*, vol. 42, no. 3–4, pp. 418–425, 2014.

[27] D. Eltigani and S. Masri, "Challenges of integrating renewable energy sources to smart grids: A review," *Renewable and Sustainable Energy Reviews*, vol. 52, pp. 770–780, 2015.

[28] A. Evans, V. Strezov, and T.J. Evans, "Assessment of utility energy storage options for increased renewable energy penetration," *Renewable and Sustainable Energy Reviews*, vol. 16, no. 6, pp. 4141–4147, 2012.

[29] K.S. Reddy, Madhusudan Kumar, T.K. M.H. Sharon, and S. Lokeswaran, "A review of integration, control, communication and metering (ICCM) of renewable energy based smart grid," *Renewable and Sustainable Energy Reviews*, vol. 38, pp. 180–192, 2014.

[30] D. Kim, A. Alaerjan, L. Lu, H. Yang, and H. Jang, "Toward interoperability of smart grids," *IEEE Communications Magazine*, vol. 55, no. 8, pp. 204–210, August 2017.

[31] T.F. Agajie, B. Khan, H.H. Alhelou, and O.P. Mahela, "Optimal expansion planning of distribution system using grid-based multi-objective harmony search algorithm," *Computers & Electrical Engineering*, vol. 87, p. 106823, 2020.

[32] Y. Kifle, B. Khan, and J. ingh, "Designing and modeling grid connected photovoltaic system (Case study: EEU Building at Hawassa City)," *International Journal of Convergence Computing*, vol. 3, no. 1, pp. 20–34, 2018.

[33] O.P. Mahela, B. Khan, H.H. Alhelou, and P. Siano, "Power quality assessment and event detection in distribution network with wind energy penetration using stockwell transform and fuzzy clustering," *IEEE Transactions on Industrial Informatics*, vol. 16, no. 11, pp. 6922–6932, November 2020, doi: 10.1109/TII.2020.2971709.

[34] O.P. Mahela, B. Khan, H. Haes Alhelou, and S. Tanwar, "Assessment of power quality in the utility grid integrated with wind energy generation," *IET Power Electronics*, vol. 13, no. 13, pp. 2917–2925, 14 October 2020, doi: 10.1049/iet-pel.2019.1351.

[35] S. Ram Ola, A. Saraswat, S.K. Goyal, V. Sharma, B. Khan, O.P. Mahela, H. Haes Alhelou, and P. Siano, "Alienation coefficient and Wigner distribution function based protection scheme for hybrid power system network with renewable energy penetration," *Energies*, vol. 13, pp. 1120, 2020.

[36] M.I. Ridwan, N.S. Miswan, M.S.M. Shokri, M.N. Noran, R.M. Lajim, and H.N. Awang, "Interoperability in smart grid using IEC 61850 standard: A power utility prospect," *2014 IEEE Innovative Smart Grid Technologies – Asia (ISGT ASIA)*, Kuala Lumpur, Malaysia, pp. 261–266, 2014.

[37] Y. Sawle, S.C. Gupta, and A.K. Bohre, "Socio-techno-economic design of hybrid renewable energy system using optimization techniques," *Renewable Energy*, vol. 119, pp. 459–472, 2018.

[38] Y. Sawle, S.C. Gupta, A.K. Bohre, and A. Novel, "Methodology for scrutiny of off-grid hybrid renewable system" *International Journal of Energy Research*, vol. 42, pp. 570–586, 2018.

[39] Y. Sawle, S.C. Gupta, and A.K. Bohre, "Review of hybrid renewable energy systems with comparative analysis of off-grid hybrid system," *Renewable and Sustainable Energy Reviews*, vol. 81, pp. 2217–2235, June 2017.

[40] Y. Sawle, S.C. Gupta, and A.K. Bohre, "PV-wind hybrid system: A review with case study," *Cogent Engineering Journal*, vol. 3, no. 1, p. 1189305, 2016.

[41] Y. Sawle, S.C. Gupta and A.K. Bohre, "Techno-economic scrutiny of HRES through GA and PSO technique," *International Journal of Renewable Energy Technology*, vol. 9, no. 1/2, pp. 84–107, 2018.

[42] Y. Sawle, S.C. Gupta, and A.K. Bohre, "Optimal sizing of standalone PV/Wind/Biomass hybrid energy system using GA and PSO optimization technique," *1st International Conference on Power Engineering, Computing and Control, PECCON-2017*, 2–4 March 2017, VIT University, Chennai Campus, Elsevier, Energy Procedia 00 (2017), Vol. 117, pp. 690–698, 2017.

[43] Y. Sawle and S.C. Gupta "Optimal sizing of photovoltaic/wind hybrid energy system for rural electrification," *Power India International Conference (PIICON), 2014 6th IEEE*, pp. 1–4. IEEE, 2014.

[44] V. Tamrakar, S.C. Gupta, and Y. Sawle, "Study of characteristics of single and double diode electrical equivalent circuit models of solar PV module," *2015 International Conference on Energy Systems and Applications*, Pune, pp. 312–317, 2015

[45] L. Daniele, M. Solanki, F. den Hartog, and J. Roes, "Interoperability for smart appliances in the IoT world. In: Groth P., et al. (eds), The Semantic Web – ISWC 2016. Lecture Notes in Computer Science, vol. 9982. Springer, Cham, 2016.

[46] T. Sato, D.M. Kammen, B. Duan, M. Macuha, Z. Zhou, J. Wu, M. Tariq, and S.A. Asfaw, *Smart Grid Standards: Specifications, Requirements, and Technologies*. Wiley, Hoboken, New Jersey, United States, 2015.

[47] Y. Yan, Y. Qian, H. Sharif, and D. Tipper, "A survey on cyber security for smart grid communications," *IEEE Communications Surveys & Tutorials*, vol. 14, no. 4, pp. 998–1010, Fourth Quarter 2012.

[48] A. Anwar and A.N. Mahmood "Cybersecurity of smart grid infrastructure," arXiv preprint arXiv:1401.3936, 2014.

[49] M.Z. Gunduz and R. Das, "Cyber-security on smart grid: Threats and potential solutions," *Computer Networks*, vol. 169, p. 107094, 2020.

Part V

Energy Management of an Active Distribution Network

7

Active Distribution Management System

Mohammad Gholami[1], *Sajjad Fattaheian-Dehkordi*[1,2], *Hesam Mazaheri*[1], *and Ali Abbaspour Tehrani-Fard*[1]

[1] *Department of Electrical Engineering, Sharif University of Technology, Tehran 1136511155, Iran*
[2] *Department of Electrical Engineering, Aalto University, Espoo 11000, Finland*

7.1 Introduction

In the power system, energy management systems (EMSs) balance the energy sources and energy consumption at the minimum possible cost [1–4]. To supply the electricity in a reliable and secure way, the operating state of the system should be observed adequately. Since a large share of energy production is related to the transmission grid, using the EMS platform, the transmission system operators (TSOs) control and keep the system secure. Unlike the EMS, which is basically an energy-focused unit, since the 1970s the distribution management system (DMS) has been developed to manage the distribution operations. In other words, the DMS is a modified version of the EMS for the distribution system and provides different functionality compared to the EMS [5].

One of the most important requirements for a smart grid transition is the DMS, which provides the basis for safe and efficient modeling, operation, and direct field activities. It also serves as an operations platform, automating tasks and filtering information for the operator. Moreover, it is the key framework for a utility that will empower the administration of a smart grid and will use its advantages. For implementation of a DMS within the utility, the integration of a wide range of components, like field equipment, communications, control centers, and supporting systems, is required. This sets up a complete solution for the utility to accomplish its business goals in delivering reliable power to its consumers [6].

7.2 The Comparison of DMS and EMS

As previously described, to reduce the cost in the power system, the EMS tries to control the energy sources and consumption. The integrated management software, which should be equipped with some tools like power flow, security assessment, system stability, and system reliability, can control energy consumption by monitoring, billing, and cost allocation. In addition, in order to collect and display data from the network in the EMS system, communication and intelligent devices like switchgear and intelligent switching controllers are used as interfaces that are connected to the computer systems through an Ethernet network. On the other hand, a DMS at low voltage levels of

Active Electrical Distribution Network: A Smart Approach, First Edition. Edited by Baseem Khan, Josep M. Guerrero, Sanjeevikumar Padmanaban, Hassan Haes Alhelou, Om Prakash Mahela, and Sudeep Tanwar.

distribution network is defined as a supervisory control of residential or commercial loads. Compared with the transmission network, the numbers of lines, transformers, and switches are higher in the distribution network. However, only a few switches are remotely controlled by DMS. The DMS monitors the system and remotely controls switches and substations in the distribution system. The main objectives of the DMS can be categorized into the following parts [7]:

- The increase of system safety by providing better visibility.
- Improving the useful life expectancy of power system assets.
- Boosting system reliability by decreasing outage times of the system.
- Increasing the efficiency of the system and utilizing the available resources in an optimum way.

There is a control room in the DMS that mainly focuses on the following items:

- Supporting the operator to keep and operate the system in a secure state.
- Loss minimization.
- Optimizing the power flow of lines.
- Coordinating the real-time functions with non-real-time data (manually operated devices) for regular management and control of the network.
- Organizing the network model database.

While DMS and EMS have much similarity in many ways, such as remotely collecting measurements of the system state and its power devices at the data collection terminals equipped with RTU, processing available data to operators via a user interface, storing and investigating data as well as historic events for later retrieval, containing investigative functions to assist operators in order to interpret the data and investigate forthcoming circumstance, and connecting to other computer systems to share data and investigate results, these systems (EMS and DMS) have fundamental differences, presented in Table 7.1 [5].

7.3 DMS Functions

The application functions that provide network model analysis and capability can be the heart of DMS. As shown in Figure 7.1, the DMS functions can be categorized into three groups: **system monitoring**, **decision support**, and **control actions**.

Also, DMS environments can be presented in the following domains:

- **Distribution operation environment:** Providing operators with system visibility, decision support, and control for managing the distribution operation.
- **Engineering study environment:** Developing historical system performance indices.
- **Operations planning environment:** Conducting operations planning such as the impact analysis of an outage.
- **Training simulator system:** Providing scenarios for training system operators.
- **Quality assurance system:** Examining new applications and updates before presenting them to the production system.

Table 7.1 The differences between the control units of distribution and transmission systems.

Feature/Control unit	DMS	EMS
Network structure	Typically radial	Typically network connected
Location of devices	Located along the length of distribution circuits	Generally located at substations
Field devices	Manually operated	Remotely controlled
Required number of the RTU	High	Low
Data amount	Low	High

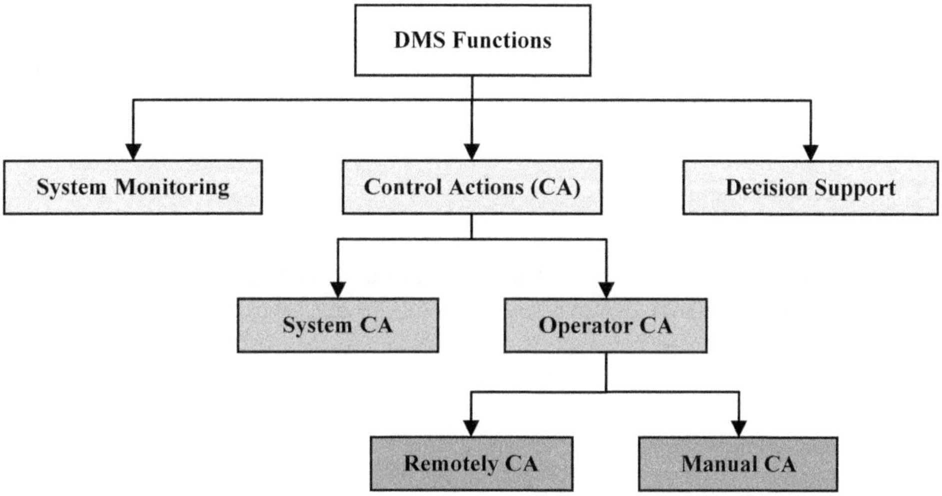

Figure 7.1 DMS functions division.

7.3.1 System Monitoring

System monitoring can help the operators to provide an accurate picture of the system condition in the distribution system. The DMS needs a lot of real-time data (such as analog and status data reported to the DMS through remote terminal units in the substations and feeders every 4 seconds) and near-real-time information (such as equipment measurements and status coming from smart meters every 5–15 minutes) about the loading and performance of the distribution system power equipment. Examples of equipment that can be monitored are distribution substation transformers and load tap changers, distribution substation switchgear, distribution substation shunt capacitor banks and shunt reactor banks, distribution substation bus meters, field re-closers, overhead and underground load-break switches with SCADA capabilities, switchable capacitor banks located in the field, field voltage regulators, faulted circuit indicators (FCIs) and other sensors located in the field, and distributed generating resources.

The main functions of DMS that can be considered in the system monitoring function group are state estimation and power flow.

7.3.1.1 State Estimation

State estimation (SE) is a procedure of exploiting noisy and existing measurements and data to extract the best estimation of system operational conditions. This function plays an inevitable role in system monitoring and controlling the system and is the main function of the EMS. However, due to the combination of distributed renewable resources directly connected to the medium and the low voltage grid of a distribution network, SE can have a key functionality in the DMS [8, 9].

7.3.1.2 Power Flow

The power flow function investigates the power flow of the network numerically and can be initiated on a periodic basis or in cases of any remarkable variation in the network and operator demand. The DMS power flow program can assist distribution operators with the electrical conditions of the network in order to understand abnormal conditions in feeders, like a voltage drop at the end part of the feeder and overload in line sections [7].

7.3.2 Decision Support

Decision support functions can provide a rich set of solutions to ensure the reliability, safety, and efficiency of the system. In detail, power system information and system models can be employed to provide decision support solutions and appropriate control actions. According to Table 7.2, decision support functions can be categorized into three application sections including assessment, optimization, and restoration applications.

7.3.2.1 Distribution System Modeler

This function obtains a detailed, up to date, three-phase electrical and connectivity model of the electric distribution system. This model is used in other DMS functions such as state estimation, power flow, and short-circuit analysis. In this function, all of the network changes and future permanent additions should be considered [7].

7.3.2.2 Load Estimation

The DMS load estimation function is employed to determine the share of each distribution service transformer from a total load of the main bus. For this purpose, historical load profiles for each distribution load and each bus are considered [10].

Table 7.2 The classification of decision support functions.

Assessment applications	Optimization applications	Restoration applications
Distribution system modeler	Volt-VAR optimization	Fault location, isolation, and service restoration
Load estimation		
Short-circuit analysis	Optimal network reconfiguration	Tagging
Short-term load forecasting		

7.3.2.3 Short-Circuit Analysis

Short-circuit analysis in a DMS structure can evaluate the voltage and current profiles of the network while a hypothesis fault with optimal location and type exists in the distribution grid. The calculated fault currents are compared with the breaking capabilities of the switchgear or the fault-current margins of the devices. During reconfiguration of the network, this function can assist the operator.

7.3.2.4 Short-Term Load Forecasting (STLF)

The STLF function tries to automatically predict the load of a system for a period of time (like a week) by employing historical load and weather data. The information on load forecasting can support operation planning applications such as optimal network reconfiguration.

7.3.2.5 Volt-VAR Optimization (VVO)

For minimization of an objective function, such as load demand or energy consumption, the VVO function evaluates the optimal control actions while feeders loading and buses voltage satisfy their acceptable ranges [11]. Such examples of control actions are control load tap changers of the substation transformer, voltage regulators, switched capacitor banks, and remotely controlled power generators connected to the distribution system. The VVO function employs the measured values of smart meters to decrease the demands on load buses by reducing their voltages in the range of minimum and maximum allowable voltages.

7.3.2.6 Optimal Network Reconfiguration (ONR)

The ONR function can propose recommended actions, typically the recommended switching actions and a switching plan, necessary to minimize an objective function (like total electrical losses, the largest peak demand, and imbalance of loads between feeders) without causing any loading or voltage constraint violations on the feeder.

7.3.2.7 Fault Location, Isolation, and Service Restoration

The FLISR function can provide a restoration when a fault happens. This functionality through decreasing the number of customer interruptions and the outage duration time tends to improve the reliability of the system and power quality [12]. For this purpose, as the first step, the section with a fault is detected and isolated. Then a new configuration for the system is employed in a way that as many customers as possible are restored to the service in a short time period. In this regard, controllable equipment, like fault detectors, automated line switches, circuit breakers, re-closers, and tie switches can be utilized. The FLISR operation has two modes, including centralized and decentralized. In the first mode, the FLISR functionality is performed by the DMS either directly or by providing the recommended control action to the operators for their automatic implementation, while the FLISR functionality in the second mode is accomplished by a peer-to-peer system without DMS intervention.

7.3.2.8 Tagging

Tagging is the DMS ability to place tags on any device in order to block specified remote control actions on the related facilities in accordance with operating procedures. One example of tags is the “Re-closer off” tag, which represents the automatic performance of reclosing in a circuit that has been disabled due to work being performed or other reasons.

7.3.3 DMS Control Actions

In the distribution network, some power system devices in the substation or field location, like circuit breakers, re-closers, disconnect switches, load tap changers, capacitor banks, shunt reactors, and voltage regulators, can be controlled by the DMS. In order to control these devices, the following control actions (CAs) can be employed:

a. System CAs: These CAs can be directly applied by the DMS with no intervention of the operator; for instance, a direct closed mode of the FLISR or a VVO function.
b. Operator advisory CAs: These CAs can be applied manually or remotely according to the recommendation provided by the decision support functions. In this type of CA, the operator can make the decision on whether the proposed recommendation should be applied or not; for example, by placing the FLISR or VVO function under the operator advisory control mode to make the final implementation decision.
c. Operator remotely CA: This system can be implemented through EMS by the operators. Opening a remotely controllable switch is an example of this CA type.
d. Operator manual CA: This type of CA can be used by the operator for equipment that cannot be operated remotely (like manually operated disconnect switches).

It should be highlighted that, based on the type of control action selected for the FLISR and VVO function, these functions can be considered in the decision support or control action function group. The interactions between mentioned groups of DMS functions (i.e. monitoring system, decision support, and control actions) are presented in Figure 7.2.

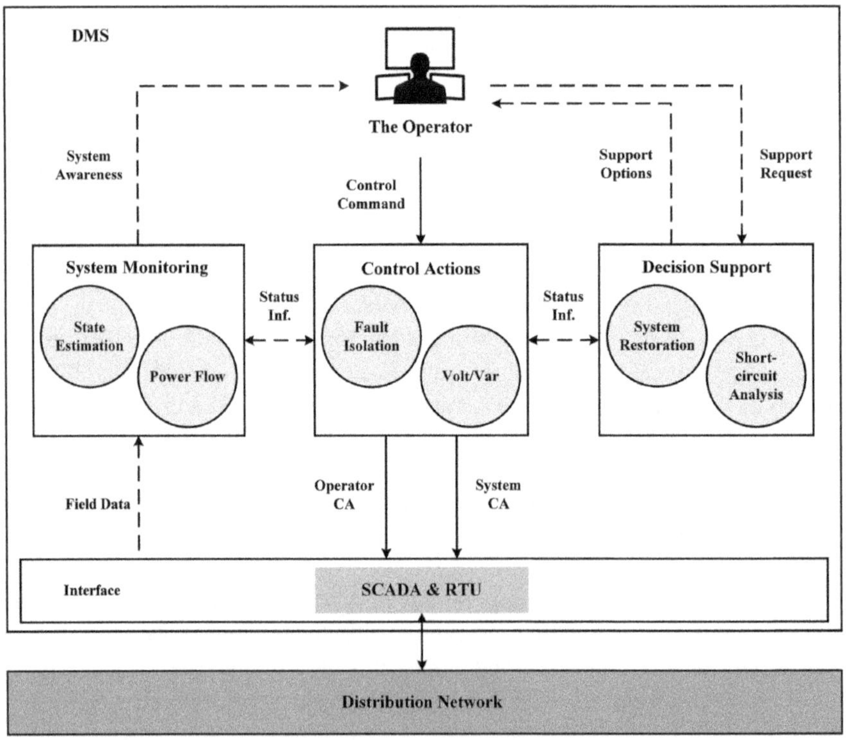

Figure 7.2 Interaction between different categories of DMS function.

7.4 DMS Architecture

As mentioned, the increasing trend in the integration of renewable energy resources (RESs), flexible resources (i.e. storage units), as well as metering and control devices has challenged the traditionally DMS entities designed for passive distribution systems. In this regard, based upon the critical role that DMS entities play in the operation and management of the distribution systems, their architecture requires using redundant designs to minimize the failure probability in the system. Moreover, different modules are defined in the conceptual design of the DMSs for smart grids in order to facilitate the incorporation of computer-based systems. It is noteworthy that the modular structure of the DMS could increase the reliability of the systems and ease the maintenance, upgrading, and expansion of the system in the future. In the following sections, the hardware and software designs of the DMS, as well as its associated characteristics and interaction with other subsystems in a power grid, are discussed.

7.4.1 Hardware Overview

As in other aspects of the power system, the state-of-the-art designs for DMS in smart grids are developed using computer-aided engineering. In this regard, the hardware design of the DMS could be categorized as below:

- Data acquisition subsystem (i.e. communication links and metering devices)
- Computer subsystems
- Man/machine subsystems

It should be noted that communication systems are an important section in DMS that provides the possibility to monitor and control different assets and subsystems in the distribution grid. In this regard, different communication infrastructures are required that enable the DMS to monitor services, local resources, protection devices, etc. Fiber optics (FO), wireless, and broadband power lines (BPLs) are viable choices to be considered in the hardware design of the DMS. The cost and acceptable latency and reliability in transferring data between subsystems are the most important criteria in selection between these potential choices.

7.4.2 Software Overview

As discussed, the system structure is developed in a modular way to increase the reliability, flexibility, and maintainability of the DMS. In this structure, each module could work independently and its required data would be provided by the common communication links. Furthermore, the structure of the DMS should be helpful for the system to be strengthened from the reliability, resiliency, flexibility, economics, and customer service perspectives. In this respect, the complete modular structure of the DMS is shown in Figure 7.3. The role of each module in the management of the distribution systems and its impacts on reliability, resiliency, flexibility, economics, and customer service quality of the system are discussed in this section.

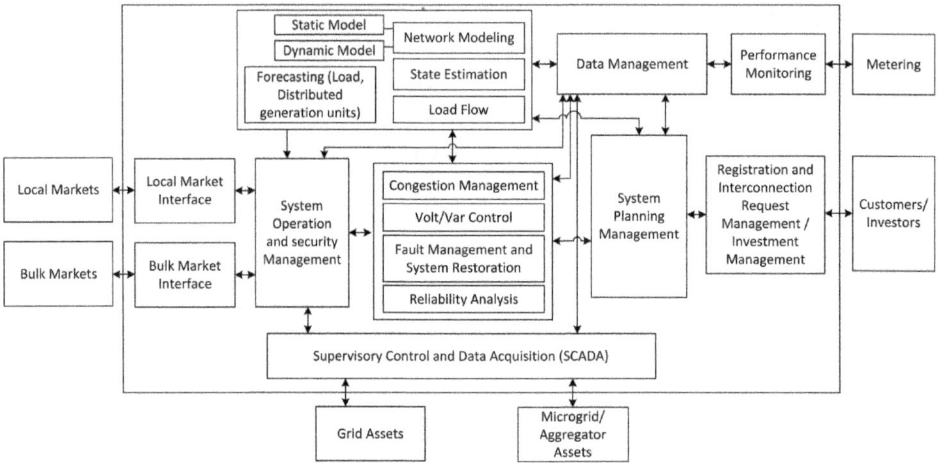

Figure 7.3 Conceptual design of the DMS.

7.4.2.1 SCADA

The supervisory control and data acquisition module has a key role in data processing, alarm analysis, as well as transferring control signals to the grid of owned assets and independent operated entities like microgrids (MGs)/aggregators in the system. This module mainly impacts the operation of the smart grid from reliability, flexibility, and resiliency perspectives.

7.4.2.2 Data Management

This module is responsible for data management in the system. Noted that DMS functioning relies on the data gathered from the current operational condition of the system as well as the historical data. In management modules, field data through the SCADA module would be stored in the data management module. Moreover, the output of the analysis modules (i.e. congestion management, volt/var control, reliability analysis, fault management, and system restoration) and pre-analysis modules (i.e. network modeling, load flow, state estimation, and forecasting) would also be kept in the data management module. This would enable the incorporation of historical data of the system in future planning and operational management of the system.

7.4.2.3 Network Modeling

Most of the system analysis tools require network modeling. In this regard, this module is responsible for providing the dynamic and static model of the network. The static model provides the network topology and connection of the assets, as well as MGs and customers, while the dynamic model considers the real-time measurement for real-time network modeling.

7.4.2.4 State Estimation

The backbone of the DMS is a state estimation function, which analyzes the network condition using real-time and pseudo-measurements. The precision of the state estimation procedure mainly depends on the precision of the metered data.

7.4.2.5 Load Flow

This function conducts the power flow in the system to determine the line flows and voltages in each node. The output of this module would be the main input data for Volt/Var control and congestion management modules.

7.4.2.6 Forecasting

DMS as a management entity provides strategies for operation and planning the system in near-future and long-term times. In this regard, the amount of load and generation by local units/MGs/aggregators should be forecasted for each time period.

7.4.2.7 Congestion Management

Privatization in the distribution systems and an increasing trend of integration local power generation units have increased the probability of congestion occurrences in the distribution systems. In this regard, this module is responsible for determining the best strategy to alleviate the congestion in case of an occurrence in the system.

7.4.2.8 Volt/Var Control

This module would be activated in the case where the voltage in a node of the grid violates the predefined acceptable intervals. The most cost-effective strategy to alleviate the issue would be given to the system operation management module to be applied in the system.

7.4.2.9 Fault Management and System Restoration

This function would be activated in the case of fault occurrences in the system and optimizes the fault clearance and system restoration process by considering the automatic or manual protection devices in the grid. This function primarily improves the system from reliability and resiliency perspectives.

7.4.2.10 Reliability Analysis

This function analyzes the reliability of the system at each point in order to consider the reliability criteria predetermined for the customers. It provides vital information for the system planning management module in order to allocate a sufficient amount of investments to improve the system reliability.

7.4.2.11 System Operation and Security Management

This is the main function in the operational management of the system, which strives to alleviate the operational and security problems of the system by considering the analysis conducted by system analysis modules. Traditionally, this function would consider the operation of protection devices, grid reconfiguration, or asset tagging in order to operate the system optimally and securely. However, integration of the distributed generations (DGs) and introduction of the prosumer (i.e. an entity that could act as a producer or consumer) in the distribution systems have dramatically shifted the traditional passive operation of the systems. In this regard, the local power generation entities (i.e. prosumers, aggregators, or microgrids (MGs)) may participate in the bulk power markets (i.e. energy and ancillary service markets). Moreover, the local markets (i.e. energy and grid ancillary services) would also be developed in the distribution systems as a result of the introduction of local power generation units (i.e. RESs and microturbines) and flexible resources (i.e.

storage units, electric vehicles). As a result, the operational management module would optimize the incentives and operational tariffs to alleviate the operational problem raised in the system in different operational time periods. Monitoring and transferring control signals would be conducted via a bulk market interface, local market interface, and SCADA subsystem.

7.4.2.12 System Planning Management

Expansion of distribution grids and integration of new resources in the system should efficiently be optimized by DMS. In this regard, the system planning management function is conceived to be the main responsible module in order to optimize the investments in the system considering the output of the analysis modules. Moreover, the integration of new customers and new resources to the grid should be authorized and managed by the planning management function. Therefore, the planning function tries to benefit the system from an economic and customer service point of view, while taking into account the reliability, flexibility, and resiliency of the system. Finally, the modifications in the system planned by this function would inform the data management system to update the data utilized by other modules, specifically the pre-analysis and analysis modules.

7.4.3 Characteristics of the DMS Architecture

The main characteristics of the DMS architecture that facilitate the employment of the other developments in the smart grid management and associated mathematical and technological methods are now discussed.

7.4.3.1 Internet of Things (IoT)

The modular computer-based structure of the DMS system, which is mainly based on the possibility to monitor and control the devices in the system, could be implemented in the system by using the IoT system. This system, by defining unique identifiers for the grid assets and using computing devices and digital machines, provides the ability to transfer data between the DMS and assets without the interaction of humans. In this regard, the cost of execution of the control and monitoring system would be decreased and the potential faults by human interaction would be avoided. The SCADA sub-system, as the module conducting supervisory control and data acquisition in the grid, would be the primary function that could be promoted by the development of IoT technology.

7.4.3.2 Machine Learning and Data Analysis Techniques

Machine learning and data analysis techniques could be employed by preanalysis, analysis, system operation management and system planning management modules to analyze the system condition and optimize the operational and planning control signals. That is why the historical data associated with metering in the system as well as analysis of the system conditions are stored in their related clusters in the data management module. This master data management module provides each function of the DMS with the required data and, therefore, would dramatically increase the speed of analysis by implementing parallel computation and distributed analysis techniques in the system.

7.4.3.3 Novel Local Market Models

Traditionally, the energy resources of the distribution systems have been managed centrally by the system operator; however, integration of distributed generation and privatization in the system have introduced the concept of decentralized operation of the energy resources. Moreover, the prosumers/MGs/aggregators could also participate in the markets run by TSO. The conceptual design of the DMS would enable the incorporation of novel market models in the distribution systems, which are based on multiagent and noncentralized operational management of the system.

7.4.3.4 Decentralized Management

The modular design of the DMS provides the possibility to develop DMS functions in a decentralized or a hybrid (i.e. a combination of centralized and decentralized) way. This is usually determined based on a cost/worth analysis considering the issues with coordination of decentralized modules, the amount of data exchange, and the required communication infrastructures.

7.4.4 DMS Interaction with Other Smart Grid Subsystems

In an electrical grid, DMS would have interaction with other control and management units. An advanced metering system, energy management system (EMS), and geographical information system (GIS) are the main units that have interaction with the DMS.

7.4.4.1 Advanced Metering System

An advanced metering infrastructure (AMI) provides the two-way information exchange between the meter and the central system. In this situation, this system enables the DMS to efficiently monitor the distribution system.

7.4.4.2 Energy Management System (EMS)

The EMS, as the unit responsible for the management of the transmission system, would have interaction with the DMS with the aim to improve customer service. The interface between the DMS and EMS is mainly dependent on the power system's overall management design. It is noteworthy that by integration of DGs and flexible resources in the distribution system, the interaction of DMS and EMS would be expanded from traditional ones to facilitate employing the energy and ancillary services from the distribution system in the transmission system management.

7.4.4.3 Geographical Information System (GIS)

The GIS system stores the physical details and updates the assets in the distribution system. In this regard, the interaction with the GIS system enables the DMS to have physical characteristics of the grid and its assets and update the network information based on the GIS database.

7.5 DMS Challenges and Its New Requirements

As previously described, the DMS is an integrated systems solution supporting the day-to-day distribution network management, related activities of the construction and maintenance, and proactively guides operators during unexpected events like storms and related restoration activities. As the utility workforce ages and utilities come under greater force to more fully employ available devices, DMS is quickly getting to be a fundamental component in keeping up and increasing delivery reliability while decreasing complexity and automating related work processes. The current acceleration in automation of the distribution and substation, as well as automated metering initiatives in the industry, have provided another impetus for DMS as a solid foundation for using these aspects of the emerging smart grid.

On the other hand, over recent years distribution networks have been encountering some new events, i.e. the existence of distributed generating (DG) units, local energy markets, the presence of new distributed energy resources (DERs), and prosumers. These events can be translated into a gradual evolution from passive to active distribution networks [13]. In this condition, the flow of distribution feeders will be bidirectional and the advantages of radial topology will be diminished. Therefore, the configuration of distribution networks should be gradually changed from a mainly radial to a more meshed topology aimed at maintaining the system reliability level. As a result, system operators and planners need to deal with highly complex active distribution networks (ADNs) as DMSs with low visibility and automation levels cannot meet the challenges of the ADN. Therefore, active DMSs with improved and efficient monitoring, control, and protection systems need to be proposed.

In order to address these issues, some new requirements and functions for operating and controlling the system are proposed to be integrated into the DMS.

7.5.1 State Estimation

Generally, due to observability issues, the low value of X/R, communication problems, high renewable energy penetration, and some other factors, SE in future distribution networks is a challenging function compared with the SE process in transmission systems. The following issues can be considered for the process of the distribution system SE algorithm:

- Utilizing pseudo-measurement (historical/forecasted data) in order to address the observability issue.
- The invalidity of conventional DC and decoupled formulation in the SE problem.
- High complexity and a long process runtime due to the intrinsic characteristics of the active distribution network, such as network topology, high DG penetration, unbalanced operation, and a high number of buses.
- Estimation errors of the SE results because of the pseudo-measurements and DG power output uncertainties.

To address these issues, some new work should be performed covering topics including selection of the state variable, the observability issue of the network (generation of pseudo-measurements), the identification of topology, renewable resource penetration, SE problem scalability, and optimal meter placement (like the phasor measurement unit (PMU) or micro-PMU with appropriate accuracy and reporting rate).

7.5.2 Load Estimation

In the past, due to the passive nature of the distribution system, the different load buses in the network were not measured and monitored. Nowadays, for grid operational planning and real-time control, estimating and predicting the load become more complex but more essential, according to the activation of the distribution grid. In order to respond to this issue, the number of measurement devices in the grid should be increased. For this purpose, due to the valuable characteristics of intelligent devices (IEDs), this equipment can be employed in the ADNs. Due to having extra functionality, like measurement of instantaneous voltage, current, and power, the IED can be replaced with the traditional electromechanical equipment, which can improve the reliability of load estimation. However, one of the problems is to collect and transmit these measured values to the DMS database in real-time for updating the algorithm of the load estimation. As a consequence, in order to support the load estimation function, some new meters and advanced communication technologies should be combined between the field and the DMS. For instance, to improve the link between field devices and RTU, different RTU devices can be joined with modbus/TCP/IP/fiber optic ports.

7.5.3 Load GIS

Due to the transformation of the power generation from a few large units to many small units, system operators are encountering some new requirements. Indeed, to minimize the distance of reactive power flow and therefore power losses, the voltage deviation and power flow imbalances should be controlled using distributed energy resources. In this situation, the location of DERs should be specified for the system operator. To address this issue, the GIS can be joined to the network model, which presents the network geographical representation to the system operator. In this situation, the vision of the operator about the network and also the ability to efficiently support the field crew will improve [14].

7.5.4 Synthetic Inertia

In the power system, after the occurrence of a disturbance, both the small and large frequency variations try to be damped by system inertia through maintaining the range of frequency change within allowable boundaries. In rotating generators, the kinetic energy mainly limits the rate of frequency change. Indeed, variation of the rotor speed in the rotating generators of the grid can cause a change in the transformation of kinetic energy into electric power. The following motion equation describes this dynamic [15]:

$$T_m - T_e = J\frac{d\omega_m}{dt} \tag{7.1}$$

where $d\omega_m/dt$ is the rate of change in rotational speed of the machine; T_m is the mechanical torque applied as input to the rotor; T_e is the electrical torque load on the machine; and J is the moment of inertia.

In future power systems, by transition from rotating machines to static generators (converters), the DMS is faced with a new challenge. Indeed, according to the increase in the number of grid-connected renewable resources (PV systems and wind turbines) with full converters in the power system, the direct-coupled inertia provided by direct-coupled machines will decrease. In this situation, the inertia response of the rotating machines simulated through a facility labeled as the synthetic inertia response can be employed. To provide synthetic inertia, some manufacturers of the wind turbine (General Electric and ENERCON) have started to combine controllers on their new wind turbine generators [16]. Moreover, the energy storage that is connected to the converter coupled generators can also provide this type of inertia. It should be highlighted that the handling of the inertia would be performed by the DMS. Therefore, a new function with the duty of providing the total available amount of inertia response and the available amount that could be sent/ received to/from the neighboring DMSs should be considered in the DMS to help the operator in real-time.

7.5.5 Energy Storage Monitoring

Based on recommendations presented for the European Energy Storage Technology Development Roadmap [17], the prices for electrical storage are predicted to decrease. In this situation, electrical storage can be employed as an economical option for providing ancillary services. On the other hand, fluctuations in the network mainly come from forecasting errors and intermittence of the renewable resources. According to the fact that electrical storage could address this issue (fluctuations), in some EU projects, like Grid4EU, this technology has been used for further investigation. In order for the operator to be able to use the energy storage systems in the network, the DMS must provide the operator with some data such as battery plans, location, and relevant energy prices.

7.5.6 Vehicles to Grid

In some countries, because of concerns about climate change and some benefits of the utilization of electric vehicles (EVs), such as its lower fuel costs and quiet operation, the application of these transportation devices has increased. For example, there are more than 13 000 EVs in Norway [18]. The EVs can use smart grids to offer an individual advantage for vehicle to grid technology (V2G). Indeed, according to the capabilities of V2G, EVs can be considered as controllable loads that respond to frequency variations. For this purpose, the operator should utilize the power participation of grid connected EVs in the primary frequency control. To address this issue, some novel algorithms should be integrated into the DMS.

7.5.7 Scaling the DMS for Supervisory Control

Due to the increase in the penetration of DG units at the distribution network level, some new requirements for controlling the system from the perspective of TSOs and DSOs should be considered. Deploying DER units increases DSO responsibility and new communications between TSO and DSO are needed to keep the system secure. To develop into more active distribution networks, the following four stages should be observed.

7.5.7.1 Increased Observability

In the near future, due to the integration of DERs and renewable resources with high levels of uncertainties, the observability of the distribution grid should be increased. For example, in Germany, based on the VDE AR-N 4105 recommendation for the distribution of voltage level grids, local controllers try to monitor and support the frequency of PV systems. According to these guidelines, by using a predefined characteristic curve, the frequencies over 50.2 Hz smoothly reduce the power of PV systems.

7.5.7.2 Active Network Management

In order to solve congestion in the transmission and distribution networks, active power management of DER units can play an important role. For example, based on German feed-in tariff law, one of the following two options should be employed in the PV units with an active power of less than 30 kWp: (1) restricting their reactive power output to 70% of their installed capacity and (2) equipping with a remote control interface to enable the DSO to send the required temporal power reduction signals. On the other hand, PV units with active power more than 30 kWp must be controlled remotely [19].

7.5.7.3 Responsibility (Control) Transferring to DSOs

Based on the activation of the distribution grid, it is predictable that the DSO for controlling the system will support the TSO more actively and, as a consequence, some responsibilities of the TSO (such as management of reactive power) can be controlled by the DSO. In accordance with this subject, the Swiss TSO proposed a concept for controlling the coordinated voltage of the 220/380 kV transmission system, which, together with the power plants connected to it, enables active participation of the related distribution networks [20].

7.5.7.4 Self-Responsible Distribution System Operation

At the fourth stage, the DSO can independently control the voltage and frequency of the related network; for instance, management of the active and reactive powers.

Based on the presented two first stages, it can be concluded that the interaction between the TSO and the DSO is increased. For instance, the TSOs can acquire the real-time measurement of medium voltage DER units in the distribution network and in case of necessity, these units can be curtailed by an official message to the DSO. In a real distribution network, the actual deployed resources can specify the need for supervision and control. In this situation, the architecture of the DMS should include features such as flexibility and modularity.

7.5.8 Multiarea Implementation of the DMS

As previously described, in the last few years, passive distribution networks are gradually transforming into active networks. However, the DMS platform is generally implemented based on a centralized control strategy, which is inefficient for ADNs. In addition, due to a large number of network buses and a growing amount of available information and remotely controlled devices in the smart grid, this strategy may be faced with the following issues:

- Requiring high computational speed processors.
- Requiring large data storage units.
- Requiring a low latency communication system to transfer large amounts of data among the field agents and control center.
- Subjecting to a single point of failure risk.

In response, decentralized strategies that are more popular in the case of large networks can be employed to eliminate the centralized strategy's weaknesses and also increase the reliability, flexibility, and efficiency of the DMS [21].

References

[1] T. Molla, B. Khan, B. Moges, H.H. Alhelou, R. Zamani, and P. Siano, "Integrated optimization of smart home appliances with cost-effective energy management system," *CSEE Journal of Power and Energy Systems*, vol. 5, no. 2, pp. 249–258, June 2019.

[2] T. Molla, B. Khan, and P. Singh, "A comprehensive analysis of smart home energy management system optimization techniques," *Journal of Autonomous Intelligence*, vol. 1, no. 1, pp. 15–21, 2018.

[3] B. Khan, H. Getachew, and H.H. Alhelou, "Components of smart grid system." In John Vacca (Ed.), *Solving Urban Infrastructure Problems Using Smart City Technologies* (pp. 385–395), Amsterdam, Netherlands: Elsevier, September 2020.

[4] B. Khan, E. Gidey, H. Getachew, and H.H. Alhelou, "Managing the generation and demand inside the smart-grid structure." In John Vacca (Ed.), *Solving Urban Infrastructure Problems Using Smart City Technologies* (pp. 433–445), Amsterdam, Netherlands: Elsevier, September 2020.

[5] J.A. Momoh, "Distribution management systems." In J.A. Momoh (Ed.), *Electric Power Distribution, Automation, Protection, and Control* (pp. 259–275). New York: CRC Press, 2007.

[6] USA Siemens, "How the distribution management system (DMS) is becoming a core function of the smart grid, reducing risks and costs by optimizing distribution network operations," *Point of View (POV) Executive White Paper Series*, 2012 [Online]. Available at: https://sie.ag/2uHECYQ.

[7] E. Vaahedi, "Distribution management system." In E. Vaahedi (Ed.), *Practical Power System Operation* (pp. 176–192). New Jersey, USA: Wiley-IEEE Press, 2014.

[8] M. Gholami, A.A.T. Fard, and M. Moeini-Aghtaie, "Linear voltage based state estimator for active distribution system including phasor measurement unit (PMU)," *2018 Electrical Power Distribution Conference (EPDC)*, IEEE, pp. 1–6, May 2018.

[9] M. Gholami, A. Abbaspour, S. Fattaheian-Dehkordi, M. Lehtonen, M. Moeini-Aghtaie, and M. Fotuhi, "Optimal allocation of PMUs in active distribution network considering reliability of state estimation results," *IET Generation, Transmission & Distribution*, vol. 14, no. 18, pp. 3641–3651, 18 September 2020.

[10] M. Rezkalla, K. Heussen, M. Marinelli, Junjie Hu, and H. W. Bindner, "Identification of requirements for distribution management systems in the smart grid context," *2015 50th International Universities Power Engineering Conference (UPEC)*, Stoke on Trent, pp. 1–6, 2015.

[11] S. Rahimi, M. Marinelli, and F. Silvestro, "Evaluation of requirements for Volt/Var control and optimization function in distribution management systems," *2012 IEEE International Energy Conference and Exhibition (ENERGYCON)*, IEEE, pp. 331–336, September 2012.

[12] M. Gholami, A. Abbaspour, M. Moeini-Aghtaie, M. Fotuhi-Firuzabad, and M. Lehtonen, "Detecting the location of short-circuit faults in active distribution network using PMU-based state estimation," *IEEE Transactions on Smart Grid*, vol. 11, no. 2, pp. 1396–1406, March 2020.

[13] C. D'Adamo, S. Jupe, and C. Abbey, "Global survey on planning and operation of active distribution networks-Update of CIGRE C6. 11 working group activities," *CIRED 2009–20th International Conference and Exhibition on Electricity Distribution Part 1*, IET, pp. 1–4, June 2009.

[14] M.M. Jensen, R. Larsen, I. Stefani, and D. Djapic, "On the implementation of an advanced distribution management system," *Nordac 2014*, pp. 1–9, 2014.

[15] P. Kundur, N.J. Balu, and M.G. Lauby, *Power System Stability and Control* (Vol. 7). New York: McGraw-Hill, 1994.

[16] C. Caerts, R. D'hulst and S. De Breucker, "Specification of smart grids high level functional architecture for frequency and voltage control," *ELECTRA Deliverable D*, 3, 2015.

[17] O. Teller, J.-P. Nicolai, M. Lafoz, D. Laing; R. Tamme, A.S. Pedersen, M. Andersson, C. Folke, C. Bourdil, M. Conte, G. Gigliucci, I. Fastelli, M. Vona, M.R. Porto, T. Hackensellner, R. Kapp, H.J. Seifert, M. Noe, M. Sander, J. Lugaro, M. Lippert, P. Hall, R. Saliger, A. Harby, M. Pihlatie, and N. Omar, "Joint EASE/EERA recommendations for a European energy storage technology development roadmap towards 2030," *European Association for Storage of Energy (EASE)*, 2013.

[18] T.E. Norbech, "Incentives and infrastructure – Crucial elements in the build-up of Norway's EV fleet," *2013 World Electric Vehicle Symposium and Exhibition (EVS27)*, Barcelona, Spain, pp. 1–4, November 17–20, 2013.

[19] J.V. Appen, M. Braun, T. Stetz, K. Diwold, and D. Geibel, "Time in the sun: The challenge of high PV penetration in the German electric grid," *IEEE Power and Energy Magazine*, 11, pp. 55–64, 2013.

[20] M. Zerva and M. Geidl, "Contribution of active distribution grids to the coordinated voltage control of the Swiss transmission system," *2014 Power Systems Computation Conference*, Wroclaw, pp. 1-8, 2014.

[21] M. Shahidehpour and Y. Wang, "Parallel and distributed processing of power systems." In M. Shahidehpour and Y. Wang (Eds.), *Communication and Control in Electric Power Systems: Applications of Parallel and Distributed Processing* (pp. 47–101). New Jersey, USA: Wiley-IEEE Press, 2003.

8

Role of Volt-VAr-W Control in Energy Management

Baidyanath Bag

Department of Electrical Engineering, National Institute of Technology Raipur, Chhattisgarh, India

8.1 Introduction

Despite strong technological innovations, especially in the area of power distribution, the weakest link in the electricity value chain is found to be the distribution sector, with significant inefficiencies present in this sector. The main reasons for this are due to high AT&C losses, large supply–demand mismatching, poor power quality, low consumer satisfaction, increased number of blackouts, etc. Further, the continuous increase in energy demand has increased the concern for adopting energy management techniques as the promising solution for the sector [1–5]. Conservation voltage reduction (CVR) is an effective technique among different available techniques for energy management in a radial distribution system.

CVR is a technique in which energy management is done through demand reduction or loss minimization. To do so, the supply voltage level across the load (constant impedance types) is being reduced to lower its consumption level, but the minimum customer utilization voltage is managed consistently as required by the level determined standard [6, 7]. CVR has induced significant interest among various utilities[1] due to the rapid reduction of demand and line losses [8]. Available research shows that a 0.3%–1% loss reduction is possible with a 1% voltage reduction. Around 3.04% of reduction in the national energy consumption can be ensured by implementing CVR in all the distribution feeders in the USA [9].

However, voltage stability is one of the critical issues faced during CVR implementation. This issue is generally addressed through integrating optimal capacitor placement (OCP), optimal DG placement, network reconfiguration, use of voltage stabilizers, many other reactive power compensation devices, etc., along with CVR. Due to the low capital required to place capacitor banks, OCPs are mostly used together with CVRs. However, the growing trend of DG has enabled the use of optimal DG placement for voltage support while implementing CVR.

The chapter contains the following sections. Section 8.2 discusses the methodology, test results, and the benefits of integrating OCP with CVR, whereas the integration of optimal DG for CVR has been discussed in Section 8.3 along with its methodology, test results, and benefits. Finally, concluding remarks have been discussed in Section 8.4.

Active Electrical Distribution Network: A Smart Approach, First Edition. Edited by Baseem Khan, Josep M. Guerrero, Sanjeevikumar Padmanaban, Hassan Haes Alhelou, Om Prakash Mahela, and Sudeep Tanwar.

8.2 OCP Integration with CVR

Problem formulation for the above stated objectives can be done while solving energy management using the concept of OCP integration with CVR. This methodology can be implemented or tested using any real-time system or a standard test system, such as IEEE test systems. To find the effectiveness of OCP integration with CVR, the following three operating scenarios are required to be tested:

1. Practical situation: Input voltage higher than rated voltage
2. Input voltage equal to rated voltage
3. Input voltage lower than rated voltage

OCP integration with CVR has been tested using the IEEE 33 bus test system for the above three operating scenarios in [10]. Following are the results and conclusions obtained in this paper. When they implemented this methodology for the second operating scenario, 17 buses faced a voltage instability issue without integrating OCP. Hence they integrated OCP with CVR to boost these bus voltages. Table 8.1 and Figure 8.1 show the test results obtained with the OCP integration while applying CVR.

Table 8.1 Capacitor banks and their impacts on bus voltages.

Capacitor banks (kVA)	108	120	135	150	300	1500
Buses	Bus operating voltages in terms of % rated bus voltage					
8	96.665	96.348	95.890	95.985	96.393	97.673
9	96.576	96.162	95.631	95.693	96.104	97.735
10	96.486	95.969	95.356	95.445	95.726	97.851
11	96.443	95.917	95.301	95.387	95.662	97.827
12	96.365	95.821	95.201	95.288	95.552	97.794
13	96.271	95.535	95.043	95.061	95.223	98.276
14	96.301	95.439	95.073	95.053	95.151	98.724
15	96.313	95.354	95.077	95.023	95.121	99.080
16	96.298	95.274	95.045	95.001	95.099	99.461
17	96.344	95.099	95.008	95.132	95.230	99.278
18	96.328	95.047	95.001	95.130	95.279	99.223
28	95.969	95.927	95.844	95.811	96.153	97.265
29	95.473	95.456	95.403	95.401	95.681	97.155
30	95.221	95.231	95.170	95.178	95.457	97.078
31	95.042	95.054	95.037	95.071	95.264	97.577
32	95.009	95.026	95.016	95.055	95.249	97.826
33	95.018	95.039	95.033	95.078	95.319	97.801

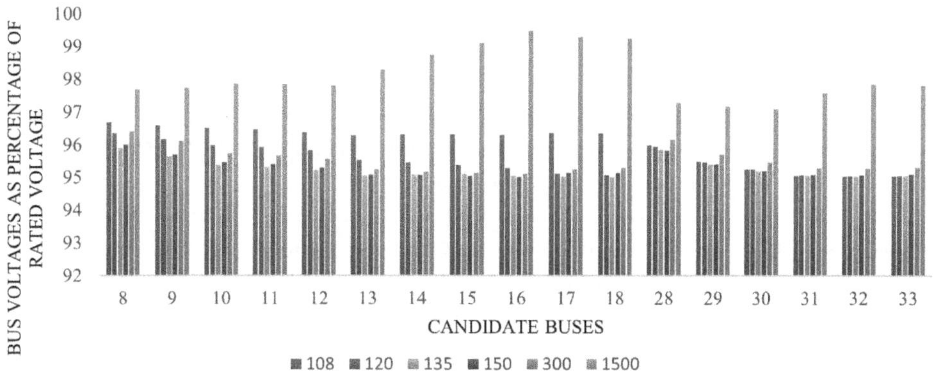

Figure 8.1 Capacitor banks and their impacts on bus voltages.

Table 8.2 Economic analysis.

Years	Cost ($)		Saving ($)		
	Installation plus purchase	Operation and maintenance	Loss reduction	Yearly profit	Accumulative profit
1	13 365.00	4455.00	16 949.24	870.76	−870.76
2	0.00	4455.00	16 949.24	12 494.24	11 623.48
3	0.00	4455.00	16 949.24	12 494.24	24 117.73
4	0.00	4455.00	16 949.24	12 494.24	36 611.97
5	0.00	4455.00	16 949.24	12 494.24	49 106.21
6	0.00	4455.00	16 949.24	12 494.24	61 600.45
7	0.00	4455.00	16 949.24	12 494.24	74 094.70
8	0.00	4455.00	16 949.24	12 494.24	86 588.94
9	0.00	4455.00	16 949.24	12 494.24	99 083.18
10	0.00	4455.00	16 949.24	12 494.24	111 577.40

As the main objective of CVR is to ensure that the maximum buses should operate at 95% of its rated voltage, 135 kVAr is the optimal size of capacitor bank to be placed during OCP implementation while under operating scenario II; this can also be seen from Table 8.1 and Figure 8.1. An economic analysis has also been carried out by the authors of this paper to find the cost benefit for their proposed methodology whose test results shown in Table 8.2.

From Table 8.2, it is clear that an accumulative profit of $111 577.40 has been obtained at the end of the planning period of 10 years through implementation of the energy management technique if applied for a second operating scenario.

When this methodology has been tested for the first operating scenario, the voltage levels of many nodes that are situated near the substation reach a value that is much higher than the rated system voltage. This hike in node voltages causes increased load drawl at such buses, thus increasing consumer meter readings and network losses. Hence this is not the recommended operating scenario for using this energy management trick. As far as operating scenario III is concerned, it is found that this methodology of energy management is not beneficial under such a scenario because the accumulated benefit is negative, it has negative branch capacity, and the network losses are very high.

Hence, it can be concluded that the operating scenario II is the best suitable operating condition for implementation of the energy management technique that is based on OCP integration with CVR. Figures 8.2 to 8.5 show some of the other major benefits that could be obtained through application of this methodology.

From Figure 8.2 it can be found that the total network loss during operating scenario II is less as compared to the other two scenarios. The variation in load demand during different operating scenarios depicted in Figure 8.3 shows that the load drawl at different buses

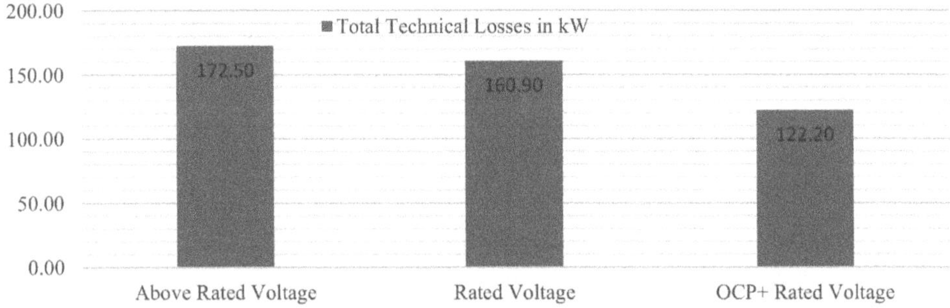

Figure 8.2 Network losses during different operating scenarios.

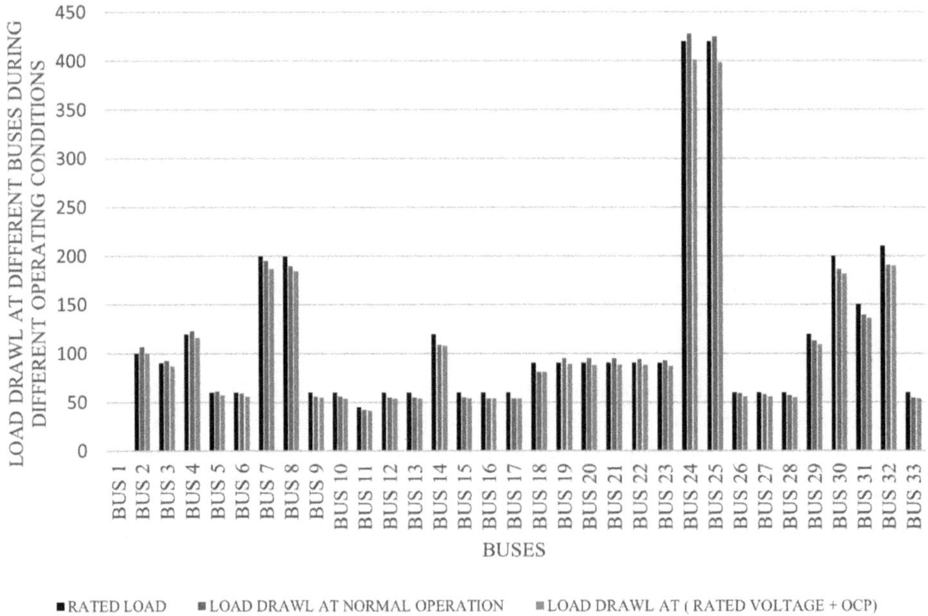

Figure 8.3 Load demand during different operating scenarios.

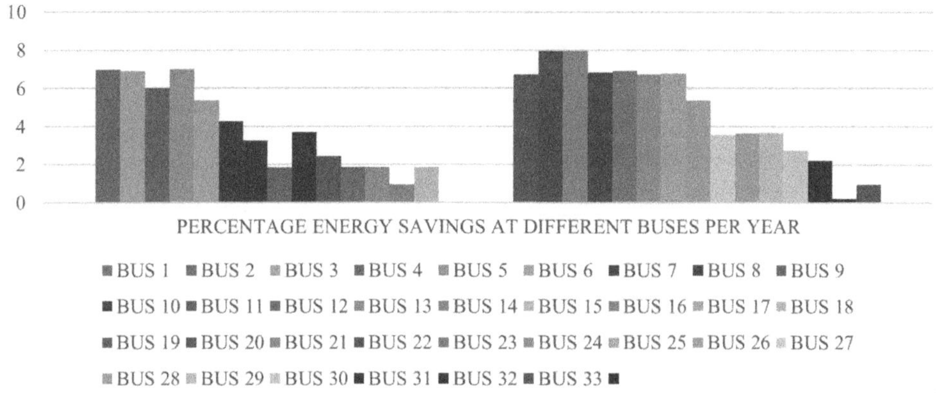

Figure 8.4 Percentage energy savings per year at different buses (operating scenario II).

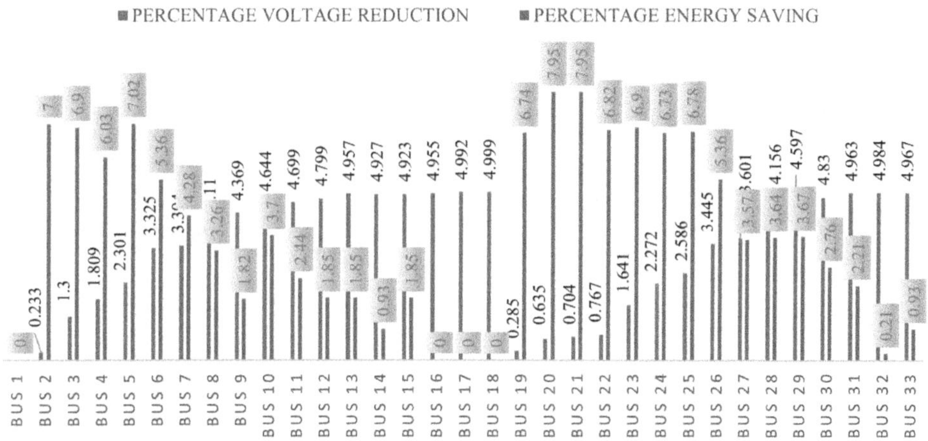

Figure 8.5 Impact of percentage voltage change on percentage energy savings per year (operating scenario II).

reduces during operating scenario II without affecting the performances of different loads. The percentage energy savings per year at different load buses depicted in Figure 8.4 also shows significant differences. The impact of percentage voltage change upon percentage energy has been depicted in Figure 8.5.

8.3 DG Integration with CVR

Like the case of OCP integration with CVR, a problem formulation for the same set of objectives is done while solving energy management using DG integration with CVR. The IEEE 33 bus test system has been used to validate the proposed methodology. A genetic algorithm is implemented in MATLAB 2019A and OpenDSS software is used for all load flow studies. To find out the effectiveness of DG integration with CVR, the same three operating scenarios as mentioned in Section 8.3 have been tested. The following results have

Table 8.3 Voltage profiles of different buses under different test cases.

Bus No.	Case 1	Case 2	Case 3
1	1.05	1	1
2	1.0472	0.99729	0.99871
3	1.0339	0.98471	0.99379
4	1.0271	0.97819	0.99178
5	1.0204	0.97178	0.99008
6	1.0037	0.95587	0.98495
7	1.0005	0.95285	0.98331
8	0.98832	0.94126	0.9801
9	0.98273	0.93593	0.97989
10	0.97757	0.93101	0.98015
11	0.9768	0.93029	0.98033
12	0.97547	0.92902	0.98079
13	0.97006	0.92387	0.98317
14	0.96806	0.92196	0.98436
15	0.96682	0.92078	0.98309
16	0.96562	0.91963	0.98187
17	0.96383	0.91794	0.98006
18	0.9633	0.91743	0.97951
19	1.0466	0.99677	0.99819
20	1.0429	0.99325	0.99467
21	1.0422	0.99256	0.99397
22	1.0415	0.99194	0.99335
23	1.0304	0.98131	0.99123
24	1.0238	0.97501	0.98587
25	1.0205	0.97188	0.9827
26	1.0019	0.95421	0.98487
27	0.99961	0.95201	0.98488
28	0.98931	0.9422	0.98384
29	0.98192	0.93517	0.9835
30	0.97873	0.93212	0.98448
31	0.97501	0.92859	0.98074
32	0.9742	0.92781	0.97992
33	0.97395	0.92757	0.97967

been obtained. In case 1, a slack bus is kept at a voltage level of 1.05 p.u., which shows the operation of the system under an above-rated voltage condition. To apply CVR, the slack bus voltage was reduced to 1 p.u. (which is the rated voltage condition) in case 2, but at this condition 17 of the buses faced a voltage instability issue, which can be seen in Table 8.3.

To mitigate the above issue of voltage stability, an optimal DG placement has been done for case 2 in case 3. The voltage profiles of different buses under three different cases have been depicted in Table 8.3 and Figure 8.6.

Out of these three cases, case 1 is the practical operating condition that represents operating scenario I, case 2 is unstable and is an invalid condition, but case 3, which represents the operating scenario II, is the optimal condition as far as test results are concerned. In case 3, three different DGs of 709.4, 1260.4, and 1180.4 kW have been placed at buses 14, 24, and 30 respectively. Total line losses taking place and the total load drawl under these three operating conditions have been depicted in Figures 8.7 and 8.8, respectively. However, as far as the operating scenario III is concerned, like the previous method, this method also does not produce economically feasible solutions and three different buses violate voltage stability conditions despite an optimal DG placement along with CVR. Hence the various test results for this operating scenario have not been discussed.

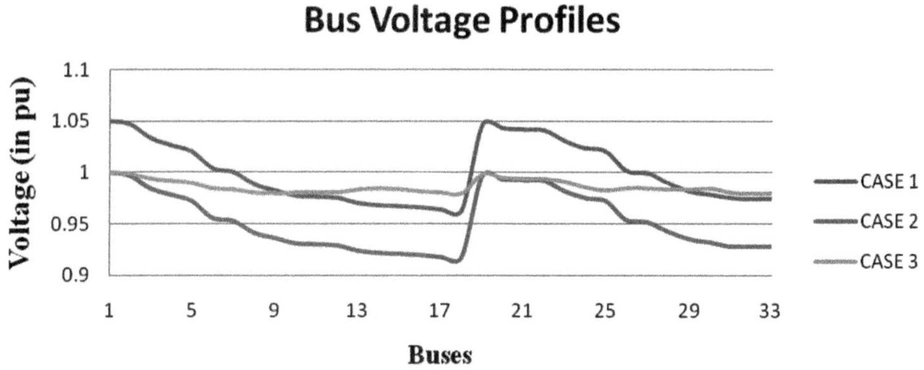

Figure 8.6 Voltage profile during different test cases.

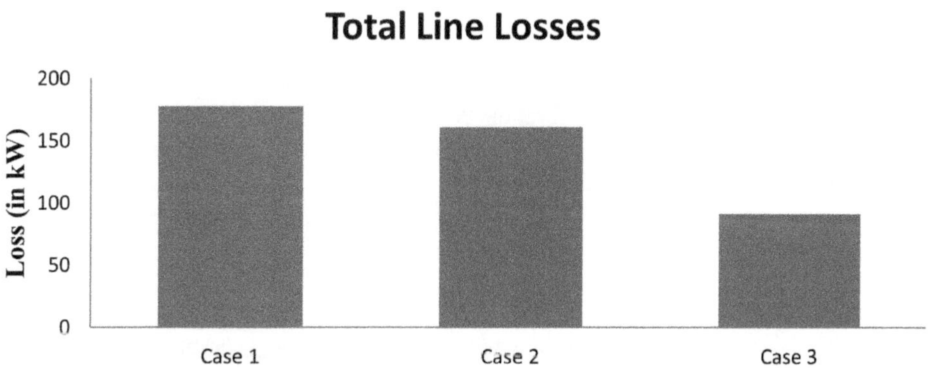

Figure 8.7 Line losses during different test cases.

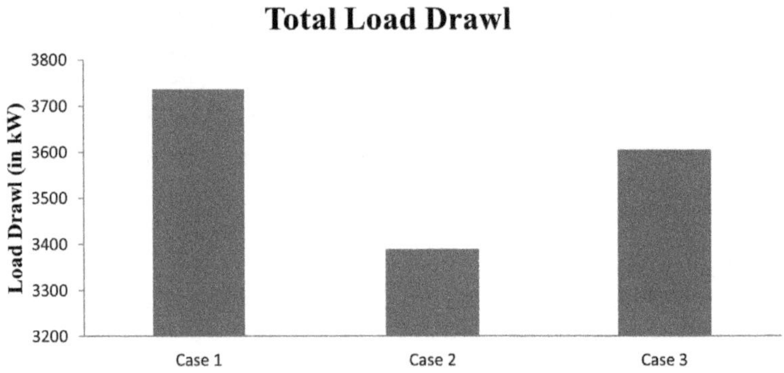

Figure 8.8 Load drawl during different test cases.

8.4 Conclusion

In this chapter, a simple and cost-effective method for energy management has been discussed. The task of energy management through this method is achieved using volt-var-w control in the power distribution sector. For controlling voltage, active power, and reactive power, conservation voltage reaction supported by optimal capacitor placement and optimal distributed generation placement are used. Different operating conditions of the power distribution network have been tested through the application of this method and the promising operating scenario for which this method of energy management is suitable has been verified from the obtained test results. As the costs of capacitor banks are cheaper and their implementation methodologies are very simple as compared to all other available means of voltage support, OCPs are the most adoptable options used with CVR. However, the growing trend of distributed generation technology and its promising benefits in the present context have enabled the use of optimal DG placement for voltage support while implementing CVR.

References

[1] B. Khan, H.H. Alhelou, and F. Mebrahtu, "A holistic analysis of distribution system reliability assessment methods with conventional and renewable energy sources," *AIMS Energy*, vol. 7, no. 4, pp. 413–429, 2019.

[2] Y. Kifle, B. Khan, and P. Singh, "Assessment and enhancement of distribution system reliability by renewable energy sources and energy storage," *Journal of Green Engineering*, vol. 8, no. 3 (2), pp. 219–262, July 2018.

[3] P. Singh and B. Khan, "Smart microgrid energy management using a novel artificial shark optimization," *Complexity*, vol. 2017, Article ID 2158926, 22 pages, 2017.

[4] T. Molla, B. Khan, B. Moges, H.H. Alhelou, R. Zamani, and P. Siano, "Integrated optimization of smart home appliances with cost-effective energy management system," *CSEE Journal of Power and Energy Systems*, vol. 5, no. 2, pp. 249–258, June 2019.

[5] B. Khan, E. Gidey, H. Getachew, and H.H. Alhelou, "Managing the generation and demand inside the smart-grid structure." In John Vacca (ed.), *Solving Urban Infrastructure Problems Using Smart City Technologies* (pp. 433–445), Amsterdam, Netherlands: Elsevier, September 2020.

[6] W. Zhaoyu and W. Jianhui, "Review on implementation and assessment of conservation voltage reduction," *IEEE Transactions on Power Systems*, vol. 29, no. 3, pp. 1306–1315, 2014.

[7] D. Kirshner, "Implementation of conservation voltage reduction at Commonwealth Edison," *IEEE Transactions on Power Systems*, vol. 5, no. 4, pp. 1178–1182, 1990.

[8] K.P. Schneider, F.K. Tuffner, J.C. Fuller and R. Singh, "Evaluation of conservation voltage reduction (CVR) on a national level," 2010. [Online]. Available at: http://www.pnl.gov/main/publications/external/technical_reports/PNNL-19596.pdf. [Accessed February 2020].

[9] T.A. Short and R.W. Mee, "Voltage reduction field trials on distributions circuits," *Proceedings of IEEE PES Transmission and Distribution Conference and Exposition (T&D)*, IEEE, pp. 1–6, 2012.

[10] B. Bag and T. Thakur, "Energy management using volt/var control of end-use appliances without affecting their performances and lives," *Journal of Renewable and Sustainable Energy*, vol. 9, no. 3, pp. 035504.1–035504.13, 2017.

9

Active Management of Distribution Networks

Hooman Firoozi[1], *Mohamad Amin Rajabi Nezhad*[2], *Hosna Khajeh*[1], *and Hannu Laaksonen*[1]

[1] *Flexible Energy Resources, School of Technology and Innovations, University of Vaasa, 65200 Vaasa, Finland*
[2] *Power Quality Research Lab, Iran University of Science and Technology, Tehran, Iran*

9.1 Overview

In conventional power systems, electrical power was unidirectionally transmitted from power plants to end-users through transmission and distribution systems. These conventional power grids have been structured as a centralized grid in which the power flows only in one way through long-distance lines. Therefore, such systems were designed hierarchically. Conventional power networks are dimensioned according to peak power. Investigations state that transmitted energy through transmission lines would not be more than 70% of the total energy capacity required for the peak demand. Moreover, the generation capacity is over-designed for about 20% to be responsible for peak demand, which occurs only 5% of the time. It is estimated that the power losses due to the transmission process are about 8% [1].

Previously distribution renewable energy sources and other distributed generation (DG) units connected to distribution networks were required to be disconnected in fault and disturbance situations. However, the technological advances and social acceptance has led to more and more customers installing DGs and other flexible energy resources such as batteries. The high injection of intermittent power produced by renewable-based DGs causes stability issues for a central system. The problem could also get worse if, for example, a considerable share of the DGs is disconnected from the grid during faults of high voltage (HV) grid or frequency disturbances. Therefore, more supportive actions are required from distribution networks. In addition, effective management of the DGs requires high cooperation between different system operators including distribution system operators (DSO) and transmission system operators (TSO).

Moreover, in the future, there is an increasing need for flexibility from distribution-network-located flexible energy resources (controllable generation units, energy storages, controllable loads, and electric vehicles) as well as those connected to transmission networks (such as traditional generators). Distribution-network-located flexible energy resources have the capability to:

a. Provide various local (corresponding voltage level) and system-wide (whole power system) flexibility services by providing active (P) and reactive (Q) power support and control.

Active Electrical Distribution Network: A Smart Approach, First Edition. Edited by Baseem Khan, Josep M. Guerrero, Sanjeevikumar Padmanaban, Hassan Haes Alhelou, Om Prakash Mahela, and Sudeep Tanwar.

b. Improve energy efficiency at the same time. For example, they can contribute to voltage control and congestion management, reduce losses, and enhance reliability of electricity supply to the end-users, for example through intended island operation.

Nevertheless, this potential of flexible energy resources and full integration of RES cannot be realized without active management of the distribution grids. In this regard, the traditional structure of distribution networks is not effective. Thus, the structure has been changed and the energy delivery procedure has become more decentralized. Moreover, for active management of the distribution network, the DSO needs to operate distribution networks in an optimal and reliable manner [2]. The DSO needs to perform the entire network energy management system using control center information systems, substation and secondary substation automation, along with customer-side equipment such as smart meters. The capabilities of an active network management (ANM) can include cost management, control, protection, fault management, etc. [3, 4].

To this end, this chapter discusses the active energy management of distribution systems with the high number of DGs. First, the chapter introduces planning and operational challenges associated with the current distribution system. Then, the active distribution system management as a potential solution is analyzed. The active management of distribution networks itself includes several solutions. These solutions are introduced in the chapter. Finally, the role of energy storage systems (ESSs) and microgrids (MGs) in ANM are discussed.

9.2 Distribution Network Challenges

Current distribution networks are faced with an increasing number of DGs connected to these networks. Although renewable-based DGs offer environmental and economic benefits for energy sectors, integrating DGs into distribution networks poses serious challenges for the distribution system operators (DSOs). The challenges can be different according to the type of DGs, the number and the location of DGs, as well as the type of distribution networks [5]. However, in general, the related challenges can be categorized into operational challenges and those related to network reinforcement and planning [6–10].

9.2.1 Operational Challenges

The production of renewable-based DGs is intermittent and volatile as it is highly dependent on environmental factors [11]. For example, the real-time production of photovoltaic (PV) panels is highly correlated with the real-time solar radiation and cloud moving patterns. In addition to intermittency, the power produced by these resources is often non-dispatchable. As a result, the unpredictable behavior of DGs along with its uncontrollability presents serious challenges for system operators, both DSOs and TSOs.

The growing number of DGs connected to distribution networks results in a bidirectional flow of power. This issue affects voltage profiles and the power flow of the distribution network. Although DGs can improve the reliability of supply and decrease power losses in distribution networks, they may jeopardize the security of this network by violating the constraints of voltage, especially during time slots with high generation. To deal with this problem, the DSO should deploy a more advanced strategy and adopt additional resources to regulate voltages of nodes in the distribution network [12]. However, the existing control actions are too costly and significantly delayed to respond to the problems arising from the intermittent power of renewable-based DGs [13]. Additionally, the existing devices deployed for voltage regulation purposes are often designed for traditional distribution networks that were operated without DGs. In this regard, considering radial networks, the voltage magnitudes were considered to reduce along the distribution feeder from the substation point to the end-user locations. However, DGs located close to the end-users disapprove of this assumption as they increase the voltages at their connection points [14].

Moreover, the presence of a large standalone DG (such as a wind turbine with a huge capacity) decreases power quality in distribution networks during times at which the DG is turned on or off. This problem increases, especially in rural areas that have weak distribution networks [12]. In addition to this, rapid voltage dips can happen to distribution networks after each failure of DGs [15]. Besides, DGs can affect the power quality by raising the fault levels and causing some system perturbations, such as harmonics or flickers.

Congestion management is another challenge for the current DSOs. The excessive installation of DG violates the physical capacity constraints, which are defined as ($P^{generation} - P^{load} > P^{max}$). In these situations, the DSO needs to curtail the generation or adopt downward flexibility in order to manage congestion in the distribution network. The same situation in the opposite direction can happen during times at which a number of electric vehicles are charging simultaneously. In these situations, the physical constraints are violated in the way presented in the following equation:

$$P^{load} - P^{generation} > P^{max} \tag{9.1}$$

In this regard, the DSO requires adopting upward flexibility by decreasing demand or increasing the production of resources located at specified nodes of the network.

Regarding the protection issues, the increasing number of DGs connected to the distribution network's feeders significantly impacts the existing applicable protection practices as they can contribute to short-circuit currents. This contribution affects the value of fault indicators utilized for finding the fault position. Hence, some important factors need to be considered before installing DGs. These factors are related to the protection of the DGs from internal faults as well as protecting the faulted network from the DGs' fault currents. The other factor is the consideration of islanding detection (anti-islanding/loss-of-mains). It should be noted that the high amount of distribution-network-located generation can increase islanding operation in distribution networks [5].

9.2.2 Network Reinforcement and Planning Challenges

The increasing penetration of DGs, especially distributed PV panels, fails to decrease the reinforcement-related needs of distribution networks. Distribution networks should be designed to accommodate peak loads occurring only in limited time slots of a year. However, the residential peak demand mostly happens in the evenings during which there is no PV production. As a result, the high installation of solar DGs is not able to reduce investment and reinforcement costs of networks. In addition to this, the high injection of DG power increases the network assets and devices, resulting in additional investment costs. For example, wider cables might be required for a distribution network with a great number of DGs in order to withstand higher voltages. This can be costly for DSOs and incurs greater connection costs for the DG developers [16].

From the planning aspect, DSOs need to take into account the effects of DGs and its required assets when they are planning their networks. Hence, the presence of DGs with different characteristics makes planning complicated and timely. The planning problem will exacerbate if the DSO does not receive information on the installation of DGs on some private networks. Hence, a decent hierarchical management method is needed so that the DSO is fully aware of the type and location of each existing DG as well as those that are going to be installed in the future. It is worth mentioning that the trend of installing DGs is not exact from the viewpoint of the passive DSO, which can lead to a more complicated planning process.

Moreover, according to the traditional centralized management method, DSOs design their networks on a top-down basis. In this way, energy delivery was performed in one direction starting from the transmission substation to the small-scale customers and end-users. As a result of this approach, there exist a few monitoring tools and devices in the lower levels as the loads were predicted and monitored in an aggregated manner. The aggregated information is also available with some time lags. However, DGs in distribution networks need to be actively monitored and controlled in real-time, which should be planned in the near future.

9.3 Active Management for Planning and Operation of Future Distribution Networks

The active management of a distribution network facilitates the cooperation between the planning and operation of the network. With the help of active management, the DSO is able to benefit from real-time flexibility, which leads to a reduction of the investments. In this way, the DSO deploys the current hosting capacity of the distribution network in a more optimal manner. However, to achieve this purpose, the network should be equipped with decent ICT technology, as well as several connections and operational requirements. DSOs also require having suitable tools for forecasting the production and loads of the network in the short term as well as the maintenance requirements of the network in the short and long terms. Figure 9.1 presents some planning-based and operation-based solutions regarding the active network management of distribution networks.

Furthermore, active DSOs can procure flexibility from small-scale flexible energy resources located in distribution networks (such as electric vehicles and batteries) so as to exploit network available capacities. In this regard, local markets for energy and flexibility should be designed to incentivize consumers and pro-active consumers (prosumers) to play more active roles and respond to the DSO's real-time flexibility need. Note that the new market structures and the management methods should be able to highly utilize distribution-network-located flexible energy resources in both operation and planning principles [17].

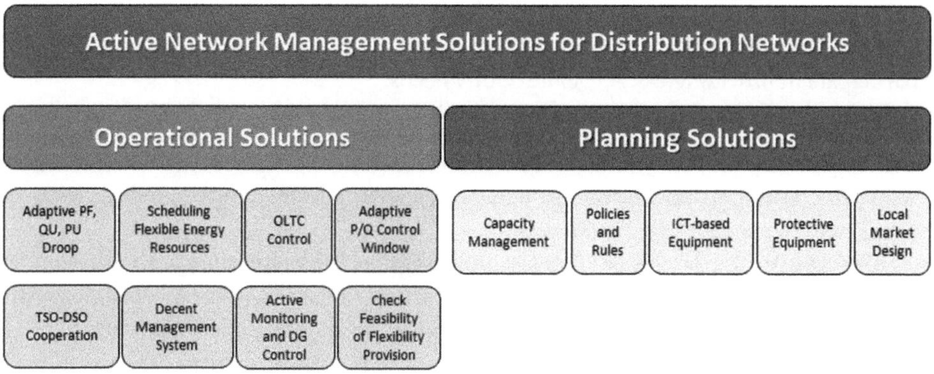

Figure 9.1 Active network management solutions for distribution system operators.

9.3.1 Planning of Active Distribution Networks

The active DSO manages the capacity of the network and aims to improve the utilization and investments on distributed energy resources and flexible energy resources at critical nodes when planning the network. For this purpose, the DSO needs to identify the areas that can be potentially overloaded. After that, it should design various incentive programs and policies to strengthen the distribution network and increase its available flexible capacities using flexible energy resources. To this end, incentive programs are designed with a target to incentivize customers and prosumers to reshape their consumption or curtail their production based on the needs of the operator.

The DSO may also establish some policies to control and manage the growth of DGs in its network. According to the IMPROGRES project [18], policies developed by the DSO can highly affect the location and the number of DGs installed in the network. The DSO must know about the current availability and the plans on the DG installation. The regulations and policies should also be compliant with the current availability, short-term operation, as well as mid-term and long-term forecasts of the network. Moreover, the DSO has to provide DG developers with information regarding expected curtailments. In this way, DG developers can conduct risk analysis prior to investing in their projects. This also explicitly helps DSO since the development and reinforcement of the network would be in line with the network requirements [16].

In terms of technical criteria, the DSO needs to ensure that the DGs fulfill the minimum technical criteria. For example, DGs and flexible energy resources must be equipped with smart meters and decent equipment. The equipment and protective relays of DGs must resist voltage dips and it should not be islanded in every critical situation. In this regard, the DSO should check each DG in terms of technical requirements.

9.3.2 Operation of Active Distribution Networks

In order to effectively operate the distribution network and increase the hosting capacity for high penetration of intermittent power of renewable DGs, a DSO needs to change into a more active operator of the network. For this purpose, it needs adequate tools and state-of-the-art ICT technologies to optimally operate the network. In addition, active DSO should create a competitive environment to motivate small-scale customers and prosumers to participate in providing DSO-level flexibility needs.

Regarding active management of the distribution system, a DSO should [16]:

1. Actively manage capacity and loss of distribution networks through active interaction with different kinds of distribution-network-located flexible energy resources such as small-scale customers and prosumers.
2. Have decent operation tools to manage and integrate distributed energy resources.
3. Be actively in touch with other operators (DSOs and TSOs) so as to trade flexibility with them. For example, a DSO can buy flexibility from other system operators when needed and sell them the surplus flexibility in case they need it. For this purpose, new commercial systems, flexibility services, and grid codes should be defined.
4. Monitor, exchange, and organize off-line and real-time information associated with its network, the flexible energy resources, and the DGs.
5. Enhance monitoring, simulation, and control of the distribution network via advanced ICT technology and effectively deploy SCADA and distribution management system (DMS).
6. Deploy configurable settings including adaptive *QU*-, *Pf*-, and *PU*-droop, adaptive OLTC control, adaptive P&Q-control window along with the flexibility of available flexible energy resources, and coordinate different services offered by different resources.
7. Actively check the feasibility of providing flexibility for other system operators.
8. Actively control voltage and manage congestion of the network using cooperative actions.

The active management methods along with compatible local flexibility market structures should activate the highest collaborative value of flexibility in distribution networks. The active management should have adaptive *QU*-, *Pf*-, and *PU*-droops, adaptive scheduling of flexible energy resources, OLTC control, and a P&Q-control window, in which P denotes active power, U is voltage, and reactive power is indicated by Q [17].

9.3.3 Voltage Control

Traditionally, DSOs control voltage by reinforcing the grid, installing preventative measures, deploying transformers with on- and off-load tap changers, and capacitor banks that inject reactive power into the grid. In this regard, the DSO determines the set points and analyzes different scenarios/ranges to control the voltage of different nodes in the network. However, with a recent increase in the number of DGs, the traditional method is no longer applicable.

In order to control voltage in the distribution network, the DSO should exploit the flexibility potential of different resources connected to the distribution network. Hence, two technical actions can be taken for voltage control in distribution networks:

1. Regarding distribution networks, active power is a more significant driver affecting the voltages changes in MV and LV networks. Accordingly, the DSO can deploy the active

power produced by flexible energy resources such as EVs and batteries in order to control the voltage.
2. Inverter-based DGs that are equipped with power electronic devices can provide reactive power for the transmission network in less than 50 milliseconds. With the help of these DGs, voltage fluctuations in distribution networks can be mitigated [19]. This means that the inverter-based DGs can supersede capacitor banks, static VAR compensators, as well as on-load tap changers in distribution networks. In addition, these devices offer a faster response and more reactive power support for the network. Unlike shunt capacitors, the inverters are capable of absorbing reactive power as well as injecting it, which can be extremely helpful for voltage control purposes [19].

9.3.4 Active Control of Inverter-Based Distributed Generation

Regulating the output active power of DGs is one of the strategies introduced to control the voltage in active distribution networks. During the moments in which, for example, power production of DGs is high while the demand is low, the problem of overvoltage would occur in the network. Therefore, reducing the output power of some DGs could be a solution for this problem.

In MGs and distribution networks, voltage source inverters (VSIs) interfacing DGs can be used, both in VCM or CCM modes, based on the type of DGs connected to the distribution network. Flexible energy resources (FERs) such as energy storages are connected by means of the VCM interface inverter strategy. DGs with intermittence characteristics such as wind and solar units are connected with an interface inverter in CCM control mode. The VCM interfaced inverters are mostly utilized to inject active/reactive powers and control the frequency/voltage, especially in the islanded MGs. However, the CCM modes are proposed for maximum power point tracking (MPPT) purposes. Furthermore, using DGs for negative-sequence current injection in order to mitigate voltage unbalance is another application that has been introduced. This application might have negative effects on the output active and reactive power injected by DGs, since the negative-sequence current injection can reduce the net capacity of the interface inverter [20, 21].

9.3.5 Adaptive Droop Control of Distributed Energy Resources

In terms of adaptive droops, for instance, the DSO can deploy less sensitive *QU*-droops for PVs during winter (very high-load situations) and more sensitive droops for summertime (very low-load situations), as illustrated in Figure 9.2, which is an example for a seasonally adaptive *QU*-droop [17, 22]. A more sensitive droop leads to smaller dead-zones. Hence, the inverter-based PV can inject more reactive power during very low-load time slots. *QU*-droop should also have seasonal settings that are compliant with the location of DGs in distribution networks. In addition to adaptive *QU*-droops, a seasonally adaptive Pf can be utilized to enable the contribution of DGs to frequency control. Finally, adaptive *PU*-droops may enable a higher participation of demand response resources that are curtailing their loads in order to contribute to frequency control in the vicinity of the PV and battery energy storage connection points.

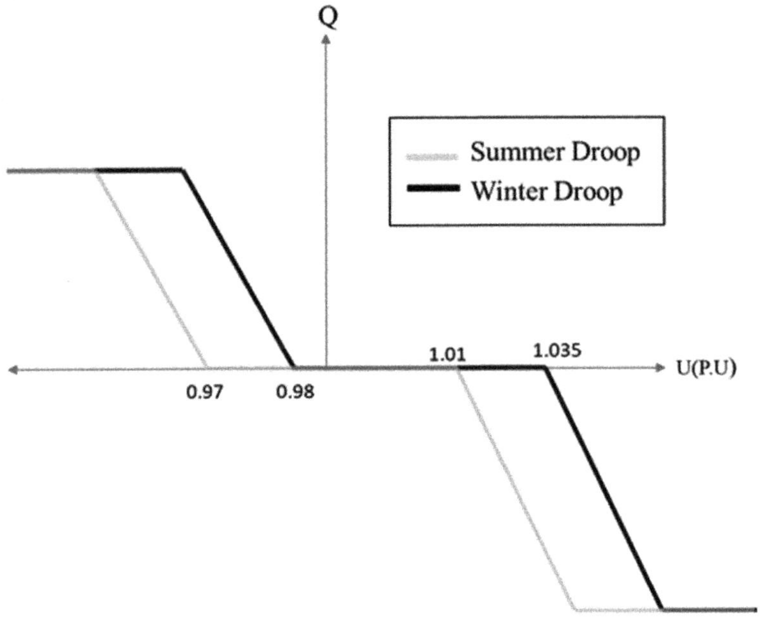

Figure 9.2 An example of a seasonally adaptive *QU*-droop.

9.3.6 Adaptive OLTC Control

The reason for the utilization of adaptive OLTC settings is threefold. First, it helps to avoid the violations of voltage constraints occurring in distribution networks. As previously mentioned, the increasing number of DGs may result in voltage limit violations. Second, during normal operation of distribution networks, the OLTC adaptive control decreases the need for voltage control. Finally, it increases the hosting capacity of distribution networks for the injection of intermittent solar power [23, 24].

As an example, the OLTC value can be set to 20.3 kV in conditions with a very high load (like winter) and equals 20.0 kV during the condition with a very low load situation (such as the summer season). However, it is worth mentioning that the *QU*-droop settings of DGs are also dependent on the OLTC set-value. Hence, careful considerations are required when adjusting these settings [17].

9.3.7 Adaptive Active and Reactive Power Control Limits Between Voltage Levels

This method is utilized to enhance the coordination of the DSO and TSO for the provision of flexibility services from resources that are located at different levels of power systems, including low voltage (LV), medium voltage (MV), and high voltage (HV) levels. The P&Q can be controlled monthly/seasonally as well as daily/hourly. In this way, the adaptive P&Q-control windows should be designed for each connection point and for each level of the power system including (HV/MV, MV/LV, and LV connection points). With the help of this method, the short-term and long-term operation of the TSO and DSO can be coordinated optimally [17].

9.3.8 Traffic Light Concept

In order to operate the distribution system, first the current state of the system should be defined and specified. These states were defined based on security standards as well as grid codes. A scheme called "traffic light" is one of the methods utilized to distinguish between different states of the system and to choose the appropriate actions accordingly [25]. This concept is illustrated in Figure 9.3. Additionally, the introduced states are defined in the following subsections.

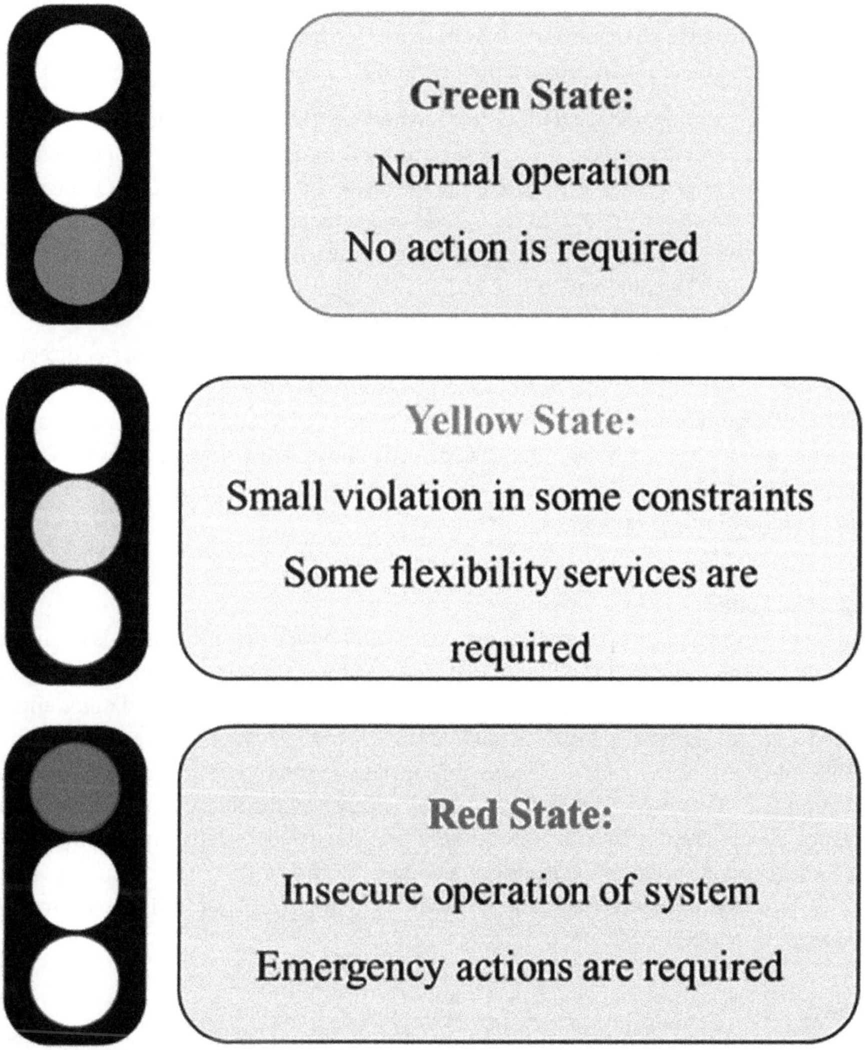

Figure 9.3 Traffic light concept.

9.3.8.1 Green State

This indicates a normal operating state in the distribution network. The green state happens if the system is stable in terms of all of its operating points. It means that the power can flow through the feeders in either direction without violating the network constraints. It is noticeable that the boundaries and constraints of the secure operation are dependent on the physical characteristics and dynamics of the network.

9.3.8.2 Yellow State

The yellow sign indicates the alert state in distribution networks. These situations happen if the voltage-related constraints are violated or when congestion occurs in the network. In these situations, the active DSO can procure its required flexibility from the local market so as to relieve the network constraints. Two management methods can be defined in these situations so that the DSO can procure the flexibility.

1. Decentralized management method: In this method, the DSO does not directly control the flexible energy resources. It forms several local markets, which are operated by an independent entity called a local market operator. In this way, the DSO submits its needed flexibility associated with each node to the local market operator. The customers also submit their available upward and downward flexibility to the local market based on their resource availability and the comfort level. Finally, the local market operator matched the bid of the DSO with the offers of the customers.
2. Centralized management method: In the centralized method, the DSO is directly in touch with the customers and flexible energy resources in the distribution network. The DSO may negotiate contracts with customers in order to directly control their resources in some time slots. In this way, the DSO directly controls the flexible resources when a yellow state happens in the system. The customers should receive appropriate monetary benefits in this situation.

9.3.8.3 Red State

This indicates the insecure operation of the system and happens if the system is in an emergency state in which the grid stability is in danger. These situations are not only about the distribution network. In other words, the total system stability may be in danger and creates a red state. In these situations, the active DSO, which is constantly in touch with the TSO, will be aware of the state and take appropriate actions to help the TSO. The emergency actions can include, for example, direct control of flexible energy resources and emergency curtailment of DGs. It is noticeable that the direct orders imposed by the TSO to the resources connected to distribution networks should be executed by the DSO. Thus, the DSO requires a decent technology facilitating the fast and reliable information exchange.

9.3.9 Adaptive Scheduling of Flexible Energy Resources

Distribution networks have a number of potent flexible energy resources that can react according to the DSO needs. For example, the charging power as well as the charging time of electric vehicles can be scheduled in such a way to control congestion and

Figure 9.4 An example for some flexible energy resources for a residential customer.

voltages in the distribution network [26]. In addition, there are other flexible appliances such as heating, ventilation, and air conditioning (HVAC) that can be controlled so as to provide DSO-level flexibility services. However, the owners' comfort level and satisfaction should be considered when scheduling these appliances. In general, the loads can be curtailed or shifted to other time slots while renewable-based resources can only be curtailed in emergency situations [27, 28]. Figure 9.4 shows some flexible energy resources regarding residential customers. It should be mentioned that the DSO cannot utilize these resources unless the customer is smart and is equipped with the decent ICT technology. Additionally, in order to highly exploit the flexibility potential of these customer-level resources, several local markets can be designed in distribution networks [29]. In local markets, customers can trade flexibility with each other and with the DSO. In this way, the customers can make benefits and minimize their costs while satisfying the flexibility needs of the DSO.

9.3.10 Increased Cooperation Between Transmission and Distribution Networks

As aforesaid, the effective operation of power systems requires appropriate cooperation between different system operators, including TSOs and DSOs. Today, there is mainly no decent ICT infrastructure for acquiring data from flexible energy resources and DGs at the customer level of power systems. Even in some cases, only TSOs can receive information on DGs to take some curtailment actions whereas DSOs do not access this information in real-time. In addition, currently, TSOs and DSOs do not exchange any information regarding the operation of their networks with each other.

To implement active management, the online information exchange between system operators is necessary for the DSO to effectively operate the distribution network with a considerable number of DGs. In this way, the available flexibility potential of distribution-network-located resources can be highly utilized in the whole system. At the TSO level, generators mainly send their schedules to the TSO in order to provide system-wide (frequency-related) services. However, regarding the DSO level, the DSO need real-time and near real-time information on DG forecast and DG schedules in order to manage the local network better. Having received the information, the DSO should send the required information with the TSO. However, the flow of information between the DSO and the TSO should be bidirectional as long as the DSO also needs to receive information on the current status of the whole system. In this regard, flexible energy resources as well as DGs connected to the distribution are able to provide system-wide ancillary services for the TSO as well as the DSO.

In addition to the decent infrastructure for exchanging information, the cooperation between DSOs and TSOs requires an appropriate coordination scheme. In this regard [30], five coordination schemes have been introduced for the TSOs and DSOs, which are described as follows:

1. Centralized market model: In this model, there exists one central market, which is operated by the TSO. Flexible resources connected to both distribution networks as well as transmission networks are able to participate in the central market. However, the feasibility of the participation of distribution-network-located resources should be analyzed by the DSO beforehand.
2. Local market model: In this model, the DSO runs some local markets in order to satisfy the flexibility needs of the DSO level. The surplus flexibility is then aggregated and offered to the central market, which is operated by the TSO.
3. Shared balancing responsibility model: In this model, the DSO is also responsible for balancing the generation and load in its region. To this end, the DSO runs local markets to provide both balancing flexibility as well as local services. The TSO also runs its central market for its own resources connected to transmission networks. In other words, the responsibilities of the system are separated.
4. Common TSO–DSO market model: The model includes some common local markets in distribution networks where both the TSO and the DSO can participate as buyers. Unlike the second model, there is no priority regarding the provision of flexibility needs. There is also a central market operated by the TSO for satisfying TSO-level flexibility needs.
5. Integrated market model: This model comprises one integrated market in which both DSO and TSO can participate as buyers. TSO-level and DSO-level resources also take part as sellers. In addition, an independent entity should play the role of a market operator.

9.4 Application of Energy Storage Systems in Active Network Management

9.4.1 Integration of Renewable Energy Resources

Power production from renewable energy resources such as wind turbines and photovoltaic (PV) systems are directly dependent on weather conditions, which mean their daily power generation has an intermittent nature. Meanwhile, the maximum generation of these resources usually takes place in off-peak times. Application of energy storage in order to store the energy surplus in off-peak moments and then discharging during peak demand times could significantly reduce the power flow, and as a result, decrease the power losses as well as a required capacity reduction for the peak demand from the sources owner's point of view. The energy storage systems (ESSs) can be considered as energy buffers that make dispatching the RES output power smoothly. Another advantage, from the market point of view, that ESS brings about to ADN, is that the stored energy in the battery at low price times (off-peak time) could be released to ADN at peak times in which the energy price is considerably high. Investigations show that the grid hosting capacity limits the high penetration amount of RESs. Hosting capacity is defined as the highest number of RESs that can be integrated into the grid while not violating power quality restrictions. Therefore, ESSs have become one of the popular pieces of equipment to increase distribution networks' hosting capacity [31].

9.4.2 Load Leveling

Distribution systems are willing to have daily load leveling as much as possible so as to reconcile their curves to the optimal and economical operating point that can cause cheaper energy as well as better quality. Therefore, the energy is stored at off-peak times and injected back to the network during peak times. Due to the changing demand based on the daily load diagram, the grid must have the proper capacity to handle the load variations constantly and to match generated and demand power, regardless of how it runs. Hence, the use of ESSs with proper management based on active network management can provide load leveling as well as stress reduction on the distribution network arising from peak loads [32, 33].

9.4.3 Peak Shaving

There are significant differences in the load profile's peaks and valleys due to demand variation. From the grid's viewpoint, it can cause several challenges about reliability and stability in grid operation. On the other hand, the industrial customers need a large amount of power during the day to run their equipment in specific short-time intervals. These customers must pay extra charges due to the peak demand they needed. In the past, distribution networks were utilizing diesel generators and gas turbines for power shortage compensation. Nowadays, optimal and effective utilization of energy storage systems have become the key solution to peak shaving. As mentioned before, the energy delivered in the peak time is more expensive. This means, from the ADN and end-user point of view, they are more interested to reduce the energy demand from the grid in such situations and use the energy stored in the ESSs if it is possible [34, 35].

There are several types of energy storage, such as compressed air, pumped hydro storage (PHS), flywheels, and battery energy storage systems (BESS). Among all these ESSs, BESS has become more popular due to its rapid response time. Figure 9.5 presents the peak shaving performed by ESSs.

Peak power, as a relative concept, is compared to a reference value that is referred to a shaved profile. This is explained as the area that exceeds the reference value. The required energy from energy storage is calculated as the maximum power that exceeds the reference value multiplied by its time duration. A principal example of the calculation method of battery energy storage required capacity is shown in Figure 9.6, given in the following equation:

$$B_{cap} = P_{bat} \times \Delta T \tag{9.2}$$

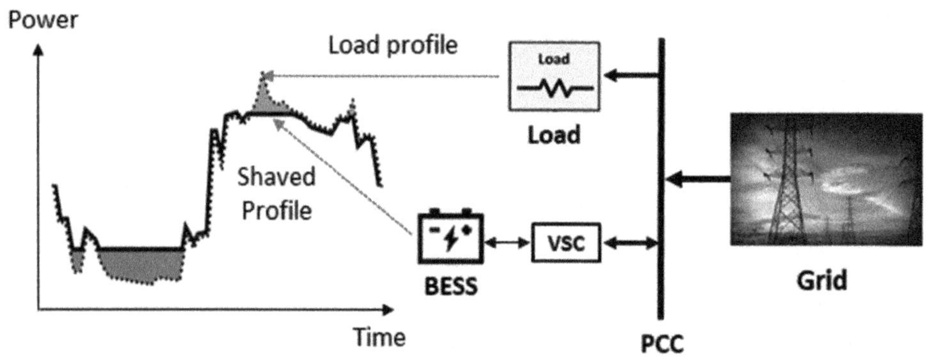

Figure 9.5 Peak shaving performed by ESSs.

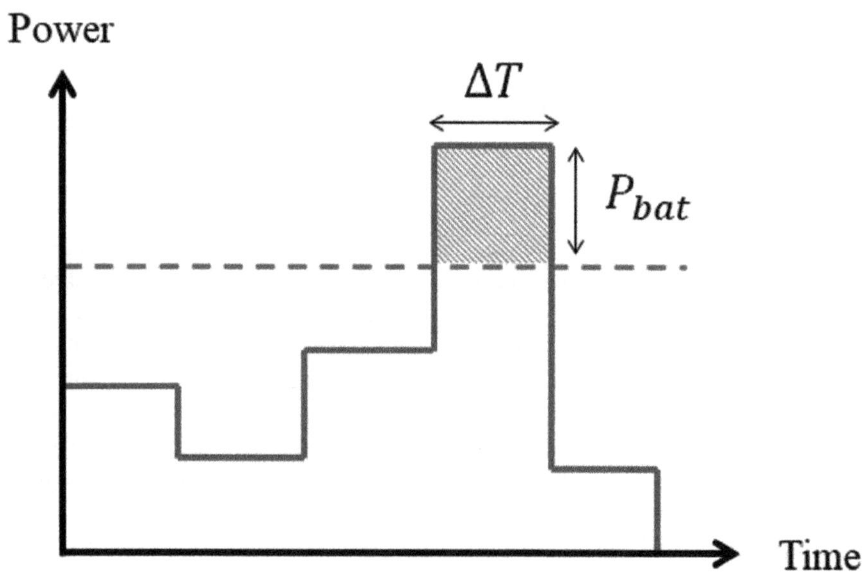

Figure 9.6 The required battery capacity calculation for peak shaving.

9.4.4 Power Quality Improvement

Integration of DGs into distribution networks has similar power quality impacts as disturbing loads. Single-phase DGs, especially in low-voltage AND, increase unbalancing issues. The other power quality issues such as harmonics might be injected into the ADN through power electronic interfaced devices that connect DGs to the networks. Due to the stochastic nature of DGs, frequency deviation and voltage fluctuation occur as a consequent. Thus, installation of RESs in the ADN with disturbing loads causes the power quality issues to become heightened.

Application of power filters such as passive and active filters and distribution static synchronous compensators have been introduced to compensate for harmonic disturbances in active distribution networks. By developing the energy storage technologies and reducing the cost of a power converter interface with this equipment, similar functionalities can be considered for ESSs. The interface inverter can be operated in both the voltage control mode (VCM) and the current control mode (CCM). For power quality issues, compensation such as harmonic, unbalance, and reactive power, the inverter works at CCM and has to inject current to the distribution network. This kind of application has attracted more attention from end-users. The output powers of ESS as well as the voltage at its terminals are controlled by regulating the output voltage of the interface inverter both in phase and amplitude in the voltage control mode. This operation mode is more typical for utility networks [20, 21].

9.4.5 Active Power Management

The combination of ESSs with wind turbines and PV systems can effectively smooth the intermittent output power of these types of renewable units in order to reduce the system-wide frequency deviation caused by high penetration of RESs into distribution networks. Several types of energy storages could be used for this purpose. With this regard, for instance, the integration of hybrid energy storages including BESS, super-capacitors, flywheels, and superconducting magnetic energy storage is believed to be beneficial. Investigations state that using BESS is a more practical and economical method because of the high cost of other types of energy storage systems and recent technology development of BESS. A simple schematic of BESS integrated into the grid is depicted in Figure 9.7.

In summary, different applications for energy storage systems are defined based on their charging/discharging rates (amount of stored energy); thus the energy storage can be divided into three different categories considering their power rate as follows:

1. ESSs that can supply the required load energy from a few minutes to several hours for energy management purposes, for example, in islanding operation conditions of MGs.
2. ESSs that have the ability to supply load energy within minutes for power shortage situations, for instance, in uninterruptible power supply (UPS) applications.
3. ESSs that can supply load energy within seconds to improve power quality or stability issues, such as harmonics unbalance, etc.

These applications of ESSs in distribution networks are summarized in Table 9.1 [36, 37]. Moreover, different types of energy storages based on their utilized technology are categorized in Figure 9.8.

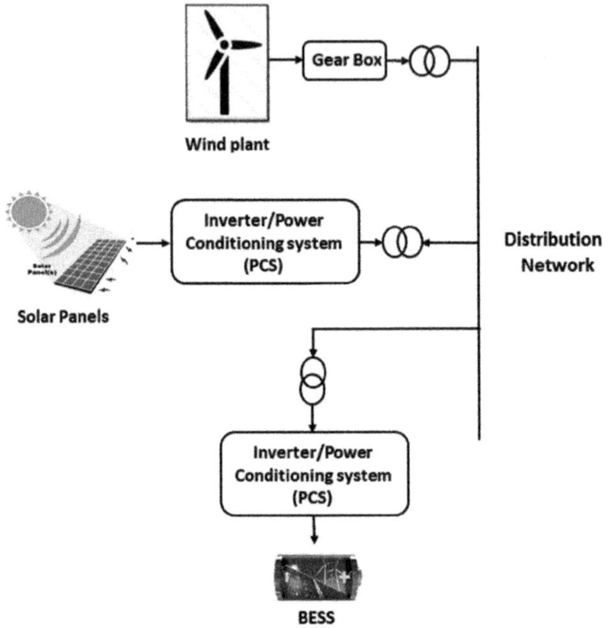

Figure 9.7 Schematic of integration of BESS to the distribution network.

Table 9.1 Various applications of ESSs in the distribution network.

Application area	Application	Goals
Grid	Load leveling	✓ Peak demand reduction ✓ Optimal utilization of transmission capacity ✓ Power losses reduction ✓ Extend the network without reinforcement
Distribution network/ customer level	Voltage regulation and disturbance compensation	✓ Power quality improvement
	Store energy in off-peak and release it in peak time	✓ Reduce energy consumption in peak time ✓ Cost reduction ✓ Voltage fluctuation compensation

9.5 Microgrids and Active Network Management

The concept active network management has been surveyed during the past decade in order to facilitate the integration and active operation of DGs into distribution networks. This concept is utilized to manage the limitation of distribution networks to integrate the large amount of DGs, instead of network reinforcement. A microgrid (MG) is a group of distribution generators, energy storages, controllable loads, etc., located in the same area,

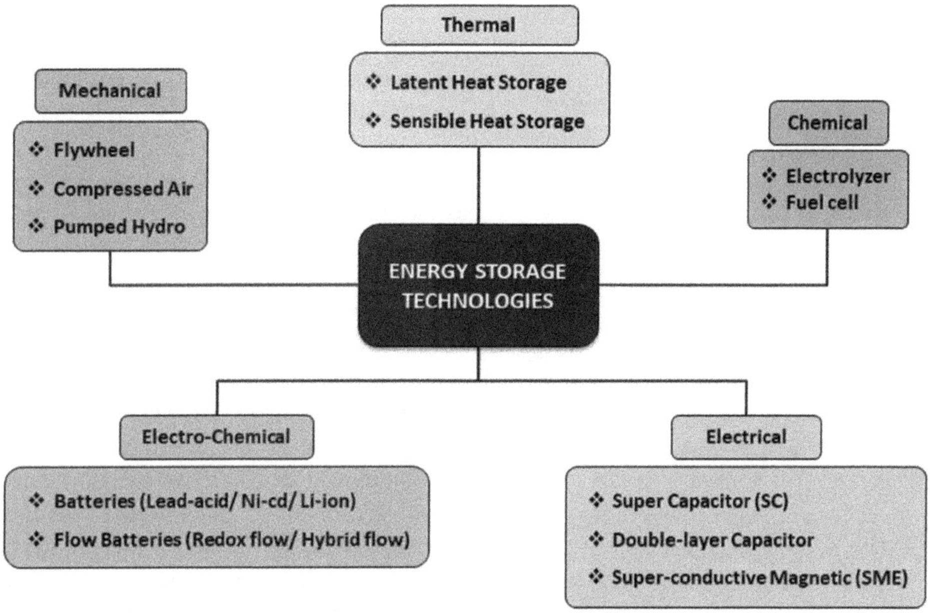

Figure 9.8 Different types of ESS technologies.

which is in turn connected to the national grid while also able to disconnect from it and operate independently. The advantages of utilization of MGs can be organized as different viewpoints. From the grid point of view, the MG is defined as a fully controllable system that can operate in both dependent (i.e. grid-connected) as well as independent (i.e. standalone) modes [38]. The integration of MGs to the distribution networks improves the monitoring and controllability of the network as well as reducing power losses stemming from transmission lines. From the distribution network point of view, improving the network quality by reducing costs and power losses during the electricity distribution can be achieved by active utilization of MGs. From an environmental point of view, improvement in sustainability by reducing carbon emissions, environmental pollution, and global warming through the utilization of low-carbon technologies could be obtained as well.

MGs can provide various technical services regarding active network management and protection of an active distribution network during both grid-connected and islanded operation modes. In accordance with this, active power control capability of DERs in ADN could effectively capture different DER uncertainty. MGs also offer excellent potential for reactive power control by proper control of DG inverters, utilization of energy storage, and FACTS devices. The control of all these devices can be performed in a centralized or decentralized manner to improve the operation of ADNs [3, 4].

9.5.1 Challenges of Microgrids

As mentioned before, one of the most important reasons to utilize the MGs was to supply the customer with high quality and without any interruption, while working in parallel with the supply grid. However, the MG should change its operation into islanded mode when a permanent fault occurs in the supply grid. In such situations, the operation of MGs in islanded mode should be based on related standards.

Islanded mode detection methods are divided into two categories, namely remote methods and local methods. Remote methods are known as centralized methods in which islanding detection is based on grid-side considerations, while local techniques are based on the local-side considerations within the MG. The local methods are divided into three types, i.e. passive, active, and hybrid techniques. In the passive method, the voltage, frequency, and current harmonic distortions are measured continually. Due to the difference between the maximum and minimum values of those measured parameters in two different modes of operation, the islanded mode method can differ from the grid-connected mode method. The active method operates with a similar principle of control mechanism. It means that this method is based on feedback techniques and analyzing the response to the intentional disturbance that injects into the grid. The hybrid method is a combination of the two mentioned methods in which the passive technique takes precedence over the active method [39].

Power interruption can occur in both operation modes of MGs. A fault in islanding operation can lead to a blackout and in this situation all DGs might be disconnected from the MG. The normalizing and restoring process of the whole system is called a black-start. Therefore, the service provider is responsible for power management, voltage controlling, and balancing, which are considered in the MG control center (MGCC). The black-start condition can be started both when the grid is accessible or is not. However, in the absence of the main grid, this process becomes more difficult because the reference voltage is not available for local DGs, and to start this process the frequency and voltage stability should be considered. This means the process requires an intelligent management, especially in the presence of critical loads in the distribution network [40].

The conventional distribution system protection is not similar to the protection of active distribution networks. In conventional networks, the fault current only supplies from the main grid in one direction, while in ADN the fault current is supplied by the nearest DGs as well. The reliability of an MG in both modes of operation depends on the applied protection system including smart measurement and protection devices, relays, etc.

In general, two types of faults may occur in the MGs. The first one is when the fault occurs in the main grid and the MG is connected to the network. In such cases, the protection relays must operate according to their protection system and isolate the MG from the faulty feeder. The MG then continues its work in islanded mode until the fault is cleared. Therefore, rapid response switching is required based on the islanding detection methods. The second mode is for proper protection of the system where the MG operates disconnected from the main grid. In this situation, the protection system should remove the smallest part of the system where the fault occurred. Hence, for the participation of MGs in active network management, the investigation of programmable protection equipment and high communication capabilities are required [4].

9.6 Summary

Current distribution networks are faced with an increasing number of DGs connected to these networks. Although renewable-based DGs offer environmental and economic benefits for energy sectors, integrating DGs into distribution networks poses serious challenges for the distribution system operators (DSOs). The active management of distribution network facilitates the cooperation between the planning and operation of the network. With the help of active management, the DSO is able to benefit from real-time flexibility, which leads to a reduction on the investments. In this way, the DSO deploys the current hosting capacity of the distribution network in a more optimal manner. In general, active network management should utilize several solutions to effectively operate the network.

In terms of planning of the network, the active DSO should manage the capacity of the network and aim to improve the utilization and investments on distributed energy resources and flexible energy resources at critical nodes. Different policies and rules should be defined so as to control the installation of DGs and flexile energy resources in order to be in line with network reinforcements. Besides, the DSO should be equipped with decent equipment, such as technological equipment, and protective and monitoring systems, in order to actively manage the network.

In terms of active operation of the network, active management of flexible energy resources is required. The DSO should be actively in touch with other system operators and trade flexibility with them while actively checking the feasibility of providing flexibility for other system operators. In addition, monitoring, simulation, and control of the distribution network should be enhanced and adaptive and configurable settings should be utilized for *QU*-, *Pf*-, and *PU*-droop, OLTC control, and P&Q-control.

The integration of renewable energy resources can be facilitated through the use of ESS. Additionally, ESS can contribute to provide some flexibility services, including peak-shaving, load leveling, and PQ improvement as well as management of the active power. MGs can also provide various technical services regarding active network management and protection of the active distribution network during both grid-connected and islanded operation modes. In accordance with this, the active power control capability of DERs in ADN could effectively capture different DER uncertainty. MGs also offer excellent potential for reactive power control by proper control of DG inverters, utilization of energy storages, and FACTS devices. However, an active DSO should be able to manage the challenges that prevailing MGs can create for the network.

References

[1] G. Pepermans, J. Driesen, D. Haeseldonckx, R. Belmans, and W. D'haeseleer, "Distributed generation: Definition, benefits and issues," *Energy Policy*, vol. 33, no. 6, pp. 787–798, 2005.

[2] A. Hirsch, Y. Parag, and J. Guerrero, "Microgrids: A review of technologies, key drivers, and outstanding issues," *Renewable and Sustainable Energy Reviews*, vol. 90, pp. 402–411, 2018.

[3] O. Palizban, K. Kauhaniemi, and J.M. Guerrero, "Microgrids in active network management – Part I: Hierarchical control, energy storage, virtual power plants, and market participation," *Renewable and Sustainable Energy Reviews*, vol. 36, pp. 428–439, 2014.

[4] O. Palizban, K. Kauhaniemi, and J.M. Guerrero, "Microgrids in active network management – Part II: System operation, power quality and protection," *Renewable and Sustainable Energy Reviews*, vol. 36, pp. 440–451, 2014.

[5] S.N. Afifi, "Impact of hybrid distributed generation allocation on short circuit currents in distribution systems," PhD Thesis, Brunel University, London, UK, 2017.

[6] T.F. Agajie, B. Khan, H.H. Alhelou, and O.P. Mahela, "Optimal expansion planning of distribution system using grid-based multi-objective harmony search algorithm," *Computers & Electrical Engineering*, vol. 87, p. 106823, 2020.

[7] O.P. Mahela, B. Khan, H.H. Alhelou, and P. Siano, "Power quality assessment and event detection in distribution network with wind energy penetration using Stockwell transform and fuzzy clustering," *IEEE Transactions on Industrial Informatics*, vol. 16, no. 11, pp. 6922–6932, November 2020.

[8] Y. Kifle, B. Khan, and P. Singh, "Assessment and enhancement of distribution system reliability by renewable energy sources and energy storage," *Journal of Green Engineering*, vol. 8, no. 3 (2), pp. 219–262, July 2018.

[9] O.P. Mahela, J. Sharma, B. Kumar, B. Khan, and H.H. Alhelou, "An algorithm for the protection of distribution feeder using Stockwell and Hilbert transforms supported features," *CSEE Journal of Power and Energy Systems*, 2020. doi: 10.17775/CSEEJPES.2020.00170.

[10] O.P. Mahela, A.G. Shaik, B. Khan, R. Mahla, and H.H. Alhelou, "Recognition of complex power quality disturbances using S-transform based ruled decision tree," *IEEE Access*, vol. 8, pp. 173530–173547, 2020, doi: 10.1109/ACCESS.2020.3025190.

[11] S. Montoya-Bueno, J.I. Muñoz-Hernández, and J. Contreras, "Uncertainty management of renewable distributed generation," *Journal of Cleaner Production*, vol. 138, pp. 103–118, 2016.

[12] N. Mahmud and A. Zahedi, "Review of control strategies for voltage regulation of the smart distribution network with high penetration of renewable distributed generation," *Renewable and Sustainable Energy Reviews*, vol. 64, pp. 582–595, 2016.

[13] R.A. Walling, R. Saint, R.C. Dugan, J. Burke, and L.A. Kojovic, "Summary of distributed resources impact on power delivery systems," *IEEE Transactions on Power Delivery*, vol. 23, no. 3, pp. 1636–1644, 2008.

[14] S.K. Ibrahim, "Distribution system optimization with integrated distributed generation," PhD Thesis, University of Kentucky, Lexington, USA, 2018.

[15] S. Ruiz-Romero, A. Colmenar-Santos, F. Mur-Pérez, and Á. López-Rey, "Integration of distributed generation in the power distribution network: The need for smart grid control systems, communication and equipment for a smart city – Use cases," *Renewable and Sustainable Energy Reviews*, vol. 38, pp. 223–234, 2014.

[16] P. Hallberg, "Active distribution system management a key tool for the smooth integration of distributed generation," *Eurelectric TF Active System Management*, vol. 2, no. 13, pp. 1–53, 2013.

[17] H. Laaksonen, C. Parthasarathy, H. Hafezi, M. Shafie-khah, and H. Khajeh, "Flexible control and management methods for future distribution networks," *CIRED Workshop*, 2020.

[18] H.K. Jacobsen and S.T. Schröder, "Curtailment of renewable generation: Economic optimality and incentives," *Energy Policy*, vol. 49, pp. 663–675, 2012.

[19] M.A.G. De Brito, L. Galotto, L.P. Sampaio, G. de A. e Melo, and C.A. Canesin, "Evaluation of the main MPPT techniques for photovoltaic applications," *IEEE Transactions on Industrial Electronics*, vol. 60, no. 3, pp. 1156–1167, 2012.

[20] S.Y. Mousazadeh Mousavi, A. Jalilian, M. Savaghebi, and J.M. Guerrero, “Power quality enhancement and power management of a multifunctional interfacing inverter for PV and battery energy storage system,” *International Transactions on Electrical Energy Systems*, vol. 28, no. 12, p. e2643, 2018.

[21] S.Y.M. Mousavi, A. Jalilian, M. Savaghebi and J.M. Guerrero, “Autonomous control of current-and voltage-controlled DG interface inverters for reactive power sharing and harmonics compensation in islanded microgrids,” *IEEE Transactions on Power Electronics*, vol. 33, no. 11, pp. 9375–9386, 2018.

[22] H. Laaksonen, C. Parthasarathy, H. Hafezi, M. Shafie-khah, H. Khajeh, and N. Hatziargyriou, “Solutions to increase PV hosting capacity and provision of services from flexible energy resources,” *Applied Science*, vol. 10, no. 15, pp. 5146, 2020.

[23] H. Laaksonen, K. Sirviö, S. Aflecht, and P. Hovila, “Multi-objective active network management scheme studied in Sundom smart grid with MV and LV network connected DER units,” *Proceedings of 25th International Conference on Electricity Distribution: CIRED 2019*, Madrid, 3–6 June 2019, 2019.

[24] H. Laaksonen, C. Parthasarathy, H. Hafezi, M. Shafie-khah and H. Khajeh, “Control and management of distribution networks with flexible energy resources,” *International Review of Electrical Engineering*, vol. 15, no. 3, p. 213, June 2020.

[25] H. Zoeller, M. Reischboeck and S. Henselmeyer, "Managing volatility in distribution networks with active network management," *CIRED Workshop 2016*, Helsinki, 2016, pp. 1–4.

[26] H. Khajeh, H. Laaksonen, A.S. Gazafroud, and M. Shafie-Khah, “Towards flexibility trading at TSO-DSO-customer levels: A review,” *Energies*, vol. 13, p. 165, 2020.

[27] H. Khajeh, A.A. Foroud, and H. Firoozi, “Robust bidding strategies and scheduling of a price-maker microgrid aggregator participating in a pool-based electricity market,” *IET Generation, Transmission & Distribution*, vol. 13, no. 4, pp. 468–477, 2018.

[28] H. Khajeh, M. Shafie-khah, and H. Laaksonen, “Blockchain-based demand response using prosumer scheduling.” In: M. Shafie-khah (Ed.), *Blockchain-Based Smart Grids* (pp. 131–144). Amsterdam, Netherlands: Elsevier, April 2020.

[29] H. Khajeh, H. Firoozi, H. Laaksonen, and M. Shafie-khah, “A new local market structure for meeting customer-level flexibility needs,” *2020 International Conference on Smart Energy Systems and Technologies (SEST)*, pp. 1–6, 2020.

[30] G. Migliavacca, M. Rossi, H. Gerard, M. Džamarija, S. Horsmanheimo, C. Madina, I. Kockar, G. Leclerq, M. Marroquin, and H. Svendsen, “TSO-DSO coordination and market architectures for an integrated ancillary services acquisition: The view of the SmartNet project,” *CIGRE*, Paris Session, 2018.

[31] H. Saboori, R. Hemmati, S.M.S. Ghiasi, and S. Dehghan, “Energy storage planning in electric power distribution networks – A state-of-the-art review,” *Renewable and Sustainable Energy Reviews*, vol. 79, pp. 1108–1121, 2017.

[32] E. Reihani, S. Sepasi, L.R. Roose, and M. Matsuura, “Energy management at the distribution grid using a battery energy storage system (BESS),” *International Journal of Electrical Power & Energy Systems*, vol. 77, pp. 337–344, 2016.

[33] I. Papic, “Simulation model for discharging a lead-acid battery energy storage system for load leveling,” *IEEE Transactions on Energy Conversion*, vol. 21, no. 2, pp. 608–615, 2006.

[34] I. Wasiak, R. Pawelek, and R. Mienski, "Energy storage application in low-voltage microgrids for energy management and power quality improvement," *IET Generation, Transmission & Distribution*, vol. 8, no. 3, pp. 463–472, 2013.

[35] A. Oudalov, R. Cherkaoui, and A. Beguin, "Sizing and optimal operation of battery energy storage system for peak shaving application," *2007 IEEE Lausanne Power Tech*, pp. 621–625, 2007.

[36] A. Ehsan and Q. Yang, "State-of-the-art techniques for modelling of uncertainties in active distribution network planning: a review," *Applied Energy*, vol. 239, pp. 1509–1523, 2019.

[37] S. Shivashankar, S. Mekhilef, H. Mokhlis, and M. Karimi, "Mitigating methods of power fluctuation of photovoltaic (PV) sources – A review," *Renewable and Sustainable Energy Reviews*, vol. 59, pp. 1170–1184, 2016.

[38] H. Firoozi and H. Khajeh, "Optimal day-ahead scheduling of distributed generations and controllable appliances in microgrid," *2016 Smart Grids Conference (SGC)*, Kerman, California, USA, 2016, pp. 1–6.

[39] M.L. Doumbia, M. Robitaille, K. Agbossou, and R. Simard, "Islanding detection method for a hybrid renewable energy system," *International Journal of Renewable Energy Research*, vol. 1, no. 1, pp. 41–53, 2011.

[40] N. Pogaku, M. Prodanovic and T.C. Green, "Modeling, analysis and testing of autonomous operation of an inverter-based microgrid," *IEEE Transactions on Power Electronics*, vol. 22, no. 2, pp. 613–625, 2007.

Part VI

Phasor Measurement Unit Placement

10

Enhancing the Performance of the State Estimation Algorithm Through Optimally Placed Phasor Measurement Units

Kinfe Negash[1], Baseem Khan[1], P. Sanjeevikumar[2], and Esayas Gidey[1]

[1] *Department of Electrical and Computer Engineering, Hawassa University, Hawassa, 05, Ethiopia*
[2] *Department of Energy Technology, Aalborg University, Esbjerg, 6700, Denmark*

10.1 Introduction

In the power system, phase angles of bus voltages are significant as the flow of real power in the transmission network is almost proportional to the sine of the phase differences of sending and receiving end bus voltages. In the power system, various planning and operational decisions are directly related to the active power flow; therefore, computing the phase differences is of great importance. Initial modern techniques measured the phase difference in a direct manner [1]. Subsequently, for estimating the local phase angle relating to the time reference, positive-going zero-crossing of a phase voltage is utilized. The main disadvantage of these techniques is that these methods are not suitable for the wide-area phasor measurement system's generalization, which is the main feature of the smart grid system. The present phasor measurement unit (PMU) technology has developed by the advanced research performed on the computer relayed transmission network. Continuously varying load in the system frequently changes the topology of the present dynamic power system network. With an increase in the load demand, the power network is loaded to a maximum extent, and is therefore vulnerable to partial or total blackout due to small or foremost outages [2].

Different techniques have been utilized by various researches to perform the optimal placement problem of the PMU. The integer linear programming (ILP) technique is utilized for optimally placing and finding the placement position of PMUs. Therefore, the multistage measurement model of the system is observable and utilized for state estimation [3–5]. Chakerabarti and Kyriakides [6] utilized an exhaustive binary search technique to determined the least number of required PMUs to make the system observable. In a situation of having greater than one placement option with a similar least quantity of PMUs, this technique picks the one that has the most desired measurement redundancy pattern. In another study [7], an integer quadratic programming technique was utilized for computing the least quantity and the optimal positions of the PMUs in order to secure absolute observability. Heuristic and meta-heuristic methods are also utilized by various researchers to find the optimal placement of PMUs [8–10]. Negash et al. [11]

Active Electrical Distribution Network: A Smart Approach, First Edition. Edited by Baseem Khan, Josep M. Guerrero, Sanjeevikumar Padmanaban, Hassan Haes Alhelou, Om Prakash Mahela, and Sudeep Tanwar.

utilized a power system analysis toolbox (PSAT) for a comparative study of an optimal placement of PMUs using various techniques like depth-first, graph theory, and simulated annealing. Sun et al. [12] optimally placed the phasor measurement unit by utilizing the sum of the variance of the robust estimators. Arpanahi et al. [13] developed the new multiobjective optimal PMU placement technique by considering various objectives such as small-signal stability, system observability, and PMU cost. Moreover, uncertainties related to the wide-area monitoring system and variations in network topology are incorporated. Wang et al. [14] developed a risk-constrained PMU placement approach by considering probability and the consequent severity of observability loss. Shalini et al. [15] developed a protection scheme with a full area back-up by utilizing a cubature Kalman filtering-based dynamic state estimation together with minimum PMU measurements to get a full system observability. A novel factor based on voltage has also been developed that utilized the dynamic state estimation data for detecting the outage line. Zhang et al. [16] presented the new faulty line detection technique on the power network, which has only sparse synchronized measurements. Guo et al. [17] developed the concept of observability propagation depth (OPD) for enhancing the optimal PMU placement (OPP) formulation with bounded observability propagation constraints. Almunif and Fan [18] utilized two mathematical approaches, namely mixed integer liner programming and nonlinear programming, to solve the optimal PMU placement problem for modeling grid observability. Manoharan et al. [19] proposed a multiobjective PMU placement problem for improving the observability and voltage stability of the system. This problem was solved by utilizing a binary coded ant lion optimization algorithm. A hybrid state estimator was proposed [20] that depends on the measured data provided by synchronized and unsynchronized sensors. For this purpose, various refresh rates of the sensors were considered. Goleijani et al. [21] proposed the artificial neural network technique for the dynamic state estimation by utilizing hybrid data obtained from supervisory control and data processing systems (SCADA) and PMUs. Tarif et al. [22] proposed a multiobjective PMU placement problem for a small signal stability assessment of the power system. For that purpose a multisignal Prony analysis of frequency signals with the magnitude and angle of the voltage signals is recorded by the PMUs. Rather et al. [23] proposed the pragmatic technique for placing the PMUs in a cost effective manner by considering realistic constraints of the Danish power system.

This work proposes a technique to place a PMU at the optimal location for the performance improvement of the state estimation algorithm to enhance the stability of the power system. The placement of the PMU is assessed on the IEEE 14-bus system and on the practical 21-bus Southern region power system of Ethiopia. The placement of the PMU is assessed at the substation level, which is connected to this grid.

This paper utilizes the integer linear programming technique for optimally placing the PMUs to enhance the conventional state estimation algorithm. The significant contributions of this chapter are as follows:

1. The integer linear programming problem is proposed for the optimal placement of the PMUs in the transmission network.
2. Information obtained from optimally placed PMUs is used to enhance the performance of the conventional state estimation algorithm.
3. A practical 21-bus Southern region power network of Ethiopia is utilized for showing the effectiveness of the developed technique.
4. A comparative analysis of conventional and improved state estimation techniques is performed in terms of a mean square error reduction.

10.2 Methodology

In this work, the optimal location of PMU is performed for the improvement of a state estimation (SE) algorithm.

10.2.1 SE with Conventional Measurements

The traditional state estimation (SE) represents a procedure to obtain all the voltage phasors of a given power network. The system is modeled on the bus-branch model, where buses represent the substations and π models indicate the transmission lines and transformers.

The classical and established SE method is the weighted least squares (WLS) model, which relies on the following measurement model [24]:

$$z = h(x) + e \tag{10.1}$$

The state vector x has a dimension of $2n - 1$, where n is the number of buses, given by

$$x = \left[\theta_2 \quad \theta_3 \cdots \theta_n \quad V_1 \quad V_2 \quad V_3 \cdots V_n\right]^{\mathrm{T}} \tag{10.2}$$

In Eq. (10.2), bus 1 is selected as the slack bus, with the phase angle set to zero, that provides a system reference. The following equation represents the measurement function $h(x)$ vector, which relates each measurement to the state variables:

$$h(x) = \begin{bmatrix} P_{km} \\ P_k \\ Q_{km} \\ Q_k \\ V_k \end{bmatrix} \tag{10.3}$$

The purpose of the WLS SE is to obtain the state variables, which minimize the following objective function:

$$J(x) = \sum_{i=1}^{m} \frac{\left(Z_i - h_i(x)\right)^2}{R_{ii}} = \left[Z - h(x)\right]^T R^{-1}\left[Z - h(x)\right] \tag{10.4}$$

A minimum is found once the first-order optimality conditions are satisfied and is achieved by a first-order approximation:

$$g(x) = \frac{\partial J}{\partial x} = -H^T(x) R^{-1}\left[Z_i - h_i(x)\right] = 0 \tag{10.5}$$

$$H(x)=\frac{\partial h(x)}{\partial x}=\begin{bmatrix} \frac{\partial P_{km}}{\partial \theta} & \frac{\partial P_{km}}{\partial V} \\ \frac{\partial P_{km}}{\partial \theta} & \frac{\partial P_{km}}{\partial V} \\ \frac{\partial Q_{km}}{\partial \theta} & \frac{\partial Q_{km}}{\partial V} \\ \frac{\partial Q_{km}}{\partial \theta} & \frac{\partial Q_{km}}{\partial V} \\ \frac{\partial V_{km}}{\partial \theta} & \frac{\partial V_{km}}{\partial V} \end{bmatrix}$$

By expanding the nonlinear function $g(x)$ by a Taylor series in the order of the state vector x^k, and by ignoring the higher-order terms, an iterative solution scheme can be obtained, given by

$$G(x^k)\Delta X^{K+1} = H^T(x^k)W[z-h(x^k)] \tag{10.6}$$

where $G\left(x^k\right)=H^T\left(x^k\right)WH\left(x^k\right)$ is the gain matrix, $W=R^{-1}$ is the weighting matrix, and $\Delta x^k = x^{k+1}-x^k$, where k is the iteration index.

10.2.2 SE with Synchronized Phasor Measurements

The PMU advent recalls as a distance relay with symmetrical components, developed in the 1970s at the Virginia Tech Laboratory. Hence, the state vector must include all the voltage angles and magnitudes, with a dimension $2n$, as presented below:

$$X=\left[\theta_1 \ \ \theta_2 \ \ \theta_3 \ \ \theta_4 \cdots \theta_n \ \ \ V_1 \ \ V_2 \ \ V_3 \ \ V_4 \ \cdots \ V_n\right]^{\mathrm{T}} \tag{10.7}$$

The rectangular coordinates present a disadvantage as it will amplify errors of PMU measurements. Such a conversion is simple to apply, as can be seen by the following set of equations:

$$I_{km}^{Re} = I_{km}\cos\left(\delta_{km}\right) \tag{10.8}$$

$$I_{km}^{Im} = I_{km}\sin\left(\delta_{km}\right) \tag{10.9}$$

where I_{km}^{Re} and I_{km}^{Im} refer to the real and imaginary parts of the phasor current measurement, I_{km} refers to the phasor current magnitude, measured by PMU, and δ_{km} refers to the phasor current angle, measured by the PMU. Thus, the measurement function equations and Jacobean matrix elements must be properly adapted to accommodate the new measurements, as follows:

$$h(x)=\begin{bmatrix} P_{km} \\ P_k \\ Q_{km} \\ Q_k \\ V_k \\ \theta_k \\ I_{km}^{Re} \\ I_{km}^{Im} \\ \end{bmatrix} \tag{10.10}$$

$$H(x)=H(x)=\frac{\partial h(x)}{\partial x}=\begin{bmatrix} \frac{\partial P_{km}}{\partial \theta} & \frac{\partial P_{km}}{\partial V} \\ \frac{\partial P_k}{\partial \theta} & \frac{\partial P_k}{\partial V} \\ \frac{\partial Q_{km}}{\partial \theta} & \frac{\partial Q_{km}}{\partial V} \\ \frac{\partial Q_k}{\partial \theta} & \frac{\partial Q_k}{\partial V} \\ \frac{\partial V_k}{\partial \theta} & \frac{\partial V_k}{\partial V} \\ \frac{\partial I_{km}^{Re}}{\partial \theta} & \frac{\partial I_{kmk}^{Re}}{\partial V} \\ \frac{\partial I_{km}^{Im}}{\partial \theta} & \frac{\partial I_{km}^{Im}}{\partial V} \end{bmatrix} \tag{10.11}$$

where $H(x)$ represents the sensivity matrix. By deriving the current phasor flow equations as a function of the power flows, it is possible to determine the partial derivatives of the Jacobean matrix:

$$I_{km}^{Re}=\frac{P_k\cos(\theta_k)+Q_{km}\sin(\theta_k)}{V_k} \tag{10.12}$$

$$I_{km}^{Re}=\frac{P_k\cos(\theta_k)-Q_{km}\sin(\theta_k)}{V_k} \tag{10.13}$$

where V_k refers to the voltage magnitude at the sending bus k, measured by the PMU, and θ_k refers to the voltage angle at the sending bus k, measured by the PMU. Figure 10.1 presents the flow chart of the utilized WLS technique [24].

10.2.3 Optimal Placement of PMU for the Performance Enhancement of SE

For secure power system operation, SE provides critical information. Security assessment of a power system is supported by the State and estimated by SE. Adequate system

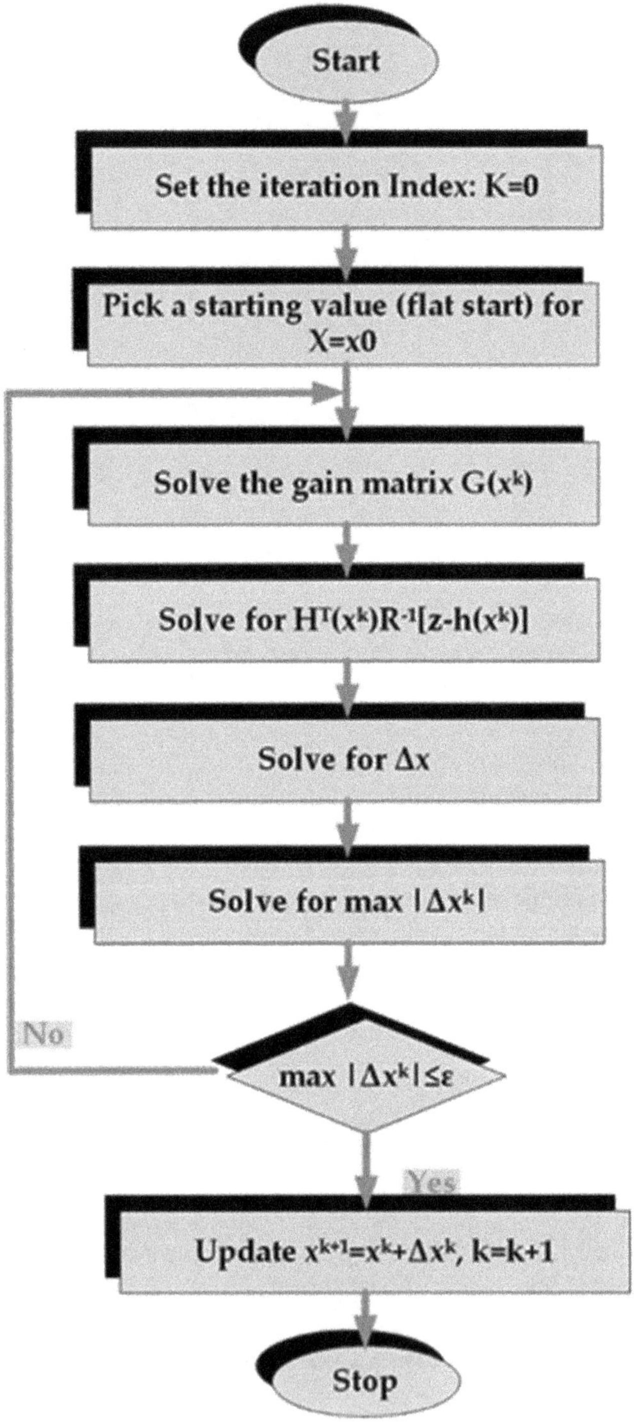

Figure 10.1 WLS algorithm.

information initiated adequate control actions for satisfactorily managing the power system operations. Since SE is highly influenced by the measurements, these should be placed at a certain level so that the detection of bad data can improve and unobservability because of failure of measurement can be minimized. There are two ways that can be used to find optimal placement for PMU for the improvement SE algorithm: optimality by observability analysis type one and observability in state estimation type two. Optimality by observability analysis type one discussed the numerical and topological approaches to determine the system observability along with further numerical methods to find the unobservable branches and observable islands and place PMU optimally to make the system observable. In this work, observability in state estimation type 2 is utilized.

10.2.3.1 System Observability in State Estimation Type 2

Consider the addition of a new measurement to the existing set of measurements. It will modify the existing gain matrix G, as follows:

$$G' = G + h_k^T h_k \tag{10.14}$$

$$L^{-1} G' L^{-T} = D + L^{-1} h_k^T h_k L^{-T} \tag{10.15}$$

Denoting $M = L^{-1} h_k$ yields the results presented in [25].

10.2.4 Problem Formulation

The angle measurements from a PMU bus and the angles at the buses adjacent to a PMU bus utilizing the subsequent current phasor measurements are utilized for calculation purposes. Therefore, the measurement vector of a PMU placed at bus i is presented by

$$T_i = \begin{bmatrix} 0 & \dots & 1 & \dots & 0 & \dots & 1 & \dots \end{bmatrix}$$

In the measurement vector T_i, elements 1 are related to the buses with angle measurements, while elements 0 are related to the buses not adjacent to bus i.

The problem formulation of optimal placement of PMU by ILP is then given by

$$\min C^T X_m \tag{10.16}$$

Subjected to $TX_m \geq b_m$

where $C = \begin{bmatrix} 1 \ 1 \ldots\ldots 1 \ 1 \end{bmatrix}^T$ represents the connectivity matrix and $X_m = \begin{bmatrix} x_1 x_2 \ \ldots\ldots \ x_m \end{bmatrix}^T$ represents an integer variable vector related to the candidate buses for PMU placement. A unitary column vector with a length of m is represented by b_m.

10.2.4.1 Solution Algorithm

Step 1: Compute the D matrix from the traditional measurements [24]. Form vector b_m in relation to D.

Step 2: Utilizing vector b_m, get all probable buses that enhance the rank of D when PMU is located at the bus.

Step 3: Utilize the developed ILP formulation for solving X, where the state estimation results will be improved.

10.2.4.2 Least-Squares Equations Formulation

The z_i measurement is a linear function of the state x_i; that is,

$$z = h(x) = h_{i1}x_1 + h_{i2}x_2 + \cdots + h_{iN_s}x_{N_s} \tag{10.17}$$

Vector h with h_{ij} coefficients is represented by

$$h = \begin{bmatrix} h_{i1} \\ h_{i2} \\ \vdots \\ h_{iN_s} \end{bmatrix} \tag{10.18}$$

Hence, z_i becomes

$$z_i = h_i^{\mathrm{T}} x \tag{10.19}$$

where

$$x = \begin{bmatrix} x_1 \\ x_2 \\ \vdots \\ x_{N_s} \end{bmatrix} \tag{10.20}$$

Finally,

$$z_i = H[X] \tag{10.21}$$

where

$$Z = \begin{bmatrix} \mathrm{Z}_1 \\ \mathrm{Z}_2 \\ \vdots \\ \mathrm{Z}_{\mathrm{N_m}} \end{bmatrix} \tag{10.22}$$

$$[H] = \begin{bmatrix} h_{11} & h_{12}\cdots & h_{1Ns} \\ h_{21} & h_{22}\cdots & \\ \vdots & \vdots & \vdots \\ h_{N_{m1}} & \cdots & h_{N_m N_s} \end{bmatrix} \tag{10.23}$$

where row i of $[H]$ is equal to the vector h_i^{T}, with N_m measurements greater than the number of states N_s. A function of the estimates of x_i is given as

$$J(X)=\sum_{i=1}^{N_m}[Z_i-h_i(x_1,x_2,\ldots,x_{N_s})]^2 \tag{10.24}$$

which implies

$$J(X)=\sum_{i=1}^{N_M}(Z_i-h_i^{\mathrm{T}}X)^2 \tag{10.25}$$

Therefore,

$$J(X)=(Z-[H]X)^{\mathrm{T}}(Z-[H]x) \tag{10.26}$$

Now, the value of x minimizes $J(x)$,

$$\frac{\partial J(X)}{\partial X_l}=0\ \text{for}\ j=1\ \cdots\ N_S \tag{10.27}$$

The gradient of $J(x)$ is given by

$$\nabla_x J(X)=\begin{bmatrix}\dfrac{\partial J(X)}{\partial X_1}\\ \dfrac{\partial J(X)}{\partial X_2}\\ \vdots\\ \\ \end{bmatrix} \tag{10.28}$$

Therefore,

$$\nabla_X\ J(x)=0 \tag{10.29}$$

where 0 is a vector of Ns, with all elements zero.
For solving, the expansion of $J(x)$ is given by

$$J(X)=(Z-[H]X)^T(Z-[H]X) \tag{10.30}$$

$$J(X)=Z^T Z-X^T[H]^T+Z^T[H]X+X^T[H]^T[H] \tag{10.31}$$

Therefore,

$$J(X)=Z^T Z-2Z^T[H]X+X^T[H]^T[H]X \tag{10.32}$$

For solving the above equation, some general relationships are required, which are as follows. The gradient is always a vector of first derivatives of a scalar function that is itself a function of a vector. Thus, by defining $F(y)$ to be a scalar function, then its gradient $\nabla_y F$ is

$$\nabla_Y F = \begin{bmatrix} \frac{\partial F}{\partial y_1} \\ \frac{\partial F}{\partial y_2} \\ \vdots \end{bmatrix} \tag{10.33}$$

Then F can be defined as

$$F = y^T b = \begin{bmatrix} y_1 & y_2 & \cdots \end{bmatrix} \begin{bmatrix} b_1 \\ b_2 \\ \vdots \end{bmatrix} \tag{10.34}$$

where b is a vector of constants bi, $i = 1,\ldots, n$. Then F can be expanded as

$$F = y_1 b_1 + y_2 b_2 + y_3 b_3 + \cdots \tag{10.35}$$

The gradient of F is given by

$$\nabla_y F = \begin{bmatrix} \frac{\partial F}{\partial y_1} \\ \frac{\partial F}{\partial y_2} \\ \vdots \end{bmatrix} = \begin{bmatrix} b_1 \\ b_2 \\ \vdots \\ b_n \end{bmatrix} = b \tag{10.36}$$

which implies that

$$F = b^T y = y^T b \tag{10.37}$$

Therefore,

$$\nabla_y \left(b^T y \right) = b \tag{10.38}$$

Now, by writing the vector b as the product of a matrix $[A]$ and a vector u,

$$b = [A] u \tag{10.39}$$

Then

$$F = y^T b = y^T [A] u \tag{10.40}$$

Therefore,

$$\nabla_y F = [A] u \tag{10.41}$$

Similarly,

$$b^T = u^T [A] \tag{10.42}$$

If F can be expressed as

$$F = b^T y = u^T [A] y \tag{10.43}$$

then

$$\nabla_y F = [A]^T u \tag{10.44}$$

Finally, if a scalar function F is quadratic, then

$$F = y^T [A] y = [y_1 \; y_2 \cdots y_n] \begin{bmatrix} a_{11} & a_{12} \cdots & \\ a_{21} & a_{22} \cdots & \\ \vdots & \vdots & \end{bmatrix} \begin{bmatrix} y_1 \\ y_2 \\ \vdots \\ y_n \end{bmatrix} = \sum_{i=1}^{n} \sum_{j=1}^{n} y_i a_{ij} y_j \tag{10.45}$$

and

$$\nabla_y F = \begin{bmatrix} \dfrac{\partial F}{\partial y_1} \\ \dfrac{\partial F}{\partial y_2} \\ \vdots \end{bmatrix} = \begin{bmatrix} 2a_{11}y_1 + & 2a_{12}y_2 + & \cdots \\ 2a_{21}y_1 + & 2a_{22}y_2 + & \cdots \\ & \vdots & \end{bmatrix} = 2[A] y \tag{10.46}$$

By utilizing the above expressions to develop the gradient of $J(x)$, i.e. $\nabla_x J$, the first term in Eq. (10.32), $Z^T Z$, is not a function of x, so ignore it. Therefore,

$$\nabla_x \left(-2Z^T [H] X\right) = -2[H]^T Z \tag{10.47}$$

Again,

$$\nabla_X \left(X^T [H]^T [H] X \right) = 2[H]^T [H] X \tag{10.48}$$

Then combining Eqs (10.47) and (10.48),

$$\nabla_X J = -2[H]^T Z + 2[H]^T [H] X \tag{10.49}$$

and

$$-2[H]^T Z + 2[H]^T [H] X = 0 \tag{10.50}$$

or

$$X = \left[[H]^T [H] \right]^{-1} [H]^T Z \tag{10.51}$$

For different weights of each measurement, $J(x)$ is given by

$$J(X) = \sum_{i=1}^{n_m} w_i \left(z_i - h_i^T X \right)^2 \tag{10.52}$$

This can be written as

$$J(X) = \left(Z - [H] X \right)^T [W] \left(Z - [H] X \right) \tag{10.53}$$

where $[W]$ is a diagonal matrix. Then

$$J(X) = Z^T [W] Z - X^T [H]^T [W] Z - Z^T [W][H] X + X^T [H]^T [W][H] X \tag{10.54}$$

Further,

$$\nabla_X J = -2[H]^T [W] Z + 2[H]^T [W][H] X \tag{10.55}$$

Which gives the weighted least squares algorithm formulation utilized in this work:

$$X = \left\{ \left\{ [H]^T [W][H] \right\}^{-1} [H]^T [W] Z \right\} \tag{10.56}$$

10.3 Results Analysis and Discussion

The IEEE 14-bus benchmark system is used as a base case and the 21-bus Southern region Ethiopian power system network is modeled in detail.

10.3.1 IEEE 14-Bus Power System

Figure 10.2 shows an IEEE 14-bus power system observability analysis. The red line in the figure indicates that the system is not fully observable with the given conventional measurement. Initially, injection measurements are included at bus 14 and a flow measurement from bus 6 to 13 (based on the IEEE 14-bus standard) is performed by using state estimation in the MATLAB environment.

Table 10.1 presents the MATLAB coded state estimation output and its voltage profile of a 14-bus system without a PMU installation. By applying the developed algorithm, the ILP model results in the following equations:

$$\min f = x_6 + x_9 + x_{12} + x_{13} + x_{14}$$

subject to

$$TX = \begin{vmatrix} 1 & 0 & 1 & 1 & 1 \\ 0 & 1 & 0 & 1 & 1 \end{vmatrix} \left[x_6 + x_9 + x_{12} + x_{13} + x_{14}\right]^T 3 \left[1 \quad 1\right]^T$$

$$X = \left[x_6 + x_9 + x_{12} + x_{13} + x_{14}\right]$$

$$X_i \in \{0, 1\}$$

The solution is

$$X = \left[0 \quad 0 \quad 0 \quad 0 \quad 1 \quad 0\right]^T$$

Now, by converting the current measurement obtained from the PMU placement into the power flows provided the results presented in Table 10.2, obtained by installing a single

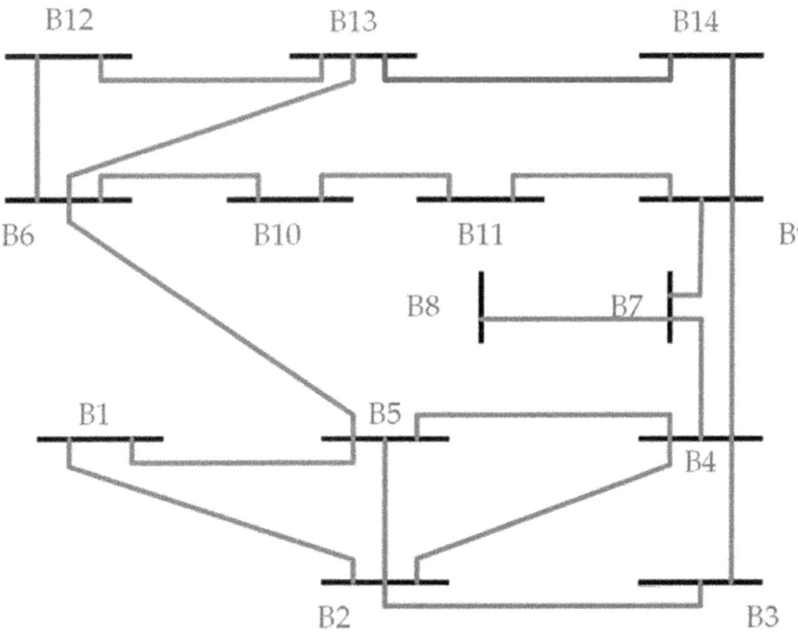

Figure 10.2 IEEE 14-bus observability analysis.

Table 10.1 Conventional state estimation results of voltage without PMU placement.

Bus	Actual value of voltage in p.u.	State estimation results of the voltage without PMU placement in p.u.
Bus 01	1.06	1.0068
Bus 02	1.045	0.9899
Bus 03	1.01	0.9518
Bus 04	1	0.9579
Bus 05	1	0.9615
Bus 06	1.07	1.0185
Bus 07	1	0.9919
Bus 08	1.09	1.0287
Bus 09	1	0.9763
Bus 10	1	0.9758
Bus 11	1	0.9932
Bus 12	1	1.0009
Bus 13	1	0.9940
Bus 14	1	–

Table 10.2 Conventional state estimation results of voltage with PMU placement.

Bus	Actual value of voltage in p.u.	State estimation results of the voltage with PMU placement in p.u.
Bus 01	1.06	1.044
Bus 02	1.045	1.028
Bus 03	1.01	0.991
Bus 04	1	0.996
Bus 05	1	1.000
Bus 06	1.07	1.054
Bus 07	1	1.029
Bus 08	1.09	1.064
Bus 09	1	1.014
Bus 10	1	1.013
Bus 11	1	1.030
Bus 12	1	1.031
Bus 13	1	1.021
Bus 14	1	–

PMU at the optimal position that improved the state estimation performance. It enhances the state estimation mean square error from 0.00144 to 0.00038, which shows around 73% improvement. Table 10.3 shows the overall performance comparison.

10.3.2 Southern Region 21-bus Network

Figure 10.3 presents the Southern region 21-bus system observability analysis. The conventional measurements are presented in Table 10.4. The red line in the figure indicates that the system is not fully observable with the given conventional measurement, which is indicated by three injection and twenty flow measurements.

Table 10.3 Mean square error comparison for state estimation performance for IEEE 14-bus system.

Bus	Actual value of voltage in p.u.	Estimated value before PMU placement		Estimated value after PMU placement	
		Voltage in p.u.	Error square	Voltage in p.u.	Error square
Bus 01	1.06	1.0068	0.00283	1.044	0.00024
Bus 02	1.045	0.9899	0.00303	1.028	0.00029
Bus 03	1.01	0.9518	0.00338	0.991	0.00036
Bus 04	1	0.9579	0.00177	0.996	1.4E-05
Bus 05	1	0.9615	0.00148	1.000	4.7E-08
Bus 06	1.07	1.0185	0.00265	1.054	0.00025
Bus 07	1	0.9919	6.5E-05	1.029	0.00086
Bus 08	1.09	1.0287	0.00376	1.064	0.00066
Bus 09	1	0.9763	0.00056	1.014	0.00019
Bus 10	1	0.9758	0.00058	1.013	0.00018
Bus 11	1	0.9932	4.6E-05	1.030	0.0009
Bus 12	1	1.0009	8.3E-07	1.031	0.00097
Bus 13	1	0.9940	3.6E-05	1.021	0.00042
Bus 14	1	–	–	–	–
Mean square error		0.00144		0.00038	

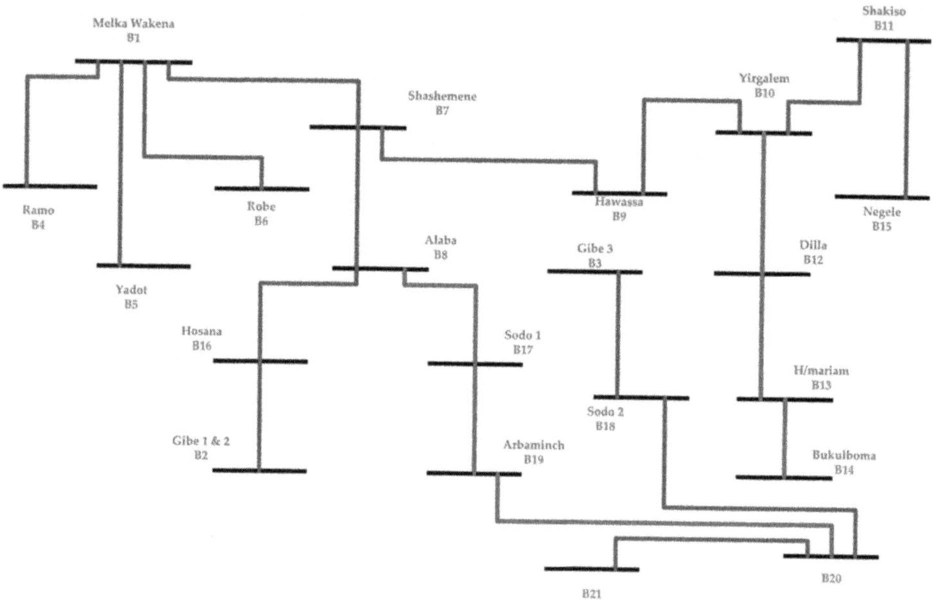

Figure 10.3 21-bus Southern region system observability analysis.

Table 10.4 21-bus Southern region conventional state estimation results without PMU placement.

Bus	Actual value of voltage in p.u.	State estimation results of the voltage without PMU placement in p.u.
Bus 01	0.9591	1
Bus 02	0.9788	1.019047
Bus 03	0.971	1.018189
Bus 04	0.9591	0.990085
Bus 05	0.9661	0.999202
Bus 06	0.9334	0.99918
Bus 07	0.9062	0.991997
Bus 08	0.9148	0.9989
Bus 09	0.8951	0.9892
Bus 10	0.8844	0.9858
Bus 11	0.8637	0.9845
Bus 12	0.887	0.9861
Bus 13	0.887	0.9867
Bus 14	0.9145	0.9873
Bus 15	0.7861	0.9845
Bus 16	0.9287	1.0087
Bus 17	0.971	1.0002
Bus 18	0.9029	1.0052
Bus 19	0.8902	1.0022
Bus 20	0.8902	1.0028
Bus 21	0.8902	–

A zero injection measurement at the bus numbers 12, 13, 14, 17, 19, and 20 is assumed to make the system state estimation solvable. The state estimation program in a MATLAB environment is running without the placement of any PMU. Table 10.4 presented the state estimation output of the Southern region of the Ethiopia 21-bus system using conventional measurements. Now by assuming all the resistance and reactance values to be 1 in the non-observable system, the Jacobean matrix is formed as

$$
H =
\begin{matrix}
4 & 0 & 0 & -1 & -1 & -1 & -1 & 0 & 0 & 0 & 0 & 0 & 0 & 0 & 0 & 0 & 0 & 0 & 0 & 0 & 0 \\
0 & 1 & 0 & 0 & 0 & 0 & 0 & 0 & 0 & 0 & 0 & 0 & 0 & 0 & 0 & -1 & 0 & 0 & 0 & 0 & 0 \\
0 & 0 & 1 & 0 & 0 & 0 & 0 & 0 & 0 & 0 & 0 & 0 & 0 & 0 & 0 & 0 & 0 & -1 & 0 & 0 & 0 \\
1 & 0 & 0 & -1 & 0 & 0 & 0 & 0 & 0 & 0 & 0 & 0 & 0 & 0 & 0 & 0 & 0 & 0 & 0 & 0 & 0 \\
1 & 0 & 0 & 0 & -1 & 0 & 0 & 0 & 0 & 0 & 0 & 0 & 0 & 0 & 0 & 0 & 0 & 0 & 0 & 0 & 0 \\
0 & -1 & 0 & 0 & 0 & 0 & 0 & 0 & 0 & 0 & 0 & 0 & 0 & 0 & 0 & -1 & 0 & 0 & 0 & 0 & 0 \\
0 & 0 & 1 & 0 & 0 & 0 & 0 & 0 & 0 & 0 & 0 & 0 & 0 & 0 & 0 & 0 & 0 & -1 & 0 & 0 & 0 \\
-1 & 0 & 0 & 1 & 0 & 0 & 0 & 0 & 0 & 0 & 0 & 0 & 0 & 0 & 0 & 0 & 0 & 0 & 0 & 0 & 0 \\
-1 & 0 & 0 & 0 & 1 & 0 & 0 & 0 & 0 & 0 & 0 & 0 & 0 & 0 & 0 & 0 & 0 & 0 & 0 & 0 & 0 \\
-1 & 0 & 0 & 0 & 0 & 1 & 0 & 0 & 0 & 0 & 0 & 0 & 0 & 0 & 0 & 0 & 0 & 0 & 0 & 0 & 0 \\
-1 & 0 & 0 & 0 & 0 & 0 & 1 & 0 & 0 & 0 & 0 & 0 & 0 & 0 & 0 & 0 & 0 & 0 & 0 & 0 & 0 \\
0 & 0 & 0 & 0 & 0 & 0 & 1 & -1 & 0 & 0 & 0 & 0 & 0 & 0 & 0 & 0 & 0 & 0 & 0 & 0 & 0 \\
0 & 0 & 0 & 0 & 0 & 0 & -1 & 1 & 0 & 0 & 0 & 0 & 0 & 0 & 0 & 0 & 0 & 0 & 0 & 0 & 0 \\
0 & 0 & 0 & 0 & 0 & 0 & 0 & 1 & 0 & 0 & 0 & 0 & 0 & 0 & 0 & -1 & 0 & 0 & 0 & 0 & 0 \\
0 & 0 & 0 & 0 & 0 & 0 & -1 & 0 & 1 & 0 & 0 & 0 & 0 & 0 & 0 & 0 & 0 & 0 & 0 & 0 & 0 \\
0 & 0 & 0 & 0 & 0 & 0 & 0 & 0 & 1 & -1 & 0 & 0 & 0 & 0 & 0 & 0 & 0 & 0 & 0 & 0 & 0 \\
0 & 0 & 0 & 0 & 0 & 0 & 0 & 0 & -1 & 1 & 0 & 0 & 0 & 0 & 0 & 0 & 0 & 0 & 0 & 0 & 0 \\
0 & 0 & 0 & 0 & 0 & 0 & 0 & 0 & 0 & 1 & -1 & 0 & 0 & 0 & 0 & 0 & 0 & 0 & 0 & 0 & 0 \\
0 & 0 & 0 & 0 & 0 & 0 & 0 & 0 & 0 & -1 & 1 & 0 & 0 & 0 & 0 & 0 & 0 & 0 & 0 & 0 & 0 \\
0 & 0 & 0 & 0 & 0 & 0 & 0 & 0 & 0 & 0 & 1 & 0 & 0 & 0 & -1 & 0 & 0 & 0 & 0 & 0 & 0 \\
0 & 0 & 0 & 0 & 0 & 0 & 0 & 0 & 0 & 0 & 0 & 0 & 0 & 0 & 0 & 0 & 0 & 1 & 0 & -1 & 0 \\
0 & 0 & 0 & 0 & 0 & 0 & 0 & 0 & 0 & 0 & 0 & 0 & 0 & 0 & 0 & 0 & 0 & 0 & 0 & 1 & -1 \\
0 & 0 & 0 & 0 & 0 & 0 & 0 & 0 & 0 & 0 & 0 & 0 & 0 & 0 & 0 & 0 & 0 & 0 & 0 & -1 & 1
\end{matrix}
$$

The corresponding gain matrix is found to be

$$G=\begin{matrix}
22 & 0 & 0 & -6 & -6 & -5 & -5 & 0 & 0 & 0 & 0 & 0 & 0 & 0 & 0 & 0 & 0 & 0 & 0 & 0 & 0\\
0 & 2 & 0 & 0 & 0 & 0 & 0 & 0 & 0 & 0 & 0 & 0 & 0 & 0 & 0 & 0 & 0 & 0 & 0 & 0 & 0\\
0 & 0 & 2 & 0 & 0 & 0 & 0 & 0 & 0 & 0 & 0 & 0 & 0 & 0 & 0 & 0 & 0 & -2 & 0 & 0 & 0\\
-6 & 0 & 0 & 3 & 1 & 1 & 1 & 0 & 0 & 0 & 0 & 0 & 0 & 0 & 0 & 0 & 0 & 0 & 0 & 0 & 0\\
-6 & 0 & 0 & 1 & 3 & 1 & 1 & 0 & 0 & 0 & 0 & 0 & 0 & 0 & 0 & 0 & 0 & 0 & 0 & 0 & 0\\
-5 & 0 & 0 & 1 & 1 & 2 & 1 & 0 & 0 & 0 & 0 & 0 & 0 & 0 & 0 & 0 & 0 & 0 & 0 & 0 & 0\\
-5 & 0 & 0 & 1 & 1 & 1 & 5 & -2 & -1 & 0 & 0 & 0 & 0 & 0 & 0 & 0 & 0 & 0 & 0 & 0 & 0\\
0 & 0 & 0 & 0 & 0 & 0 & -2 & 3 & 0 & 0 & 0 & 0 & 0 & 0 & 0 & -1 & 0 & 0 & 0 & 0 & 0\\
0 & 0 & 0 & 0 & 0 & 0 & -1 & 0 & 3 & -2 & 0 & 0 & 0 & 0 & 0 & 0 & 0 & 0 & 0 & 0 & 0\\
0 & 0 & 0 & 0 & 0 & 0 & 0 & 0 & -2 & 4 & -2 & 0 & 0 & 0 & 0 & 0 & 0 & 0 & 0 & 0 & 0\\
0 & 0 & 0 & 0 & 0 & 0 & 0 & 0 & 0 & -2 & 3 & 0 & 0 & 0 & -1 & 0 & 0 & 0 & 0 & 0 & 0\\
0 & 0 & 0 & 0 & 0 & 0 & 0 & 0 & 0 & 0 & 0 & 0 & 0 & 0 & 0 & 0 & 0 & 0 & 0 & 0 & 0\\
0 & 0 & 0 & 0 & 0 & 0 & 0 & 0 & 0 & 0 & 0 & 0 & 0 & 0 & 0 & 0 & 0 & 0 & 0 & 0 & 0\\
0 & 0 & 0 & 0 & 0 & 0 & 0 & 0 & 0 & 0 & 0 & 0 & 0 & 0 & 0 & 0 & 0 & 0 & 0 & 0 & 0\\
0 & 0 & 0 & 0 & 0 & 0 & 0 & 0 & 0 & 0 & -1 & 0 & 0 & 0 & 1 & 0 & 0 & 0 & 0 & 0 & 0\\
0 & 0 & 0 & 0 & 0 & 0 & 0 & -1 & 0 & 0 & 0 & 0 & 0 & 0 & 0 & 3 & 0 & 0 & 0 & 0 & 0\\
0 & 0 & 0 & 0 & 0 & 0 & 0 & 0 & 0 & 0 & 0 & 0 & 0 & 0 & 0 & 0 & 0 & 0 & 0 & 0 & 0\\
0 & 0 & -2 & 0 & 0 & 0 & 0 & 0 & 0 & 0 & 0 & 0 & 0 & 0 & 0 & 0 & 0 & 3 & 0 & -1 & 0\\
0 & 0 & 0 & 0 & 0 & 0 & 0 & 0 & 0 & 0 & 0 & 0 & 0 & 0 & 0 & 0 & 0 & 0 & 0 & 0 & 0\\
0 & 0 & 0 & 0 & 0 & 0 & 0 & 0 & 0 & 0 & 0 & 0 & 0 & 0 & 0 & 0 & 0 & -1 & 0 & 3 & -2\\
0 & 0 & 0 & 0 & 0 & 0 & 0 & 0 & 0 & 0 & 0 & 0 & 0 & 0 & 0 & 0 & 0 & 0 & 0 & -2 & 2
\end{matrix}$$

Using the proposed algorithm, the D matrix is developed and the diagonal of the D matrix is Diag $D=\left[22,2,2,1.36,1.07,0.5,3,1.67,2.4,2.33,1.29,0,0,0,2.22,2,0,1,0,2,0\right]$

This implies that to make the system observable and the state estimation solvable, the PMU is placed at buses 12, 13, 14, 17, 19, and 21 and/or at the adjacent buses 8, 10, and 20. Therefore, the rank of D will be increased. Considering all the possible buses (8, 10, 12, 13, 14, 17, 19, 20, 21), the T matrix is given by

$$T=\begin{matrix}
0&0&0&0&0&0&1&1&0&0&0&0&0&0&0&1&1&0&0&0&0\\
0&0&0&0&0&0&0&0&1&1&1&1&0&0&0&0&0&0&0&0&0\\
0&0&0&0&0&0&0&0&0&1&0&1&1&0&0&0&0&0&0&0&0\\
0&0&0&0&0&0&0&0&0&0&0&1&1&1&0&0&0&0&0&0&0\\
0&0&0&0&0&0&0&0&0&0&0&0&1&1&0&0&0&0&0&0&0\\
0&0&0&0&0&0&0&1&0&0&0&0&0&0&0&0&1&0&1&0&0\\
0&0&0&0&0&0&0&0&0&0&0&0&0&0&0&0&1&0&1&1&0\\
0&0&0&0&0&0&0&0&0&0&0&0&0&0&0&0&0&1&1&1&1\\
0&0&0&0&0&0&0&0&0&0&0&0&0&0&0&0&0&0&0&1&1
\end{matrix}$$

Further,

$$b_m = \begin{bmatrix} 0 & 0 & 1 & 1 & 1 & 1 & 1 & 0 & 1 \end{bmatrix}$$

Now by applying the proposed algorithm in the following ILP model,
$\min f = x_8 + x_{10} + x_{12} + x_{13} + x_{14} + x_{17} + x_{19} + x_{20} + x_{21}$
subject to

$$= \begin{vmatrix} 0 & 1 & 1 & 1 & 0 & 0 & 0 & 0 & 0 \\ 0 & 0 & 1 & 1 & 1 & 0 & 0 & 0 & 0 \\ 0 & 0 & 0 & 1 & 1 & 0 & 0 & 0 & 0 \\ 1 & 0 & 0 & 0 & 0 & 1 & 1 & 0 & 0 \\ 0 & 0 & 0 & 0 & 0 & 1 & 1 & 1 & 0 \\ 0 & 0 & 0 & 0 & 0 & 0 & 0 & 1 & 1 \end{vmatrix} [x_8 + x_{10} + x_{12} + x_{13} + x_{14} + x_{17} + x_{19} + x_{20} + x_{21}]^T 3 \begin{bmatrix} 1 & 1 & 1 & 1 & 1 & 1 \end{bmatrix}^T$$

The solution is given by

$$X = \begin{bmatrix} 0 & 0 & 0 & 1 & 0 & 0 & 1 & 0 & 0 \end{bmatrix}^T$$

Table 10.5 21-bus Southern region conventional state estimation results without PMU placement.

Bus	Actual value of voltage in p.u.	State estimation results of the voltage with PMU placement in p.u.
Bus 01	0.9591	0.95929
Bus 02	0.9788	0.97878
Bus 03	0.971	0.97097
Bus 04	0.9591	0.94909
Bus 05	0.9661	0.9587
Bus 06	0.9334	0.95852
Bus 07	0.9062	0.95065
Bus 08	0.9148	0.958
Bus 09	0.8951	0.948
Bus 10	0.8844	0.944
Bus 11	0.8637	0.943
Bus 12	0.887	0.943
Bus 13	0.887	0.942
Bus 14	0.9145	0.941
Bus 15	0.7861	0.943
Bus 16	0.9287	0.968
Bus 17	0.971	0.954
Bus 18	0.9029	0.957
Bus 19	0.8902	0.954
Bus 20	0.8902	0.955
Bus 21	0.8902	-

Table 10.6 Mean square error comparison for the state estimation performance evaluation for the 21-bus Southern region system.

Bus	Actual value of *voltage* in p.u.	Estimated value before PMU placement		Estimated value after PMU placement	
		Voltage in p.u.	Error square	Voltage in p.u.	Error square
1	0.9591	1	0.0016728	0.95929	3.5995E-08
2	0.9788	1.019047	0.00161978	0.97878	5.706E-10
3	0.971	1.018189	0.00222682	0.97097	1.0108E-09
4	0.9591	0.990085	0.00096008	0.94909	0.00010013
5	0.9661	0.999202	0.00109576	0.9587	5.4817E-05
6	0.9334	0.99918	0.00432707	0.95852	0.00063122
7	0.9062	0.991997	0.0073611	0.95065	0.00197542
8	0.9148	0.9989	0.00707123	0.958	0.00184948
9	0.8951	0.9892	0.00885007	0.948	0.00276649
10	0.8844	0.9858	0.0102766	0.944	0.00357031
11	0.8637	0.9845	0.01458155	0.943	0.00625571
12	0.887	0.9861	0.00981833	0.943	0.00317973
13	0.887	0.9867	0.00993785	0.942	0.00304241
14	0.9145	0.9873	0.00529677	0.941	0.0006908
15	0.7861	0.9845	0.03937621	0.943	0.02458305
16	0.9287	1.0087	0.00639271	0.968	0.00154218
17	0.971	1.0002	0.00085073	0.954	0.00030365
18	0.9029	1.0052	0.01046715	0.957	0.00296609
19	0.8902	1.0022	0.01254589	0.954	0.00410989
20	0.8902	1.0028	0.01267295	0.955	0.00417624
21	0.8902	–	–	–	–
Mean square error		0.0079715		0.00294275	

Now solving the state estimation by optimally placing the PMUs, the obtained results are presented by Table 10.5.

Table 10.6 shows the overall performance comparison for the 21-bus Southern region system. The optimal placing of PMUs enhanced the state estimation mean square error from 0.0079715 to 0.00294275, which shows around 63% improvement.

10.4 Conclusion

The inclusion of synchronized phasor measurements is discussed by addressing the use of voltage and current phasor measurements at the bus and branch levels. The theory of observability analysis, including the topological and numerical approaches, are also discussed. The ILP approach is investigated as it is utilized for the proposed method, since it is used to identify observable and unobservable portions of the system, forming the super-nodes in the proposed methods.

The proposed algorithms are illustrated on the IEEE 14-bus system as a base case and 21-bus Southern region of Ethiopia system as a practical case study using MATLAB and PET software. Computed results show that the proposed algorithm offered PMU placement solutions that improve robustness of the traditional state estimation in such a way that improves the state estimation result by minimizing the mean square error from 0.00144 to 0.00038 in the IEEE 14-bus analysis and from 0.007972 to 0.002943 in the 21-bus Southern region of Ethiopia system, which show improvements of 73% and 63%, respectively.

References

[1] K. Negash, B. Khan, and E. Yohannes, “Artificial intelligence versus conventional mathematical techniques: A review for optimal placement of phasor measurement units,” *Journal of Technology and Economics of Smart Grids and Sustainable Energy*, vol. 1, p. 10, 2016.

[2] J. Aghaei, A. Baharvandi, A. Rabiee, and M. Akbari, “Probabilistic PMU placement in electric power networks: An MILP-based multiobjective model,” *IEEE Transactions on Industrial Informatics*, vol. 11, no. 2, pp. 332–341, April 2015.

[3] D. Dua, S. Dambhare, R.K. Gajbhiye, and S.A. Soman, “Optimal multistage scheduling of PMU placement: An ILP approach,” *IEEE Transactions on Power Delivery*, vol. 23, no. 4, pp. 1812–1820, October 2008.

[4] S. Razavi, H. Falaghi, and M. Ramezani, “A new integer linear programming approach for multi-stage PMU placement,” *2013 Smart Grid Conference (SGC)*, Tehran, pp. 119–124, 2013.

[5] B. Gou, “Generalized integer linear programming formulation for optimal PMU placement,” *IEEE Transactions on Power Systems*, vol. 23, no. 3, pp. 1099–1104, August 2008.

[6] S. Chakrabarti and E. Kyriakides, “Optimal placement of phasor measurement units for power system observability,” *IEEE Transactions on Power Systems*, vol. 23, no. 3, pp. 1433–1440, August 2008.

[7] S. Chakrabarti, E. Kyriakides, and M. Albu, “Uncertainty in power system state variables obtained through synchronized measurements,” *IEEE Transactions on Instrumentation and Measurement*, vol. 58, no. 8, pp. 2452–2458, August 2009.

[8] Q. Li, R. Negi, and M.D. Ilić, “Phasor measurement units placement for power system state estimation: A greedy approach,” *2011 IEEE Power and Energy Society General Meeting*, Detroit, MI, USA, pp. 1–8, 2011.

[9] M. Hajian, A.M. Ranjbarb, T. Amraee, and B. Mozafari, “Optimal placement of PMUs to maintain network observability using a modified BPSO algorithm,” *Electrical Power and Energy Systems*, vol. 33, pp. 28–34, 2011.

[10] J. Wang, C. Li, and J. Zhang, “Optimal phasor measurement unit placement by an improved PSO algorithm,” *2012 Asia-Pacific Power and Energy Engineering Conference*, Shanghai, pp. 1–4, 2012.

[11] K. Negash, B. Khan, D. Tesfaye, and M. Gayathri, “Optimal placement of phasor measurement unit for system observability,” *Proceedings of IEEE International Conference on Computing Methodologies and Communication (ICCMC 2017)*, Erode India, pp. 73–77, 18–19 July 2017.

[12] L. Sun, T. Chen, X. Chen, W.K. Ho, K.-V. Ling, K.-J. Tseng, and G.A.J. Amaratunga, "Optimum placement of phasor measurement units in power systems," *IEEE Transactions on Instrumentation and Measurement*, vol. 68, no. 2, pp. 421–429, February 2019.

[13] M.K. Arpanahi, H.H. Alhelou, and P. Siano, "A novel multiobjective OPP for power system small signal stability assessment considering WAMS uncertainties," *IEEE Transactions on Industrial Informatics*, vol. 16, no. 5, pp. 3039–3050, May 2020.

[14] S. Wang, R. Huang, X. Ke, J. Zhao, R. Fan, H.Wang, Z. Huang, A. Sathanur, and D. Vrabie, "Risk-oriented PMU placement approach in electric power systems," *IET Generation, Transmission & Distribution*, vol. 14, no. 2, pp. 301–307, 31 January 2020.

[15] S.R. Samantaray and A. Sharma, "Enhancing performance of wide-area back-up protection scheme using PMU assisted dynamic state estimator," *IEEE Transactions on Smart Grid*, vol. 10, no. 5, pp. 5066–5074, September 2019.

[16] C. Zhang, X. Wang, Y. Liu, D. Shi, and Y. Li, "Fast faulted line identification method for wide-area backup protection with PMU optimal placement strategy," *IET Generation, Transmission & Distribution*, vol. 13, no. 9, pp. 1599–1611, 7 May 2019.

[17] X. Guo, C. Liao, and C. Chu, "Enhanced optimal PMU placements with limited observability propagations," *IEEE Access*, vol. 8, pp. 22515–22524, 2020.

[18] A. Almunif and L. Fan, "Optimal PMU placement for modeling power grid observability with mathematical programming methods," *International Transactions on Electrical Energy Systems*, vol. 30, p. e12182, 2020.

[19] H. Manoharan, S. Srikrishna, G. Sivarajan, and A. Manoharan, "Economical placement of PMUs considering observability and voltage stability using binary coded ant lion optimization," *International Transactions on Electrical Energy Systems*, vol. 28, p. e2591, 2018.

[20] N.M. Manousakis and G.N. Korres, "A hybrid power system state estimator using synchronized and unsynchronized sensors," *International Transactions on Electrical Energy Systems*, vol. 28, p. e2580, 2018.

[21] S. Goleijani and M.T. Ameli, "Neural network-based power system dynamic state estimation using hybrid data from SCADA and phasor measurement units," *International Transactions on Electrical Energy Systems*, vol. 28, p. e2481, 2018.

[22] T. Tarif, A.A. Ladjici, Y. Amrane, and Y. Chabane, "Multi-objective optimal PMUs placement for online assessment of small-signal stability," *International Transactions on Electrical Energy Systems*, vol. 29, p. e12117, 2019.

[23] Z.H. Rather, Z. Chen, P. Thøgersen, and P. Lund, "Pragmatic approach for multistage phasor measurement unit placement: A case study of the Danish power system and inputs from practical experience," *International Transactions on Electrical Energy Systems*, vol. 26, pp. 2532–2551, 2016.

[24] N.M. Manousakis and G.N. Korres, "A weighted least squares algorithm for optimal PMU placement," *IEEE Transactions on Power Systems*, vol. 28, no. 3, pp. 3499–3500, August 2013.

[25] B. Gou and R.G. Kavasseri, "Unified PMU placement for observability and bad data detection in state estimation," *IEEE Transactions on Power Systems*, vol. 29, no. 6, pp. 2573–2580, November 2014.

Part VII

Smart Microgrid Integration and Optimization

11

Smart Microgrid Integration and Optimization

M. Thirunavukkarasu and Yashwant Sawle

Vellore Institute of Technology, Vellore, Tamilnadu, India

11.1 Introduction

New challenges are emerging from the trend toward distributed renewable energies. Typical weather-based renewable energy sources contribute to difficult to manage intermittent generation of energy. They are, though, an essential aspect of potential smart grids and hence there are a lot of measures being used to render these sources more effective. In addition to integrating distributed renewable resources, future smart grids will also need to integrate information and communication systems for coordination and monitoring [1]. For the future, control and performance can be improved by breaking the smart grid into subsystems. These subsystems are known by the name of smart microgrids and consist of small-scale energy consumers and producers that can be controlled. A smart microgrid can either be attached to the mainline grid or it can operate in a so-called island mode [2]. Smart means modern, safe, sleek, trendy, or workable in automation; a grid is an energy supply network that supplies power at certain levels. The intelligent grid has no description that is acknowledged unanimously. This can be defined in basic terms as well as in more nuanced ways. Stated quite clearly, it is an intellectual system that is able to collect, interact, and decide a conclusion. Only energy can be generated or relayed via a conventional grid. The intelligent grid switches the actual system to one that works in a collaborative, sensitive, and ecological responsiveness [3].

A smart grid comes with techniques for smart measurement, a digital remote sensor, and smart control automation with analytical instruments to process, track, and manage the two-way energy flow during power service [4]. Smart grids give reliable distributed generation power with a voltage control ability for vastly higher DRES permeation into the generation system. Smart grids provide sensitive frequency-regulated loads to improve the stability of the system. A smart grid system is always in service with an architectural relationship to goals and strengths. The execution of a smart grid will enhance the strength, self-healing, and grid digital system integration [5]. Small-scale integration renewable energy sources (RESs) trigger problems such as voltage spikes, harmonic disturbances, and involve grid synchronization of sources [6]. Hence the smart grid network helps to improve these concerns by avoiding outages and encouraging customers to monitor the use of energy. The photovoltaic institutional power grid combined with battery storage and

Active Electrical Distribution Network: A Smart Approach, First Edition. Edited by Baseem Khan, Josep M. Guerrero, Sanjeevikumar Padmanaban, Hassan Haes Alhelou, Om Prakash Mahela, and Sudeep Tanwar.

portable generators has tremendous potential for a functioning grid-powered microgrid as well as an islanding mode microgrid. In the last five years, solar photovoltaic (PV) systems linked across a low tension distributed network are steadily growing at a combined average growth rate of 88% [7]. An intelligent grid infrastructure may be used to build a broad network of smart grids, which ultimately link to all other microgrids. These intelligent grids have an enormous capacity and can be a reliable solution in developed countries that lack architecture to transfer and distribute power. In the United States, transport emissions account for only 20% of carbon dioxide overall while power production accounts for 40% of carbon dioxide pollution because of the rising electricity demand. The effective supplies of electric power and the subsequent removal of greenhouse gases and contaminants such as nitrogen and sulfur oxides are considered as a smart grid central function to address the problem [8]. A smart infrastructure also supports the consumer by predicting their demand and the right economy technology use of energy consumption.

11.2 Architecture and Traits of the Smart Grid

In order to understand the design and conception of an intelligent grid, we need to comprehend its distinction from the conventional grid. Table 11.1 provides this relationship. The National Institute of Standards and Technology (NIST) has provided a conceptual smart grid model that describes the formulation of scheduling, expansion, interconnected stakeholders, and needed resources [9]. These stakeholders are grouped into seven mapping domains, as outlined in Table 11.2 by NIST.

In order to define new technical standards relevant to smart grids as well as to recognize existing technological standards applicable in smart grids, a smart grid architecture model (SGAM) was initially developed under the European Commission (EC) M/490 Directive

Table 11.1 Difference between a conventional and an intelligent grid.

Terms	Conventional grid	Intelligent grid
Network	Interconnected network	Radial network
Generation	Centralized for power generation	Distributed system of power generation
Communication	One-way communication	Two-way real-time communication
Sensors	Limited number of essential sensors are used	Several sensors are implicated
Operation	Mechanical	Digital
Control and monitor	Manual and limited control	Automatic and wide range of control
Security	No security or privacy concern	Provides cyber security and privacy concerns
Terms	Conventional grid	Intelligent grid

Table 11.2 Stakeholders of a smart grid.

Stakeholders	Summary
Generation	The generation of energy. Generating firms are interested as competitors in bulk electricity.
Transmission	Power is transmitted. It provides corporation or competitors liable for the transfer of electricity produced for delivery.
Distribution	Electricity is distributed and tracked to end users by energy providers, which include the energy distributor and the consumer.
Customer	The end users utilize electric power. It may be residential, corporate, or commercial.
Operations	It consists of the administrative or energy flow control authorities to perform power management activities.
Markets	Stakeholders are using intelligent grid assets. Both managers and buyers are market operators.
Service provider	Offers aid to all parties interested in generation, production, and transfer of electrical power.

for the European Standardization Bodies CEN (Comitè Europèen de Normalization), CENELEC (European Committee for Electrotechnical Standardization), and ETSI (European Telecommunications Standards Institute). In view of the distribution and the number of experts, the consensus on terminology, technologies, context, and topic areas for debate seem to have become quite an issue [10]. The smart grid architecture model was then used as a method for the classification of reference to address this fatal issue (i.e. that the issue itself is already a problem formulation) [11–20]. The smart grid suggested by the federal Department of Energy (DOE) is divided into ten areas: network transmission framework, evaluation of system management, activity of infrastructure, distribution control, integration of green renewable sources, energy conservation, distributed generation and storing, necessity signals and alternatives of demand participation and smart machines, plug-in hybrid electric vehicles (PHEVs), and stockpiling, illustrated in Figure 11.1.

Figure 11.2 illustrates how the second architectural structure is split into knowledge and expertise levels, and modern technologies and inventions that imply production of energy, storage, delivery, and end consumer aspects of the electrical grid.

11.2.1 Layer Architecture of the Smart Grid

The structure of the smart grid consists of power layers, a market and service layer, an application layer, and a control and functional layer, as shown in Figure 11.3. Furthermore, the control layer reflects the communication systems and expands widely across the entire smart phase terrain zone.

11.2.1.1 Power Layer

The first layer is the electrical grid's physical framework, which comprises all the elements that relate to the power cycle, including generation, transmission, distribution, distributed

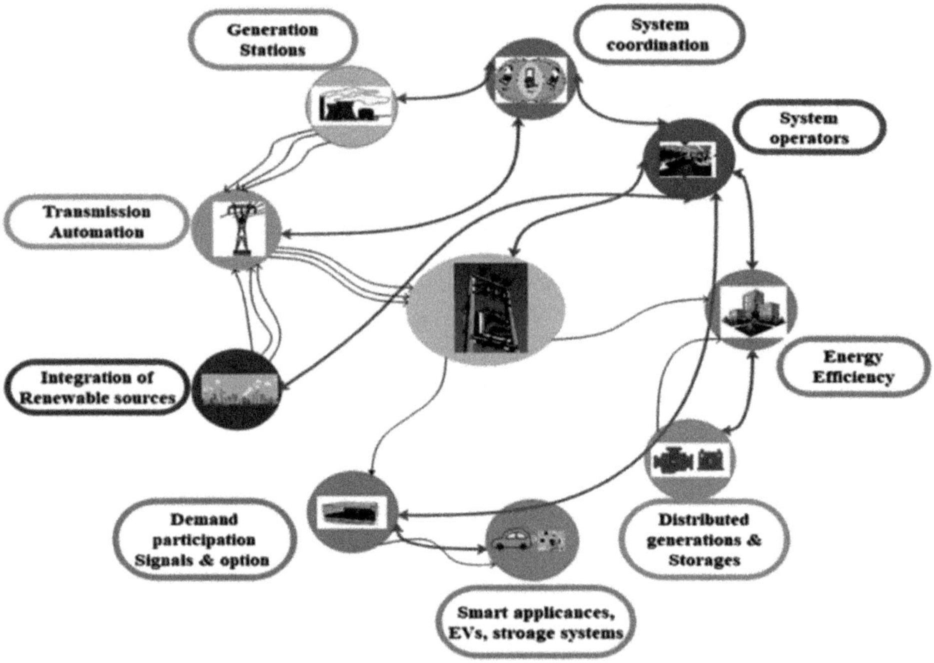

Figure 11.1 Architecture of smart grid by Federal Department of Energy.

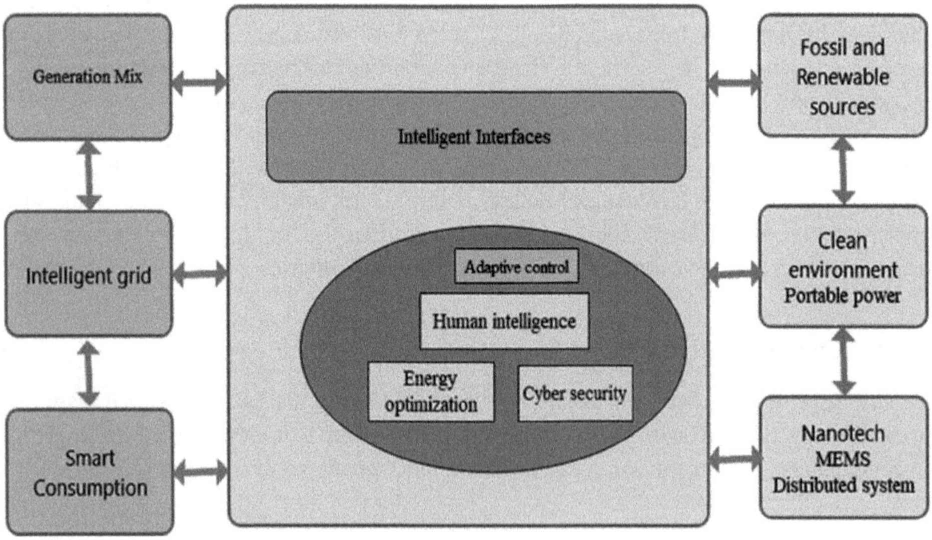

Figure 11.2 The smart grid – Architecture 2.

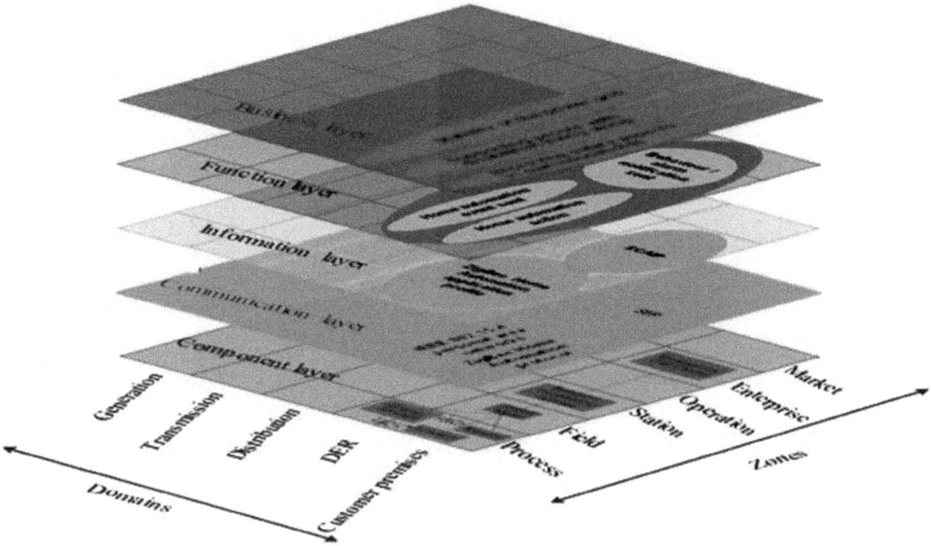

Figure 11.3 Smart grid multilayered architecture.

energy resources (DER), storage devices, and end users. The power layer is the independent identity of the energy storage system (ESS) and, depending on their condition, it may be a supplier or a customer.

11.2.1.2 Operation and Access Control Layer

All information needs to be shared in this layer to monitor distributed generation (DG), configured dispatchable loads in potential systems, and other intelligent devices. It supports energy transport control activities and model power system operations such as grid balancing, fault detection, and real-time control. There is also an access control actuator attached to the mounted smart meter and household gadgets so as to diminish vitality use and total burdens by means of the AMI systems. A wide collection of nodes in the area should be linked to the service and control layer and utilities should use and acknowledge different networks such as WAN, FAN, NAN, HAN, BAN, and IAN.

11.2.1.3 Market and Service Layer

This layer enables market sharing offers and requirements to be matched inside the power grid. The retail sector and the wholesale business are the two available market sectors. The vast number of retail consumer's usage of ICT equipment details and reviews information and picks a compatible retail sector tariff offered by businesses in the area of electricity. The wholesale sector is a deregulated system, where efficient energy providers allow their consumers to have the requisite capacities. The service layer executes consumer and utilities services. The service provider is the cornerstone of the power system's performance inside a smart grid, which should be linked to the controller, the market, and the buyer.

11.2.1.4 Application Layer

The application layer covers all the systems, which adds the best functionality of the intelligent framework in all the systems and layers. The application layer offers clients different smart grid network applications and also offers utilities depending on the current database.

11.3 Smart Microgrids

A modern concept that has emerged from electrical grids is the intelligent system. In addition to electricity delivery, smart grids provide significant complexity of the distribution of knowledge in the integrated energy supply network, but electrical grids deal primarily with power transport in wide regions. This intelligent grid is fitted with servers, robotics, control, and other latest technology systems capable of remotely reacting to ever-changing energy needs. A modern grid system, namely the intelligent microgrid network, has been with further progress of technology in the field of distributed energy resources management. An intelligent microgrid is a limited variant of a smart grid that can be customized to fulfill the energy needs of a given area utilizing, for example, intelligent controls, optimization methods, management of capital, and electricity marketing. The concept of an intelligent microgrid is to manage, under precisely specified electrical limits, a network of interconnected loads and distributed power resources as a single controllable unit in terms of the grid. A smart microgrid is a revamped power grid form, concentrating not only on the output of bulk electricity but also on smart meters, smart vehicles, green, and energy-efficient technologies. This modern intelligent grid conducts a two-way flow of energy and knowledge, not just on a one-way delivery of electricity, such as the conventional grid, but on one that leads to efficient and more sophisticated power distribution. India's Smart Grid Interface Vision Integration, with its intelligent city mission goals to achieve smart solutions for energy management (smart measurement, renewable energy sources, efficient and green buildings), aims to achieve intelligent energy-intensive solutions for waste management and water management in metropolitan centers [21]. Figure 11.4 depicts high-level interdependencies among intelligent cities, smart grids, and smart microgrids.

11.3.1 Architecture of the Smart Microgrid System

Power generation units are normally situated in isolated places in densely developed countries. Since power is conveyed over a great distance, this is first raised to a higher voltage and then transport through high tension transmission lines. Users other than the local electricity distributors receive electricity from these lines. The voltage from the transmission to the distribution level is decreased upon arrival at the substation. In addition, the voltage is once more converted into an essential service voltage from the distribution level. The voltage is reduced from the transmission to the distribution level when it comes to the sub-station. In addition, the voltage is again converted to an essential service voltage from the distribution level. There are many benefits arising from an intelligent microgrid system. Intelligent grids incorporate elements of the conventional power grid with automated

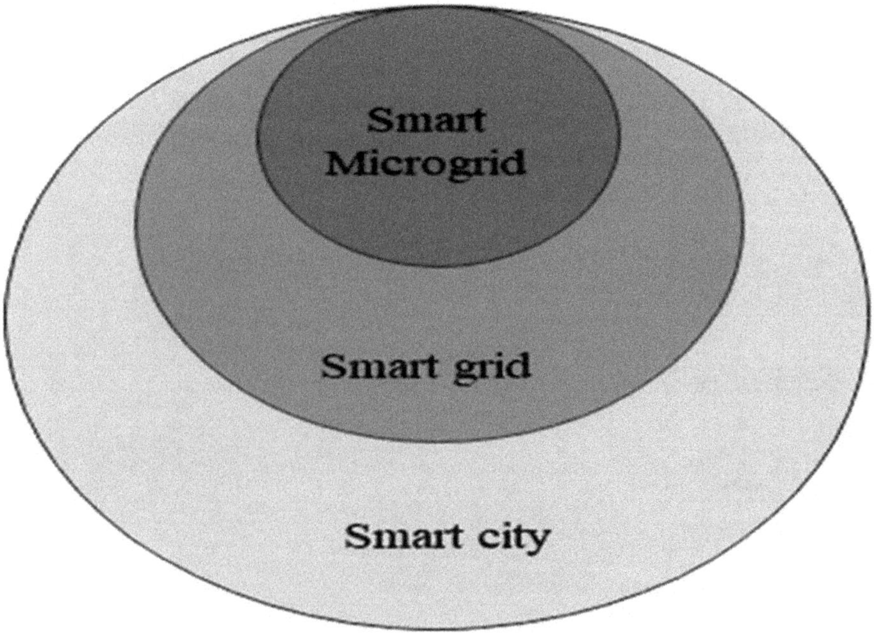

Figure 11.4 Interdependencies between smart cities, smart grid, and smart microgrid.

processes, which enable them to transmit power centrally through information to the benefit of everyone. Figure 11.5 displays the overall design of smart microgrid systems, which involves multiple elements for various uses.

Smart microgrids are mainly characterized by: maximum use of renewable and other forms of energy; load priority is important or not critical; profiling of raw material and load; control and prediction control; electrical and physical signal acquisition and monitoring in real time; and minimize and respond quickly to network disruptions by automatically connecting/disconnecting the system components [22].

11.3.2 Operating Modes of a Smart Microgrid

In the smart microgrid, the grid tied in normal modes, the grid tied on the emergency mode, and the grid in the off mode are basically three modes of operation available. Economic optimization is taken into consideration in the grid tied normal mode of operation. In that respect, DER, load, and storage controls take account of energy expenditure and are considered to fulfill the capacity demands in the finest potential or economic energy source. No economic optimization will be considered in the grid-linked emergency mode. This is the operational mode during peak periods or during variability in the grid. During these hours the source available will not take into account the energy price during these hours, but will meet the load requirement. The microgrid is disconnected from the centralized grid when the last mode of operation is the off grid mode. This mode is regarded

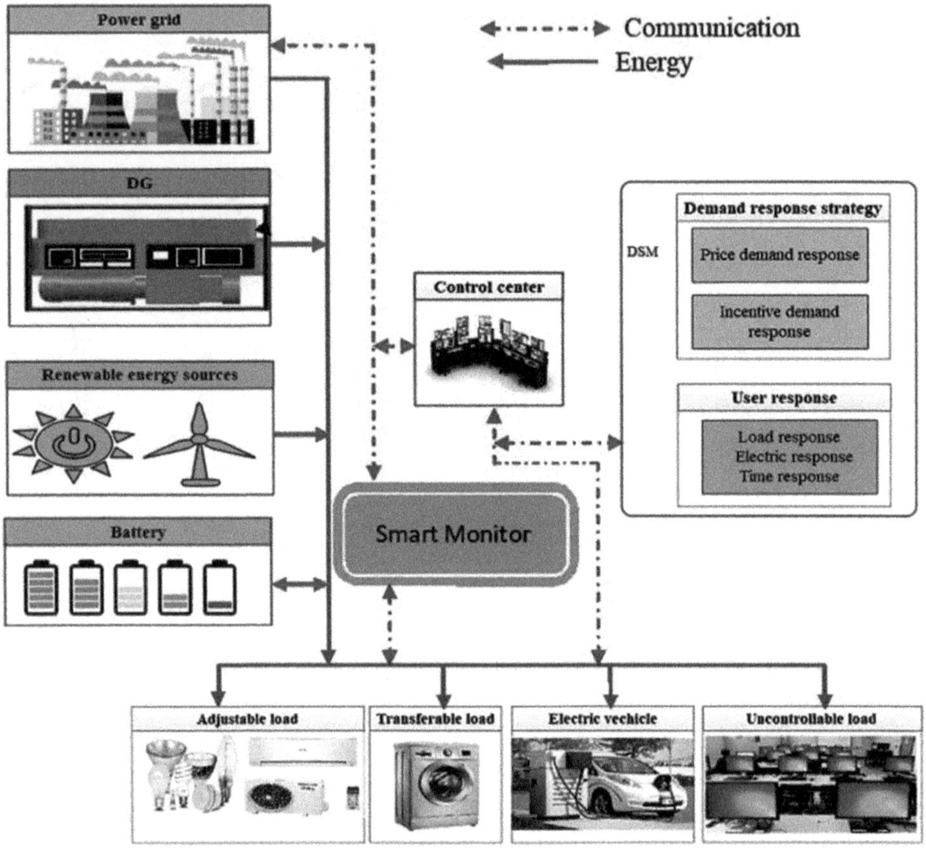

Figure 11.5 Architecture of smart microgrid system.

during high hours, low-energy hours in grid conditions, power cuts, and a storm, or worse. In this mode the loads connected to microgrids are entirely dependent on the DERs and the energy storage devices.

11.3.3 Topologies for an Intelligent Microgrid

Radial: A radial grid is the simple structure called a radial network and involves a series of radial networks and sub-networks that commence from a source of power and distribute power through networks with increasingly lower voltage levels, and simply end at communities, households, and industries.

Mesh: A network of the mesh includes the radial structure but also consists of redundant lines that connect nodes, as well as the main lines, to retrieve power if a main line fails.

Ring grids: The interconnection in ring networks and loading feed power are operated by secondary grid nodes. There are backward and forward routes to access any power flow node in ring grid topology. The benefits of ring operation include better features, stability of voltage, and low loss of power [23].

11.4 Demand Response

The reliability of the power grid, efficiency indices, and service quality indices may be influenced by degradation of the RES [24]. In this relationship, a microgrid was implemented in order to integrate a range of DER technology and loads into a limited distributed power system as an active and regulated infrastructure [25]. In the view of the demand response (DR) program, microgrids have a great potential to facilitate the active involvement of end-user consumers [26]. The DR program may not only reduce energy costs but also offers the highest level of reliability and safety standards by using the microgrid core controller (MGCC) [27]. A significant function of the smart grid has been established to enable active consumer engagement in the demand response [28]. This function enables the equilibrium between generation and demand to be managed efficiently, while maintaining the device frequency and voltage within necessary limits. To be specific, demand will be successful if intermittent renewables are exploited more effectively. The primary determinant of generation excess or deficiency in a power system depends on the frequency drop up or down [29]. This frequency variance can be regulated by a demand response. The MGCC takes advantage of the potential of the DR program in order to provide microgrid frequency protection in a more effective and cost-efficient way. The demand response applies to shifts in the usage of energy by end consumers in response to increases in power costs over time or to attempts to limit energy use in periods of elevated market rates [30]. The demand response may be divided into two pricing and incentive-based product responses, with each of these grouped into several subgroups, driven on the consumer's interest in modifying their consumer behavior. Figure 11.6 illustrates the classification of the different DR programs.

DRPs can also be found in the microgrid along with DG units as the key components. In other words, DR is one of the most useful options to achieve optimal activity [31]. Therefore, synchronized DR controls will greatly facilitate the availability of auxiliary frequency control resources and the enhancement of the stability of the microgrid in an intelligent microgrid. DR systems are typically graded according to costs and reward. The demand bidding and auxiliary service systems were planned to work within the microgrid frequency

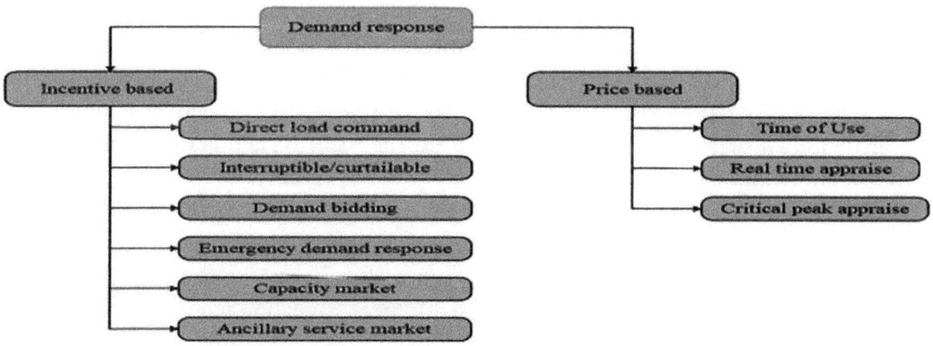

Figure 11.6 Classifications of demand response.

management system and focused on the dispatchability and reliability characteristics of incentive-based services [32].

11.5 Demand Side Management

Options of demand side control and energy conservation solutions are considered to find reliable ways to adjust customer demand to reduce high generator running costs and to incorporate deferred power. Demand side management (DSM) is another technology that is becoming increasingly common. This kind of regulation is counter to the conventional principle of power production, which implies that the utilities ought to have adequate resources to satisfy the demands of consumers, often called supply-side management. The main task of management on the demand side is to redistribute usage. Consumers are advised to limit their usage at peak hours and to postpone their electricity requirements over valley times. A broad range of specific features are essential to enforce demand management: frequent and reliable data analysis is required for access to current data on consumption; operational limits for warning and data storage; the exchanging of information needed in two directions to send messages to smart meters and to obtain the measuring data; and home system control as a prerequisite to enable operation [33].

DSM encourages consumers to make better electricity use choices and helps electricity suppliers to minimize demand for full load, overall expense, and the global amount of CO_2 emissions [34]. DSM techniques can be bifurcation in two specific ways. First, the activity of the different appliances in the domestic, business, and industrial industries is regulated. The second approach is to integrate distributed renewable energies and design algorithms to enable RESs generated electricity to fulfill appliance requirements. If electricity demand for consumers are impacted by DSM then the cumulative demand for peak load can be lowered, the demand profile can be restructured, and grid reliability can be improved. DSM's prime objective is to optimize the usage of green energy options for electrical energy supplies, decreasing the cost of power, minimizing energy imports from the central grid, or decreasing demand for peak loads [35]. Authors suggest a forecasting strategy for a load demand strategy that extends to an intelligent home that utilizes green resources to fulfill a portion of the charge demand and can monitor linked equipment. The authors conclude that green energy rates and its power supply are valid for the next 24 hours and the output of electricity from renewable energy supplies must be rated according to the appliances and scheduled appliances. The demand side management framework is structured as a conditional linear problem and the target is to reduce regular electricity costs by optimizing the use of renewable energy [36]. Cicek and Delic [37] analyzed the usage of demand side management in a smart grid in order to improve social security, taking into consideration wind power production (wind farm) and the consequent volatility.

By adding demand side control at the power plant delivery point will often reduce the difference between supplies and customers. Each of these processes is the "demand response," which allows a consumer to volunteer load control. Consumer/customer participation in peak demand and generation-based market transition would reduce the disparity between demand and supply. The integration of a demand response is complicated since it requires intellectual coordination, negotiation, and regulation.

11.6 Agents in Smart Grids

Agents are knowledgeable individuals in certain contexts, versatile and autonomous, who make sound decisions based on their combined knowledge and their previous experience. Smart companies tend to be the most fit for tracking, regulating, marketing activities, and productive energy use. They have a marketplace platform for electric vehicles and a demand response (DR). Figure 11.7 shows anatomy of an intelligent agent. The smart agent can be defined by examining various factors, whereas the most appropriate concept of the agent is "A smart decision-making device situated in some situations that can function in a manner which is versatile and autonomous to achieve its design goals as the situation shifts" [38]. Smart agents are independent, sensitive, constructive, and social in characteristics.

The smart agent typically consists of four components: an input interface, an output interface, a decision system, and a communication network, as seen in Figure 11.8.

Intelligent agents were graded using numerous viewpoints and the ranking is outlined in the tree diagram in Figure 11.9 [39]. A multiagent method (MAS) is an agent unit operating and collaborating collectively to accomplish a shared aim. Agent-based schemes may have one specific agent, as with user interface agents or program secretaries [40], but when they organize a colony of agents in a far better way to pursue a purpose the term MAS is used for they are achieving their target [41]. The exchange of knowledge between agents takes place in all languages of the agent. Smart grid intelligent agents can help to sustain a complex dynamic scenario for reliable and secure service when functioning as a distributed control system, and also disturbance evaluation, power network restoration, secondary voltage management, and power grid visualization can be effectively carried out through an MAS.

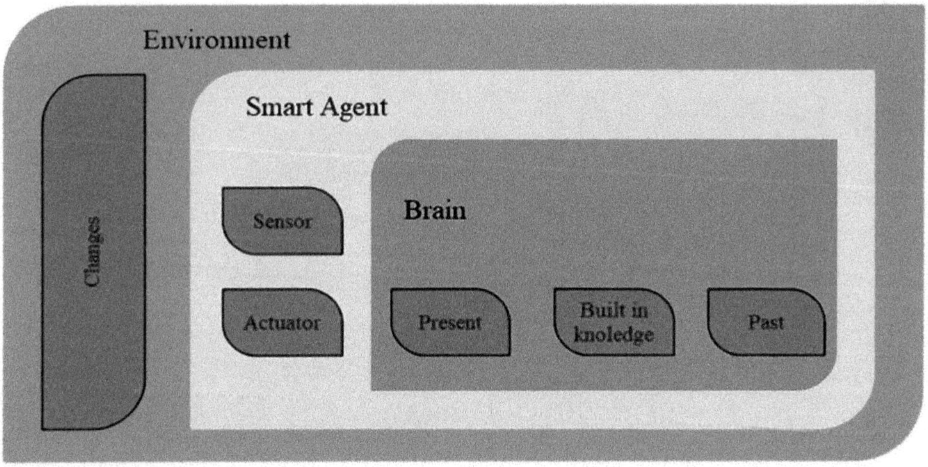

Figure 11.7 The smart agent.

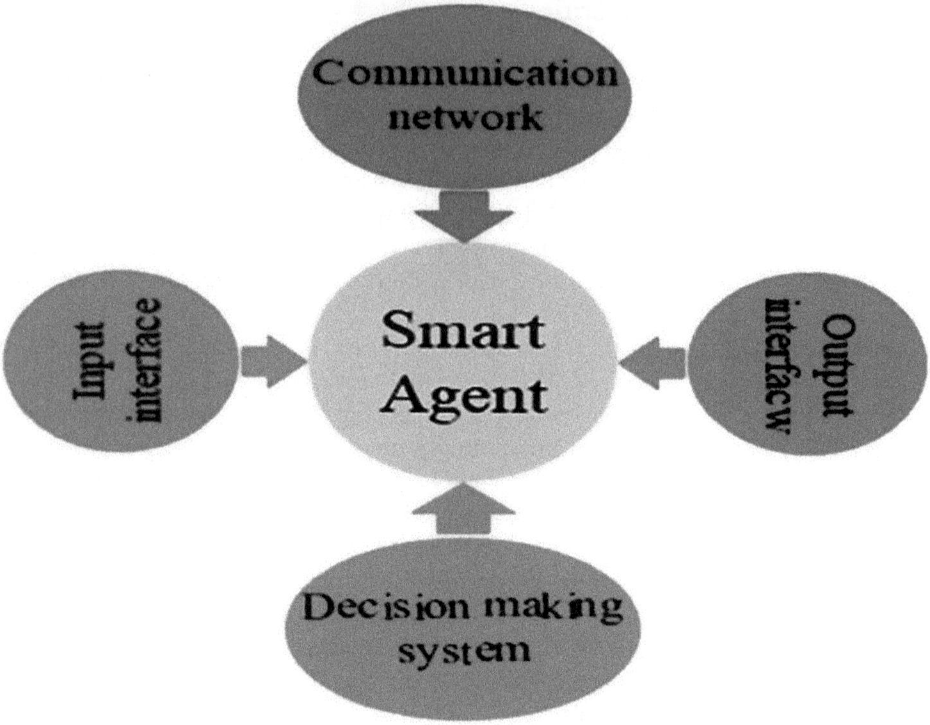

Figure 11.8 Smart agent model.

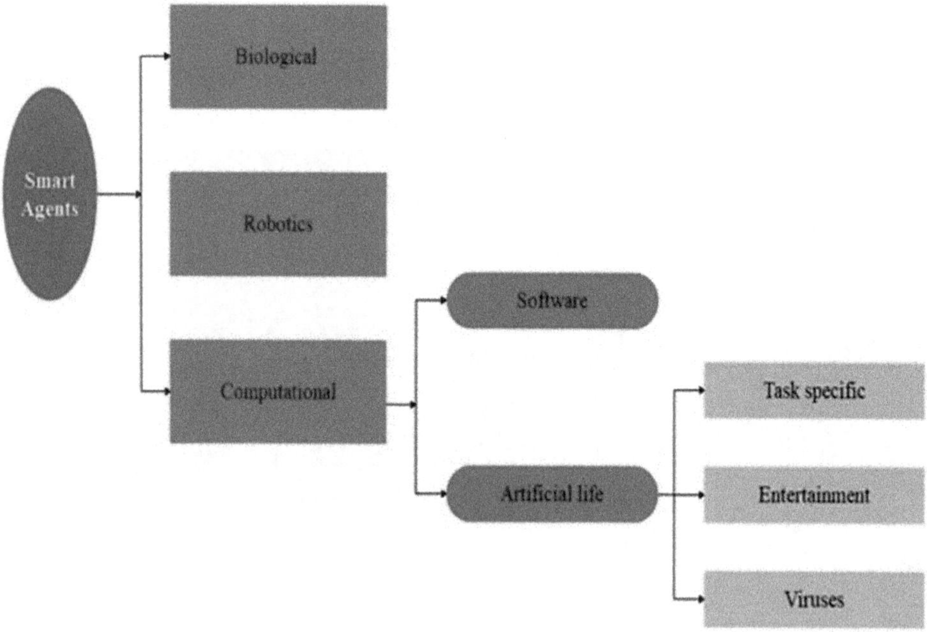

Figure 11.9 Classification of smart agents.

11.6.1 Smart Grid Agent-Based Modeling

The agent-oriented smart grid model will determine the effect of projected energy dynamics on the grid in operation. A smart grid focused on intelligent agents has two classes of agents, one for the electricity market problems, energy flow control, price allocation, and electricity scheduling, and the other on issues related to network protection, stability, fault management, and performance problems [42]. Agents could be named as device agents, distributed energy resource (DER) agents, consumer agents, intelligence control agents, intelligence response control agents, and graphical user interface (GUI) agents according to their application in the power system.

11.6.2 Role of Agents in the Smart Grid

11.6.2.1 Regulation of the Smart Grid

Enhancing the DER penetration will result in a reverse power flow that transforms the distribution network to become an active device based on power sources and loads with a current flow and voltage levels. For technological and economic challenges in the intelligent network real and reactive power variation has become a significant concern in the distribution network. Furthermore, Denmark has found a device operating solution focused on three distinct energy factors for distributed combined heat and power (CHP) production [43]. Due to distributed control, any generator in the microgrid may function independently. A multiagent framework for distributed control of microgrids may therefore be built [44].

Moreover, as the generation is progressively dispersed and controlled by multiple individuals, it is difficult to explicitly track each DG through the control order issued by the device operator. Therefore it needs re-regulation of the infrastructure for energy management [45]. In [46], the mobile agent is the foundation of a modern paradigm for the management of supply and demand balancing, which facilitates bilateral trading between customers.

11.6.2.2 Fault Management and Self-Healing

Many articles have already proposed fault handling of power systems using intelligent agents; for example, Tarhuni et al. [47] have detailed decentralized fault handling controls in the medium voltage (MV) network. State evaluation is addressed by a decomposition method [48], which requires a decentralized optimization process for solving the problem of state estimation. By coordination with the agent-based architecture, Lu et al. [49] present an assessment by the state of the power network. The introduction of intelligent agents in the security of the power structure has been illustrated in [50] by a modern strategy. In order to restore the power grid, Eriksson et al. [51] have built a smart agent model for the self-healing network by isolating the defective component. The technical facets of the multiagent system-managed power grid are concentrated in [52].

11.6.2.3 Energy Balance Management

It is a great challenge to manage energy production and usage any time electricity is to be used at the moment it is produced. By using intelligent energy management (IEM), the electricity market will be more mature, connecting the Regional Transportation Organization (RTO) with end consumers and a home power management system at the consumer level [53].

11.7 Integrating Renewable Energy Sources in a Smart Microgrid

Owing to its high uncertainty, green power integration in intelligent grids has been very complicated in real-time balancing energy demand and power output in the networks. Therefore, it is of considerable significance to forecast energy rates correctly in the short term. In recent decades, the conventional power grid has experienced major shifts that are also influenced by the advent of technologies such as solar as well as distributed electricity. Incorporating the effective renewable energy source (RES) framework in smart grids decreased transmission losses and (conceivably) the need for construction of new transmission lines. However, wind and solar RESs have a stochastic and erratic effect on the reliability of the energy grid. Since the distributed electricity output by a RES fluctuates and there are also difficulties in forecasting wind and solar production for the next day, the existence of variability and randomness of RES power in a current smart grid stability study has become a tough process. Smart grids should therefore be viewed as a significant obstacle in the efficient service of potential power grids to be resilient to these uncertainties [54].

The modern electricity system is a power web. Sources for efficiency, reliability, and economy are integrated into the mainstream system [55]. Tiny forms of electricity such as photovoltaic systems, fuel cells and wind turbines, electronic control equipment and controls, data and coordination, as well as monitoring networks, are also part of the scheme [56]. The distribution and storage of electric resources are rapidly playing a significant part in non-conventional renewable energies and systems (RESSs), such as not being limited to biomass, biogas, and geothermal energy. The goal is to build a robust electricity grid that lowers prices for remote power and encourages renewable and sustainable growth activities [57]. In addition, the incorporation of a limited and co-located output into a global power grid with a large share of renewable energy sources has been remedied for complications of power system quality [58]. All this is integrated through a distributed generation system to reduce environmental consequences and enhance supply stability. On the other side, the constant and autonomous usage of green energy technologies and devices raises threats for voltage variations and power variability, inconsistency rates, harmonics, inefficiencies, and other crucial issues. Small distributed energy sources can be incorporated with a smart grid framework into an urban network to manage them in real-time and to promote their connectivity with other infrastructures [59]. The grid incorporation of green energy supplies with all technological ramifications therefore plays a vital role in smart cities' sustainability growth. There are major roles in sustainable development of smart cities through the incorporation of grid energy resources with all the strategic consequences. It is important to build a control scheme that allows connection of new sources without disrupting current networks, allows rapid and smooth attachment and disconnection of the RESS from the main grid, and allows autonomous management of frequency and power for complex load demands [60].

Compared to external grids, the microgrid is a complex DG-based energy supply network with integration status, islanded or utility grid connected. More and more severe harmonic pollution occurs through the extensive use of electronic power systems. Micronetwork harmonics are inevitable because of distributed power production uncertainties. Prior to integration, it is crucial to resolve harmonic issues. Smart distribution networks can be an effective way to solve the problems, such as being able to resist cascading failures and accidents with blackout and to address the DG integration and wide-ranging blackout problems in order to increase the reliability and security of distribution networks. Two independent systems are difficult to adjust in frequency, voltage, or the angle of phase before integration to a complete conformity. A certain linkage between the two grids needs to be identified in order to ensure the most efficient and stable integration [61].

The smart grid combines the conventional electricity infrastructure of the twentieth century with the new twenty-first century telecommunications and IT technology. This interface facilitates the effective use of capital to maximize energy usage, deploy and maintain distributed sources, and exchange the power generated. A vast number of utilities worldwide have begun installing green energy sources, including solar and wind power, near consumptive locations. Residential owners have begun installing intelligent household equipment and green energy in their premises so they can produce and use electricity effectively. The grid operators face crucial challenges for balancing the supply and demand in the power sector as intelligent network capabilities expand. Grid operators must find ways to balance supply and consumption when taking into account the distinctive and stochastic behavior of intelligent grid services. In particular, the probability of future risks of potential losses due to non-optimal energy system planning leads to rising expenditure decisions. The goal of traditional integrated resource planning (IRP), subject to planning and organizational restrictions, is to reduce investment costs. In terms of interval stochastic, new energy sources under uncertainties are modeled. Uncertainties including load growth and price of fuel have been taken into consideration [62]. However, only the impacts of smart grid resources are considered in order to narrow the scope of this analysis to the influence of smart grid resources. A Benders's decomposition technique is used to simultaneously evaluate deterministic and stochastic parameters [63].

The smart grid deployment map is the starting point for the smart grid journey and any utility attempting to do this may not have the same smart grid technology in all jurisdictions. However, regardless of the existing base lines, strategic goals, and organizational capacities of a utility, the ideal way for addressing smart grid system integration is to evaluate the intelligent grid capabilities with respect to its core policy making and the interface requirements of the "input customer/control provider." In this analysis, the data collection, command and control, interface protocol, and communication specifications will be defined, and which domain those roles will belong to and with which layer they will be connected to or to reside will be discussed. The final goal must be to promote the deployment of the interconnected network domains for a utility in the intelligent grid integration map, to provide a distributed control system (using an intelligent agent system) to run through several utility network domains, to enable end-to-end connectivity, and to ensure sharing of information between all utilities [64].

Pudjianto et al. [65] presented the concept of a virtual power plant (VPP) and illustrate the incorporation of RERs into VPPs. Key developments have been explored recently in

[66] for the convergence of various forms of RERs. Specifically, the study focused on the vigorous scheduling/dispatch issue for multitime frames. Kowalczyk et al. [67] proposed the control mechanisms built to use renewable energy sources for a microgrid management system. The authors developed a load management control strategy and conducted a virtual analysis using MATLAB/Stimulant. In comparison, a new class of acidic, advanced lead batteries invented by the Commonwealth Science and Industrial Association, used the innovation ‘Ultra Battery’ to handle the instability and changing of electricity demands, as one possible solution for endogenous renewable power generation systems. This technology was designed to integrate microgrid systems with a storage system, offering many advantages, such as allowing secure and dispatchable power for renewable sources [68].

11.8 Modeling of Smart Microgrid Components

At the current time, the economic activities of the micronetwork with increasing incorporation of renewable energy has become a significant issue in the world of the smart grid. In various literature groups there are many meta-heuristic techniques of optimization available. The discovery of most fitting optimization strategies is one of the most challenging activities to reduce the expense of the microgrid. One of the toughest challenges to reduce microgrid costs is to choose the most effective optimization strategy.

11.8.1 Objective Function

11.8.1.1 Overall Cost

The following formulation illuminates how to articulate the economic dilemma of the intelligent microgrid. The overall expense of the intelligent microgrid is minimized by

$$\min C(X) = \sum_{t=1}^{NT} C_t + OM_{DG} + \text{CTCPD} + C_f \sum_{i=1}^{N} (aP_{\text{DiG}(i)}^2 + bP_{\text{DiG}(i)} + c) \tag{11.1}$$

where CTCPD is the cumulative total cost per day. The combined average daily cost of batteries reflects the amount of BES, BEV, and PHEV’s overall daily expense:

$$CTCPD = TCPD_{BES} + TCPD_{BEV} + TCPD_{PHEV} \tag{11.2}$$

where

$TCPD_{BES}$ is the complete cost every day of battery energy storage,

$TCPD_{BEV}$ is the total cost per day of battery electric vehicle,

$TCPD_{PHEV}$ is the complete cost every day of a plug-in hybrid electric vehicle, and C_t is the overview of the grid, with running and combustion costs for DG, BES, BEV, PHEV, and FC and MT and FCEV are start-up costs:

$$\begin{aligned} C_t = {} & \text{Cost}_{\text{grid,t}} + \text{Cost}_{\text{DG,t}} + \text{Cost}_{\text{BES,t}} + \text{Cost}_{\text{BEV,t}} + \\ & \text{Cost}_{\text{PHEV,t}} + \text{SUC}_{\text{FC,t}} + \text{SUC}_{\text{MT,t}} + \text{SUC}_{\text{FCEV,t}} \end{aligned} \tag{11.3}$$

where
DG is distributed generation,
BES is battery energy storage,
BEV is battery electric vehicle,
PHEV is plug-in hybrid electric vehicle,
FC is fuel cell,
MT is microturbine, and
FCEV is fuel cell electric vehicle.

11.8.1.2 Cost of the Grid Supply

Grid supply costs are described by

$$\mathrm{Cost}_{grid,t}\begin{cases}\mathrm{Bid}_{grid,t}P_{grid,t}\ \text{if}\ P_{grid,t}>0\\(1-\mathrm{tax})\mathrm{Bid}_{grid,t}P_{grid,t}\ \text{if}\ P_{grid,t}<0\\0\ \text{if}\ P_{grid,t}=0\end{cases}\tag{11.4}$$

where
$\mathrm{Bid}_{grid,t}$ is bid of utility and
$P_{grid,t}$ is power created by the grid.

11.8.1.3 DG Operation and Maintenance Cost

The continuous maintenance and running costs of the distributed generators are represented by

$$\mathrm{OM_{DG}}=\left(\mathrm{OM_{MT}}+\mathrm{OM_{FC}}+\mathrm{OM_{FCEV}}+\mathrm{OM_{PVi}}+\mathrm{OM_{WTi}}\right)\mathrm{NT}\tag{11.5}$$

where
$\mathrm{OM_{MT}}$ is operational and maintenance cost of microturbine.
$\mathrm{OM_{FC}}$ is operational and maintenance cost of fuel cell,
$\mathrm{OM_{FCEV}}$ is operational and maintenance cost of fuel cell electric vehicle,
$\mathrm{OM_{PVi}}$ is operational and maintenance cost of photovoltaic,
$\mathrm{OM_{WTi}}$ is operational and maintenance cost of wind turbine, and
NT is operating time in hours.

11.8.2 Constraints

The limit is a mathematical term that restricts or limits the variables of the problem. The following equations demonstrate all the constraints and conditions of the system considered.

11.8.2.1 Electrical Demand Balance

This balance demands that the overall power production should be equal to the load and device failure. The energy demand balance equation is described by

$$P_{\mathrm{grid},t} + P_{\mathrm{DiG},(i)} + P_{MT,t}u_{MT,t} + P_{FC,t}u_{FC,t} + P_{FCEV,t}u_{FCEV,t} + P_{BES,t}u_{BES,t} + P_{PHEV,t}u_{PHEV,t} + P_{BEV,t}u_{BEV,t} = P_{Demand,t},\ t = 1,2,\ldots,\ \mathrm{NT} \tag{11.6}$$

where

$P_{grid,t}$ is power created by the grid,

$P_{\mathrm{DiG},(i)}$ is power generated by diesel generator,

$P_{MT,t}$, $P_{FC,t}$, $P_{FCEV,t}$, $P_{BES,t}$, $P_{PHEV,t}$, and $P_{BEV,t}$ are the generated powers of microturbine, fuel cell, fuel cell electric vehicle, battery energy storage, plug-in battery electric vehicle, and battery electric vehicle, respectively,

$u_{MT,t}$, $u_{FC,t}$, $u_{FCEV,t}$, $u_{BES,t}$, $u_{PHEV,t}$, and $u_{BEV,t}$ are the on/off status values of microturbine, fuel cell, fuel cell electric vehicle, battery energy storage, plug-in battery electric vehicle, and battery electric vehicle, respectively, and

NT is operating time in hours.

11.8.2.2 Dispatchable DG Constraints

The power output limits of different DGs are defined by

$$\begin{aligned} P_{FC,min} \le P_{FC,t} \le P_{FC,max}, t = 1,\ldots,\mathrm{NT} \\ P_{MT,min} \le P_{MT,t} \le P_{MT,max}, t = 1,\ldots,\mathrm{NT} \\ P_{FCEV,min} \le P_{FCEV,t} \le P_{FCEV,max}, t = 1,\ldots,\mathrm{NT} \end{aligned} \tag{11.7}$$

where

$P_{FC,min}$, $P_{MT,min}$, and $P_{FCEV,min}$ are the minimum values of generated power for fuel cell, microturbine, and fuel cell electric vehicle, respectively.

$P_{FC,max}$, $P_{MT,max}$, and $P_{FCEV,max}$ are the maximum generated power values for fuel cell, microturbine, and fuel cell electric vehicle, respectively.

NT is operating time in hours.

11.8.2.3 BES Constraints

This is the maximum and minimum rate for charging and discharging battery energy storage.

11.8.2.3.1Discharging Mode of Battery Energy Storage

The mode of discharge is illustrated by

$$C_{BES,t+1} = \max\left\{\left(C_{BES,t} - \frac{\Delta t P_{BES,t}}{\eta_{discharge}}\right), C_{BES,min}\right\}, t = 1,\ldots,\mathrm{NT} \tag{11.8}$$

$$\underline{P_{BES,t}} \le P_{BES,t} \le \overline{P_{BES,t}}, t = 1,\ldots,\mathrm{NT} \tag{11.9}$$

where

$C_{BES,t}$ is energy stored in battery energy storage,

$C_{BES,\min}$ is minimum capacity of battery energy storage,
$\Delta t P_{BES,t}$ is multiplication of deviation in time Δt with capacity of the battery energy storage system $P_{BES,t}$,
$\eta_{discharge}$ is discharging efficiencies of various batteries,
NT is operating time in hours, and
$\underline{P_{BES,t}}$ and $\overline{P_{BES,t}}$, are maximum and minimum discharging rates of battery energy storage at time t.

11.2.3.2 Charging Mode of Battery Energy Storage

The charging mode is illustrated by

$$C_{BES,t+1} = \min\left\{\left(C_{BES,t} - \Delta t P_{BES,t}\eta_{discharge}\right), C_{BES,\max}\right\},\ t = 1,\ldots,\ NT \tag{11.10}$$

$$\underline{P_{BES,t}} \leq P_{BES,t} \leq \overline{P_{BES,t}},\ t = 1,\&.,\ \mathrm{NT} \tag{11.11}$$

where
$C_{BES,t}$ is energy stored in battery energy storage,
$C_{BES,max}$ is maximum capacity of battery energy storage,
$\Delta t P_{BES,t}$ is multiplication of deviation in time Δt with capacity of the battery energy storage system $P_{BES,t}$,
$\eta_{discharge}$ is discharging efficiency of various batteries,
NT is operating time in hours, and
$\underline{P_{BES,t}}$ and $\overline{P_{BES,t}}$, are maximum and minimum discharging rates of battery energy storage at time t, where

$$\overline{P_{BES,t}} = \min\left\{\frac{P_{BES,\max}\left(C_{BES,t} - C_{BES,\min}\right)\eta_{discharge}}{\Delta t}\right\},\ t = 1,\ldots,\ \mathrm{NT} \tag{11.12}$$

$$\underline{P_{BES,t}} = \max\left\{\frac{P_{BES,\min}\left(C_{BES,t} - C_{BES,\max}\right)}{\eta_{discharge}\Delta t}\right\},\ t = 1,\ldots,\ \mathrm{NT} \tag{11.13}$$

11.8.2.4 Grid Constraints

The supply of the grid should be within the defined limit presented by

$$P_{grid,min} \leq P_{grid,t} \leq P_{grid,max} \tag{11.14}$$

where
$P_{grid,t}$ is power generated by the grid and
$P_{grid,min}$ and $P_{grid,max}$ are the maximum and minimum power generated by the grid.

11.8.2.5 Diesel Generator Constraints

The diesel generator should generate electricity within the limit as defined by

$$0 \leq P_{\mathrm{Dig}(i)} \leq P_{\mathrm{Dig}(i)}^{max} \tag{11.15}$$

where
$P_{\text{DiG (i)}}$ is the power generated by diesel generator.

11.8.2.6 Battery Electric Vehicle (BEV) Constraints

These specify the maximum and minimum rates of BEV charging and discharging.

11.8.2.6.1 Discharging Mode of a Battery Electric Vehicle

The discharging mode of a BEV is illustrated by

$$C_{BEV,t+1} = \max\left\{\left(C_{BEV,t} - \frac{\Delta t P_{BEV,t}}{\eta_{discharge}}\right), C_{BEV,min}\right\},\ t = 1,\ldots.,\ \text{NT} \tag{11.16}$$

$$\underline{P_{BEV,t}} \le P_{BEV,t} \le \overline{P_{BEV,t}},\ t = 1,\&.,\ \text{NT} \tag{11.17}$$

where
$C_{BEV,t}$ is energy stored in battery electric vehicle,
$C_{BEV,min}$ is minimum capacity of battery electric vehicle,
$\Delta t P_{BEV,t}$ is multiplication of deviation in time Δt with capacity of the battery energy storage system $P_{BEV,t}$,
$\eta_{discharge}$ is discharging efficiencies of various batteries,
NT is operating time in hours, and
$\underline{P_{BEV,t}}$ and $\overline{P_{BEV,t}}$,are maximum and minimum discharging rates of battery electric vehicle at time t.

11.8.2.6.2 Charging Mode of a Battery Electric Vehicle

The charging mode of a BEV is illustrated by

$$C_{BEV,t+1} = \min\left\{\left(C_{BEV,t} - \Delta t P_{BEV,t}\eta_{discharge}\right), C_{BEV,max}\right\},\ t = 1,\ldots.,\ \text{NT} \tag{11.18}$$

$$\underline{P_{BEV,t}} \le P_{BEV,t} \le \overline{P_{BEV,t}},\ t = 1,\ldots.,\ \text{NT} \tag{11.19}$$

where
$C_{BEV,t}$ is energy stored in battery electric vehicle,
$C_{BEV,max}$ is maximum capacity of battery electric vehicle,
$\Delta t P_{BEV,t}$ is multiplication of deviation in time Δt with capacity of the battery energy storage system $P_{BEV,t}$,
$\eta_{discharge}$ is discharging efficiencies of various batteries,
NT is operating time in hours, and
$\underline{P_{BEV,t}}$ and $\overline{P_{BEV,t}}$, are maximum and minimum discharging rates of battery electric vehicle at time t, where

$$\overline{P_{BEV,t}} = \min\left\{\frac{P_{BEV,max}\left(C_{BEV,t} - C_{BEV,min}\right)\eta_{discharge}}{\Delta t}\right\},\ t = 1,\ldots.,\ \text{NT} \tag{11.20}$$

$$\underline{P_{BEV,t}} = \max\left\{\frac{P_{BEV,min}\left(C_{BEV,t} - C_{BEV,max}\right)}{\eta_{discharge}\Delta t}\right\},\ t = 1,\ldots.,\ \text{NT} \tag{11.21}$$

11.8.2.7 PHEV Constraints

These specify the maximum and minimum rates of PHEV charging and discharging.

11.8.2.7.1 Discharging Mode of a Plug-in Battery Electric Vehicle

The discharging mode of a PHEV is illustrated by

$$C_{PHEV,t+1} = \max\left\{\left(C_{PHEV,t} - \frac{\Delta t P_{PHEV,t}}{\eta_{discharge}}\right), C_{PHEV,\min}\right\},\ t = 1,\ldots.,\ \mathrm{NT} \tag{11.22}$$

$$\underline{P_{PHEV,t}} \le P_{PHEV,t} \le \overline{P_{PHEV,t}},\ t = 1,\ldots.,\ \mathrm{NT} \tag{11.23}$$

where

$C_{PHEV,t}$ is energy stored in plug-in battery electric vehicle,
$C_{PHEV,min}$ is minimum capacity of plug-in battery electric vehicle,
$\Delta t P_{PHEV,t}$ is multiplication of deviation in time Δt with capacity of the battery energy storage system $P_{PHEV,t}$,
$\eta_{discharge}$ is discharging efficiencies of various batteries,
NT is operating time in hours, and
$\underline{P_{PHEV,t}}$ and $\overline{P_{PHEV,t}}$, are maximum and minimum discharging rates of plug-in battery electric vehicle at time t.

11.8.2.7.2 Charging Mode of a Plug-in Battery Electric Vehicle

The charging mode of a PHEV is illustrated by

$$C_{PHEV,t+1} = \min\left\{\left(C_{PHEV,t} - \Delta t P_{PHEV,t}\eta_{discharge}\right), C_{PHEV,\max}\right\},\ t = 1,\ldots.,\ \mathrm{NT} \tag{11.24}$$

$$\underline{P_{PHEV,t}} \le P_{PHEV,t} \le \overline{P_{PHEV,t}},\ t = 1,\ldots.,\mathrm{NT} \tag{11.25}$$

where

$C_{PHEV,t}$ is energy stored in plug-in battery electric vehicle,
$C_{PHEV,max}$ is maximum capacity of plug-in battery electric vehicle,
$\Delta t P_{PHEV,t}$ is multiplication of deviation in time Δt with capacity of the battery energy storage system $P_{PHEV,t}$,
$\eta_{discharge}$ is discharging efficiencies of various batteries,
NT is operating time in hours,
$\underline{P_{PHEV,t}}$ and $\overline{P_{PHEV,t}}$, are maximum and minimum discharging rates of plug-in battery electric vehicle at time t, where

$$\overline{P_{PHEV,t}} = \min\left\{\frac{P_{PHEV,max}\left(C_{PHEV,t} - C_{PHEV,min}\right)\eta_{discharge}}{\Delta t}\right\},\ t = 1,\ldots.,\ \mathrm{NT} \tag{11.26}$$

$$\underline{P_{PHEV,t}} = \max\left\{\frac{P_{PHEV,min}\left(C_{PHEV,t} - C_{PHEV,max}\right)}{\eta_{discharge}\Delta t}\right\},\ t = 1,\ldots.,\ \mathrm{NT} \tag{11.27}$$

11.8.2.8 Operating Reserve Constraints

Reliability is accomplished through the acquisition of energy storage in smart microgrids, e.g. operating reserves, BES, and EVT. The addition of the standby generation power of switched on BES, EVTs, FC, MT, DiG, and Grid is the working reserve at each point of time. In less than 10 minutes, it may be supplied to the SMG and is described by

$$\begin{aligned}&\mathrm{P}_{\mathrm{grid},max}+\mathrm{P}_{\mathrm{DiG},(\mathrm{i})}+P_{MT,max}u_{MT,t}+P_{FC,max}u_{FC,t}+\overline{P_{FCEV,t}}u_{FCEV,t}+\overline{P_{BES,t}}u_{BES,t}+\\&\overline{P_{PHEV,t}}u_{PHEV,t}+\overline{P_{BEV,t}}u_{BEV,t}\geq \mathrm{OR}_{\mathrm{t}}+P_{Demand,t},t=1,2,\&.,\mathrm{NT}\end{aligned} \tag{11.28}$$

where

OR_t is essential backup generation in kW and

$P_{Demand,t}$ is demand power in kW at t step time.

11.9 Optimization Techniques

Currently, in the smart networks setting, the economic functioning of microgrids under enhanced incorporation of renewable energy has become a major issue. Renewable energy supplies are intermittently dependent on the atmosphere, rendering interconnections more complicated and challenging. Different energy storage facilities are used to address intermittent and weather-related problems, described above with clean energy goods. The implementation of the microgrid is therefore a perfect option for incorporation of renewable energy sources into an intelligent grid environment. It provides multiple benefits, such as minimizing energy loss, efficiency, and improving energy conservation. Therefore, existing trends for smart research include automated and self-organizing control systems, green energy prediction, microgrid stabilization by storage add-ons, or constructing DC microgrids to properly handle the existence of PV-derived energy and batteries.

Through implementing smart grid technologies, microgrid systems and a reliable economic power grid are also an important part of demand-side management through the implementation of distributed generation and production. Given the widespread implementation of microgrid technologies and a demand response, it has become an essential topic for utilities as to what activities should be carried out in the form of a demand response. An intelligent grid is a modern model of a grid that combines technologies for producing renewable energy supplies, an EMS system, and an infrastructure for transmission and distribution. Not only does this increase energy efficiency in a cleaner manner, but it also increases the power grid's stability and reliability. It can also significantly limit energy use and environmental impacts [69]. The interdisciplinary new science of operational research adopts the philosophy and special methods to find the best solution of deep technical and scientific challenges. The effort to find the right solution to a given problem is the method of optimizing. There are several solutions to the optimization problem and sometimes the criterion is to decide the right solution. The techniques of optimization aim to find the optimal or closest equivalent parameter(s) from the original parameter(s) to optimize the stated problem.

11.9.1 Overview of the Optimization Platform

Many optimization methods have been used to work with challenges of optimization and these strategies may be split into conventional mathematical optimization and sophisticated computer optimization techniques. To evaluate the optimum response, traditional approaches, such as direct and gradient-based strategies, use a deterministic methodology. The gradient-based optimization methods use gradient/derivative information in order to find the best solution. Linear programming is a linear objective function optimization approach, subject to linear equality and inequality constraints. The nonlinear program, subject to inequality restrictions, is concerned with nonlinear objective functions. Integer programming is a program that defines some or all of the variables as integers in mathematical optimization. The drawback of traditional approaches is that as the quest space size grows it becomes complex in locating the right value [70]. Because of its ease, its broad implementation, its time-saving and its global viewpoint, and computer intelligence, gradient-free optimization techniques have in many places been primarily used to solve the difficult issues. Heuristic and meta-heuristic algorithms may be categorized as intelligence algorithms. Researchers developed various meta-heuristic approaches to address the issue of smart grid reduction of costs. Two major classifications of population-dependent meta-heuristic optimization techniques include swarm intelligence (SI) and evolutionary algorithms (EA). A variety of well-known strategies for evolution are: the Evolution Approach, Genetic Algorithm, Differential Evolution, Evolution Programming, and several more. A number of techniques based of swarm intelligence include the following: Firefly, Shuffled Frog Leaping, Particle Swarm Optimization, Ant Colony Enhancing, Artificial Bee Colony, etc. In addition to the meta-heuristic optimization methods listed above, numerous natural methods based on natural phenomena are also available, such as Gravitational Search, Harmony Search, Flower Pollination, Biogeography-Based Optimization, etc.

11.9.1.1 Genetic Algorithm

The genetic algorithm (GA) is a global technique of quest focused on Darwin's evolutionary theory. Genetic algorithms are algorithms for optimization, searching, and learning influenced by natural and genetic advances. This method resembles the natural selection process, where the best people are picked for reproduction to create offspring for the next generation. They create offspring that inherit the traits of their parents and are in the next generation. If parents are stronger and the likelihood to live is stronger, the children may live even longer. This method goes on before the optimum condition is sought. Figure 11.10 displays the fundamental GA flow chart.

11.9.1.2 Particle Swarm Optimization

Particles reflect options that are feasible. PSO is founded on swarms and utilizes the best methods for a collection of particles. The best experience is then gained through particles and swarms, in which the algorithm learns the best position and pace for each particle. The particle movement is accompanied as the particles pass to the end criteria, which is an appropriate ideal answer or a certain amount of iterations. For each individual particle the best information is retained in the particle memory and the best worldwide information is recognized as the best global particle. The PSO limits are premature convergence and a fair

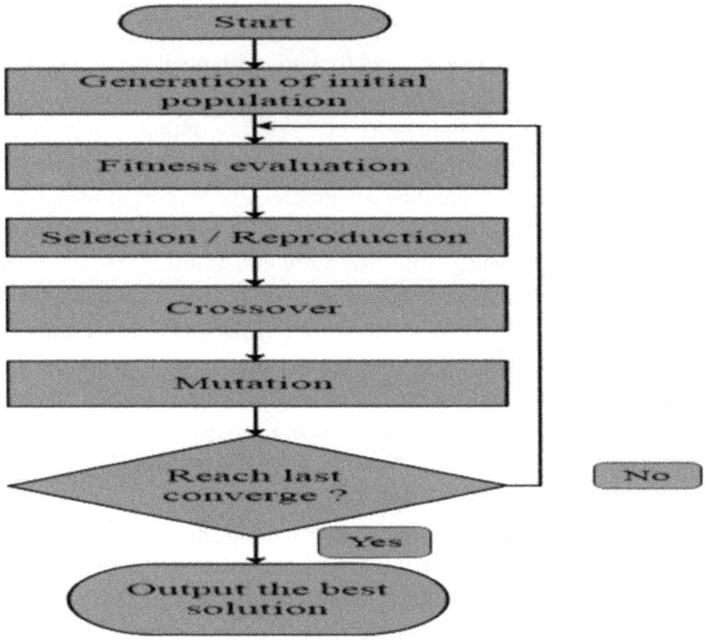

Figure 11.10 Basic flow chart of a genetic algorithm.

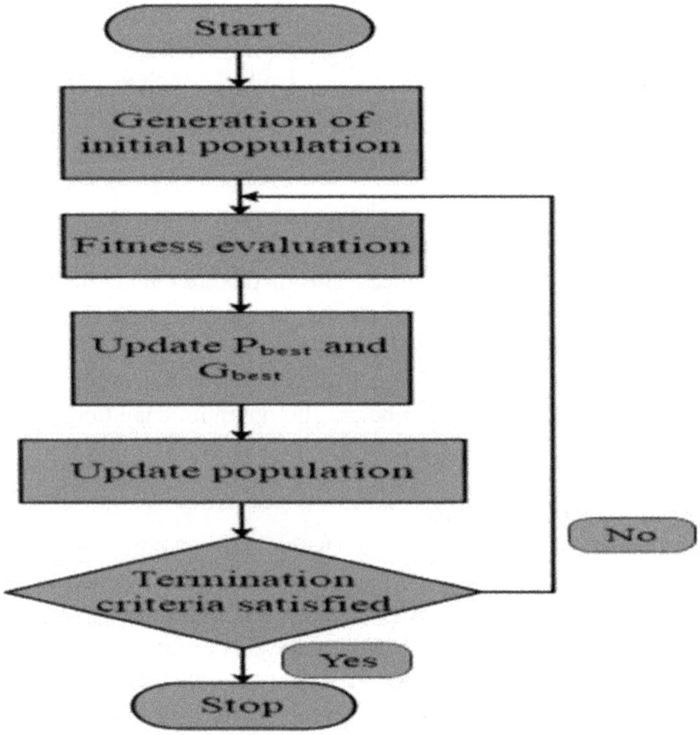

Figure 11.11 Basic flow chart of a particle swarm optimization.

risk of being trapped in an optimum locality. Figure 11.11 illustrates the basic PSO flow chart.

11.9.1.3 Harmony Search Algorithm

The harmony search algorithm (HSA) is a meta-heuristic system inspired by the population of the musical quest for an optimal state of harmony. The following measures are considered in this algorithm: initializing harmonic memory, improvising a new harmony, and upgrading the new memory. The harmonic database is a file that stores all the parameters and fitness values. Figure 11.12 illustrates the basic HSA flow chart.

11.9.1.4 Tabu Search Algorithm

The Tabu search algorithm (TSA) is a local-search meta-heuristic algorithm. Using motion, it looks for new ideas from the supplied adjacent system and solution. It substitutes the incumbent solution iteratively with the strongest next solution before the end conditions are fulfilled and the solution in the neighborhood is strengthened. Figure 11.13 illustrates the fundamental TSA flowchart.

Pei et al. suggested a modern method of microgrid interconnection and coordination that proves that microgrid groups will further optimize subsidiary networks by interconnecting microgrid groups [71]. In order to accomplish optimization schemes under

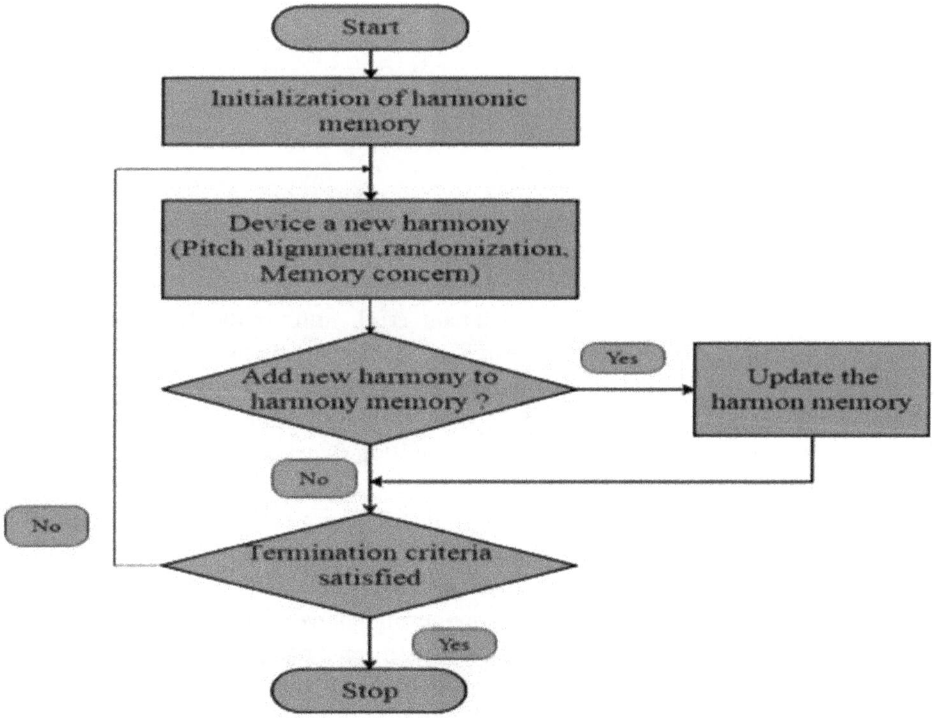

Figure 11.12 Basic flow chart of a harmony search algorithm.

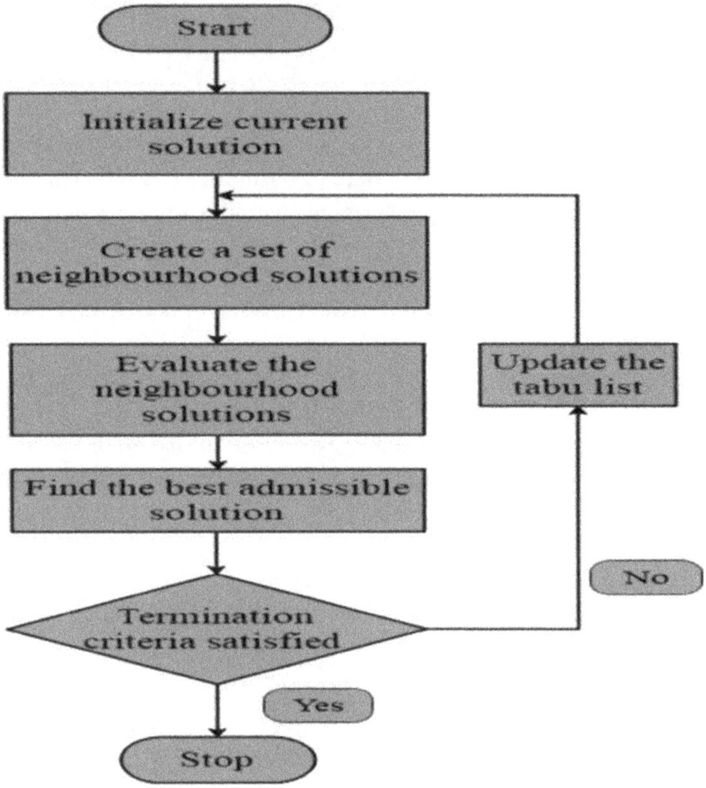

Figure 11.13 Basic flow chart of a Tabu search algorithm.

expensive and realistic constraints, Fu et al. developed a microgrid market optimization model taking into account the costs of infrastructure, operating, repair, and fuel, and environmental costs for conversion [72]. For the cost-efficient working of MG, which included solar PV, wind turbine, and fuel cell, Niknam et al. utilized the Honeybee Mating Optimization [73]. The comparative research on meta-heuristic microgrid optimization strategies was proposed in [7486]. Another powerful cost optimization approach is load balancing and control shifting. De Angelis et al. developed frameworks for minimizing energy costs using linear programming, where the strategy used rearranged the functions to reduce energy price periods [87].

In the perfect positioning and scale of DG in distribution networks, the meta-heuristic approach is often utilized. Figure 11.14 demonstrates the application of an overall optimization strategy to solve the optimum positioning and size of the DG problem.

The designs of two or more algorithms of optimization have superior benefits relative to individual implementation. Consequently, there has been a rising trend in hybrid deployment. Hybrid algorithms, however, can be the hybrid of two or more strategies of meta-heuristic hybridism or a combination of conventional and meta-heuristic approaches to optimization.

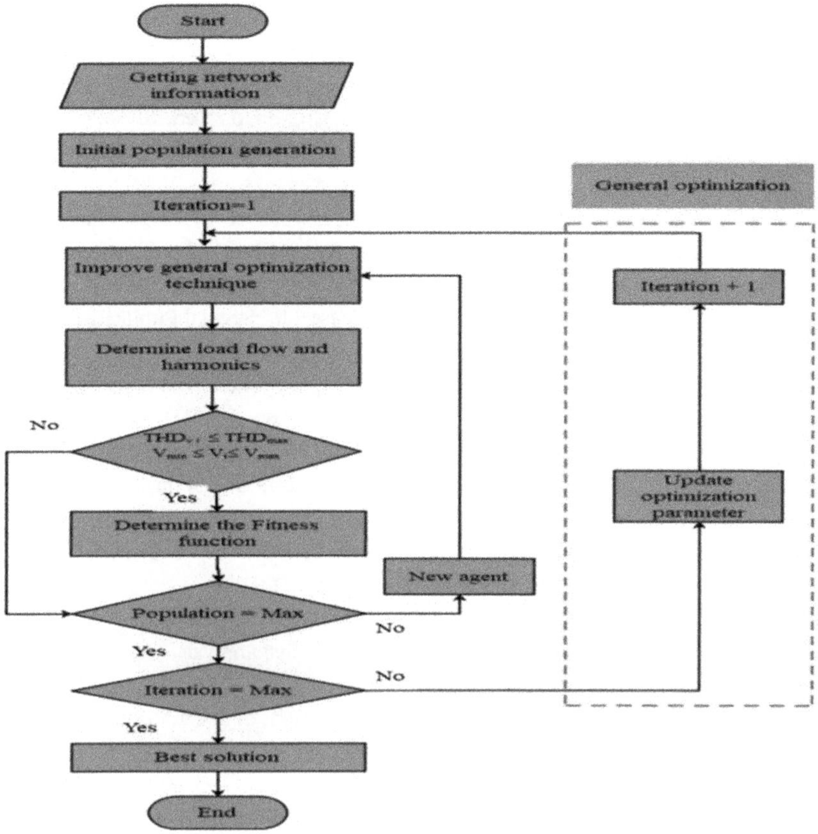

Figure 11.14 Basic flow chart for optimum position and size of the DG.

References

[1] H. Farhangi, "The path of the smart grid," *IEEE Power and Energy Magazine*, vol. 8, no. 1, pp. 18–28, January 2010.

[2] R.H. Lasseter, "Smart distribution: Coupled microgrids," *Proceedings of the IEEE*, vol. 99, no. 6, pp. 1074–1082, June 2011.

[3] Litos Strategic Communication, The Smart Grid: *An Introduction*. Washington, DC, 2008.

[4] H. Louie, M. Burns, and C. Lima, "An introduction and user's guide to the IEEE smart grid web portal," *IEEE PES Innovative Smart Grid Technologies Conference Europe (ISGT Europe)*, pp. 1–5, 2010.

[5] Y. Cunjiang, Z. Huaxun, and Z. Lei, "Architecture design for smart grid," *Energy Procedia*, vol. 17, pp. 1524–1528, 2012, doi: 10.1016/j.egypro.2012.02.276.

[6] G.M. Shafiiulla, A.M.T. Oo, D. Jarvis, A.B.M.S. Ali, and P. Wolfs, "Potential challenges: Integrating renewable energy with smart grid," *20th Australasian Universities Power Engineering Conferences* (AUPEC), pp. 1–6, 2010.

[7] A. Sharma, M. Kolhe, K.M.S.Y. Konara, K. Muddineni, A. Mudgal, S. Garud, and N. Ulltveit-Moe, "Performance assessment of building integrated photovoltaic and battery energy system: A case study of TERI-retreat faculty in India," *4th International Conference on Smart and Sustainable Technologies (SpliTech)*, Split, Croatia, pp. 1e5, 2019a.
[8] US Department of Energy, "The Smart Grid: An Introduction," Available: https://www.energy.gov/oe/downloads/smart-grid-introduction-0 [Accessed 6 June 2020].
[9] C. Greer, D.A. Wollman, D.E. Prochaska, P.A. Boynton, J.A. Mazer, C.T. Nguyen, G.J. FitzPatrick, T.L. Nelson, G.H. Koepke, A.R. Hefner Jr., V.Y. Pillitteri, T.L. Brewer, N.T. Golmie, D.H. Su, A.C. Eustis, D.G. Holmberg, and S.T. Bushby, "NIST framework and roadmap for smart grid interoperability standards, release 3.0," Special Publication (NIST SP) – 1108r3, NIST Pubs, October 1, 2014.
[10] J. Momoh, *Smart Grid: Fundamentals of Design and Analysis*, 1st ed. Hoboken, NJ: Wiley-IEEE Press, 2012.
[11] B. Khan and P. Singh, "Selecting a meta-heuristic technique for smart micro-grid optimization problem: A comprehensive analysis," *IEEE Access*, vol. 5, pp. 13951–13977, 2017.
[12] P. Singh and B. Khan, "Smart microgrid energy management using a novel artificial shark optimization," *Complexity*, vol. 2017, p. 22, 2017, Article ID 2158926.
[13] B. Khan and P. Singh, "Optimal power flow techniques under characterization of conventional and renewable energy sources: A comprehensive analysis," *Journal of Engineering*, vol. 2017, pp. 16, 2017, Article ID 9539506.
[14] Y. Kifle, B. Khan, and P. Singh, "Assessment and enhancement of distribution system reliability by renewable energy sources and energy storage," *Journal of Green Engineering*, vol. 8, no. 3(2), pp. 219–262, July 2018.
[15] O.P. Mahela, B. Khan, H.H. Alhelou, and P. Siano, "Power quality assessment and event detection in distribution network with wind energy penetration using stockwell transform and fuzzy clustering," *IEEE Transactions on Industrial Informatics*, vol. 16, no. 11, pp. 6922–6932, November 2020, doi: 10.1109/TII.2020.2971709.
[16] O.P. Mahela, B. Khan, H.Alhelou, and S. Tanwar, "Assessment of power quality in the utility grid integrated with wind energy generation," *IET Power Electronics*, vol. 13, no. 13, pp. 2917–2925, 14 October 2020, doi: 10.1049/iet-pel.2019.1351.
[17] T.F. Agajie, B. Khan, H.H. Alhelou, and O.P. Mahela, "Optimal expansion planning of distribution system using grid-based multi-objective harmony search algorithm," *Computers & Electrical Engineering*, vol. 87, pp. 106823, 2020.
[18] Y. Kifle, B. Khan, and J. Singh, "Designing and modeling grid connected photovoltaic system: (Case Study: EEU Building at Hawassa city)," *International Journal of Convergence Computing*, vol. 3, no. 1, pp. 20–34, 2018.
[19] T. Molla, B. Khan, B. Moges, H.H. Alhelou, R. Zamani, and P. Siano, "Integrated optimization of smart home appliances with cost-effective energy management system," *CSEE Journal of Power and Energy Systems*, vol. 5, no. 2, pp. 249–258, June 2019.
[20] G.S. Yogee, O.P. Mahela, K.D. Kansal, B. Khan, R. Mahla, H. Haes Alhelou, and P. Siano, "An algorithm for recognition of fault conditions in the utility grid with renewable energy penetration," *Energies*, vol. 13, pp. 2383, 2020.
[21] Ministry of Urban Development (MOUD), Government of India (GOI), "Smart city mission statement and guidelines," June 2015.

[22] W. Bower, D. Ton, R. Guttromson, S. Glover, J. Stamp, D. Bhatnagar, and J. Reilly, “The advanced microgrid integration and interoperability,” Sandia Report, SAND2014-1535, February 2014. Sandia National Laboratories, Albuquerque, New Mexico and Livermore, California, USA.
[23] S.N. Singh, V.S. Prathiba, and N. Katiki, “Smart micro grid model for Rural India,” Inn*ovative Engineering Technologies (ICIET2015)*, Bangkok, Thailand, 7–8 August 2015.
[24] F. Katiraei, R. Iravani, N. Hatziargyriou, and A. Dimeas, “Microgrids management,” *IEEE Power Energy Magazine*, vol. 6, no. 3, pp. 54–65, 2008.
[25] N. Hatziargyriou, H. Asano, R. Iravani, and C. Marnary, “Microgrids,” *IEEE Power Energy Magazine*, vol. 5, no. 4, pp. 78–94, 2007.
[26] D.E. Olivares, C.A. Canizares, and M. Kazerani, “A centralized energy management system for isolated microgrids,” *IEEE Transactions on Smart Grid*, vol. 5, no. 4, pp. 1864–1875, 2014.
[27] T.L. Vandoorn, J.C. Vasquez, J. De Kooning, J.M. Guerrero and L. Vandevelde, “Microgrids: Hierarchical control and an overview of the control and reserve management strategies,” *IEEE Industrial Electronics Magazine*, vol. 7, no. 4, pp. 42–55, December 2013.
[28] US Department of Energy, Office of Electric Transmission and Distribution, “Grid 2030: A national vision for electricity’s second 100 years,” July 2003. Available at: www.ferc.gov/eventcalendar/files/20050608125055-grid-2030.pdf.
[29] D. Trudnowski, M. Donnelly, and E. Lightner, “Power-system frequency and stability control using decentralized intelligent loads,” *Proceedings of IEEE PES Conference and Exhibition Transmission and Distribution*, pp. 1453–1459, 2005.
[30] US Department of Energy. Benefits of demand response in electricity markets and recommendations for achieving them. Report to US Congress, 2006.
[31] M.H. Albadi and E.F. El-Saadany, “A summary of demand response in electricity markets,” *Electric Power Systems Research*, vol. 78, pp. 1989–1996, 2008.
[32] Available: http://www.ferc.gov/industries/electric/indus-act/demand-response/demres-adv-metering.asp.
[33] M.S. Hoosain and B.S. Paul, “Smart homes: A domestic demand response and demand side energy management system for future smart grids,” *2017 International Conference on the Domestic Use of Energy (DUE)*, Cape Town, South Africa, pp. 285–291, 2017.
[34] T. Remani, E.A. Jasmin, and T.P.I. Ahamed, “Load scheduling problems under demand response schemes: A survey,” *2015 IEEE International Conference on Signal Processing, Informatics, Communication and Energy Systems (SPICES)*, pp. 1–5, February 2015.
[35] T. Logenthiran, D. Srinivasan, and T.Z. Shun, “Demand side management in smart grid using heuristic optimization,” *IEEE Transactions on Smart Grid*, vol. 3, no. 3, pp. 1244–1252, June 2012.
[36] M. Jaradat, M. Jarrah, Y. Jararweh, M. Al-Ayyoub, and A. Bousselham, “Integration of renewable energy in demand-side management for home appliances,” *2014 International Renewable and Sustainable Energy Conference (IRSEC)*, pp. 571–576, October 2014.
[37] N. Cicek and H. Delic, “Demand response management for smart grids with wind power,” *IEEE Transactions on Sustain Energy*, vol. 6, no. 2, pp. 625e34, 2015.
[38] M. Wooldridge, “Agent-based software engineering,” *IEE Proceedings – Software Engineering*, vol. 144, no. 1, pp. 26–37, February 1997.

[39] S. Franklin and A. Graesser, "Is it an agent, or just a program? A taxonomy for autonomous agents," *Proceedings of the Third International Workshop on Agent Theories, Architectures, and Languages*, Budapest, Hungary, Springer-Verlag, 1996.

[40] P. Maes, "Agents that reduce work and information overload," *Communications of the ACM*, vol. 31, no. 7, pp. 31–40, 1994.

[41] R.V. Guha and D.B. Lenat, "Enabling agents to work together," *Communications of the ACM*, vol. 37, no. 7, pp. 127–142, 1994.

[42] M. Shahidehpour and Y. Wang, Communication and Control in Electric Power Systems: Applications of Parallel and Distributed Processing. Hoboken, NJ: John Wiley & Sons, pp. 349–389 (Ch. 1 and Ch. 10), 2003.

[43] S.B. Ghosh, P. Ranganathan, S. Salem, T. Jingpeng, D. Loegering, and K.E. Nygard, "Agent-oriented designs for a self healing smart grid," *2010 First IEEE International Conference on Smart Grid Communications*, pp. 461–466, 2010.

[44] Z. Jian, A. Qian, J. Chuanwen, W. Xingang, Z. Zhanghua, and G. Chenghong, "The application of multi agent system in microgrid coordination control," *2009 International Conference on Sustainable Power Generation and Supply*, Nanjing, pp. 1–6, 2009.

[45] K. Kawamata, T. Tsuji, and T. Oyama, "A study on supply and demand control method based on mobile-agent technology in microgrid with renewable energy," *2011 International Conference on Proceedings of IEEE Clean Electrical Power (ICCEP)*, pp. 720–724, 14–16 June 2011.

[46] Y. Kusakabe, R. Hara, and T. Oyama, "A study in distributed autonomous control of distributed generations using the concept of token," *Proceedings of the International Conference on Electrical Engineering (ICEE)*, 2005 *(CDROM, DG1-07)*.

[47] N.G. Tarhuni, N.I. Elkalashy, T.A. Kawady, and M. Lehtonen, "Autonomous control strategy for fault management in distribution networks," *Electric Power Systems Research*, vol. 121, pp. 252–259, April 2015.

[48] X. Gonzalez, J.M. Ramirez, J.A. Marmolejo, and G. Caicedo, "Methodology for multi area state estimation solved by a decomposition method," *Electric Power Systems Research*, vol. 123, pp. 92–99, June 2015.

[49] Z. Lu, J. Zhang, T. Feng, and H. Cheng, "Distributed agent-based state estimation considering controlled coordination layer," *Electric Power Systems Research*, vol. 54, pp. 569–575, January 2014.

[50] J.A. Hossack, J. Menal, S.D.J. McArthur, and R.J. Mcdonald, "A multiagent architecture for protection engineering diagnostic assistance," *IEEE Transactions on Power Systems*, vol. 18, no. 2, pp. 639–647, May 2003.

[51] M. Eriksson, M. Armendariz, O.O. Vasilenko, A. Saleem, and L. Nordstrom, "Multi agent-based distribution automation solution for self-healing grids," *IEEE Transactions on Industrial Electronics*, vol. 62, no. 4, pp. 2620–2628, 2015.

[52] M. Nordman and M. Lehtonen, "An agent concept for managing electrical distribution networks," *IEEE Transactions on Power Delivery*, vol. 20, no. 2, pp. 696–703, April 2005.

[53] C. Ziyuan, D. Yizhou, Y. Ming, and M. Steurer, "A secure and distributed control network for the communications in smart grid," *2011 IEEE International Conference on Proceedings of IEEE Systems, Man, and Cybernetics (SMC)*, pp. 2652–2657, 9–12 October 2011.

[54] T.M. Lawrence, M.-C. Boudreau, L. Helsen, G. Henze, J. Mohammadpour, D. Noonan, and R.T. Watson, "Ten questions concerning integrating smart buildings into the smart grid," *Building and Environment*, vol. 108, pp. 273–283, 2016, doi: 10.1016/j.buildenv.2016.08.022.

[55] National Renewable Energy Laboratory (NREL). "Using distributed energy resources." Available at: http://www.nrel.gov.

[56] EPRI White Paper, "Integrating distribution resources into electric utility distribution system," *Technology Review*, 2011.

[57] P. Mazidi and G. Sreenivas, "Reliability assessment of a distributed generation connected distribution system," Presented at *The International Conference on Power System Operation and Energy Management*, Bangalore, India, pp. 1–7, 2013.

[58] M.H. Bollen and F. Hassan, Integration of Distributed Generation in the Power System. John Wiley & Sons, 2011.

[59] C.H. Lo and N. Ansari, "The progressive smart grid system from both power and communications aspects," *Communications Surveys & Tutorials, IEEE*, vol. 14, no. 3, pp. 799–821, 2012.

[60] Lasseter, R. Akhil, A. Marnay, C. Stephens, J. Dagle, J. Guttromson, and R.A.S. Meliopoulous, "Technical report. Consortium for electric reliability technology solutions," 2004.

[61] B. Chen, L. Cheng, Z. Wu, L. Lin, R. Wang, and Z. Chen, "Study on the reliability of integration between smart micro-grid and distribution network," *2012 Asia-Pacific Power and Energy Engineering Conference*, Shanghai, pp. 1–5, 2012, doi: 10.1109/APPEEC.2012.6307230.

[62] A. Ioannou, G. Fuzuli, F. Brennan, S.W. Yudha, and A. Angus, "Multi-stage stochastic optimization framework for power generation system planning integrating hybrid uncertainty modelling," *Energy Economics*, vol. 80, pp. 760–776, 2019.

[63] J.Á. López, K. Ponnambalam, and V.H. Quintana, "Generation and transmission expansion under risk using stochastic programming," *IEEE Transactions on Power Systems*, vol. 22, pp. 1369–1378, 2007.

[64] H. Farhangi, "A road map to integration: Perspectives on smart grid development," *IEEE Power and Energy Magazine*, vol. 12, no. 3, pp. 52–66, May–June 2014, doi: 10.1109/MPE.2014.2301515.

[65] D. Pudjianto, C. Ramsay and G. Strbac, "Virtual power plant and system integration of distributed energy resources," *IET Renewable Power Generation*, vol. 1, no. 1, pp. 10–16, March 2007.

[66] H. Chen, P. Xuan, Y. Wang, K. Tan, and X. Jin, "Key technologies for integration of multitype renewable energy sources – Research on multitimeframe robust scheduling/dispatch," *IEEE Transactions on Smart Grid*, vol. 7, no. 1, pp. 471–480, January 2016.

[67] A. Kowalczyk, A. Wlodarczyk, and J. Tarnawski, "Microgrid energy management system," *2016 21st International Conference on Methods and Models in Automation and Robotics, MMAR 2016*, pp. 157–162, 2016.

[68] J. Wood, "Integrating renewables into the grid: Applying ultra battery technology in MW scale energy storage solutions for continuous variability management," *2012 IEEE International Conference on Power System Technology (POWERCON 2012)*, pp. 1–4, 2012.

[69] M. Kumar, S.C. Srivastava, and S.N. Singh, "Control strategies of a DC micro-grid for grid connected and islanded operations," *IEEE Transactions on Smart Grid*, vol. 6, pp. 1588–1601, 2017.
[70] C. Kanzow, N. Yamashita, and T. Fukushima, "Levenberg–Marquardt methods with strong local convergence properties for solving nonlinear equations with con-vex constraints," *Journal of Computational and Applied Mathematics*, vol. 172, pp. 375–397, December 2004.
[71] W. Pei, Y. Du, H. Li, Y. Yang, W. Deng, and Z. Qi, "Novel solution and key technology of interconnection and interaction for large scale micro-grid cluster integration," *High Voltage Engineering*, vol. 41, pp. 3193–3203, 2015.
[72] Y. Fu, Y. Jiang, Z. Li, and C. Wei, "Optimal economic dispatch for micro-grid considering shiftable loads," *Proceedings of CSEE*, vol. 34, pp. 2612–2620, 2014.
[73] T. Niknam, A. Kavousifard, S. Tabatabaei, and J. Aghaei, "Optimal operation management of fuel cell/wind/photovoltaic power sources connected to distribution networks," *Journal of Power Sources*, vol. 196, no. 20, pp. 8881–8896, 2011.
[74] S. Sharma, S. Bhattacharjee, and A. Bhattacharya, "Grey wolf optimisation for optimal sizing of battery energy storage device to minimise operation cost of microgrid," *IET Generation, Transmission & Distribution*, vol. 10, pp. 625–637, 2016.
[75] Y. Sawle, S.C. Gupta, and A.K. Bohre, "Socio-techno-economic design of hybrid renewable energy system using optimization techniques," *Renewable Energy*, vol. 119, pp. 459–472, 2018.
[76] Y. Sawle, S.C. Gupta, A.K. Bohre, "A NovelMethodology for scrutiny of off-grid hybrid renewable system," *International Journal of Energy Research*, vol. 42, pp. 570-586, 2018.
[77] Y. Sawle, S.C. Gupta, and A.K. Bohre, "Review of hybrid renewable energy systems with comparative analysis of off-grid hybrid system," *Renewable and Sustainable Energy Reviews*, vol. 81, pp. 2217–2235, June 2017.
[78] Y. Sawle, S.C. Gupta, and A.K. Bohre, "PV-wind hybrid system: A review with case study," *Cogent Engineering Journal*, vol. 3, no. 1, pp. 1189305, 2016.
[79] Y. Sawle and S.C. Gupta, "Optimization of a standalone hybrid renewable energy system for telecom base station," *Progress in Petrochemical Science Crimson Journals*, vol. 2, no. 5, pp. 1–10, 2018.
[80] Y. Sawle and S.C. Gupta, "A novel system optimization of a grid-independent hybrid renewable energy system for telecom base station," *International Journal of Soft Computing, Mathematics and Control (IJSCMC)*, vol. 4, no. 2, pp. 49–57, May 2015.
[81] V. Tamrakar, S.C. Gupta, and Y. Sawle, "Single-diode and two-diode PV cell modeling using Matlab for studying characteristics of solar cell under varying conditions," *Electrical & Computer Engineering: An International Journal (ECIJ)*, vol. 4, no. 2, pp. 67–77, June 2015, ISSN: 2201-5957.
[82] V. Tamrakar, S.C. Gupta and Y. Sawle, "Single-diode PV Cell modeling and study of characteristics of single and two-diode equivalent circuit," *Electrical and Electronics Engineering: An International Journal (ELELIJ)*, vol. 4, no. 3, pp. 13–24, August 2015, ISSN: 2200-5846.
[83] Y. Sawle and M. Thirunavukkarasu, "Design, analysis and optimal sizing of standalone PV/diesel/battery hybrid energy system using HOMER, Recent Trends on Renewable

Energy, Smart Grid and Electric Vehicle Technologies," 9 July 2020, Organized by VIT Vellore (Accepted, Scopus).

[84] Y. Sawle, S.C. Gupta, and A.K. Bohre, "Optimal sizing of standalone PV/wind/biomass hybrid energy system using GA and PSO optimization technique," *1st International Conference on Power Engineering, Computing and Control, PECCON-2017, 2–4 March 2017, VIT University, Chennai Campus, Elsevier, Energy Procedia 00 (2017)*, vol. 117, pp. 690–698, 2017.

[85] Y. Sawle and S.C. Gupta, "Optimal sizing of photovoltaic/wind hybrid energy system for rural electrification," *Power India International Conference (PIICON), 2014 6th IEEE*, IEEE, pp. 1–4, 2014.

[86] V. Tamrakar, S.C. Gupta, and Y. Sawle, "Study of characteristics of single and double diode electrical equivalent circuit models of solar PV module," *2015 International Conference on Energy Systems and Applications*, Pune, pp. 312–317, 2015.

[87] F. De Angelis, F.M. Boaro, D. Fuselli, S. Squartini, F. Piazza, and Q. Wei, "Optimal home energy management under dynamic electrical and thermal constraints," *IEEE Transactions on Industrial Informatics*, vol. 9, no. 3, pp. 1518–1527, August 2013.

12

Control Algorithms for Energy Storage Systems to Reduce Distribution Power Loss of Microgrids

Yun Yang[1], *Siew-Chong Tan*[2], *and Shu-Yuen Ron Hui*[3]

[1] *Research Assistant Professor, Tang Ping Yuan Building CF615, Department of Electrical Engineering, The Hong Kong Polytechnic University, Hong Kong SAR, China and Honorary Research Assistant Professor, Department of Electrical and Electronic Engineering, The University of Hong Kong, Hong Kong SAR, China*
[2] *Professor, Chow Yei Ching Building 719, Department of Electrical and Electronic Engineering, The University of Hong Kong, Hong Kong SAR, China*
[3] *Chair Professor, Chow Yei Ching Building 601E, Department of Electrical and Electronic Engineering, The University of Hong Kong, Hong Kong SAR, China, and Department of Electrical and Electronics Engineering, Imperial College London, UK*

12.1 Introduction

Massive carbon emissions by traditional electric utilities with combustion of fossil fuels have stimulated investigations of renewable energy sources (RESs) in microgrids [1]. However, the definition of microgrids is not unified and its concept may vary across different areas. Nevertheless, the general consensus is that microgrids are (i) low-voltage (LV) power distribution networks in small-scale applications, such as a data center, a smart building, an island, and a suburban locality, etc. [2]; (ii) active power distribution networks that can be fully isolated from large power grids and operate in standalone modes [3]; and (iii) power grids that facilitate the integrations of distributed energy resources (DERs) and energy storage systems (ESSs) to guarantee sustainable and price-competitive electricity [4].

Typical microgrids are either AC microgrids, DC microgrids, or their hybrids. AC microgrids are known for their easy integrations to the existing power grids [5]. However, the control strategies of AC microgrids are more complicated than their DC counterparts [6–10]. DC microgrids can normally operate with higher efficiencies because of less power conversion stages [11–15]. Nevertheless, many components, such as the protection circuits, etc., of DC microgrids are more expensive than that of their AC counterparts [16].

In both AC and DC microgrids, due to the high penetrations of intermittent RESs, ESSs such as batteries, flywheels, and ultracapacitors are required in the system to achieve a real-time power balance [17]. Among them, battery storages are the most popular in emerging microgrid applications due to the fact that their power densities are relatively higher than other types of ESS [18]. They are usually installed and distributed to supplement DER with short-term overloads and low-voltage transients. Thus, batteries in microgrids are also named as distributed battery systems (DBSs). Generally, external power converters are required for a DBS to be integrated into microgrids [13]. However, some grid-connected converters are embedded to evolve the loads into smart loads (SLs) [19]. The emerging electric spring (ES) is incorporated in non-critical loads (NLs) to become SLs, which can be

Active Electrical Distribution Network: A Smart Approach, First Edition. Edited by Baseem Khan, Josep M. Guerrero, Sanjeevikumar Padmanaban, Hassan Haes Alhelou, Om Prakash Mahela, and Sudeep Tanwar.

used as an alternative ESS to an existing DBS. For ES installations, the loads are classified as NLs and critical loads (CLs). NLs include lighting systems, refrigerators, and water heating systems that can tolerate significant voltage fluctuations, while CLs are sensitivity loads that need compliance to strict voltage regulations, i.e. medical equipment.

In standalone microgrids, high distribution power loss is a prominent issue that can deteriorate efficient operations of microgrid systems [14, 20–24]. Not only will the distribution power loss degrade the power transfer efficiency, but it will also increase the cooling system costs for the excessive heat dissipation [25]. Particularly in emerging data centers, more than 50% energy loss is wasted on the transmission/distribution lines [26]. To this end, several researchers have investigated different control strategies to lower the distribution power loss of DERs [27–36]. Optimal dispatches of DER systems are adopted to reduce the distribution power loss based on power flow models [27–30]. Model-based optimization strategies are adopted for online optimization of distribution power loss of DER systems in [31–36]. A comprehensive review of microgrid optimization techniques with storage systems is presented in [37, 38]. This chapter presents four control algorithms, i.e. an adaptive differential evolution (ADE)-based hierarchical control [39], a particle swarm optimization (PSO)-based hierarchical control [40], and two predictive control-based hierarchical controls [41, 42] to lower the line loss of microgrids with DBS and ES.

For standalone DC and AC microgrids with DBS, the distribution power loss reductions can be modeled as optimization problems without multiple unknown system parameters. Therefore, heuristic algorithms like ADE and PSO are especially suitable for solving these problems. The adopted ADE algorithm gains the merit of overcoming the drawbacks of the conventional evolutionary algorithm, i.e. the Genetic Algorithm (GA) with (i) global optimum, (ii) lower randomness, (iii) fewer parameter tunings without depending on an external probability density function, and (iv) better exploration performance by dynamically adjusting the differential weigh and crossover rate. For the modeled DBS in standalone DC microgrids, the objective function comprising the terms of both power loss on the distribution lines and the deviations of bus voltages may result in multiple local optimal points. The ADE algorithm performs well in searching out the global optimal point with relatively fast convergence. The PSO algorithm is known for its ability to optimize multiple coupled parameters of a system with fast convergence and high reliability and accuracy. For the modeled DBS in standalone AC microgrids, multivariables are indirectly coupled in the objective function. Hence, the PSO algorithm is able to solve this issue. In regard to distributed ES in standalone DC and AC microgrids, the distribution power loss models are more explicit. Thus, model predictive control (MPC) or model-based predictive control can be used to optimize the objective functions. MPC is an intuitive control scheme that can be easily extended to multivariable systems by simply reconfiguring the objective function. It also has the advantage of easy incorporation of variable constraints. For ES in standalone DC microgrids, the power loss on the distribution lines and the deviations of bus voltages are explicit functions of bus voltages and line impedances. Two constraints of the weighting factor and the bus voltages are included in the objective function. Therefore, the designed centralized model prediction control (CMPC) can be used to optimize this objective function. A similar predictive control is designed for ES in standalone AC microgrids. Compared to the CMPC, the predictive control for AC microgrids is simplified, which can be implemented using an inexpensive digital controller of a power flow router (PFR).

12.2 ADE-Based Hierarchical Control for DBS in Standalone DC Microgrids

In this section, an ADE-based hierarchical control is presented to achieve power loss reduction of the distribution cables in standalone DC microgrids. The secondary layer of the hierarchical control based on the ADE algorithm provides bus voltage references for the local control in the primary layer.

12.2.1 Overview of ADE-Based Hierarchical Control

The circuitry of a DBS with the proposed ADE-based hierarchical control in a standalone DC microgrid is depicted in Figure 12.1. Based on V_{bus}, I_{line}, and state of charge (SoC), $V^{*}_{\text{ref}i}$ for the ith DBS can be calculated. The reference is further adopted by the conventional cascaded inner-loop controller to regulate the grid-connected converter. The flowchart of the ADE-based hierarchical control algorithm is given in Figure 12.2. $V_{\text{b}i}$, $I_{\text{b}i}$, and $T_{\text{b}i}$ of the ith battery pack are used to estimate SoCi(0). Based on the Coulomb-counting method, the SoC of the battery can be calculated as

$$\text{SoC}_i = \text{SoC}_i(0) - \frac{1}{C_{ei}} \int I_{\text{b}i} \, dt \tag{12.1}$$

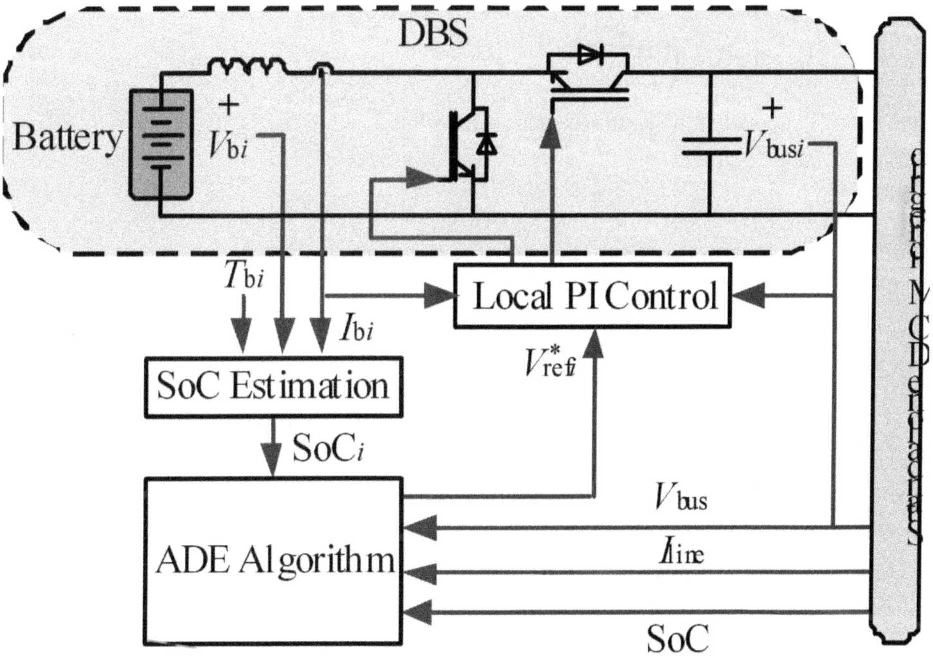

Figure 12.1 A typical DBS controlled by the ADE-based hierarchical control in a standalone DC microgrid.

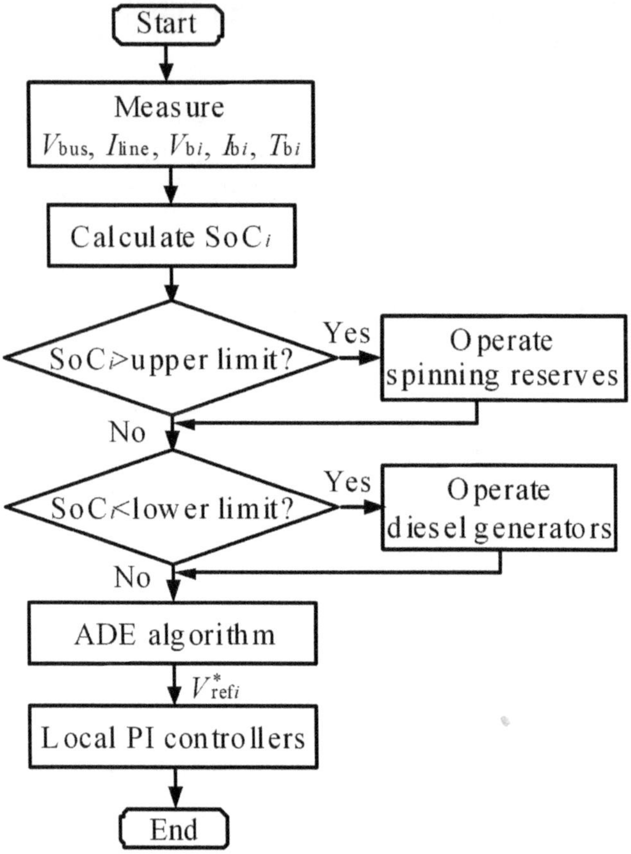

Figure 12.2 Flowchart of the proposed control algorithm.

where C_ei is the battery capacity. The SoC of the battery needs to be regulated within the upper and lower limits (i.e. $SoC_{max} = 90\%$ and $SoC_{min} = 20\%$).

12.2.2 Proposed ADE Algorithm

The ADE algorithm consists of initialization, iteration, and final stages. The pseudocode of the algorithm is given as follows.

ADE algorithm

Input: dimension dim(V_{bust}), population size P_{size}, crossover rate bound (P_{min}, P_{max}), mutation rate bound (F_{min}, F_{max}), maximum generation max_{gen}, maximum deviation rate max_{dr}

Output: optimal solutions V_{busi}, optimal fitness J

1. Begin
2. G←1 // g: generation number
3. **Initialization**

4. $V_{busi}(g)$ = initialization();
5. **Iteration**
6. While the terminal criteria not met do
   J = fitness ($v_{busi}(g)$);
7. //J: fitness of the objective function
   //**mutation operation**
8. $V_{busmi}(g)$ = mutation ($V_{busi}(g)$)
   J = fitness ($V_{busmi}(g)$)
9. //**crossover operation**
10. $V_{busti}(g)$ = crossover ($V_{busi}(g)$,$V_{busmi}(g)$)
    J = fitness ($V_{busti}(g)$)
11. //**selection operation**
12. $V_{busi}(g)$ = selection ($V_{busi}(g)$, $V_{busi}(g)$)
13. **End while**
14. **The final stage**
15. Output optimal solution V_{busi}; optimal fitness J
16. **End**

Details of the adopted ADE algorithm can be found in [39].

12.2.3 Simulation-Based Case Studies

The effectiveness of the ADE-based hierarchical control in distribution power loss mitigation is validated in a five-bus 48 V DC microgrid (as shown in Figure 12.3) using the RTDS.

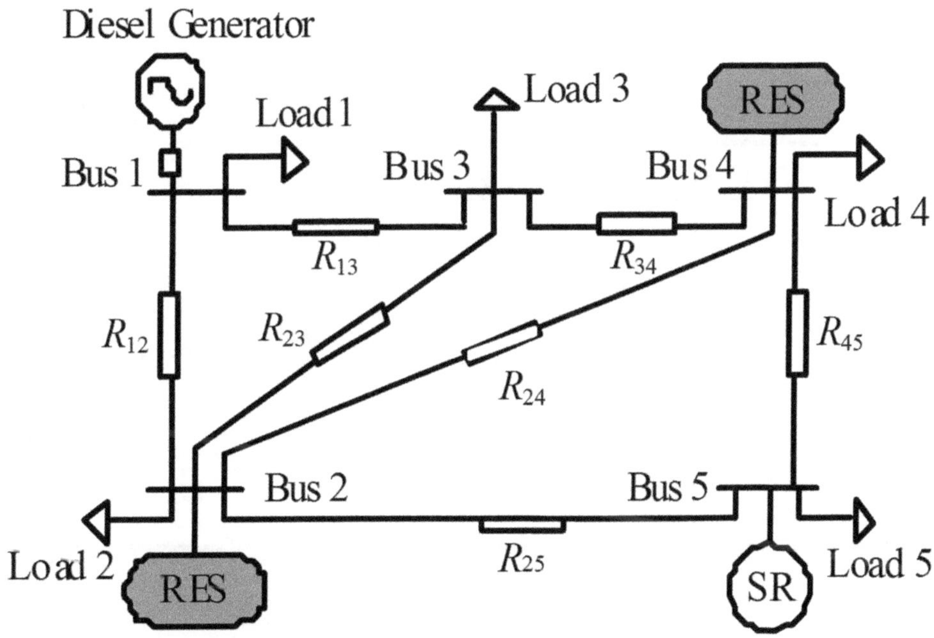

Figure 12.3 Topology of the five-bus 48 V DC microgrid.

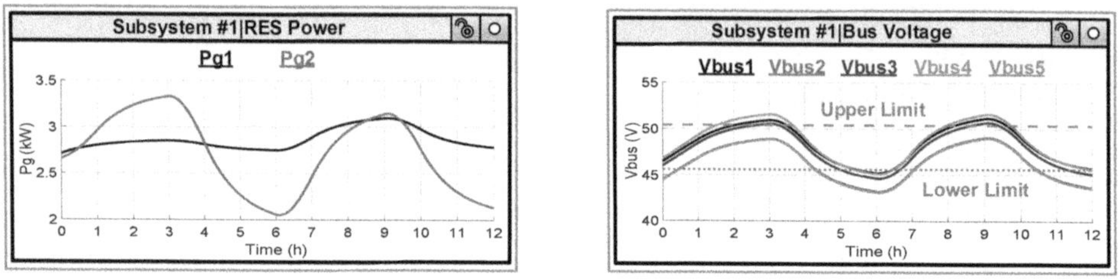

(a) (b)

Figure 12.4 Waveforms of RES power and bus voltages without DBS.

Two RESs are integrated at buses 2 and 4 of the microgrid. A diesel generator and spinning reserves are connected to the buses 1 and 5, respectively. The main parameters of the microgrid, DBS, and adopted controllers are provided in [39]. Without DBS, all the bus voltages will fluctuate when the two RES powers vary in a 12-hour period. The fluctuated power and voltages are presented in Figure 12.4. The bus voltages violate the limits.

Then, three DBS with the initial SoC of 0%, 40%, and 60% are connected to the buses 2, 4, and 5, respectively. An outer SoC control loop is included in the cascaded inner-loop local control to ensure that the SoCs of the batteries are within the boundaries. Figure 12.5 shows the bus voltage waveforms. Obviously, the bus voltages are well-regulated. Then, the ADE-based hierarchical control is applied. Figures 12.6 and 12.7 show the bus voltage waveforms and the SoCs of the batteries in 12 hours. Apparently, all the voltages and SoCs

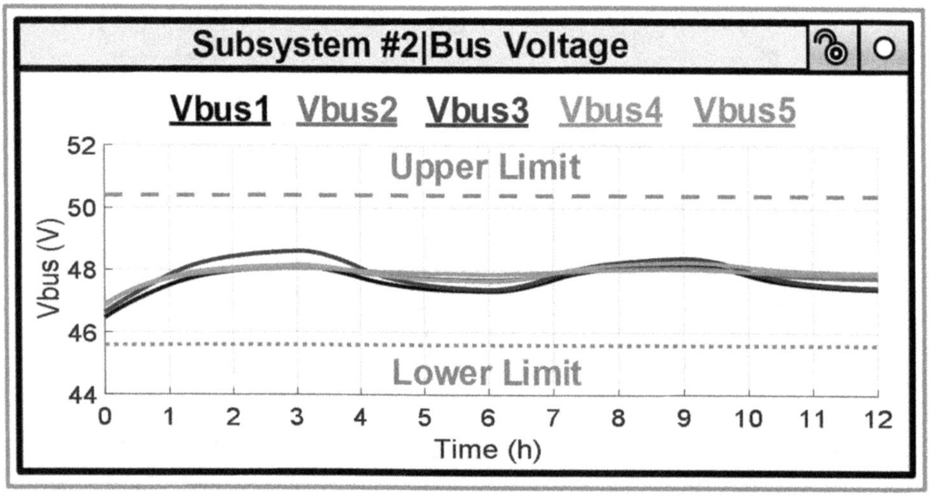

Figure 12.5 Bus voltage waveforms for the DBS controlled by the PI controllers.

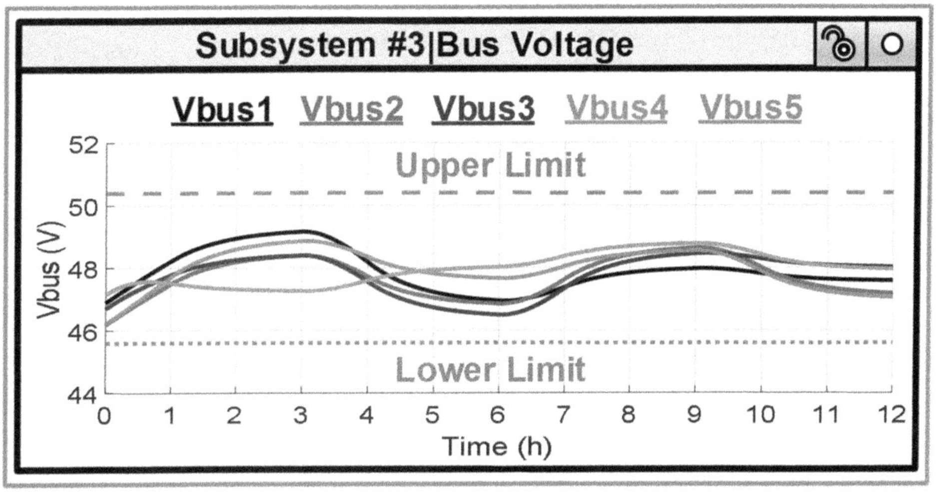

Figure 12.6 Bus voltages waveforms for the DBS controlled by the ADE-based hierarchical control.

are controlled within the limits. The comparative power loss waveforms between the conventional control (i.e. Ploss1) and the proposed ADE-based hierarchical control (i.e. Ploss2) are presented in Figure 12.8. The distribution energy loss over the 12-hour period is improved by about 29.85% by the proposed control (i.e. 2.35 kWh for the proposed control and 3.35 kWh for the conventional control).

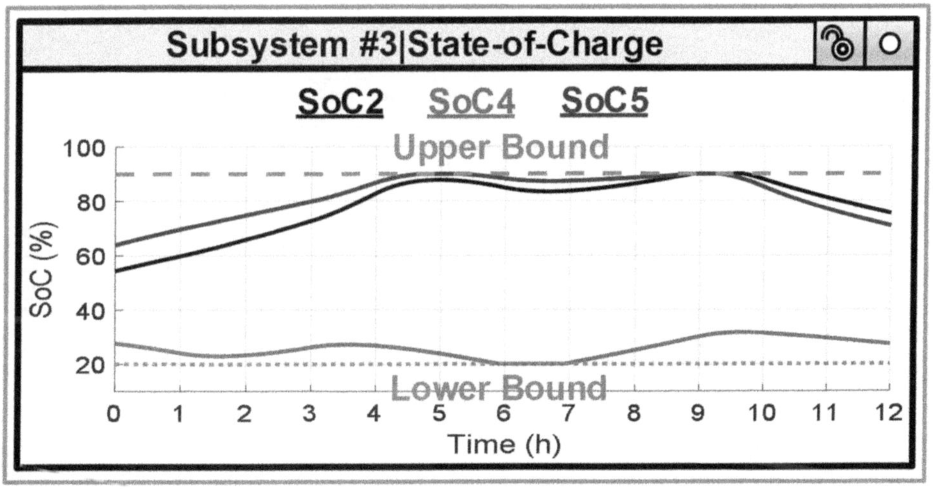

Figure 12.7 SoC of the batteries for the DBS controlled by the ADE-based hierarchical control.

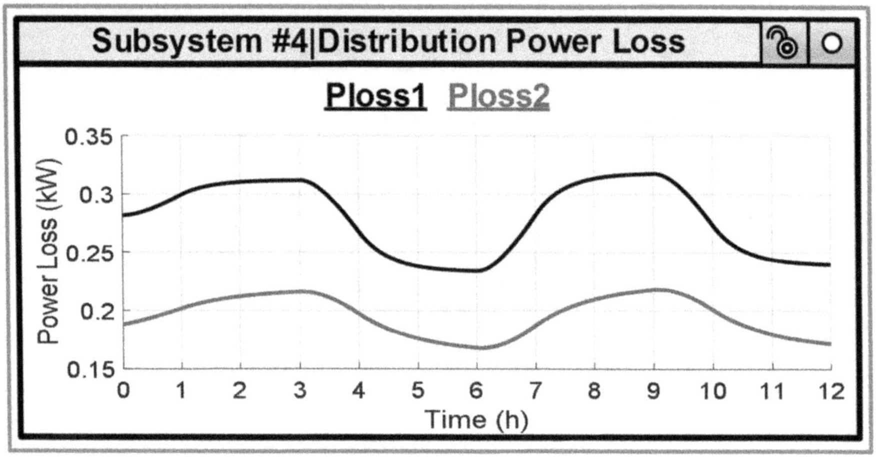

Figure 12.8 Power loss waveforms of the two control schemes.

12.3 PSO-Based Hierarchical Control for DBS in Standalone AC Microgrids

In this section, the discussion is on a PSO-based hierarchical control that is adopted in standalone AC microgrids to reduce both power loss on distribution lines and deviations of bus voltages. The block diagram of the PSO-based hierarchical control is shown in Figure 12.9.

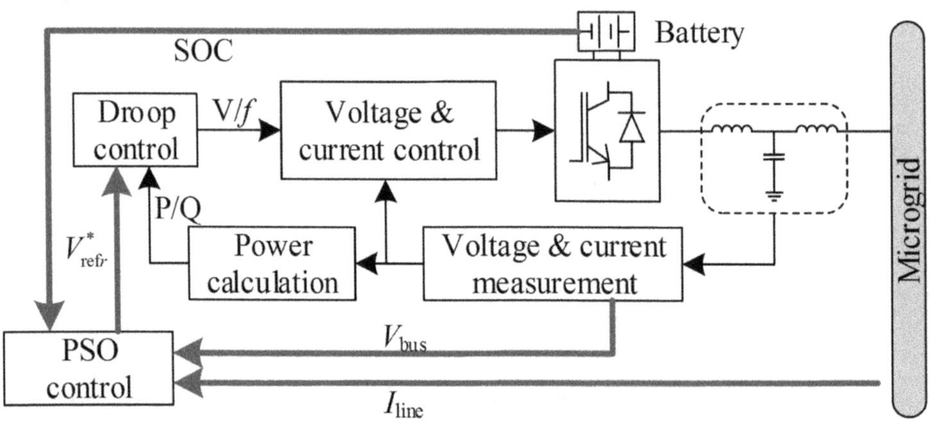

Figure 12.9 Block diagram of the PSO-based hierarchical control.

12.3.1 Overview of PSO-Based Hierarchical Control

The flow chart of the PSO-based hierarchical control is depicted in Figure 12.10. Here, r, v_{br}, i_{br} I_{ab}, and V^*_{refr} are the number, output voltage of DBS, output current of DBS, root-mean-square (RMS) value of the current from the bus a to the bus b, and the reference of the bus voltage. The SoC of the batteries can be estimated by the Coulomb-counting method in Eq. (12.1). The PSO algorithm comprises five steps. The first step is the initialization. During this step, initial positions of the particles are generated. The second step is the fitness function calculation. During this step, the fitness function is given as

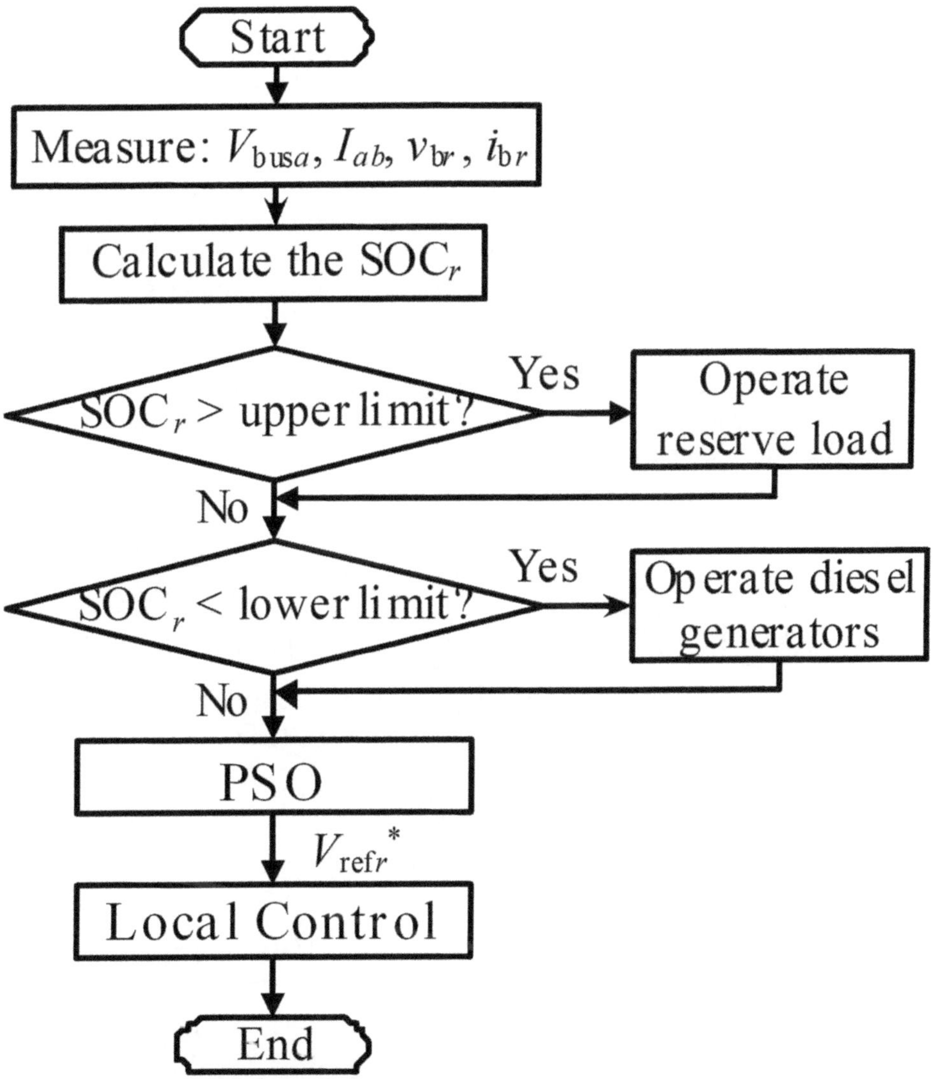

Figure 12.10 Flowchart of the PSO-based hierarchical control.

$$J = \alpha \sum_{a=1}^{m} \left| V_{\text{nom}} - V_{\text{busa}} \right| + (1 - a) \sum_{a,b=1}^{m} I_{ab}^{2} R_{ab} \tag{12.2}$$

The third step is to update inertia weight, velocity, and position. The fourth step is to find the best positions of particles and population and the optimal objective function. In the last step, the algorithm iterates from the second step to the fourth step until the stop criterion is satisfied. Details of the PSO algorithm can be found in [40].

12.3.2 Simulation-Based Case Studies

Simulation is conducted in a five-bus AC microgrid, as illustrated in Figure 12.11. A diesel generator, two RES, and reverse loads are connected to bus 1, buses 2 and 4, and bus 5, respectively. The main specifications of the AC microgrid and the PSO algorithm are provided in [40]. The PSO-based hierarchical control with different weighting factors (i.e. $\alpha = 0.9$, $\alpha = 0.15$, and $\alpha = 0.09$) are adopted for the DBS at buses 2 and 4. The corresponding bus voltages for $\alpha = 0.9$ are shown in Figure 12.12 (the bus voltage waveforms for $\alpha = 0.15$ and $\alpha = 0.09$ are shown in [40]). All bus voltages are controlled within the boundaries. Comparisons of the power loss on distribution lines and deviations of the bus voltages in the three cases are provided in Figure 12.13. The steady-state bus tracking performance $\alpha = 0.9$ is better than the other two cases. However, the power loss reduction of $\alpha = 0.09$ is more significant than the other two cases. For $\alpha = 0.15$, both the power loss on the distribution lines and voltage deviations of the bus voltages are mitigated. The SoCs of the BDS at buses 2 and 4 in a 12-hour period are presented in Figure 12.14. Apparently, the SoCs of the batteries are within the limits.

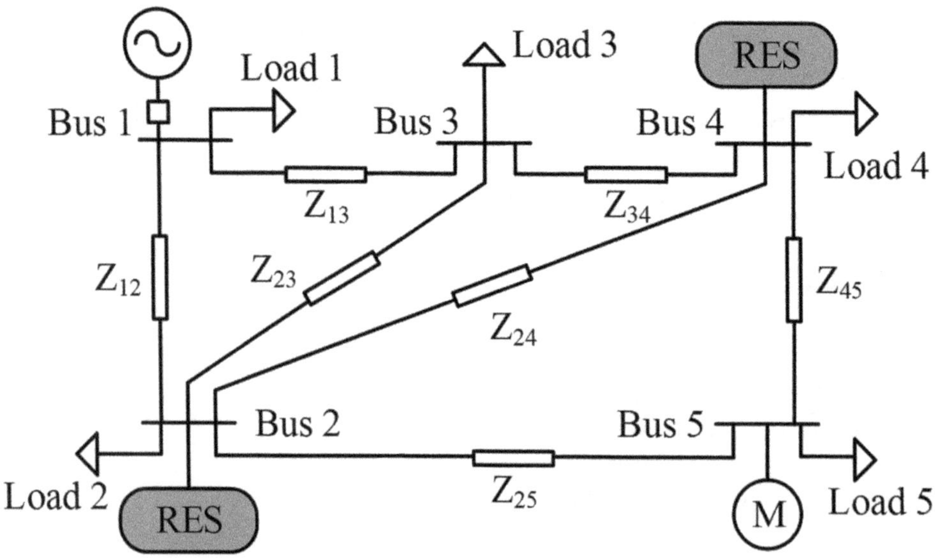

Figure 12.11 Structure of the five-bus AC microgrid.

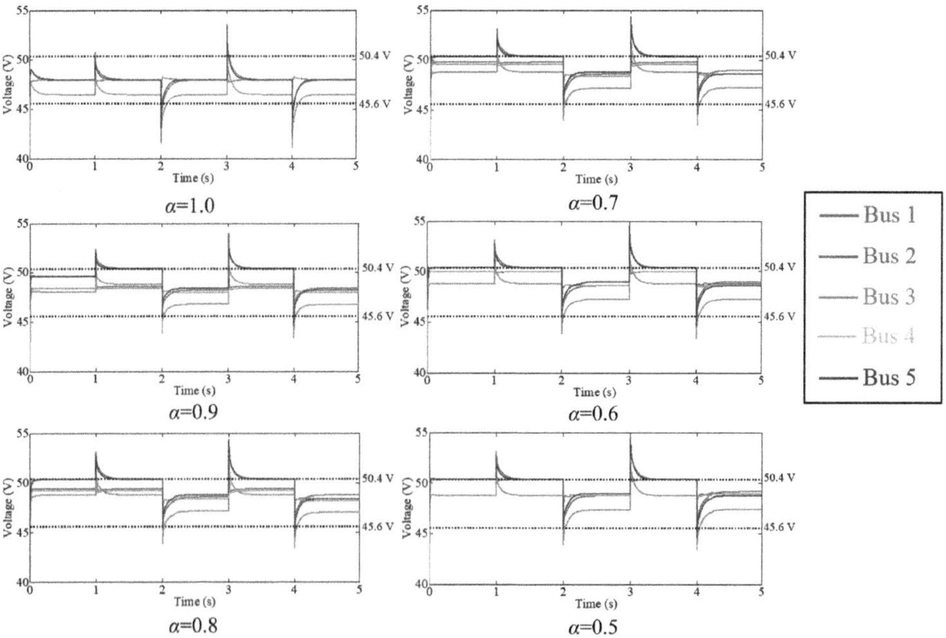

Figure 12.12 (a) bus voltages, (b) references, and (c) fitness values for the weighting factor $\alpha = 0.9$.

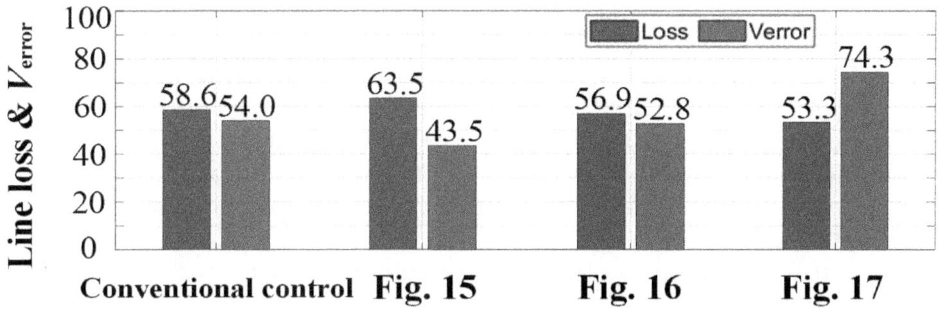

Figure 12.13 Comparative bar charts of the PSO control with different weighting factors.

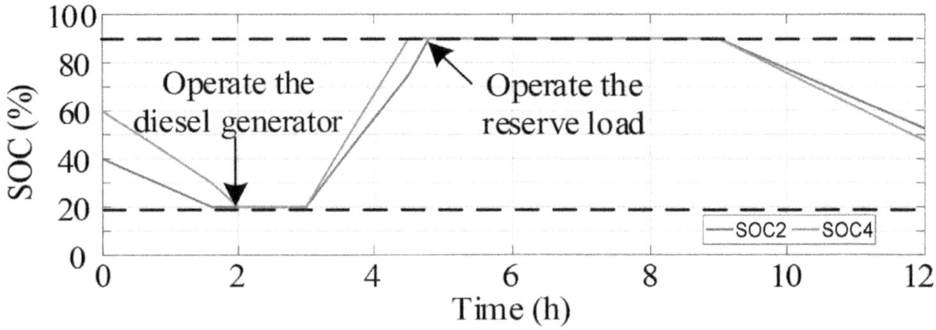

Figure 12.14 SoC of the batteries for the DBS at buses 2 and 4.

12.4 A CMPC-Based Hierarchical Control for ES in Standalone DC Microgrids

A CMPC with adaptive weighting factors is presented in this section. It is adopted to lower the power loss on the distribution lines of DC microgrids with the DCES. Details of the topologies, equivalent circuits, and local control strategies of the ES are provided in [41]. Simulation-based case studies are conducted to verify the proposed control scheme.

12.4.1 Overview of CMPC-Based Hierarchical Control

The flow chart of the proposed algorithm is depicted in Figure 12.15. Based on the measured and searched bus voltages, the distribution power loss at the next sampling time (i.e. $\hat{P}_{loss}$) is initially predicted. The bus voltages are searched by a fixed step change of $\Delta V'_{busi}$ from $(1-\eta)V_{nom}$ to $(1+\eta)V_{nom}$. Meanwhile, the bus voltage deviations are calculated by the term $\hat{V}_e^2$. Then, the objective function can be formed by summing $\hat{V}_e^2$, $\hat{P}_{loss}$, and the weighting factor α, which is furthered optimized to derive the optimal bus voltage references in several iterations. The weighting factor α is adaptive. Details of the algorithm of obtaining the adaptive weighting factor algorithm and the comprehensive control block diagram of the CMPC-based hierarchical control are shown in [41].

12.4.2 Simulation-Based Case Studies

Case studies are conducted on a 48 V five-bus DC microgrid without DBS (as shown in Figure 12.16), the specifications of which are provided in [41]. The sampling frequency in the simulation is 1 kHz. The bus voltage tolerance is ± 5%. Two RESs (i.e. P_{g1} and P_{g2}) are installed at buses 1 and 5. Four DCES are installed at buses 1, 2, 4, and 5. Initially, the weighting factor α is manually tuned from 1 to 0.5 with an interval of 0.1 without using the adaptive algorithm. (Note: $\alpha = 0.5$ is the minimum weighting factor.) The resulting bus

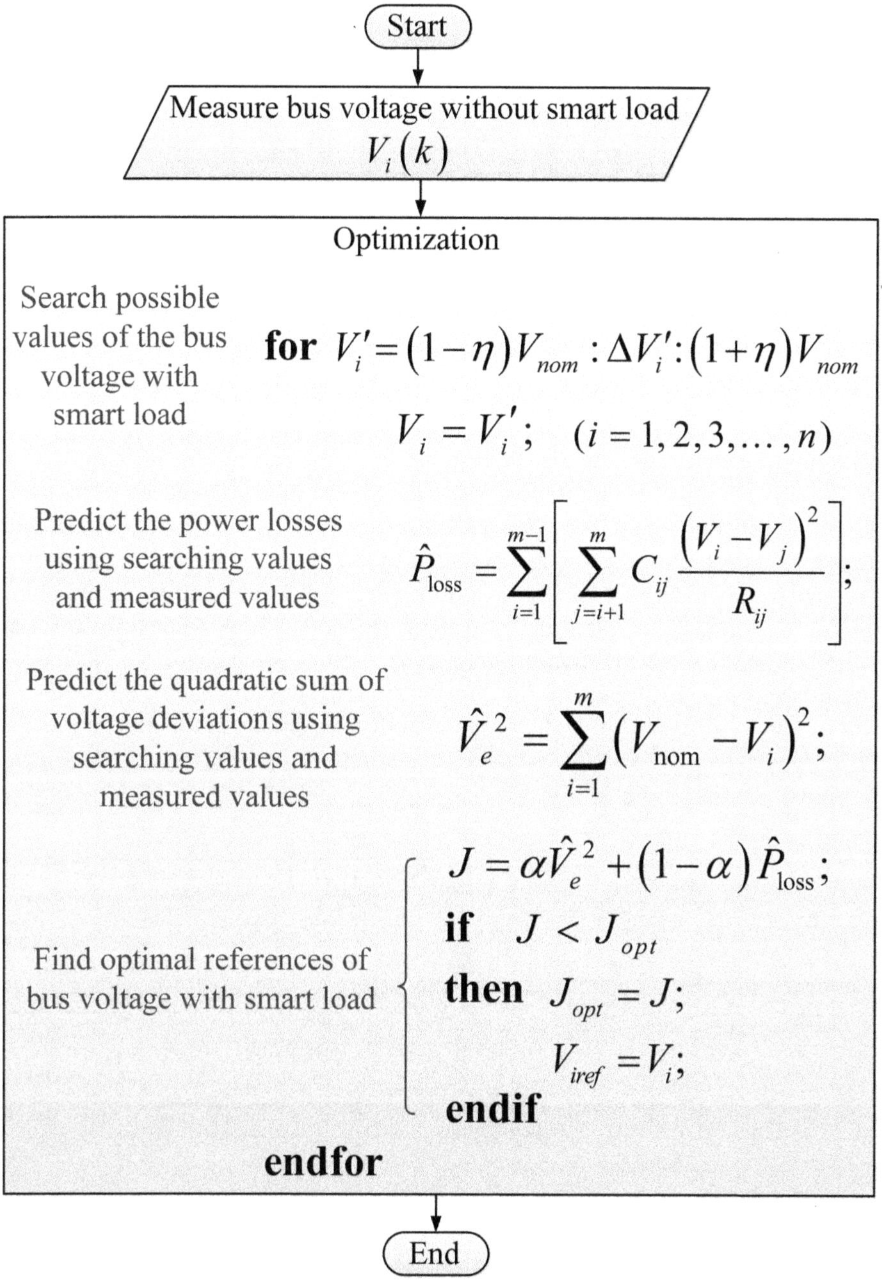

Figure 12.15 Flowchart of the CMPC algorithm.

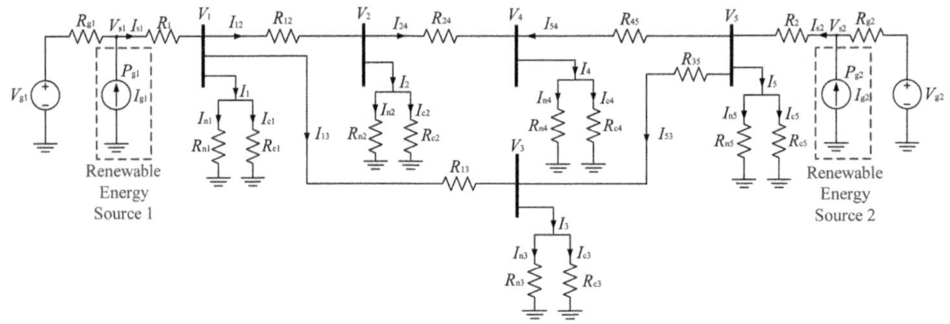

Figure 12.16 The 48 V five-bus DC microgrid without DBS in simulation.

voltages and the corresponding distribution power loss are shown in Figures 12.17 and 12.18. Apparently, the distribution power loss can be significantly reduced from $\alpha = 1$ to $\alpha = 0.9$. The reductions are still furthered but not that significant when α is changed from 0.9 to 0.8, 0.8 to 0.7, until 0.6 to 0.5. Compared to the conventional control strategy, the energy saving is about 58.5% when the weighting factor is 0.5. However, when the adaptive weighting factor algorithm is activated, the power loss on the distribution lines are further decreased. With the same RES power fluctuations, the weighting factor can be adaptively changed, as shown in Figure 12.19. The comparisons of the energy saving on the distribution lines between the CMPC with a non-adaptive weighting factor and the adaptive

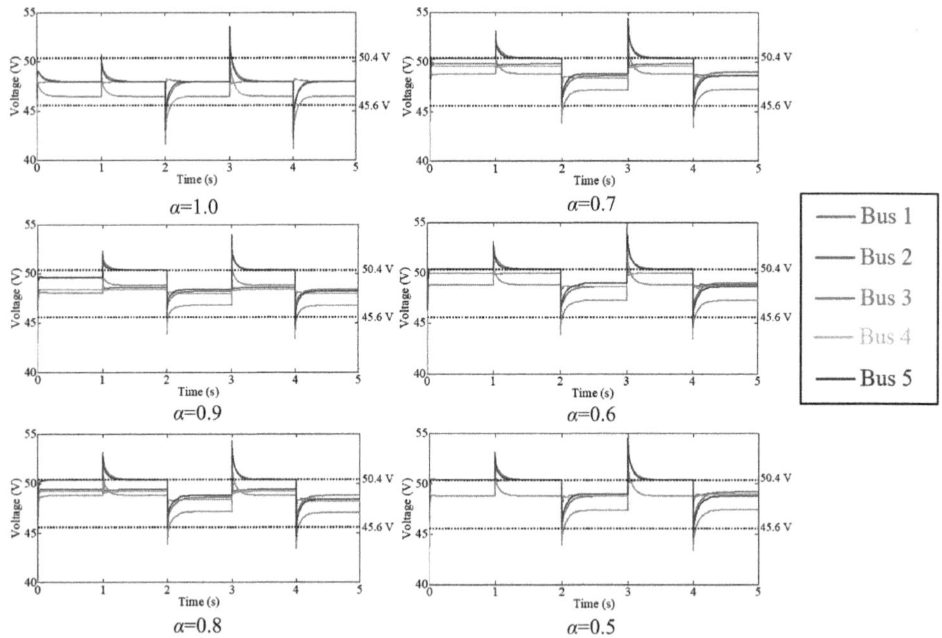

Figure 12.17 Waveforms of the bus voltages of the DC microgrid with four DCES.

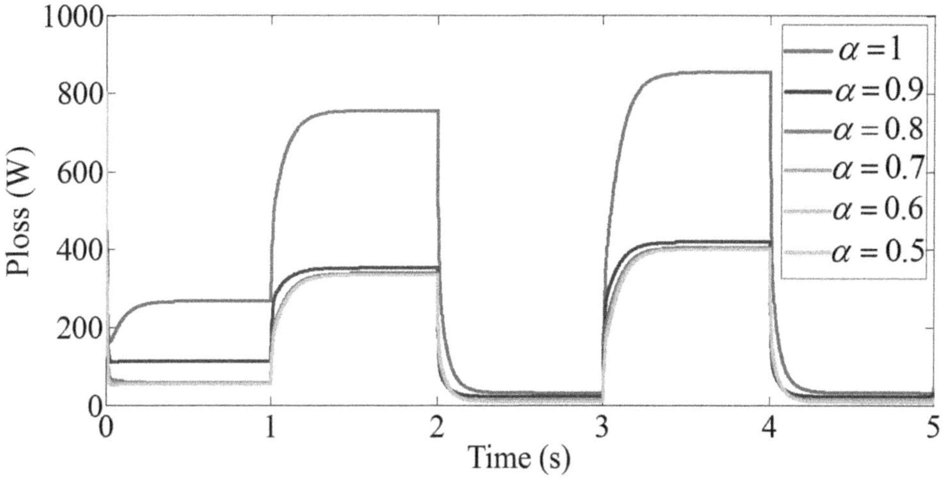

Figure 12.18 Distribution power loss for different weighting factors.

weighting factor are shown in Figure 12.20. More about 0.8% of energy can be saved by changing from the non-adaptive weighting factor to the adaptive weighting factor.

12.5 A Predictive Control-Based Hierarchical Control for ES in Standalone AC Microgrids

A predictive control-based hierarchical control is presented in this section to reduce the power loss on the distribution lines of standalone AC microgrids with distributed ESs. The adopted predictive control is simple and can be implemented in inexpensive PFR.

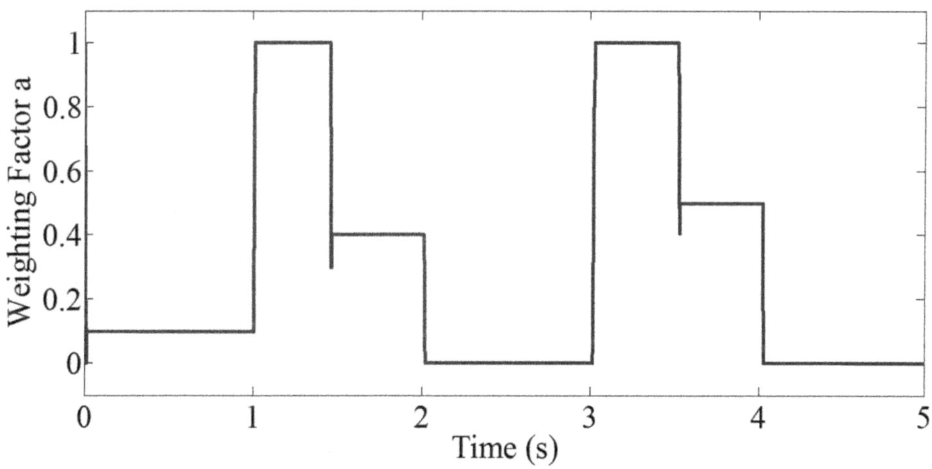

Figure 12.19 Adaptive weighting factor.

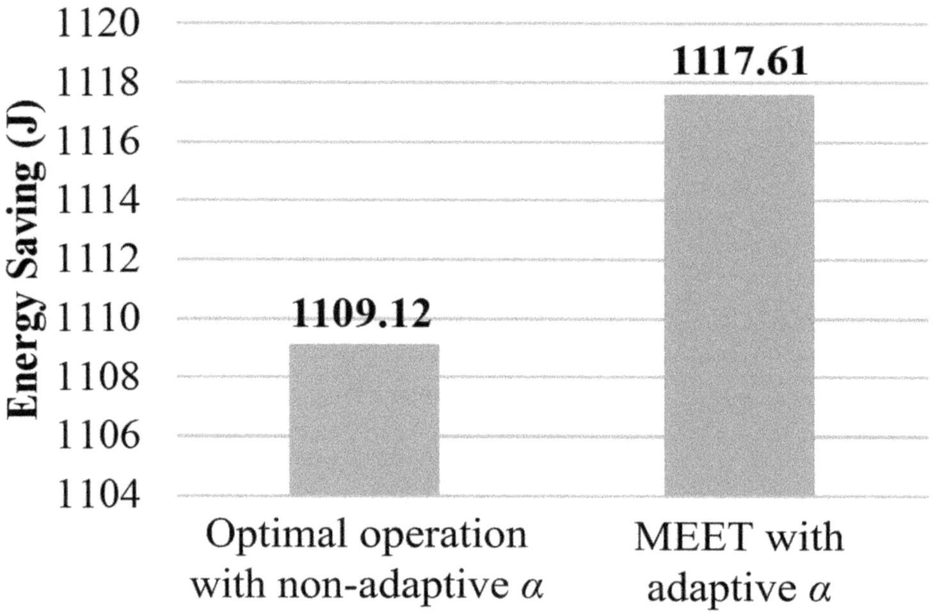

Figure 12.20 Comparisons of the distribution energy saving between the non-adaptive and adaptive weighting factors.

Experimental results validate the proposed control scheme to effectively regulate the bus voltages within the boundaries and simultaneously mitigate the distribution power loss of a 3-bus standalone AC microgrid.

12.5.1 Overview of Predictive Control-Based Hierarchical Control

The flow chart of the predictive control is similar to the CMPC in Figure 12.15. The objective function of the predictive control is

$$J=(1-\lambda)\sum\Delta\hat{P}_{ij}(k+1)+\lambda\sum\hat{V}_{\text{bei}}(k+1) \tag{12.3}$$

where $\sum\Delta\hat{P}_{ij}(k+1)$ and $\sum\hat{V}_{\text{bei}}(k+1)$ are the predicted power loss on the distribution lines and deviations of the bus voltages at the next sampling time. The weighting factor λ ($0 \le \lambda \le 1$) is automatically tuned by an adaptive weighting factor algorithm. The optimized bus voltage references are further adopted for a local ES to adjust the bus voltages. The comprehensive control block diagram is plotted in Figure 12.21. Here, M is the modulation signal. When ES operates in the inductive mode, M is positive ($M > 0$). Conversely, when ES operates in the capacitive mode, M is negative ($M < 0$). Details of the predictive control-based hierarchical control, including the adaptive weighting factor algorithm, are provided in [42].

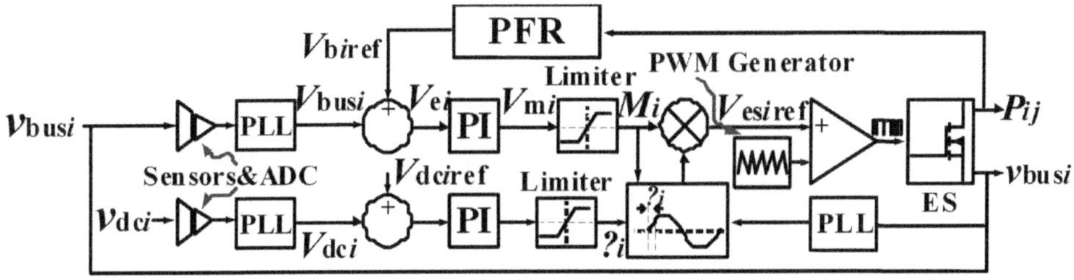

Figure 12.21 Comprehensive control block diagram of the predictive control-based hierarchical control.

12.5.2 Experiment-Based Case Studies

Experiments are carried out on a three-bus standalone 50 Hz AC microgrid with the nominal bus voltage of 220 V (RMS). The structure of the microgrid is shown in Figure 12.22. The parameters of the components in the microgrid and the controllers are provided in [42]. A Siglent SDM3045X digital multimeter is used to measure and store all the bus voltages.

The PV power fluctuates from the rated power of 2 kW during the period from 0 to 5 minutes, to 2.25 kW during the period from 5 to 11 minutes, to 1.82 kW during the period from 11 to 17 minutes, to 2.135 kW during the period from 17 to 23 minutes, and to 1.76 kW during the period from 23 to 28 minutes. When the ES is not activated, the bus voltage at bus 1 will exceed the bus voltage limits during the periods from 5 to 11 minutes and from 23 to 28 minutes. However, all the bus voltages can be controlled within the limits by activating ES. Figure 12.23 shows the waveforms of the bus voltage of bus 1 (i.e. V_{bus}), the output voltage of ES1 (i.e. V_{es}), and the voltage of NL1 (i.e. V_n) during the period from 5 to 11 minutes (in Figure 12.23a, without ES, $V_{bus}(\text{rms}) = 243\text{ V} > 242\text{ V}$, $V_{es}(\text{rms}) = 0$, $V_n(\text{rms}) = 242\text{ V}$; in Figure 12.23b, with ES, $V_{bus}(\text{rms}) = 220\text{ V} < 242\text{ V}$, $V_{es}(\text{rms}) = 110\text{ V}$, $V_n(\text{rms}) = 188\text{ V}$).

Then, the weighting factor λ is altered from 1.0 to 0.8. All three bus voltages are controlled within the boundaries. Figure 12.24 shows the measured bus voltages (at bus 1) for the two weighting factors. The comparisons of the power loss on the distribution lines for the two weighting factors are shown in Figure 12.25. The corresponding energy loss over the 28-min period can be further calculated. Due to the energy loss on the

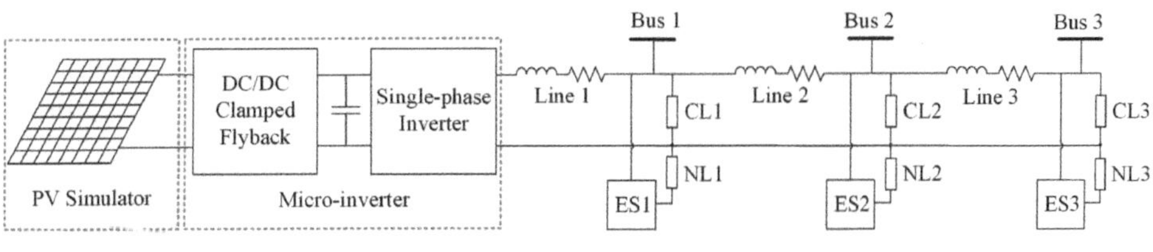

Figure 12.22 Architecture of the three-bus microgrid in the experiment.

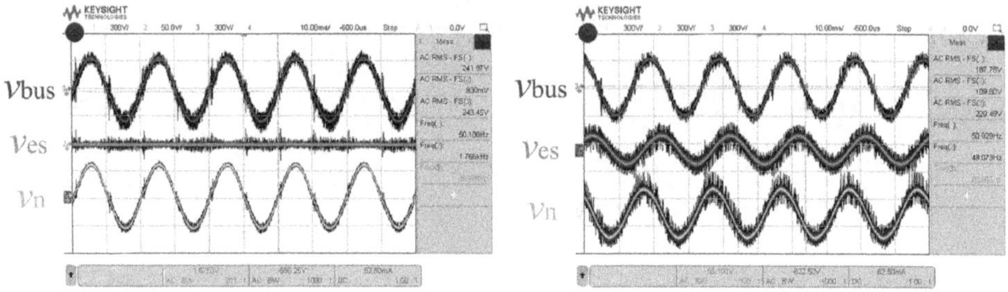

Figure 12.23 Bus 1 voltage waveforms without and with ES during the period from 5 to 11 minutes.

distribution lines, $\lambda = 0.8$ and $\lambda = 1$ are 16.27 kJ and 19.95 kJ, respectively. The energy loss can be mitigated about 17.94%.

12.6 Summary

The distribution power loss of microgrids will not only degrade the power transmission efficiency but also increases the expense of the cooling system. This chapter presents four hierarchical control strategies that can be adopted to mitigate the power loss on the distribution lines of both standalone AC and DC microgrids with the DBS and ES. For the DBS in DC microgrids, the distribution power loss model is complex and the objective function may have multiple local optimal solutions. Therefore, an ADE algorithm with the merits of a global optimum and less tuning parameters is adopted as the secondary control for the local control of DBSs. Compared to the conventional single-layer control without power loss optimization, the ADE-based hierarchical control can save more than 29% of distribution energy loss. For DBS in AC microgrids, multivariables are indirectly coupled in the objective function. A PSO algorithm can optimize the objective function with high accuracy and fast convergence. The PSO-based hierarchical control saves more than 9% distribution power loss than the conventional droop control without power loss optimization. For ESs in DC and AC microgrids, due to the distribution power model, which can be modeled based on the line impedances and bus voltages, a CMPC and a predictive control are designed with consideration of variable constraints. The CMPC-based hierarchical control saves more than 58% than the conventional single-layer control without power loss optimization. The algorithm of the predictive control in AC microgrids is simpler than that of the CMPC in DC microgrids. Compared to the conventional droop control without power loss optimization, the predictive control-based hierarchical control is validated to save more than a 17% distribution energy loss in a practical setup.

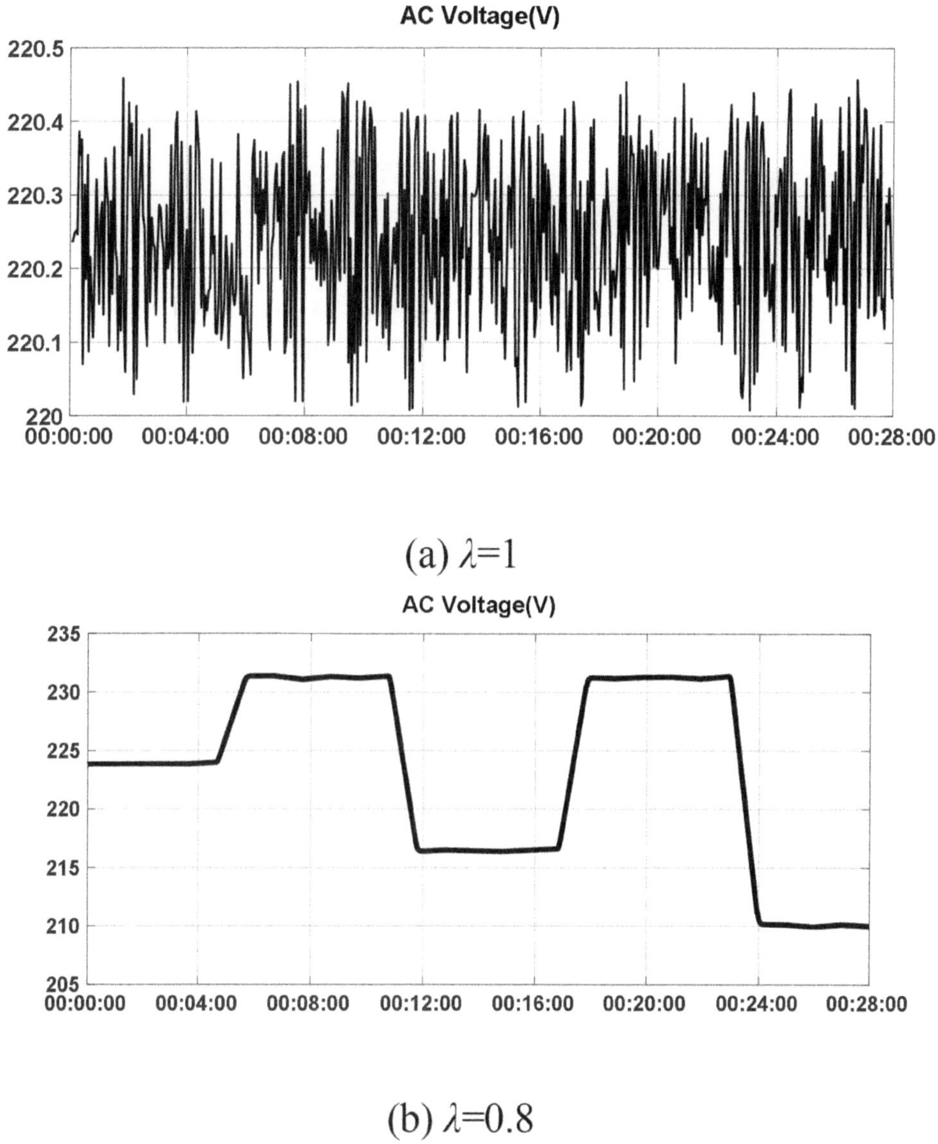

Figure 12.24 Bus 1 voltages for (a) $\lambda = 1$ and (b) $\lambda = 0.8$.

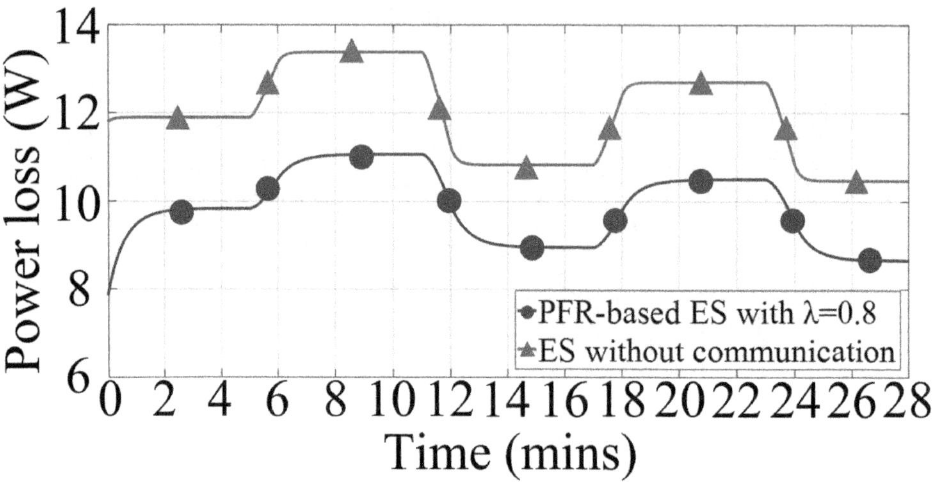

Figure 12.25 Waveforms of the power loss on the distribution lines in the experiment.

References

[1] A. Ipakchi and F. Albuyeh, "Grid of the future," *IEEE Power and Energy Magazine*, vol. 7, no. 2, pp. 52–62, March 2009.

[2] N. Hatziargyriou, H. Asano, R. Iravani, and C. Marnay, "Microgrids," *IEEE Power and Energy Magazine*, vol. 5, no. 4, pp. 78–94, July 2007.

[3] S. Chowdhury, S.P. Chowdhury, and P. Crossley, *Microgrids and Active Distribution Networks*. United Kingdom: Institution of Engineering & Technology, 2009.

[4] D. Boroyevich, I. Cvetkovic, D. Dong, R. Burgos, W. Fei, and F. Lee, "Future electronic power distribution systems – A contemplative view," *12th International Conference on Optimization of Electrical and Electronic Equipment*, pp. 1369–1380, 2010.

[5] B.C. Beaudreau, *World Trade: A Network Approach*. iUniverse, 2004.

[6] M.H. Nehrir, C. Wang, K. Strunz, H. Aki, R. Ramakumar, J. Bing, Z. Miao, and Z. Salameh, "A review of hybrid renewable/alternative energy systems for electric power generation: Configurations, control, and applications," *IEEE Transactions on Sustainable Energy*, vol. 2, no. 4, pp. 392–403, October 2011.

[7] Y.A.I. Mohamed and E.F. El-Saadany, "Adaptive decentralized droop controller to preserve power sharing stability of paralleled inverters in distributed generation microgrids," *IEEE Transactions on Power Electronics*, vol. 23, no. 6, pp. 2806–2816, November 2008.

[8] Y.W. Li and C.-N. Kao, "An accurate power control strategy for power electronics-interfaced distributed generation units operating in a low voltage multibus microgrid," *IEEE Transactions on Power Electronics*, vol. 24, no. 12, pp. 2977–2988, December 2009.

[9] Y. Xu and Z. Li, "Distributed optimal resource management based on the consensus algorithm in microgrid," *IEEE Transactions on Industrial Electronics*, vol. 62, no. 4, pp. 2584–2592, April 2015.

[10] Y. Han, H. Li, P. Shen, E.A.A. Coelho, and J.M. Guerrero, "Review of active and reactive power sharing strategies in hierarchical controlled microgrids," *IEEE Transactions on Power Electronics*, vol. 32, no. 3, pp. 2427–2451, March 2017.

[11] "DC for Efficiency," [Online]. Available at: https://library.e.abb.com/public/1afa6036874fd0bb85257d5000710a17/DC%20for%20efficiency.pdf.

[12] T. Dragičević, X. Lu, J.C. Vasquez, and J.M. Guerrero, "DC microgrids – Part II: A review of power architectures, applications, and standardization issues," *IEEE Transactions on Power Electronics*, vol. 31, no. 5, pp. 3528–3549, May 2016.

[13] J.M. Guerrero, J.C. Vasquez, J. Matas, L.G. Vicuna, and M. Castilla, "Hierarchical control of droop-controlled AC and DC microgrid – A general approach toward standardization," *IEEE Transactions on Industrial Electronics*, vol. 58, no. 1, pp. 158–172, January 2011.

[14] D.J. Becker and B.J. Sonnenberg, "DC microgrids in buildings and data centers," *2011 IEEE 33rd International Telecommunications Energy Conference (INTELEC)*, Amsterdam, Netherlands, pp. 1–7, October 2011.

[15] S. Mudaliyar and S. Mishra, "Coordinated voltage control of a grid connected ring DC microgrid with energy hub," *IEEE Transactions on Smart Grid*, vol. 10, no. 2, pp. 1939–1948, March 2019.

[16] R.M. Cuzner and G. Venkataramanan, "The status of DC micro-grid protection," *Proceedings of IEEE Industrial Applications Society Annual Meeting*, pp. 1–8, 2008.

[17] J.P. Barton and D.G. Infield, "Energy storage and its use with intermittent renewable energy," *IEEE Transactions on Energy Conversion*, vol. 19, no. 2, pp. 441–448, June 2004.

[18] H. Zhou, T. Bhattacharya, and D. Tran, "Composite energy storage system involving battery and ultracapacitor with dynamic energy management in microgrid applications," *IEEE Transactions on Power Electronics*, vol. 26, no. 3, pp. 923–930, March 2011.

[19] S.Y.R. Hui, C.K. Lee, and F.F. Wu, "Electric spring – A new smart grid technology," *IEEE Transactions on Smart Grid*, vol. 3, no. 3, pp. 1552–1561, September 2012.

[20] L. Meng, T. Dragičević, J.C. Vasquez, and J.M. Guerrero, "Tertiary and secondary control levels for efficiency optimization and system damping in droop controlled DC–DC converters," *IEEE Transactions on Smart Grid*, vol. 6, no. 6, pp. 2615–2626, June 2015.

[21] M.K. Zadeh, R. Gavagsaz-Ghoachani, S. Pierfederici, B. Nahid-Mobarakeh, and M. Molinas, "Stability analysis and dynamic performance evaluation of a power electronics-based DC distribution system with active stabilizer," *IEEE Journal of Emerging and Selected Topics in Power Electronics*, vol. 4, no. 1, pp. 93–102, March 2016.

[22] J. Ma, F. He, and Z. Zhao, "Line loss optimization based OPF strategy by hierarchical control for DC microgrid," *Proceedings of 2015 IEEE Energy Conversion Congress & Exposition (ECCE)*, Montreal, Canada, p. 6212, September 2015.

[23] Y. Jiang, Y. Yang, S.C. Tan, and S.Y.R. Hui, "Adaptive current sharing of distributed battery systems in DC microgrids using adaptive virtual resistance-based droop control," *Proceedings of 2019 IEEE Energy Conversion Congress & Exposition (ECCE)*, Baltimore, USA, pp. 4262–4267, September 2019.

[24] K. Natori, T. Tanaka, Y. Takahashi, and Y. Sato, "A study on high-efficiency floating multi-terminal power flow controller for next generation DC power networks," *Proceedings of 2017 IEEE Energy Conversion Congress & Exposition (ECCE)*, Cincinnati, USA, pp. 2631–2637, October 2017.

[25] E. Candan, P.S. Shenoy, and R.C. Pilawa-Podgurski, “A series-stacked power delivery architecture with isolated differential power conversion for data centers,” *IEEE Transactions on Power Electronics*, vol. 31, no. 5, pp. 3690–3703, May 2016.

[26] G. Gough, *Energy Efficient Servers: Blueprints for Data Center Optimization*. Berkeley, CA: ApressOpen, 2015.

[27] M. Farasat and A. Arabali, “Voltage and power control for minimising converter and distribution losses in autonomous microgrids,” *IET Generation, Transmission & Distribution*, vol. 9, no. 13, pp. 1614–1620, October 2015.

[28] M. Hussain, “Optimal placement of distributed energy resources in a DC microgrid with constant power loads to minimize bus voltage deviations and line losses,” *Global Power, Energy and Communication Conference*, Cappadocia, Turkey, pp. 424–429, June 2019.

[29] A. Garcés, “On the convergence of Newton's method in power flow studies for DC microgrids,” *IEEE Transactions on Power Systems*, vol. 33, no. 5, pp. 5770–5777, September 2018.

[30] D. Zhang, L. Wang, J. Jiang, and W. Zhang, “Optimal power management in DC microgrids with applications to dual-source trolleybus systems,” *IEEE Transactions on Intelligent Transportation Systems*, vol. 19, no. 4, pp. 1188–1197, April 2018.

[31] K. Rouzbehi, J.I. Candela, A. Luna, G.B. Gharehpetian, and P. Rodriguez, “Flexible control of power flow in multiterminal DC grids using DC–DC converter,” *IEEE Journal of Emerging and Selected Topics in Power Electronics*, vol. 4, no. 3, pp. 1135–1144, September 2016.

[32] T. Hong and F. de Leon, “Centralized unbalanced dispatch of smart distribution DC microgrid,” *IEEE Transactions on Smart Grid*, vol. 9, no. 4, pp. 2852–2861, July 2018.

[33] H. Su, C. Deng, F. Guo, X. Chen, and C. Qi, “Distributed load sharing and transmission power loss optimisation for DC microgrids,” *IET Control Theory*, vol. 13, no. 17, pp. 2930–2939, April 2019.

[34] J. Beerten, S. Cole, and R. Belmans, “Generalized steady-state VSC MTDC model for sequential AC/DC power flow algorithms,” *IEEE Transactions on Power Systems*, vol. 27, no. 2, pp. 821–829, May 2012.

[35] J. Cao, W. Du, H.F. Wang, and S.Q. Bu, “Minimization of transmission loss in meshed AC/DC grids with VSC-MTDC networks,” *IEEE Transactions on Power Systems*, vol. 28, no. 3, pp. 3047–3055, August 2013.

[36] J.H. Teng, S.-H. Liao, W.-H. Huang, and C.C. Chiang, “Smart control strategy for conversion efficiency enhancement of parallel inverters at light loads,” *IEEE Transactions on Industrial Electronics*, vol. 63, no. 12, pp. 7586–7596, December 2016.

[37] B. Khan and P. Singh, “Selecting a meta-heuristic technique for smart micro-grid optimization problem: A comprehensive analysis,” *IEEE Access*, vol. 5, pp. 13951–13977, 2017.

[38] P. Singh and B. Khan, “Smart microgrid energy management using a novel artificial shark optimization,” *Complexity*, vol. 2017, pp. 22, 2017, Article ID 2158926.

[39] J. Deng, Y. Mao, and Y. Yang, “Distribution power loss reduction of standalone DC microgrids using adaptive differential evolution-based control for distributed battery systems,” *Energies*, vol. 13, no. 9, pp. 2129, January 2020.

[40] Y. Jiang, Y. Yang, S.C. Tan, and S.Y.R. Hui, “Mitigating distribution power losses of standalone AC microgrids using particle-swarm-optimization control for distributed

battery systems," *2019 IEEE Applied Power Electronics Conference and Exposition (APEC)*, Anaheim, USA, pp. 3224–3229, March 2019.

[41] Y. Yang, S.C. Tan, and S.Y.R. Hui, "Mitigating distribution power loss of DC microgrids with DC electric springs," *IEEE Transactions on Smart Grid*, vol. 9, no. 6, pp. 5897–5906, November 2018.

[42] Y. Yang, Y. Qin, S.C. Tan, and S.Y.R. Hui, "Reducing distribution power loss of islanded AC microgrids using distributed electric springs with predictive control," *IEEE Transactions on Industrial Electronics*, vol. 67, no. 10, pp. 9001–9011, October 2020.

13

Higher Levels of Wind Energy Penetration into the Remote Grid

Challenges and Solutions

Gajendra Singh Chawda[1], Abdul Gafoor Shaik[1], and Om Prakash Mahela[2]

[1] *Department of Electrical Engineering, Indian Institute of Technology, Jodhpur, Rajasthan, India*
[2] *Power System Planning Division, Rajasthan Rajya Vidyut Prasaran Nigam Ltd., Jaipur, Rajasthan, India*

13.1 Introduction

The rise and rapid spread of COVID-19 that originated in late 2019 had shifted into a world-wide pandemic by the first half of 2020, building an extensive wellness and financial emergency. The electrical power supply and demand have been extensively interrupted. In this environment, there is curiosity in knowing what happened in the renewable energy (RE) area last year. However, it is also essential to focus on the present scenario of the power sector.

The renewable energy contribution to the global power sector made a tremendous improvement in 2019 with an installed power capacity of 200 gigawatts (GW). Presently, wind energy (WE) and solar-photovoltaic (SPV) energy sources have come to emerge as mainstream power specialists. These RE sources proved to be new renewable energy sources and generated more cost-effective electric power than current coal-fired conventional electric power plants. Also, the estimated global share of traditional electrical power is 72.7% and non-conventional electric power is 27.3%, in which wind energy contributed 5.9% of total electric power share of world-wide electric power production by the end of 2019 [1]. A block diagram of the percentage sharing by traditional and renewable energy sources is depicted in Figure 13.1.

Presently, in rural areas (remote locations) the preferred low-cost RE source in the specific wind energy (WE) source has features like small installation as well as running costs, maintenance-less operation, low power loss, and the ability to meet load demand with high-quality power [2]. Therefore, continuous enhancements are added for WE sources. In this regards, onshore wind energy sources are more suitable and globally used. In contrast, wind energy sources are still in the cradle stage and not ideal for every environmental condition. The major differences found in the literature are technical challenges and the initial and running costs with ecological suitability. Worldwide wind power generation is the second-largest yearly addition, with offshore WE source estimated to be a record 10% of fresh installations and 621 GW onshore. In fact,, the extensive WE power generation increased by 19% with 60 GW in 2019, which is the second-largest yearly addition, giving a total of 650 GW. Wind power generation is expected to be 57% of Denmark's electrical power production in 2019, with the contributions of Ireland with 32%, Uruguay with 29.5%,

Active Electrical Distribution Network: A Smart Approach, First Edition. Edited by Baseem Khan, Josep M. Guerrero, Sanjeevikumar Padmanaban, Hassan Haes Alhelou, Om Prakash Mahela, and Sudeep Tanwar.

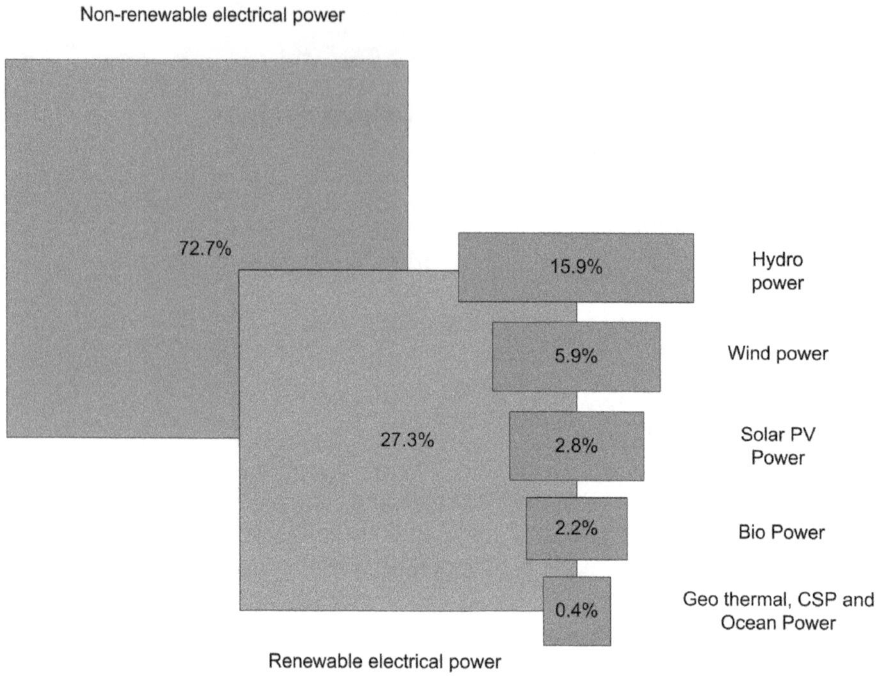

Figure 13.1 Estimated RE share of global electricity production, end 2019.

Portugal with 26.4%, as well as numerous other regions and countries. The total wind power generation capacity is 5.9% of global electric power generation [3].

WE sources are vital to fulfill these conditions, as they can provide continuous electric power day and night without any specific energy storage device [4, 5]. This source also accommodates grid flexibility and minimizes costs and delivers technical advantages due to co-localization. Mahela et al. [6] made an assessment of power quality in the utility grid integrated with wind energy generation. Gebru et al. [7] analyzed a low-voltage ride through capability of doubly-fed induction generator-based wind turbines on a wind farm. Rathore et al. [8] presented a wavelet-alienation neural-based protection scheme for a STATCOM compensated transmission line, integrated with renewable energy sources. Ola et al. [9] proposed an alienation coefficient and a Wigner distribution function-based protection scheme for a hybrid power system network with renewable energy penetration. Yogee et al. [10] presented an algorithm for the recognition of fault conditions in the utility grid with renewable energy. However, wind power generation faced many problems due to its intermittent nature and variable wind speed [11]. Also, the output power of this source depends on the input velocity of the wind.

However, the principal issue of the WE source is that output power variations may occur when integrated with the grids, especially a remote grid, due to grid impedance [12] and connected loads [13]. This penetration poses numerous operational and control challenges that hinder the safe and stable performance of the distribution system. However, most of the literature available in this context generally discussed technical, economic, and social challenges. This chapter aims to summarize and provide an overview of emerging challenges and solutions associated with higher levels of wind integration. Thus, the reader will obtain sufficient information on this topic.

Considering the importance of WE integration, the developments of WE source from late 2019 and the first half of 2020 are presented briefly and then an overview is provided of the emerging challenges and solutions associated with WE integration into the remotely located grid. This followed by a summary of the findings of wind integration into the remote grid.

13.2 Present Scenario of WE Integration

The WE electricity power generation is continuously growing worldwide. Most of the necessary parts for grid development have been provided for various countries and regions. As discussed earlier, globally offshore power generation accounts for a record 10% of fresh installations, i.e. 650 GW, and is listed as the second-largest RE power generation in the globe, just after solar-PV [14]. The yearly extension of wind power generation worldwide increases its integration levels into the entire power network and adds this renewable power into the current electric power. However, the top six countries, China, United States of America, United Kingdom, Germany, India, and Spain have maximum annual investment and maximum power generation, while Denmark and Eritrea are exceptional cases as they have already reached 50% of WE power. In this connection, China with 210 GW by 2020, United Kingdom with 39 GW by 2030, Germany with 71 GW by 2030, and India with 60 GW by 2022 have set a goal for WE to share targets as of the end of2019, illustrated in Table 13.1. The US retains its second position after China in yearly wind electricity generation [15–17].

Several countries and certified agencies have reached relatively high WE penetration levels in their distribution networks. However, these electricity grids are not defined as to whether they are stiff or weak. Also, there is no document available that shows whether these high penetration levels of the WE sources have been achieved on what loads (i.e. the percentage of nonlinear static and nonlinear dynamic). From the literature and documents available for researchers, the present scenario of higher levels of wind energy penetration is as illustrated in Figure 13.2.

It is imperative to know whether WE penetration levels are also estimated by independent system operators (ISOs). In 2018, WE penetration (represented as a percentage of connected loads) had been recorded as 23.9% in the Southwest Power Pool (SPP), 18.6% in the Electric Reliability Council of Texas (ERCOT), 7.3% in both the Midcontinent Independent System Operator (MISO) and the California Independent System Operator (CAISO), 2.8% in ISO New England (ISO-NE), 2.7% in the PJM Interconnection (PJM), and 2.5% in the New York Independent System Operator (NYISO). However, the WE capacity in the global power market significantly increased, with the most impressive numbers in SPP (25%), the Mountain region (16%), and the Midwest (16%). The smaller number is observed in the Northwest (7%), with no plan in the Southeast by the end of 2018 [18–21].

Table 13.1 International comparison of wind power capacity.

Annual capacity in MW-2019		Cumulative capacity in MW-2019	
Name of country	**Power in MW**	**Name of country**	**Power in MW**
China	26 155	China	236 402
United States	9137	United States	105 591
United Kingdom	2393	Germany	61 406
India	2377	India	37506
Spain	2319	Spain	25850
Germany	2189	United Kingdom	23 340
Sweden	1588	France	16 645
France	1336	Brazil	15 452
Mexico	1281	Canada	13 413
Argentina	931	Italy	10 406
Rest of the world	10 639	Rest of the world	104671
Total	**60 639**	**Total**	**650 682**

Source: GWEC, AWEA, WindIQ, Berkeley lab.

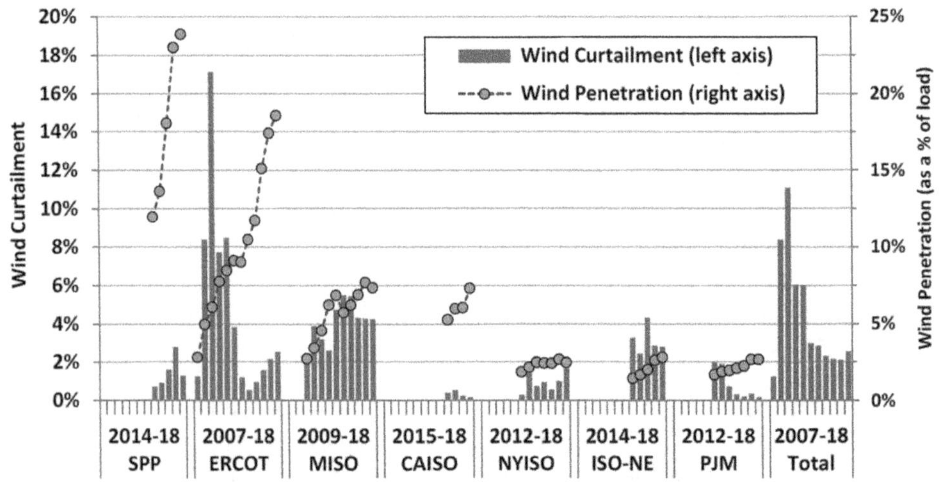

Figure 13.2 Wind energy penetration percentages by independent system operators.

Multiple attempts have been made to extend general representations of the four types used for the Wind Energy Generating (WEG) arrangements. These wind energy generating methods are as follows: type 1 (limited variable speed), type 2 (limited variable speed), type 3 (variable speed with partial power electronics conversion), type 4 (variable speed with full power electronics conversion), and type 5 (variable speed with full control) [22, 23]. Figure 13.3 presents the wind energy generation system with type 1 and type2 configurations. Figures 13.4 and 13.5 present the wind energy generation system for type 3 and type 4.

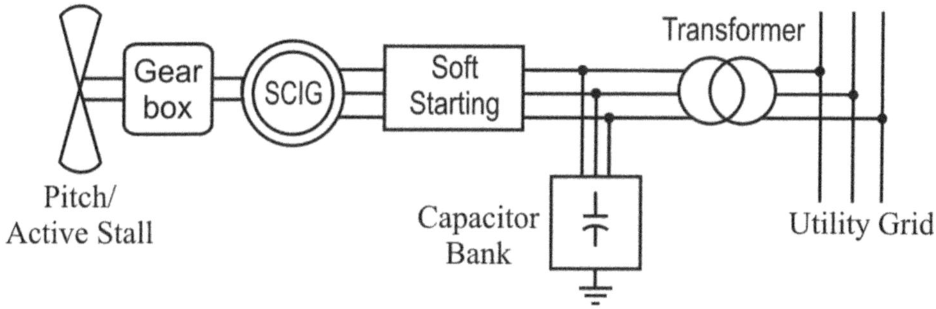

Figure 13.3 Wind energy generation system with type 1 and type 2 configurations.

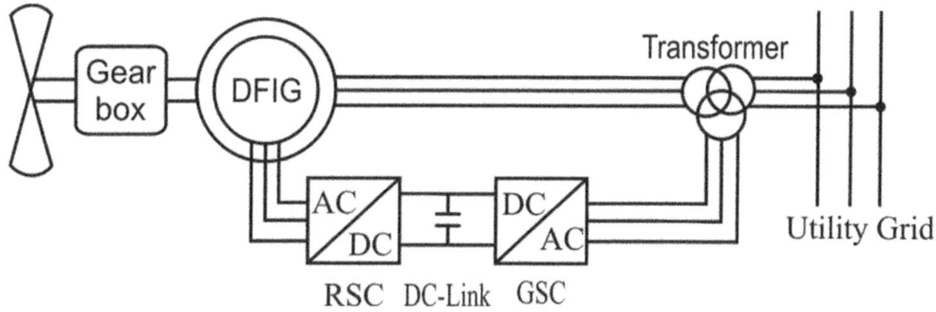

Figure 13.4 Wind energy generation system type 3.

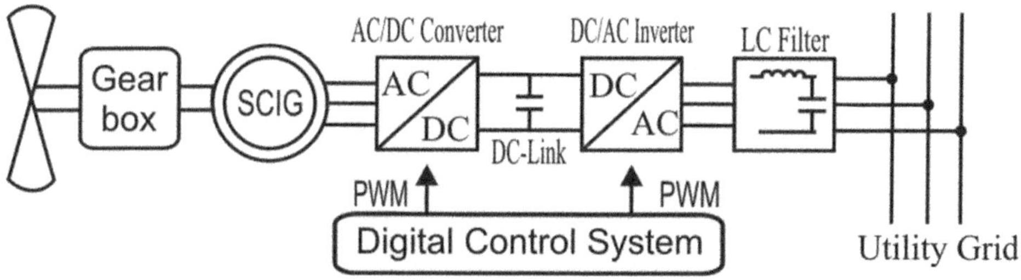

Figure 13.5 Wind energy generation system type 4.

13.2.1 Type 1 WEG System

- Squirrel-cage induction generator (SCIG) combined straight to the transformer.
- The turbine speed is fixed to the grid frequency.
- The turbine shaft rotates faster than the grid frequency and produces a negative slip for the generation of active power.
- The operating speed of the turbine under controlled conditions is nearly a linear function of torque when the speed is fixed.
- The mechanical inertia of the drivetrain can restrict the rate of change in electrical output during variable speeds.

Merits: Simple, low cost, and low maintenance
Demerits: High mechanical stress, weak voltage stability, and challenging to control

13.2.2 Type 2 WEG System

- Wound rotor induction generators (WRIGs) are connected directly to the WEG, similar to type 1. However, one additional variable resistor is added in the rotor circuit.
- This additional resistor helps to control the rotor currents very quickly to maintain continuous power.
- It can also affect the machine's dynamic response under grid interruptions.

Merits: Simple, low cost and low maintenance
Demerits: High starting inrush, weak voltage stability, and challenging to control

13.2.3 Type 3 WEG System

- The doubly-fed induction generator (DFIG) design is a next-level WEG system, by combining variable frequency AC excitation to the rotor circuit.
- The new rotor excitation is furnished through slip rings by VSC, which regulates the current by adjusting the magnitude and phase angle.
- This rotor-side converter (RSC) is combined with a grid side converter (GSC), which transfers electric power instantly to the grid.

Merits: Good conversion efficiency, reasonable control of powers
Demerits: Weak voltage stability, weak fault ride-through capability, challenging to control, sizable short-circuit contribution, damage during improper synchronization, required specific reactive power support, high stress on the rotor as well as gearbox under unbalancing

13.2.4 Type 4 WEG System

- The full AC–AC converter design is also considered as a next-level WEG system as it provides flexibility in smooth operation.
- The output power of the RSC is directly sent through the DC link capacitor to the GSC.
- The turbine is permitted to rotate at its optimal aerodynamic speed for smooth operation.
- The rotor drives a gearbox in the geared system to increase generator shaft speed.
- The gearbox is eliminated when direct driven for generating specified frequency well below that of the grid.
- The type 4 WEG system can be used in wound rotor synchronous machines, permanent magnet synchronous machines, or squirrel cage induction machines, based on design consideration of the specific applications.

Merits: Voltage stability and fault ride-through capability, flexibility and fast operation, control on short-circuit currents, frequency and voltage

Demerits: Fault current contribution is restricted, and reactive power support is comparable to the synchronous machine

All of these WEG systems are generally used as wind energy sources, either in isolated mode or grid-connected mode. Many issues have been addressed in the literature, especially when higher levels of these sources were integrated into remotely located grids. These issues are power forecasting, low voltage ride-through, protection, the strength of the AC grid, reactive power capability, harmonics, voltage stability, power quality disturbances, frequency stability, small-long signal stability, and other challenges [24, 25].

13.3 Emerging Challenges Associated with WE Integration

This section covers the emerging challenges associated with higher levels of WE integration into the remotely located grids that must consider maintaining the security and power quality of the distribution network.

13.3.1 Power Forecasting Challenge

Forecasting of the output power of the WES has been observed very helpfully for distribution power system supervision. It has been perceived that there is no ideal approach for forecasting WE output power due to the unpredictable nature of the wind. However, during higher levels of WE penetration, the forecast performs a notable role in decreasing the expenses of electric power generation. Moreover, unpredictable disturbances of ramps can reduce the security of the remotely located network. In this regard, some methods for forecasting data using historical data have been suggested in the literature. The forecast methods are the deterministic forecast (DF) method and the uncertainty analysis (UA) method [26].

Four types of strategies are included in the deterministic forecast (DF) method. These are physical DF, statistical DF, intelligent DF, and hybrid DF. The physical DF strategy utilized climate data like temperature, altitude, terrain, etc. These data are collected from the multiple research stations scattered in geographic regions or virtually designed using dynamic equations. Also, the statistical DF is a time series based strategy and utilized historical statistics, probabilities, and stochastic data. In continuation, intelligent DF strategies generally used historical weather data for future forecasting of WE output. However, the uncertainty analysis (UA) utilized three approaches, such as probability prediction, risk indices, and generation scenarios, for predicting the output power of the WE sources in remote areas. Moreover, mathematical functions have been developed using power curves for a precise explanation of the wind speed versus the output power of the WE [27].

13.3.2 Low-Voltage Ride-Through Challenge

The ability of the wind energy generation (WEG) to continue integrated into the distribution grid in the event of undervoltage issues is termed the low-voltage ride-through (LVRT). A shut-down of high-capacity WEG farms becomes critical and influences transient and steady-state stability of remotely located grid controls. Consequently, the study of LVRT is a basis for WEG under the faulty events of the grid [28]. This system needs to resist voltage interruptions without getting disconnected from the remote grid. Concurrently, the international standards state that the requirement of voltage to attain 80% of the nominal value within 0.5 seconds and 95% within 15 seconds after the fault removal at the point of connection [29]. This is one of the major challenges associated with high WE integration, which needs to be more focused and addressed effectively to the perceived objectives mentioned above. Thus there is a continuous requirement for new research works for specific designs under such conditions with matching grid codes and incorporated LVRT schemes.

13.3.3 Power Quality Challenges

13.3.3.1 Voltage/Reactive Power Support

The induction generators are generally utilized in the WEG systems, and reactive power is required for excitation. Thus, reactive power support like synchronous generators has not been encouraged. It has been recognized that when modern technologies are associated with higher levels of generations and loads, maintaining the voltage is always a major challenge. Therefore, during operation of higher levels of WE generation this is a major cause of high voltage fluctuations as well as voltage flickers. In addition, the strength of the AC grid also provides a considerable impact during more elevated levels of WE integration, as it requires continuous voltage support at the point of interconnection. In the past, these issues have been investigated with the wind side converter. However, due to the rating issue of these sources, the wind side converter rating is also recognized less. Therefore, the capacity of reactive power injection of this converter is found limited to meeting the reactive power demand of non-ideal grids. Hence, this may also affect the WE penetration levels [30]. However, the proportion of WE in the distribution system should not exceed 20% [31].

It has been observed that to maintain the rated power of the WE source an additional inverter is essential to fulfilling the reactive power demand. Therefore, the necessity of an additional inverter, named the D-FACTS device, is increased, in order to support the voltage of weak grids during the higher levels of WE penetration.

13.3.3.2 Frequency Support

The continuous expansion of WEG technology and the capacity of the grid-tied WE source are progressively increasing, which dangerously influences the frequency stability of the distribution grids. The DFIG is utilized world-wide due to its simple operation, good controllability, and cost. However, the WEG turbine converter decouples the fan rotor speed from the system frequency. Therefore, the system's equivalent inertia is continuously reducing due to the WEG turbine not providing system inertia. In addition, higher levels of WE penetration also dangerously influence the frequency stability of the remotely located grid due to interfacing converters. This is one of the critical reasons why most of the country has a limited penetration rate.

One of the solutions found in the literature is the inertial frequency response (IFR). The IFR generally receives strength from the rotating masses to resist a frequency variation from the acceptable range of frequencies. In continuation, the primary control, the automatic governing system, is initiated to maintain the frequency variations within the standard range. Later, the secondary control also helps to restore the system frequency under the stipulated limit. Energy storage devices, DFACTS devices, their control, along with load control, can also be suitable alternate solutions for a frequency degradation challenge [32].

13.3.3.3 Harmonics

Integration of higher levels of the WEG system into the grid, which is mainly a remotely located grid, injects current and voltage harmonics due to power electronics-based devices and interfacing converters. Ensuring that these harmonics are under the permissible limit is one of the most critical challenges today. The IEC 61000-3-2, 1159 and IEEE-519-2014 international standards address the allowable limits of harmonics distortion. The estimation of harmonics at a remote location is rigid when higher levels of WEG are integrated, as at this point of connection the grid is weak and voltage as well as current harmonics are already present in the system. Also, at this point of contact, the wave shape is not sinusoidal and current is not in-phase with the voltage, which indicates a low power factor. Hence it has been found that there are always harmonic voltages present in the grid. These harmonics include the integer harmonic of fifth and seventh order, which have a major effect on the estimations [33]. It has been recognized from the literature that the self-commutated pulse width modulation inverter is included in the variable-speed WEG turbines. The advantage of this system is that it controls active and reactive power as per the given rating. The disadvantage of this system is that it produces a harmonic current. Hence, researchers have started to implement some filters, such as the LC resonant passive filter. However, these filters have the disadvantages of considerable size, fixed compensation, and resonance. In continuation, DC offset and harmonic levels have been significantly mitigated at higher levels of WE penetration and provide smooth grid synchronization in remote grids using DSTATCOM [34].

One more solution found in the literature is the active utilization of D-FACTS devices. These devices are capable of mitigating all types of PQ issues, especially harmonics [33].

13.3.3.4 Flicker

Flicker has widely been considered as a power quality issue in grid-tied WEG systems. It is generated by a grid-tied variable-speed WEG system with interfacing converters, variations in output power, wind shear, the strength of grid, wind tower shadow and characteristics of the WEG. Fluctuations in voltage cause it due to load flow variations in the grid. It may restrict the higher penetration levels of the WEG system in a grid-connected mode. However, a variable-speed WEG system has exhibited excellent performance for flicker emission in association with a fixed-speed WEG system. It has been recognized from the literature that flickers can generate in a fixed-speed WEG system when it reaches maximum speed. However, a large-size WEG turbine generates lower flicker compared to a small-size WEG turbine. Meanwhile, the level of flicker can be observed by amplitude, shape, and frequency components of the fluctuated voltage waveform based on the flicker meter defined in IEC 61000-4-15.

The literature also suggested some solutions to overcome this PQ issue. These include reactive power compensation, active power curtailment, voltage and frequency regulation, and maintenance of the DC link voltage of the grid side converter.

13.3.4 Angular Stability Challenge

The angular stability is also termed as inter-area oscillations, and it is the strength of the interfaced devices in the distribution system to stay synchronized during specified or unspecified disturbances. It has been observed from the literature that small-signal stability is continuously increased during the higher levels of WE penetration and may introduce angular instability due to a reactive power issue. Also, some of the researchers concluded that the higher the WE penetration levels into the grid the lower the inertia and may lead to the system collapsing. Hence this specific challenge needs to be more focused at higher levels of WE integration [35].

13.3.5 Protection Challenge

Protection is one of the significant challenges when higher levels of the WE integrate to the remotely located grids. In this connection, the short circuit is a notable cause of distribution network failure as it provides high stress on the connected distribution elements. It has been recognized that types of the WEG system contribute a short-circuit current (SCC) based on the system design and its configurations. These configurations and design consider the number of WEG systems connected in parallel, their types, type of connected grid, and loads or isolated modes, etc. In addition, one of the most frequent challenges found in the literature is under-reach and over-reach issues of relays during high WE integration [36]. However, the additional protection challenges associated with the remote grids are false tripping/mal-operation, loss of coordination, blinding security, auto recloser difficulties, and grid disconnection issues, which are well discussed in [36].

Protection issues, as mentioned above, need to consider new protection schemes, design, implementation methods, unique and fast measurement units, coordination of each connected unit, and most crucially modern protection plans. These points could be the solution to protection challenges, as mentioned above.

13.3.6 Environmental Challenge

The output power of the WEG system depends on the input speed of the wind, and the speed of the wind highly relies on the weather and climate variability. Therefore, active power penetration from the WEG system is affected. The intermittent nature and seasonal variability of wind speed also significantly affect the system operation. It has been recognized from the literature that in multiple areas/regions in the globe there is a shift in the wind speed pattern from higher to smaller wind velocities (WVs). Hence, the appearance of WVs that are much smaller than the cut-in rate become more expected, which causes lower WE output power.

The presently available temporary comprehensive assessment is not enough for climate projections. Hence, there is a strong need for a specific design of wind blades, structure, types, concrete base, and interfacing converters and its control based on the wind speed data of different regions/areas around the globe. One more solution found in the literature is wind speeds at varying lengths near to the typical turbine hub length can be offered in future regional climate projections.

13.4 Smart Solutions Associated with WE Integration

The WE sources have accelerated the development toward greener energy sources. The rising number of WES necessitates fresh approaches for the operation and control of the remotely located grid to support or enhance the reliability and quality of power. The power electronic-based technology plays an essential role in a grid-tied WEG system and it is rapidly being used in remote locations because of the involvement of real-time workstation-based controllers. These factors collectively have led to the advancement of cost-efficient and grid-favorable converters. Moreover, continuous research is being carried out and researchers have addressed some possible solutions, which are depicted in Figure 13.6.

13.4.1 Without Energy Storage-Based Solution

A grid-tied DFIG-based WEG system has been classified into two sections for easy explanation of the mechanical and electrical section, which is illustrated in Figure 13.7. The mechanical section of the connected system includes an aerodynamic design with rotor blades and a drive train system. However, the electrical section includes a generator, power electronic-based back-to-back converters, and a control structure for the control of crowbar, pitch angle, generator torque, power optimization, frequency, converter, etc. Also, RSC and GSC side control has been managed by a suitable control technique and improves the performance of these converters [37, 38]. Hence, grid operators may consider this type of configuration for addressing various kinds of issues and provide possible practical solutions to allow higher levels of WE integration into remote location grids.

13.4.2 With Energy Storage-Based Solution

One of the vital solutions for the challenges mentioned above is the battery energy storage system (BESS). It is connected with the GSC and injunction with a DC-link capacitor of the back-to-back converters, as depicted in Figure 13.8. The benefits of connection of this additional BESS at a DClink is to provide auxiliary services for enhancing the dispatching ability.

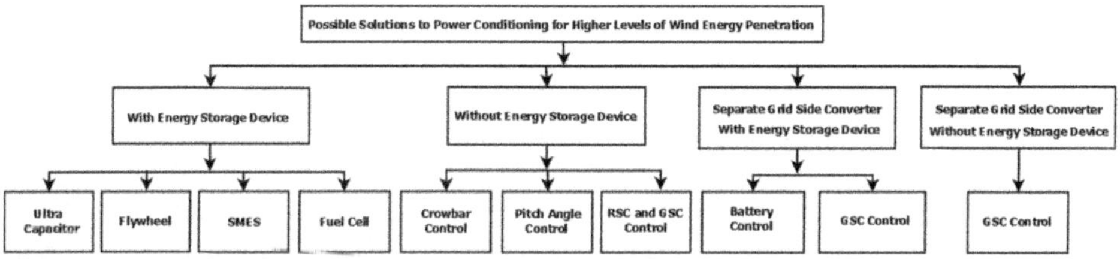

Figure 13.6 Emerging possible solution of challenges associated with grid-tied DFIG-based WEG system.

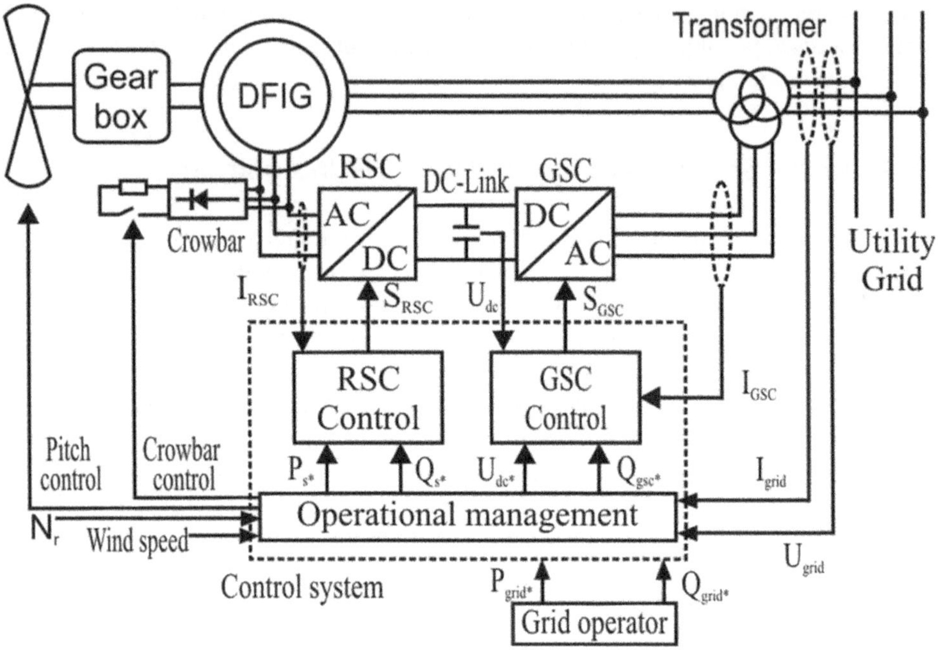

Figure 13.7 First possible solution of challenges associated with grid-tied DFIG-based WEG system.

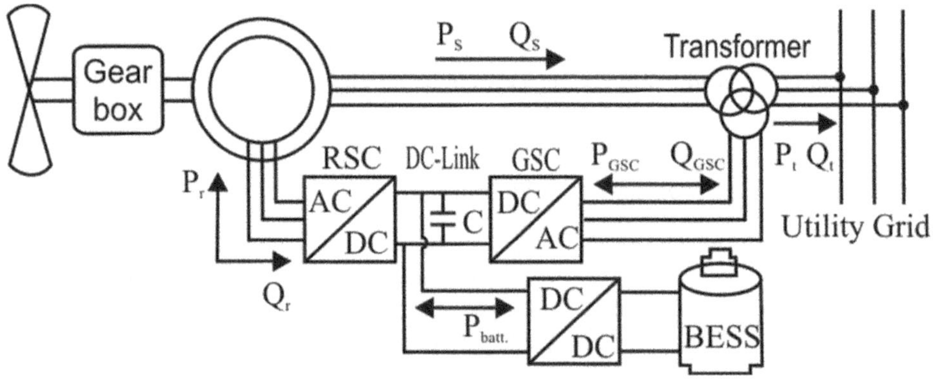

Figure 13.8 Second possible solution of challenges associated with grid-tied DFIG-based WEG system.

This modified configuration functions as a "shock absorber" for the connected system and also includes grid reliability, flexibility, and efficiency. Also, most importantly, it helps to integrate higher levels of WE sources into the remotely located grid without any compromise and mitigates peak load demand and price of electrical power during running conditions. It has been observed that the pumped hydroelectric energy storage system (PHESS) and compressed air energy storage system (CAESS), which are BESS technologies, are vital for large-scale applications. However, the superconducting magnetic energy storage system (SMES), supercapacitor (SC), and flywheel energy storage system (FESS) have also been right choices for some remote located WE applications.

13.4.3 Additional or Separate Grid Side Converter-Based Solution

It has been recognized from the literature that, in one solution, RSC and GSC side controls were designed based on conventional methods, like vector control, and these controls did not consider the strength of the grid and connected distribution loads. Hence these controls are unable to adopt the changes associated with grid strength and connected loads as well as fluctuations in the corresponding converters. Also, reactive power management, DC link control and voltage-frequency stability were a significant concern during synchronization and higher penetration levels of WE sources. Therefore, the concept of a separate grid side converter comes into the picture, and recently researchers have focused on it. The various benefits of the shunt connection of this converter have been investigated in [39]. From this research, higher levels of WE integration can be enhanced by mitigating power quality issues. It has also been found that integration of a separate grid side converter increases the initial cost of the system due to the need to rate the converter for a specific application. This particular issue and issues discussed above are solved by the control algorithm employed. Chawda and Shaik [39] looked at the control algorithms for this device that were designed in such a way that they adopted the changes associated with the WE output power, loads, strength of the grid, and DC-link voltage. In this way, not only did it mitigate the power quality issues but also reduced the cost of this converter. The solution of challenges associated with higher levels of WE penetration is depicted in Figure 13.9.

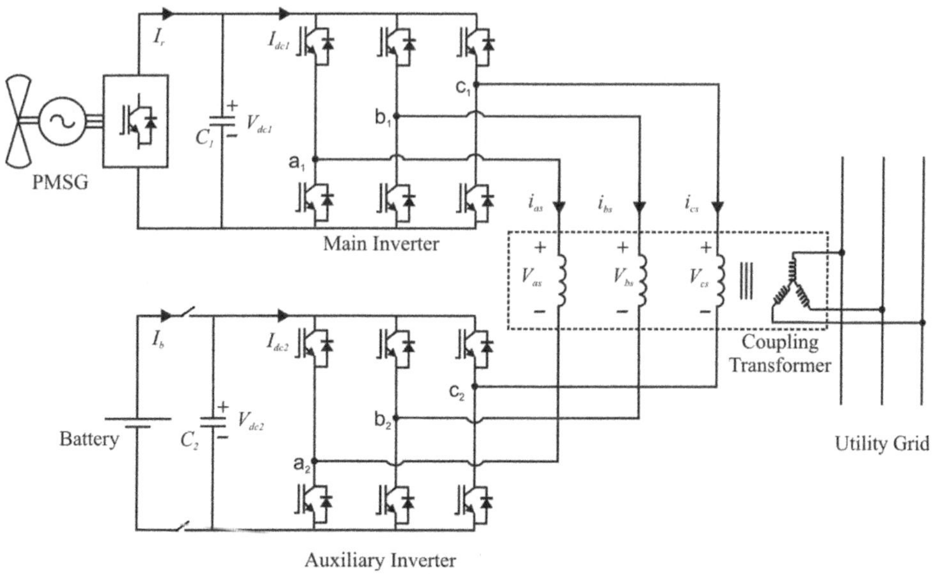

Figure 13.9 Third possible solution of challenges associated with grid-tied DFIG-based WEG system.

In addition, energy storage can also be added to the separate grid side control to enhance the capability and control of the DC-link capacitor with the additional facility of storing the power, which is also illustrated in Figure 13.9, by connecting the switch. Also, the battery energy storage system can be used as a separated source with the interfacing of DC–DC and DC–AC converters, but the DC to AC extra converter cost would be added with an additional control strategy [38, 40]. Hence recommendation leans toward the separated grid side converter for enhancing WE integration levels in remotely located grids by mitigating various power quality issues.

13.4.4 Other Possible Solutions

The other possible solutions are well discussed in [36]. Supercapacitor (SC)-based techniques for grid-tied WE applications includes a direct DC-link connection of SC, connection to a DC-link through the DC–DC converter, a common AC bus connection, and a combined BESS and SC connection to the DC-link. All of these combinations can be designed and implemented based on the application. It has also been noted that possible solutions are not limited, and research on other possible solutions is in the development stage for the promotion of higher penetration levels of green energy [41–43].

13.5 Conclusions

This chapter provides emerging challenges and solutions associated with higher levels of WE integration into remotely located grids. Remotely located grids are generally considered as weak grids, where the integration of higher levels of WE sources is a major challenge due to various power quality issues. The discussed solutions may have potential to mitigate PQ issues related to a remote grid with WE integration. It has been found that power electronics-based DFACTS devices, their control and energy storage systems, are capable of alleviating these issues and establishing an emerging solution. Also, there is a necessity for new grid codes to enhance the WEG system capability in remote areas and a requirement to ensure that they do not adversely affect the operation of the power system.

References

[1] R. Secretariat, "Renewables 2020 global status report," Report, Paris: REN12, 2020.

[2] O. Ellabban, H. Abu-Rub, and F. Blaabjerg, "Renewable energy resources: Current status, future prospects and their enabling technology," *Renewable and Sustainable Energy Reviews*, vol. 39, pp. 748–764, 2014.

[3] R.H. Wiser and M. Bolinger, "2018 wind technologies market report," U.S. Department of Energy, Office of Energy Efficiency and Renewable Energy, 2019, Available at: https://www.energy.gov/eere/wind/downloads/2018-wind-technologies-market-report (Accessed: 16 January 2020).

[4] Z. Chen, J.M. Guerrero, and F. Blaabjerg, "A review of the state of the art of power electronics for wind turbines," *IEEE Transactions on Power Electronics*, vol. 24, no. 8, pp. 1859–1875, 2009.

[5] A.Q. Al-Shetwi, M. Hannan, K.P. Jern, M. Mansur, and T. Mahlia, "Grid-connected renewable energy sources: Review of the recent integration requirements and control methods," *Journal of Cleaner Production*, vol. 253, p. 119831, 2020.

[6] O.P. Mahela, B. Khan, H. Haes Alhelou, and S. Tanwar, "Assessment of power quality in the utility grid integrated with wind energy generation," *IET Power Electronics*, vol. 13, no. 13, pp. 2917–2925, 14 October 2020, doi: 10.1049/iet-pel.2019.1351.

[7] F.M. Gebru, B. Khan, and H.H. Alhelou, "Analyzing low voltage ride through capability of doubly fed induction generator based wind turbine," *Computers & Electrical Engineering*, vol. 86, p. 106727, 2020.

[8] B. Rathore, O.P. Mahela, B. Khan, H. Haes Alhelou, and P. Siano, "Wavelet-alienation-neural based protection scheme for STATCOM compensated transmission line," *IEEE Transactions on Industrial Informatics*, June 2020. doi: 10.1109/TII.2020.3001063.

[9] S. Ram Ola, A. Saraswat, S.K. Goyal, V. Sharma, B. Khan, O.P. Mahela, H. Haes Alhelou, and P. Siano, "Alienation coefficient and wigner distribution function based protection scheme for hybrid power system network with renewable energy penetration," *Energies*, vol. 13, p. 1120, 2020.

[10] G.S. Yogee, O.P. Mahela, K.D. Kansal, B. Khan, R. Mahla, H. Haes Alhelou, and P. Siano, "An algorithm for recognition of fault conditions in the utility grid with renewable energy, penetration," *Energies*, vol. 13, p. 2383, 2020.

[11] G.S. Chawda and A.G. Shaik, "Performance evaluation of adaline controlled dstatcom for multifarious load in weak AC grid," *2019 IEEE PES GTD Grand International Conference and Exposition Asia (GTD Asia)*, Bangkok, Thailand, pp. 356–361, 2019, doi: 10.1109/GTDAsia.2019.8715994.

[12] G.S. Chawda and A.G. Shaik, "Fuzzy logic based control algorithm for DSTATCOM connected to weak AC grid," *2018 2nd International Conference on Power, Energy and Environment: Towards Smart Technology (ICEPE)*, Shillong, India, pp. 1–6, 2018, doi: 10.1109/EPETSG.2018.8659093.

[13] O.P. Mahela, B. Khan, H.H. Alhelou and P. Siano, "Power quality assessment and event detection in distribution network with wind energy penetration using Stockwell transform and fuzzy clustering," *IEEE Transactions on Industrial Informatics*, vol. 16, no. 11, pp. 6922–6932, November 2020, doi: 10.1109/TII.2020.2971709.

[14] G.S. Chawda, O.P. Mahela, N. Gupta, M. Khosravy, and T. Senjyu, "Incremental conductance based particle swarm optimization algorithm for global maximum power tracking of solar-PV under nonuniform operating conditions," *Applied Sciences*, vol. 10, no. 13, p. 4575, 2020.

[15] IRENA, *Wind Energy*, 2019. [Online]. Available at: https://www.irena.org/wind.

[16] R.H. Wiser, M. Bolinger, B. Hoen, D. Millstein, J. Rand, G.L. Barbose, N.R. Darghouth, W. Gorman, S. Jeong, A.D. Mills, and B. Paulos, "Wind technology data and trends: Land-based focus, 2020 update," Lawrence Berkeley National Laboratory,2020.

[17] Y. Sun, S.V. Wachche, A. Mills, and O. Ma, *2018 Renewable Energy Grid Integration Data Book*. No. NREL/BK-6A20-74823; DOE/GO-102020-5208. National Renewable Energy Lab. (NREL), Golden, CO, United States, 2020.

[18] American Wind Energy Association (AWEA), *U.S. Wind Industry Annual Market Report: Year Ending 2018*. Washington, D.C.: American Wind Energy Association, 2019. Available at: http://www.awea.org/market-reports.

[19] American Wind Energy Association (AWEA). *U.S. Wind Industry First Quarter 2019 Market Report*. Washington, D.C.: American Wind Energy Association, 2019. Available at: http://www.awea.org/market-reports.

[20] IEA, *World Energy Outlook 2019 Access to Electricity Database*, https://iea.blob.core.windows.net/assets/ecaa2844-dce8-4710-80fb-4085a58f292f/WEO2019-Electricity-database.xlsx; A. Contejean, IEA, Paris, personal communication with REN21, 20 May 2020.

[21] Wind Europe, "Wind in Europe in 2019," Brussels, p. 7, 2020, Available at: https://windeurope.org/wp-content/uploads/files/about-wind/statistics/WindEurope-Annual-Statistics-2019.pdf.

[22] T. Ackermann, *Wind Power in Power Systems*. Hoboken, NJ: Wiley, 2012. [Online]. Available at: https://books.google.com.sa/books?id=y7430s86pQAC.

[23] A. Ellis, Y. Kazachkov, E. Muljadi, P. Pourbeik, and J. Sanchez-Gasca, "Description and technical specifications for generic WTG models: A status report," *Proceedings of the IEEE/PES Power Systems Conference and Exposition*, March 2011, pp. 1–8.

[24] S.D. Ahmed, F.S.M. Al-Ismail, M. Shafiullah, F.A. Al-Sulaiman, and I.M. El-Amin, "Grid integration challenges of wind energy: A review," *IEEE Access*, vol. 8, pp. 10857–10878, 2020, doi: 10.1109/ACCESS.2020.2964896.

[25] G.S. Chawda, A.G. Shaik, M. Shaik, S. Padmanaban, J.B. Holm-Nielsen, O.P. Mahela, and P. Kaliannan, "Comprehensive review on detection and classification of power quality disturbances in utility grid with renewable energy penetration," *IEEE Access*, vol. 8, pp. 146807–146830, 2020.

[26] C. Monteiro, R. Bessa, V. Miranda, A. Botterud, J. Wang, G. Conzelmann, and I. Porto, "Wind power forecasting: State-of-the-art 2009," Argonne Nat. Lab., Argonne, IL, USA, Tech. Report ANL/DIS-10-1, Nov. 2009, doi: 10.2172/968212.

[27] S. Zolfaghari, G.H. Riahy, and M. Abedi, "A new method to adequate assessment of wind farms' power output," *Energy Conversion and Management*, vol. 103, pp. 585–604, October 2015, doi: 10.1016/j.enconman.2015.07.001.

[28] O.P. Mahela, N. Gupta, M. Khosravy, and N. Patel, "Comprehensive overview of low voltage ride through methods of grid integrated wind generator," *IEEE Access*, vol. 7, pp. 99299–99326, 2019.

[29] P.-H. Huang, M.S. El Moursi, W. Xiao and J.L. Kirtley, "Fault ride-through configuration and transient management scheme for self-excited induction generator-based wind turbine," *IEEE Transactions on Sustainable Energy*, vol. 5, no. 1, pp. 148–159, 2013.

[30] G.S. Chawda and A.G. Shaik. "Power quality mitigation in weak AC grid with low X/R ratios using distribution static compensator controlled by LMF algorithm," *2020 IEEE Region 10 Symposium (TENSYMP) 5–7 June 2020*, IEEE, Dhaka, Bangladesh, pp. 1–4, 2020.

[31] B. Kroposki, "Integrating high levels of variable renewable energy into electric power systems," *Journal of Modern Power Systems and Clean Energy*, vol. 5, no. 6, pp. 831–837, 2017.

[32] A.B. Attya, J.L. Dominguez-Garcia, and O. Anaya-Lara, "A review on frequency support provision by wind power plants: Current and future challenges," *Renewable and Sustainable Energy Reviews*, vol. 81, pp. 2071–2087, 2018.

[33] G.S. Chawda, A.G. Shaik, O.P. Mahela, S. Padmanaban, and J.B. Holm-Nielsen, "Comprehensive review of distributed FACTS control algorithms for power quality enhancement in utility grid with renewable energy penetration," *IEEE Access*, vol. 8, pp. 107614–107634, 2020, doi: 10.1109/ACCESS.2020.3000931.

[34] G.S. Chawda and A.G. Shaik, "Smooth grid synchronization in weak AC grid with high wind energy penetration using distribution static compensator," *2019 2nd International Conference on Smart Grid and Renewable Energy (SGRE)*, Doha, Qatar, pp. 1–6, 2019, doi: 10.1109/SGRE46976.2019.9020671.

[35] D. Gautam, V. Vittal, and T. Harbour, "Impact of increased penetration of DFIG-based wind turbine generators on transient and small signal stability of power systems," *IEEE Transactions on Power Systems*, vol. 24, no. 3, pp. 1426–1434, 2009.

[36] V. Telukunta, J. Pradhan, A. Agrawal, M. Singh, and S.G. Srivani, "Protection challenges under bulk penetration of renewable energy resources in power systems: A review," *CSEE Journal of Power and Energy Systems*, vol. 3, no. 4, pp. 365–379, 2017.

[37] A. Urtasun, P. Sanchis, I.S. Martín, J. López and L. Marroyo, "Modeling of small wind turbines based on PMSG with diode bridge for sensorless maximum power tracking," *Renewable Energy*, vol. 55, pp. 138–149, 2013.

[38] O.P. Mahela and A.G. Shaik, "Comprehensive overview of grid interfaced wind energy generation systems," *Renewable and Sustainable Energy Reviews*, vol. 57, pp. 260–281, 2016.

[39] G.S. Chawda and A.G. Shaik, "Adaptive reactive power control of dstatcom in weak ac grid with high wind energy penetration," *2019 IEEE 16th India Council International Conference (INDICON)*, IEEE, pp. 1–4, 2019.

[40] S.D.G. Jayasinghe, D. Mahinda Vilathgamuwa, and U.K. Madawala, "Direct integration of battery energy storage systems in distributed power generation," *IEEE Transactions on Energy Conversion*, vol. 26, no. 2, pp. 677–685, 2011.

[41] S.R. Sinsel, R.L. Riemke, and V.H. Hoffmann, "Challenges and solution technologies for the integration of variable renewable energy sources – A review," *Renewable Energy*, vol. 145, pp. 2271–2285, 2020.

[42] H.S. Hayajneh, "A solution to maximize renewable energy generations, incentivize battery deployment and promote green transportation," PhD Dissertation, Texas A&M University-Kingsville, TX, USA, 2020.

[43] O. Benzohra, S.S. Echcharqaouy, F. Fraija, and D. Saifaoui, "Integrating wind energy into the power grid: Impact and solutions," *Materials Today: Proceedings*, 2020.

14

Internet of Things and Machine Learning for Improving Solar-PV Plant Efficiency

Forecasting Aspects

Pankaj Kumar[1], *Gajendra Singh Chawda*[2], *and Om Prakash Mahela*[3]

[1] *Department of Electrical Engineering, Rajasthan Technical University, Kota, Rajasthan, India*
[2] *Department of Electrical Engineering, Indian Institute of Technology, Jodhpur, Rajasthan, India*
[3] *Power System Planning Division, Rajya Vidyut Prasaran Nigam Ltd, Jaipur, Rajasthan, India*

14.1 Introduction

With the growth of technology and research, conventional ways are being modified, and the whole world is gaining a new shape. The smart grid is one of the examples of technology advancement and research growth for efficient operation with improved overall efficiency. Forecasting is the most critical issue in the power system to maintain a healthy equilibrium between power generation and the connected load. The Internet of Things (IoT) and machine learning (ML) are one of the technology combinations that can help real-time power generation issues as reflected by the existing loads [1]. IoT will help in providing real-time information in terms of different parameters related to power generation, monitoring of the system's condition in real-time, fault conditions in terms of location and type and aspects of forecasting, leading to efficiency improvement [2]. The renewable energy contribution in the global power sector has made tremendous progress in 2019 with an installed power capacity of 200 Gigawatts (GW). Presently, solar-PV based renewable energy sources increased by 12% in 2019 to a record 115 GW (direct current), for a total of 627 GW [3]. In this context, India also contributes to a massive amount of solar-PV installation in various places in India. As per the annual investment/production in 2019, India is placed at third position and secures fifth position in total capacity or generation. The present installed capacity of the solar-PV is 30 GW. Recently India has started a National Action Plan on Climate Change with a National Solar Mission and has set a binding target of 100 GW by the end of 2022 [4]. Hence, renewable energy penetration [5], their issues [6], and solutions [7] need to be taken care of. It has been recognized from the literature that a solar-PV plant generates big data (BD) in the form of plant parameters, irradiations, temperature, and relative humidity, etc. The Government of India has established a total of 121 Solar Radiation Resource Assessment (SRRA) stations across the country [8]. A total of 37 parameters are collected and transmitted to the main server at the Centre for Wind Energy Technology (C-WET), Chennai, India, through an Internet facility. Meteorological data are used to predict the weather conditions, responsible for solar PV output [9]. BD is the most critical set of hidden information in terms of different dependent and independent variable parameters. This hidden data information is used by ML to extract the required information on efficient operation for the system [10]. BD cannot be categorized by data volume; it can be of any size, like terabyte, petabyte, or exabyte. These

Active Electrical Distribution Network: A Smart Approach, First Edition. Edited by Baseem Khan, Josep M. Guerrero, Sanjeevikumar Padmanaban, Hassan Haes Alhelou, Om Prakash Mahela, and Sudeep Tanwar.

big data need to be studied for setting active steps of the operation to enhance overall efficiency under the off-grid condition and grid integration. For this purpose, BD is used for analytics to find the meaningful patterns in the existing data to be used by machine learning (ML) and artificial intelligence (AI) to interpret the past and present optimization, and predictions for the future. Here, IoT facilitates the collection of the aggregate data needed for ML and AI, where machine learning is an integral part of AI [11]. Together these technologies are revolutionizing the power sector. The Internet of Things (IoT), machine learning, and artificial intelligence boost operational performance and, most importantly, develop intelligent machines to aid in day-to-day functioning for efficient control and performance [12]. In [13], a comprehensive review is given of the impact of partial shading on various PV array configurations and different modeling approaches. In [14], recognition of power quality issues associated with grid-integrated solar-photovoltaic plant in experimental framework is presented. In [15], a protection scheme for a power system with solar energy penetration is presented. In [16], an algorithm for the protection of distribution feeder using Stockwell and Hilbert transforms supported features is presented. In [17], an algorithm for the recognition of fault conditions in the utility grid with renewable energy penetration is presented. There have been critical research and development going on over the exploitation of solar-PV sources of energy for technological development, environmental awareness, and better production for efficiency improvement. Hence this chapter provides an overview of forecasting aspects of improving plant efficiency of the solar-PV system using the Internet of Things and machine learning techniques.

The layout of this chapter is as follows: Section 14.2 explains the forecasting methods used in the literature and factors affecting and challenges associated with the solar-PV forecasting. The basics of the Internet of Things and machine learning techniques and their applications are described in Sections 14.3 and 14.4. Section 14.5 elaborates on forecasting and efficiency improvement. Real-time application of the solar-PV system and explanation of SRRA data collection are in Section 14.6. Finally, a conclusion is drawn in Section 14.7.

14.2 Forecasting

The prediction of load demand is most important to ensure and map a healthy operating condition for the power generation system. Also, a prediction about the load demand for a day–week–month and the year is exceedingly tricky and depends upon the number and types of consumers [18]. Moreover, forecasting indeed improves the efficiency of solar-photovoltaic-based power generation systems by managing the existing power generation in advance with the load demand, leading to cost reduction. Forecasting needs a massive amount of data over a long period with real-time information of the system to extract meaningful information in different curve shapes and comes here with IoT–ML as the smart solution [19]. Nowadays, the accuracy of forecasting depends upon the IoT and ML. IoT application is highly capable of extracting real-time data and can store it through wired and wireless networks for further use [20]. Forecasting is categorized as short-term forecasting, medium-term forecasting, and long-term forecasting. Future load demand defines the power to be generated to meet the equilibrium position among power generation and load demand. Therefore, forecasting is related to power generation issues expressed by the connected loads and weather conditions in an area. Short-term forecasting represents the prediction of load demand for the next hour or the next day. However, medium-term forecasting represents the prediction of load demand for up to one year. It is an essential category of forecasting, which involves load switching patterns and planning for maintenance. Long-term forecasting represents the prediction of power generation for load demand planning for a couple of years.

14.2.1 Factors Affecting Forecasting

1. Time Factor
   - Hours of the day
   - Day of the week
   - Time of the year
2. Weather conditions
   - Ambient temperature
   - Irradiations
   - Cloud conditions
   - Relative humidity
   - Wind speed
   - Atmospheric pressure, etc.
3. Class of customers
   - Industrial
   - Agriculture
   - Commercial
   - Residential
4. Special events
   - Festival like Diwali, International Sports Events, etc.
5. Economical indication
   - Gross National Product (GNP)
   - Per Capita Income
   - Gross Domestic Product (GDP), etc.
6. Use of new technology
   - Electric vehicles
   - Power electronics devices, etc.
7. Generation cost and population, etc.

14.2.2 Significant Challenges Associated with Forecasting

1. Forecasting depends upon weather conditions, which arc highly unexpected.
2. Different regions have different weather conditions.

3. Difficulty in getting accurate data on consumer behavior.
4. The complex nature of loads makes forecasting difficult.
5. Security and privacy issues in the IoT.

14.3 Internet of Things

The IoT is a set of sensors and devices having wired and wireless network communication through the Internet to send and receive real-time information without human intervention. Measuring and communicating the data over the system for monitoring and control in real-time is performed by the IoT application [21]. IoT helps establish a data network to feed the system with necessary decisions so that overall efficiency increases. It provides the self-decision-making capability to the system, which can monitor to perform the task of self-healing [22]. IoT is making improvements on a large scale to connect the different power devices. In this regard, Figure 14.1 shows the various components of the IoT. IoT-based systems can be wired or wireless and can work in online and offline modes [23]. In addition, the sensor senses the parameter to produce BD in a raw form. Encryption means changing the original shape of the data or locking it, and decryption is the reverse process to that of the encryption. The data is retrieved in its original form by decryption. Encryption is the part of security issues in which information is protected during travel through any channel. Upon receiving the data, the decryption process converts the data into its original form. Encryption and decryption are like locking and unlocking the data with a unique key to maintain privacy.

14.4 Machine Learning

Machine learning is a programmed algorithm-based decision-making tool that can work with or without human intervention. It is seen as a subset of artificial intelligence and can improve through experience. Machine learning has three different types, as shown in Figure 14.2.

14.4.1 Supervised Learning

In this type of machine learning, a labeled dataset is used to train the machine. A labeled dataset means that the output is known for every input used to prepare it [24]. Such a dataset is termed as a training dataset. In supervised learning, the machine acts on every input as per the input–output relation provided by the training dataset.

Regression, Classification, Naive Bayesian Model, Random Forest Model, Neural Networks, and Support Vector Machine are types of supervised learning [25, 26].

Regression:	Single output using training data.
Classification:	Grouping the data into classes.
Naive Bayesian Model:	Used for the large finite dataset.
Random Forest Model:	Operates by building a broad set of decision trees and outputs a classification of the individual trees.

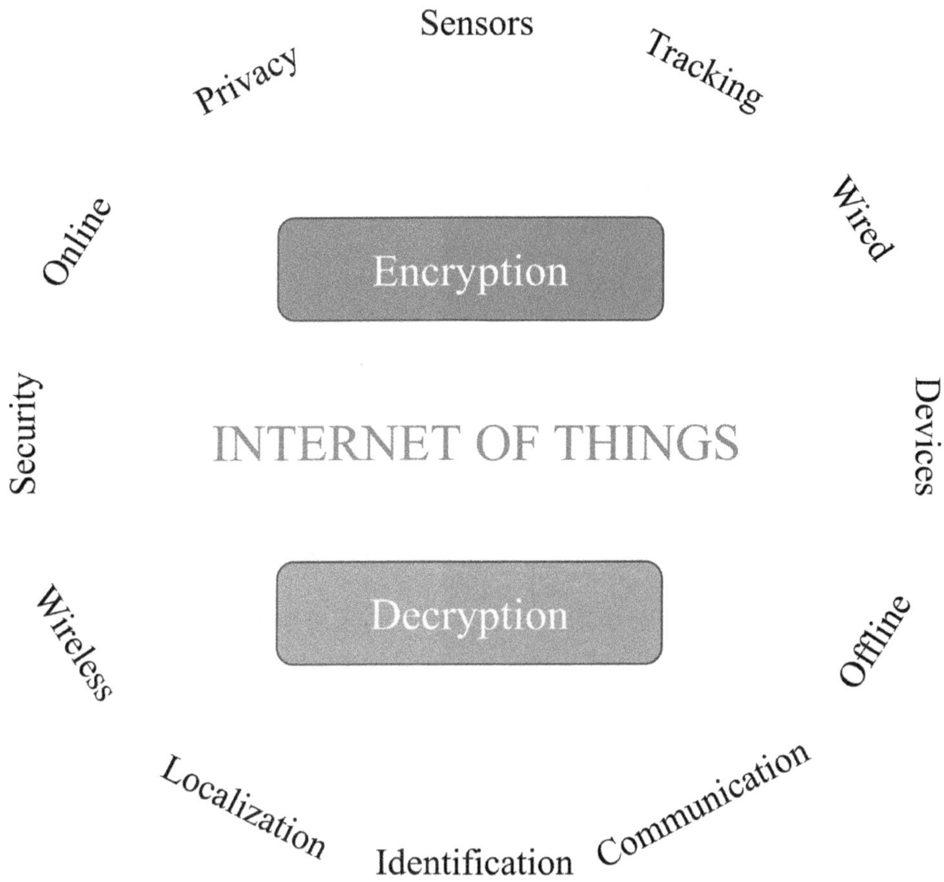

Figure 14.1 Components of Internet of Things (IoT).

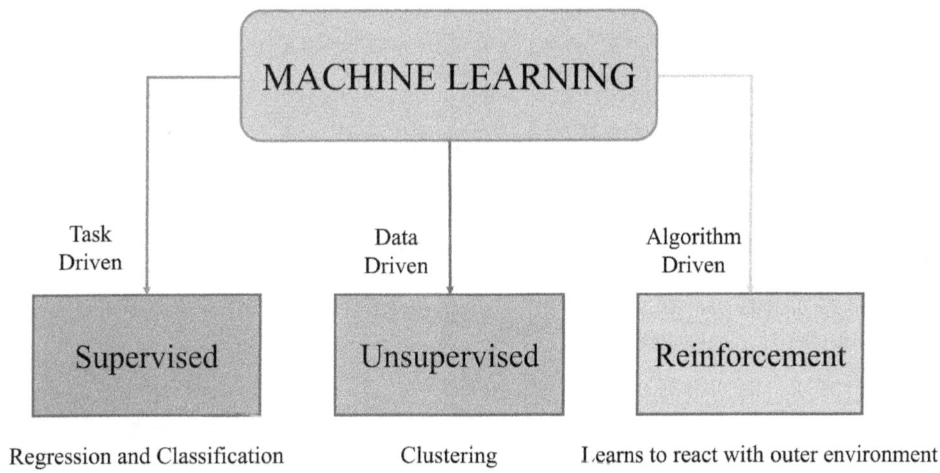

Figure 14.2 Types of machine learning.

Neural Networks: The designed algorithm is used to shape raw input, recognize patterns to interpret sensory data as per the labeled dataset used for learning.

Support Vector Machine: These are supervised learning models with associated learning algorithms that analyze data utilized for classification and regression analysis.

Note: Neural networks are simply architectures and can be used for both supervised and unsupervised learning. It depends whether there is a supervision signal or whether they are only used to extract data patterns.

14.4.2 Unsupervised Learning

This deals with the unlabeled dataset and has minimum human intervention. In this machine learning process, the machine learns to find hidden patterns and information not extracted previously. No labeled dataset is provided and the model finds a different structure on its own. Parametric Unsupervised Learning and Non-parametric Unsupervised Learning are two categories of Unsupervised Learning. Parametric unsupervised learning is based on the parametric distribution of the data. However, non-parametric unsupervised learning is based upon the cluster formation. Here, in non-parametric unsupervised learning, the dataset is grouped into clusters. Each cluster defines different information contained by data of each cluster. Clustering is done to find the structures in a set of data received from a group of sensors over a said period.

14.4.3 Reinforcement Learning

In this type of machine learning, the algorithm learns by interacting with the related environment. It is based on the reward and punishment system. The algorithm steps are marked by reward for correct performance and marked by the penalty for incorrect interpretation. The goal is to get top tips and to minimize the penalties without human intervention. It works on its own to solve the problem. Reinforcement learning is a long-term process as it requires a more extended period to analyze the system's environment and minimize the errors towards the goal [27]. Also, state-action-reward-state-action (SARSA), Q-learning, and Deep Q networks are types of reinforcement learning.

14.5 Forecasting and Efficiency Enhancement

Forecasting and efficiency improvement revolves around data collection. Data collected in real-time from experience and present work is utilized to extract hidden information and then go through the ML process to estimate forecast and efficiency. All the data collection processes are completed under the term IoT application and go through the necessary doors for final task completion.

14.5.1 Door 1: System Consideration

Under this door, we go through the solar PV system for the study of different types of parameters and dependencies on sunlight and all other environmental conditions. Sunlight is a combination of different wavelengths, absorbed by the solar-photovoltaic system as fuel to output electrical energy. Solar-photovoltaic output depends on environmental conditions like ambient temperature, wind speed, clouds, relative humidity, etc. Hence, we create a map of requirements in terms of the necessary sensors and their placement in the system to carry out the information in real-time.

14.5.2 Door 2: Data Collection – IoT

The Internet of Things is implemented within the system to collect valuable real-time information in terms of different electrical and non-electrical parameters. Accuracy and precision are essential to have high-quality data in ready-to-use formats. Pyranometer, Shaded Pyranometer, Pyrheliometer, Solar Tracker, Ultrasonic Wind Sensor, Rain Gauge, Pressure Sensor, Temperature Sensor, Relative Humidity Sensor, Data Logger and Modem, and the Global Positioning System (GPS) are used in a solar PV system for gathering real-time information.

14.5.3 Door 3: Machine Learning

Here the focus is on using ML to do even more with the data. ML helps in refining the process by removing unnecessary and repeated information. The system can perform more complicated procedures with refined details and can have the right information and decisions to improve overall efficiency and forecasting.

14.5.4 Door 4: Learning about System Dependencies – Forecasting

Under this, the refined and the most required data are analyzed to measure the impact on the system's characteristics. This information is used to learn by the system dependencies for proper planning. This information is plotted in different shapes and patterns, and is then analyzed to predict future values, i.e. forecasting.

14.5.5 Door 5: Improving Performance

Refined data contains all the necessary information required to maximize the output of the solar-PV system. Accurate prediction of future trends helps to improve overall efficiency.

14.6 Real-Time Application in a Solar-PV System

Forecasting revolves around collecting related data, modification, and refining into a usable form to extract the exact future predictive information [28]. For this, the system requires several sensors to measure the different parameters, such as horizontal global irradiance, horizontal diffuse irradiance, direct normal solar irradiance, sun movement, ambient temperature, wind speed, rain, clearness index, humidity, and air pressure. These parameters define the solar-photovoltaic-based power generation's output and hence are termed efficiency parameters. Maximum power tracking [29], employed to increase the power output, can also benefit through such sets of efficiency parameters. This will help in enhancing renewable energy penetration and maintaining smooth grid synchronization [30].

Therefore, the efficiency of the solar-photovoltaic-based power generation can be improved and can be predicted for the future using such parameters. The IoT is a tool to collect and store the sensed data in terms of different parameters in a remote location, where it is used in the ML process for extracting valuable information, which has a significant impact on overall efficiency improvement and future trend predictions.

Solar radiations vary from morning to evening, with latitude, longitude, altitude, and season. Hence, it becomes necessary to know inadequate details of solar resources quite accurately when making decisions for the solar power plant in order to achieve higher efficiency in power generation.

In Figure 14.3, the IoT setup and ML implementation are shown to work for forecasting and efficiency enhancement of the solar-photovoltaic-based power generation. The system is observed and analyzed through sight inspection to know the count and placement of the sensors and IoT devices. All sensors and tools are in a communication link with each other. Wired and wireless communications are two paths used to transmit and receive data from sensors and devices in real-time [2]. The data are encrypted for the security point of the issue before transmitting over a channel. Upon receiving the information on another site, it is necessary to decrypt it; otherwise, it will not work. Encryption and decryption are a process of locking the data and providing a key for authentication to the user to decrypt it. After collection of the dataset, it is processed using ML techniques to get the best minimum data for delivering a maximum impact on efficiency and forecasting. Control decisions are designed based upon data received in real-time to optimize the solar PV plant in accordance with load demand and environmental conditions [31]. In reality, by this method, losses are minimized by making accurate future predictions, which helps in optimizing the solar-photovoltaic-based power plant generation in advance for an efficient, secure, and reliable operation.

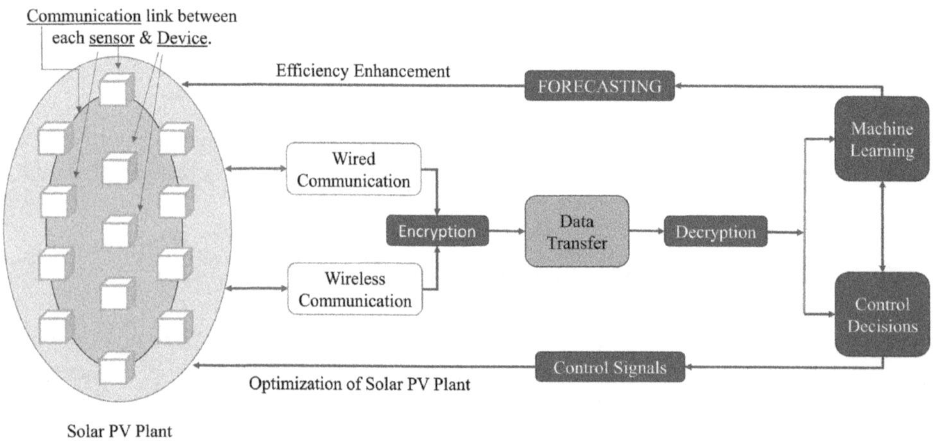

Figure 14.3 IoT and machine learning implementation in Solar PV system.

The Ministry of New and Renewable Energy in the Indian Government has established a network of solar radiation resource assessment (SRRA) stations with an autonomous R&D Institute, C-WET, in Chennai, India. C-WET, Chennai, has a rich experience in developing a wind atlas for the nation. Every second, data on 37 parameters for both solar and meteorological data are collected from each SRRA station and received at the central control server of C-WET, Chennai, through the Internet. SRRA stations are established in two phases in different states of India, as mentioned in Table 14.1 [32]. To date, a total of 51 SRRA stations in phase I, 60 SRRA stations in phase II, a 06 Advanced Measurement Station (AMS), and a 04 SRRA station for the Maharashtra Energy Development Agency (MEDA) have been established across the country [32].

Table 14.1 SRRA Stations established in India.

		SRRA station				
S. No.	**State/UT**	**Phase I**	**Phase II**	**MEDA**	**AMS**	**Total**
1	Assam	–	2	–	–	2
2	Andhra Pradesh	5	1	–	–	6
3	Andaman and Nicobar Islands (UT)	–	1	–	–	1
4	Arunachal Pradesh	–	2	–	–	2
5	Bihar	–	3	–	–	3
6	Chhattisgarh	1	1	–	–	2
7	Chandigarh (UT)	–	1	–	–	1
8	Dadra and Nagar Haveli (UT)	–	1	–	–	1
9	Gujarat	11	1	–	1	13
10	Haryana	1	1	–	1	3
11	Himachal Pradesh	–	2	–	–	2
12	Jammu and Kashmir	1	2	–	–	3
13	Jharkhand	–	3	–	–	3
14	Karnataka	5	1	–	–	6
15	Kerala	–	2	–	–	2
16	Madhya Pradesh	3	4	–	–	7
17	Manipur	–	1	–	–	1
18	Mizoram	–	1	–	–	1
19	Maharashtra	3	6	6	–	15
20	Meghalaya	–	1	–	–	1
21	Nagaland	–	1	–	–	1
22	Odisha	–	4	–	–	4

(*Continued*)

Table 14.1 (*Continued*)

		SRRA station				
S. No.	**State/UT**	**Phase I**	**Phase II**	**MEDA**	**AMS**	**Total**
23	Punjab	–	2	–	–	2
24	Pondicherry (UT)	1		–	–	1
25	Rajasthan	12	1	–	–	13
26	Sikkim	–	1	–	–	1
27	Tamil Nadu	7	1	–	1	9
28	Tripura	–	1	–	–	1
29	Telangana	1	2	–	–	3
30	Uttar Pradesh	–	5	–	–	5
31	Uttarakhand	–	2	–	–	2
32	West Bengal	–	3	–	1	4
	Total	51	60	6	4	121

14.6.1 SRRA Station (Jodhpur, India)

All the instruments and sensors are placed on two different towers, one of 1.5 m and one of 6 m height. However, a pyranometer with the solar tracker, a pyranometer with a shading disc, and a pyrheliometer are placed on a 1.5 m tower for measuring irradiation, based on global, direct, and diffused irradiation. Instruments and sensors for measuring relative humidity, wind speed, wind direction, rain, ambient temperature, atmospheric pressure, and a data acquisition system with an Internet connection are placed on the 6 m tower. All the sensors are traceable to the World Meteorological Organization (WMO) and the World Radiometric Reference (WRR) with high accuracy to ensure the excellent quality of the recorded data. Data are received at C-WET Chennai from each SRRA station, through GPRS [8]. Table 14.2 presents the solar-PV measuring instruments. Tables 14.3 and 14.4 present the meteorological measuring instruments and general instruments [32].

14.6.2 SRRA Data Collection and Prediction

In every small fraction of time, the SRRA station collects 37 different parameters through various sensing processes. The Internet of Things is the most crucial part of this process, facilitating the sensing, communication between devices, and transfer of data information through the wired or wireless method. Data collected from the sensors is named as raw data. Before processing ML's raw data, a copy of this is saved at a separate memory location called a mirror. After performing the ML task, the refined data is then available for future prediction, i.e. forecasting, and to optimize the solar-photovoltaic power generation as per load demand and weather conditions as per future predictions, hence improving the overall efficiency of the system. Finally, the data are presented monthly, hourly, a couple of

Table 14.2 Solar-PV measuring instruments.

S. No.	Sensor/instrument	Parameter/purpose	Image
(a)	Pyranometer	Global horizontal irradiance (W/m^2) Average global energy (kWh/m^2 and MJ/m^2)	
(b)	Shaded pyranometer	Diffused horizontal irradiance (W/m^2) Average diffused energy (kWh/m^2 and MJ/m^2)	
(c)	Pyrheliometer	Direct normal irradiance (W/m^2) Average direct energy (kWh/m^2 and MJ/m^2)	
(d)	Solar tracker	With and without shading disc	

Table 14.3 Meteorological measuring instruments.

S. No.	Sensor/instrument	Parameter/purpose	Image
(a)	Ultrasonic wind sensor	Wind speed (m/s) Wind direction (degree)	
(b)	Rain gauge	Rainfall (mm)	
(c)	Pressure sensor	Atmospheric pressure – QFE and QNH (mb/hpa)	
(d)	Temperature and relative humidity sensor	Temperature (°C) Dew point	

Table 14.4 General instruments.

S. No.	Sensor/instrument	Parameter/purpose	Image
(a)	GPS	For synchronization of sun tracker with sun movement	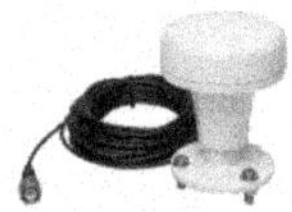
(b)	Data logger and modem	For logging sensor data and transfer	
(c)	GPRS	For transmission of data through mobile sim-card having Internet facility	
(d)	Cleaning switch	Cleaning status tracking	
(e)	External battery	Storage	

minutes, and one minute formats. The collected data from the SRRA station are transmitted wirelessly via the Internet to the Chennai-based server [8]. Figure 14.4 presents the SRRA data collection and processing methodology.

Six different parameters, horizontal global, horizontal diffuse, extra-terrestrial, clearness index, ambient temperature, and wind velocity, have been collected by the SSRA station, Jodhpur, India, in the year 2019, shown in Table 14.5 [33]. The collection of data demands a more extended period, and as of now in this chapter, only the available data from the National Institute of Solar Energy (NISE) is used. This data is used to generate the graphs representing the variation of parameters with time and weather conditions, as shown in Figures 14.5 and 14.6.

In Figure 14.5, solar irradiations are available between March and October. In the May–June month maximum, solar irradiation is available for the solar-photovoltaic panel. On the other part, if we look at Figure 14.6, the temperature rises from March to October and is maximum in the month of May–June. The power generation of the solar-photovoltaic system increases with irradiation, but the temperature rise has a reverse impact on the solar-photovoltaic system's power output. Excess temperature reduces the power output and hence reduces the power conversion efficiency. By getting a proper knowledge of such

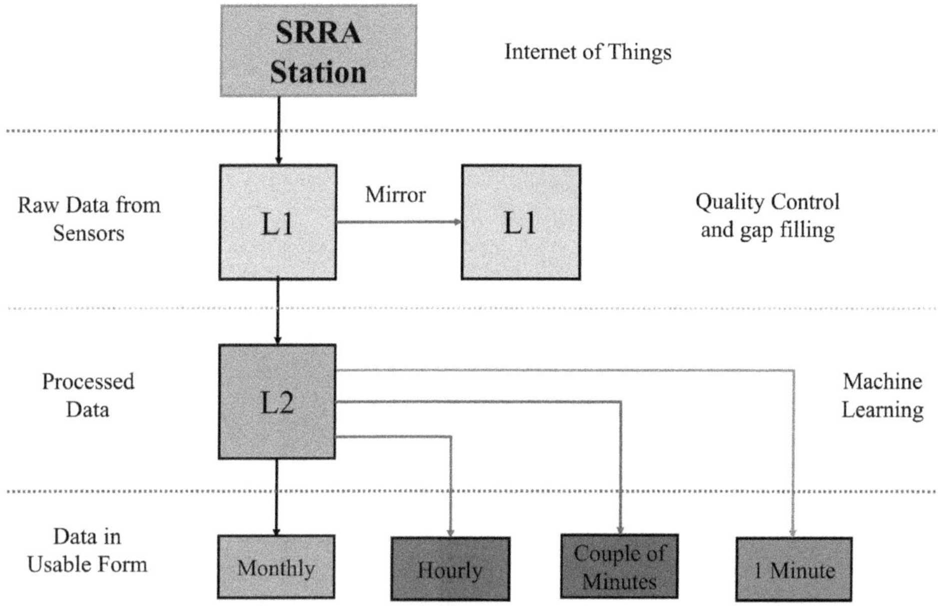

Figure 14.4 SRRA data collection and processing.

Table 14.5 Meteo data for Jodhpur, India.

Geographical Site:
Jodhpur, Country: India, Meteonorm 7.2 Station
Latitude – 26.30° Longitude – 73.02° Altitude – 217 m
Legal time Time zone UTC/GMT + 5.5 hours

Monthly meteo values	Horizontal global (kWh/m²)	Horizontal diffuse (kWh/m²)	Extra-terrestrial (kWh/m²)	Clearness index (ratio)	Ambient temperature (°C)	Wind velocity (m/s)
January	133.8	34.7	202.9	0.659	17.9	0.9
February	144.6	37.4	218.5	0.662	20.8	0.9
March	188.4	56.4	284.0	0.663	26.8	1.0
April	202.0	71.6	311.7	0.684	31.5	1.2
May	218.4	85.3	343.2	0.636	34.8	2.1
June	187.2	98.0	338.8	0.553	33.6	2.0
July	167.7	101.7	346.3	0.484	31.1	1.6
August	162.5	85.5	330.3	0.492	29.8	1.3
September	172.1	67.5	289.1	0.595	30.0	1.0
October	162.0	49.0	256.3	0.632	28.6	0.4
November	126.3	39.5	206.5	0.612	23.1	0.5
December	119.2	27.9	191.0	0.624	18.6	0.8
Year –2019	1984.2	754.5	3318.8	0.598	27.2	1.1

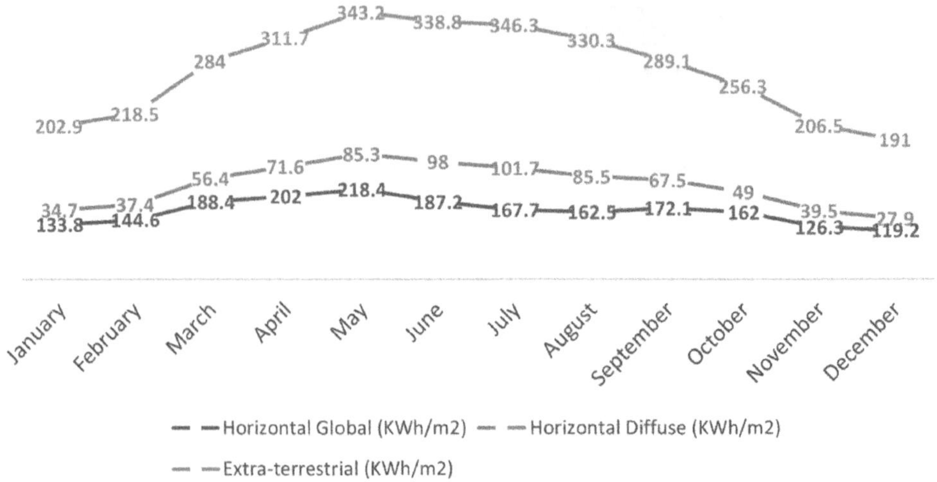

Figure 14.5 Variation of solar irradiations with time and weather change.

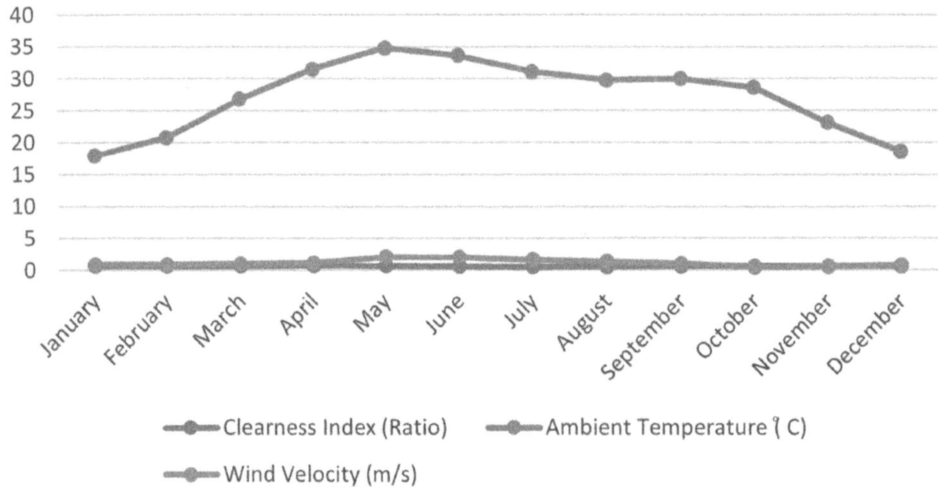

Figure 14.6 Change in clearness index, ambient temperature, and wind velocity in a year.

parameters, the structure can be optimized to meet the load demand in advance for future applications. Therefore, we can forecast the solar photovoltaic system's behavior based upon the necessary information produced by the IoT application and ML to get refined information for optimizing and enhancing system efficiency.

Hourly data is constructed with the help of monthly data collected from the selected site at Jodhpur, India. Figure 14.7 shows the sun path at Jodhpur, India, having latitude 26.3000°N,

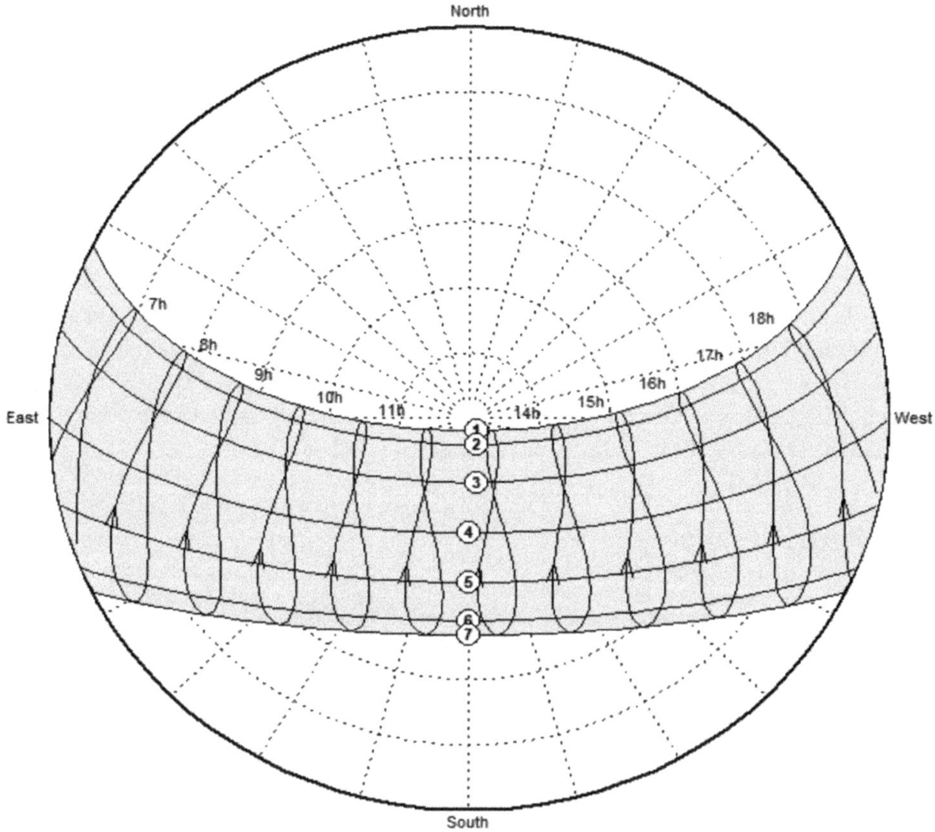

Figure 14.7 Sun paths at Jodhpur (latitude 26.3000°N, longitude 73.0200°E, altitude 217 m) – Legal time, polar coordinates.

longitude 73.0200°E, and altitude 217 m in polar coordinates [33]. The sun path represents a lot of information about the impact of the sun on Jodhpur. It can help in providing the solar azimuth and altitude for a given location. Sunlight is concentrated on Earth during summer and during winter, spread over a large area. The Earth's surface receives the same amount of sun in spring and autumn. Therefore, based on this information, a prediction about the future availability of solar-PV plant can be done in an efficient way.

14.7 Conclusion

IoT–ML has gained a significant important position in technological research among researchers at every level. Even in forecasting, IoT and ML are in use to enhance the solar-PV plant's efficiency. Solar-PV depends totally on weather conditions of the region where the solar-PV plant is situated. Therefore, it is necessary to have real-time information of such environment-based parameters in order to predict future trends. Therefore, IoT and ML have been introduced to solar-PV for enhancing the overall efficiency and preparing for mitigating future issues in advance. IoT provides real-time data/information every second, every day, for a year or more. Now, such data/information is further refined through the ML application to get the most valuable and influencing information used for predicting future requirements and issues. Therefore, the Government of India has established 121 SRRA stations in different states to collect a large set of data/information in terms of various weather parameters. We have selected the meteorological data of Jodhpur, India, collected by SRRA, a 7.2 station having geographic coordinates of latitude 26.3000°N, longitude 73.0200°E, and altitude 217 m. After refining the data collected from the SRRA station, we conclude that sunlight was concentrated on Earth during summer and during winter, spread over a large area. Earth's surface received the same amount of sun in spring and autumn. Therefore, we can easily predict and plan the power production in future. This will help the power sector in smooth functioning with higher reliability.

References

[1] R.M. Pratapa and A. Jaya Laxmi, "IOT based online load forecasting using machine learning algorithms," *Procedia Computer Science*, vol. 171, pp. 551–560, 2020.

[2] T. Rieger, S. Regier, I. Stengel, and N.L. Clarke, "Fast predictive maintenance in Industrial Internet of Things (IIoT) with Deep Learning (DL): A review", *CERC*, pp. 69–80, 2019.

[3] R. Secretariat, "Renewables 2020 global status report," Report Paris: REN12, 2020.

[4] Available at: https://mnre.gov.in/solar/current-status.

[5] G.S. Chawda and A. Gafoor Shaik, "Adaptive reactive power control of DSTATCOM in weak AC grid with high wind energy penetration," *2019 IEEE 16th India Council International Conference (INDICON)*, Rajkot, India, pp. 1–4, 2019, doi: 10.1109/INDICON47234.2019.9029010.

[6] G.S. Chawda, A.G. Shaik, M. Shaik, S. Padmanaban, J.B. Holm-Nielsen, O.P. Mahela, and P. Kaliannan, "Comprehensive review on detection and classification of power quality disturbances in utility grid with renewable energy penetration," *IEEE Access*, vol. 8, pp. 146807–146830, 2020.

[7] G.S. Chawda, A.G. Shaik, O.P. Mahela, S. Padmanaban and J.B. Holm-Nielsen, "Comprehensive review of distributed FACTS control algorithms for power quality enhancement in utility grid with renewable energy penetration," *IEEE Access*, vol. 8, pp. 107614–107634, 2020, doi: 10.1109/ACCESS.2020.3000931.

[8] Available at: https://niwe.res.in/NIWE_OLD/html/srra_station.html.

[9] E.B. Abrahamsen, O.M. Brastein, and B. Lie, "Machine learning in python for weather forecast based on freely available weather data," *Proceedings of the 59th Conference on Simulation and Modelling (SIMS 59)*, 26–28 September 2018, Oslo Metropolitan University, Norway, 2018.

[10] H. Wang, Z. Lei, X. Zhang, B. Zhou, and J. Peng, "A review of deep learning for renewable energy forecasting," *Energy Conversion and Management*, vol. 198, p. 111799, 2019.

[11] A.E. Braten and F.A. Kraemer, "Towards cognitive IoT: Autonomous prediction model selection for solar-powered nodes," *2018 IEEE International Congress on Internet of Things (ICIOT)*, IEEE, pp. 118–125, 2018.
[12] F.A. Kraemer, D. Palma, A.E. Braten, and D. Ammar, "Operationalizing solar energy predictions for sustainable, autonomous IoT device management," *IEEE Internet of Things Journal*, vol. 7, no. 12, pp. 11803–11814, December 2020.
[13] R.K. Pachauri, O.P. Mahela, A. Sharma, J. Bai, Y.K. Chauhan, B. Khan, and H.H. Alhelou, "Impact of partial shading on various PV array configurations and different modeling approaches: A comprehensive review," *IEEE Access*, vol. 8, pp. 181375–181403, 2020, doi: 10.1109/ACCESS.2020.3028473.
[14] O.P. Mahela, A.G. Shaik, N. Gupta, M. Khosravy, B. Khan, H.H. Alhelou, and S. Padmanaban, "Recognition of power quality issues associated with grid integrated solar photovoltaic plant in experimental framework," *IEEE Systems Journal*, October 2020. doi: 10.1109/JSYST.2020.3027203.
[15] S. Ram Ola, A. Saraswat, S.K. Goyal, S.K. Jhajharia, B. Khan, O.P. Mahela, H. Haes Alhelou, and P. Siano, "A protection scheme for a power system with solar energy penetration," *Applied Sciences*, vol. 10, p. 1516, 2020.
[16] O.P. Mahela, J. Sharma, B. Kumar, B. Khan, and H.H. Alhelou, "An algorithm for the protection of distribution feeder using Stockwell and Hilbert transforms supported features," *CSEE Journal of Power and Energy Systems*, July 2020. doi: 10.17775/CSEEJPES.2020.00170.
[17] G.S. Yogee, O.P. Mahela, K.D. Kansal, B. Khan, R. Mahla, H. Haes Alhelou, and P. Siano, "An algorithm for recognition of fault conditions in the utility grid with renewable energy penetration," *Energies*, vol. 13, p. 2383, 2020.
[18] K. Zaouali, R. Rekik, and R. Bouallegue, "Deep learning forecasting based on auto-lstm model for home solar power systems," *2018 IEEE 20th International Conference on High Performance Computing and Communications; IEEE 16th International Conference on Smart City; IEEE 4th International Conference on Data Science and Systems (HPCC/SmartCity/DSS)*, IEEE, pp. 235–242, 2018.
[19] R.R. Batcha and M. Kalaiselvi Geetha, "A survey on IoT based on renewable energy for efficient energy conservation using machine learning approaches," *2020 3rd International Conference on Emerging Technologies in Computer Engineering: Machine Learning and Internet of Things (ICETCE)*, IEEE, pp. 123–128, 2020.
[20] N. Tang, S. Mao, Y. Wang, and R.M. Nelms, "Solar power generation forecasting with a LASSO-based approach," *IEEE Internet of Things Journal*, vol. 5, no. 2, pp. 1090–1099, 2018.
[21] N.S. Deshmukh and D.L. Bhuyar. "A smart solar photovoltaic remote monitoring and controlling," *2018 Second International Conference on Intelligent Computing and Control Systems (ICICCS)*, IEEE, pp. 67–71, 2018.
[22] IRENA, "Innovation landscape brief: Internet of Things", International Renewable Energy Agency, Abu Dhabi, 2019.
[23] L. Li, K. Ota, and M. Dong, "When weather matters: IoT-based electrical load forecasting for smart grid," *IEEE Communications Magazine*, vol. 55, no. 10, pp. 46–51, 2017.
[24] Y. Zhou, S. Zheng, and G. Zhang, "Machine-learning based study on the on-site renewable electrical performance of an optimal hybrid PCMs integrated renewable

system with high-level parameters' uncertainties," *Renewable Energy*, vol. 151, pp. 403–418, 2020.

[25] A. Javed, B.K. Kasi, and F.A. Khan, "Predicting solar irradiance using machine learning techniques," *2019 15th International Wireless Communications & Mobile Computing Conference (IWCMC)*, IEEE, pp. 1458–1462, 2019.

[26] F. Jawaid and N. Khurum, "Predicting daily mean solar power using machine learning regression techniques," *2016 Sixth International Conference on Innovative Computing Technology (INTECH)*, IEEE, pp. 355–360, 2016.

[27] M. Adibi and J. van der Woude, "A reinforcement learning approach for frequency control of inverted-based microgrids," *IFAC – PapersOnLine*, vol. 52, no. 4, pp. 111–116, 2019.

[28] S. Balaji and S. Karthik, "Energy prediction system using internet of things," *2020 6th International Conference on Advanced Computing and Communication Systems (ICACCS)*, IEEE, pp. 1131–1135, 2020.

[29] G.S. Chawda, O.P. Mahela, N. Gupta, M. Khosravy, and T. Senjyu, "Incremental conductance based particle swarm optimization algorithm for global maximum power tracking of solar-PV under nonuniform operating conditions," *Applied Sciences*, vol. 10, no. 13, p. 4575, 2020.

[30] G.S. Chawda and A.G. Shaik, "Smooth grid synchronization in weak AC grid with high wind energy penetration using distribution static compensator," *2019 2nd International Conference on Smart Grid and Renewable Energy (SGRE)*, Doha, Qatar, pp. 1–6, 2019. doi: 10.1109/SGRE46976.2019.9020671.

[31] H. Mortaji, S.H. Ow, M. Moghavvemi, and H.A.F. Almurib, "Load shedding and smart-direct load control using internet of things in smart grid demand response management," *IEEE Transactions on Industry Applications*, vol. 53, no. 6, pp. 5155–5163, 2017.

[32] Available at: https://www.indiawaterportal.org/sites/indiawaterportal.org/files/srra_india_0.pdf.

[33] Available at: https://nise.res.in/departments/solar-resource-assessment-division.

15

Modular Design of Nonlinear Controllers for Photovoltaic Distributed Generation Systems

Meher Preetam Korukonda[1], *Man Mohan Garg*[2], *Laxmidhar Behera*[1]

[1] *Department of Electrical Engineering, Indian Institute of Technology Kanpur, India*
[2] *Department of Electrical Engineering, Malaviya National Institute of Technology Jaipur, India*

15.1 Introduction

Managing the power flow with advanced control strategies has become unavoidable in solar photovoltaic distributed generation (SPVDG) systems and are generally sorted by adopting a hierarchical control structure as in [1, 2]. The major control challenges for a PVDG system with battery energy storage are maximum power point tracking and DC voltage control.

Neely et al. [3] and Tahim et al. [4] propose a model predictive control technique based on a boost converter setup. However, the controller design is linear in nature, which makes the system operate only in a limited manner. It is limited in the sense that it does not cater for large changes in operating points. Zeng et al. [5] describe a passivity-type controller for a similar boost converter setup but the system is very sensitive to exact parameter tuning, which may create problems during implementation. In [6], a comprehensive review related to the microgrid optimization techniques is presented. In [7], recognition of power quality issues associated with grid-integrated solar photovoltaic plant in an experimental framework is presented. In [8], a comprehensive review related to the impact of partial shading on various PV array configurations and different modeling approaches is presented. In [9], recognition of islanding and operational events in the power system with renewable energy penetration using a Stockwell transform-based method is presented. In [10], an algorithm for recognition of fault conditions in the utility grid with renewable energy penetration is presented. Alnejaili et al. [11] adopt a sliding mode-based design to appropriately charge or discharge the battery to maintain voltage. However, many precautions need to be taken to select a sliding surface that can avoid unnecessary switching during sudden disturbances.

In its daily course, an SPVDG encounters a large region of operation and thus a far advanced and complex control strategy is justified. A backstepping controller is a good choice of controller design as it operates in a larger range of operation, especially when the system is underactuated, as in the SPVDG case. Moreover, a detailed stability analysis can be carried out with this kind of controller, which is very desirable for the system under consideration. A backstepping type of control has been designed for a shipboard power system to control voltage [12]. Xu et al. [13] and Roy and Mahmud [14] applied a similar backstepping technique in microgrids for voltage control.

Active Electrical Distribution Network: A Smart Approach, First Edition. Edited by Baseem Khan, Josep M. Guerrero, Sanjeevikumar Padmanaban, Hassan Haes Alhelou, Om Prakash Mahela, and Sudeep Tanwar.

However, all these works have been developed to control only a specific subsystem rather than the overall system. This chapter develops a generic modular backstepping-based design that can be applied for achieving multiple control goals in the SPVDG system. Also, a plug-and-play approach, such as the one developed in this work, also allows independent operation of each device in the system, assuring greater flexibility and improving reliability simultaneously. Finally, the use of such nonlinear techniques also improves the speed of response in the case of transients. It is also very easy to scale up to higher-order systems with many sub-systems. The system's stability is shown by constructing a series of Lyapunov functions. The overall system then guarantees good dynamic performance along with flexibility.

The following is the organization of this chapter. Section 15.2 describes the SPVDG system under consideration, its mathematical modeling, and its control hierarchy. Section 15.3 discusses the generation of secondary-level references for the system. Section 15.4 gives the complete stability proof of the piecewise backstepping control along with the designed controllers and observers. Section 15.5 shows the results that are obtained due to the occurrence of different intermittencies while Section 15.6 concludes the chapter.

15.2 System Description and Modeling

This section explains the details of different building blocks of the SPVDG system, its control structure, and the large signal modeling of the overall system.

15.2.1 Basic Blocks

The SPVDG system under consideration consists of a PV array and a battery energy storage system (BESS) feeding to a load. The PV array continuously extracts power from the available solar energy and transfers it to the load using a DC–DC converter. The BESS is connected to the grid via a bidirectional DC–DC converter. This is essential to modulate power flow in both directions as demanded by load conditions in the SPVDG system. Both of the converters are controlled using a duty cycle to regulate the desired flow of power in the PVDG system.

15.2.2 Hierarchical Control Structure

As evident from Figure 15.1, the hierarchy of control in the SPVDG system is present in two levels of the system, namely, primary and secondary. The secondary level controller handles the overall energy management function of the SPVDG system [15]. It is responsible for ensuring maximum power extraction from the PV panel and also plans the power flow of the battery depending on the imbalance between the PV power generation and load consumption. The secondary controller achieves this operation by setting the reference values for all the important states in the SPVDG system, like inductor currents and capacitor voltages. The major function of the primary controller is to bring the system states to the reference values set by the secondary controller. A properly designed primary controller performs this function, even in the presence of large disturbances affecting the system time to time.

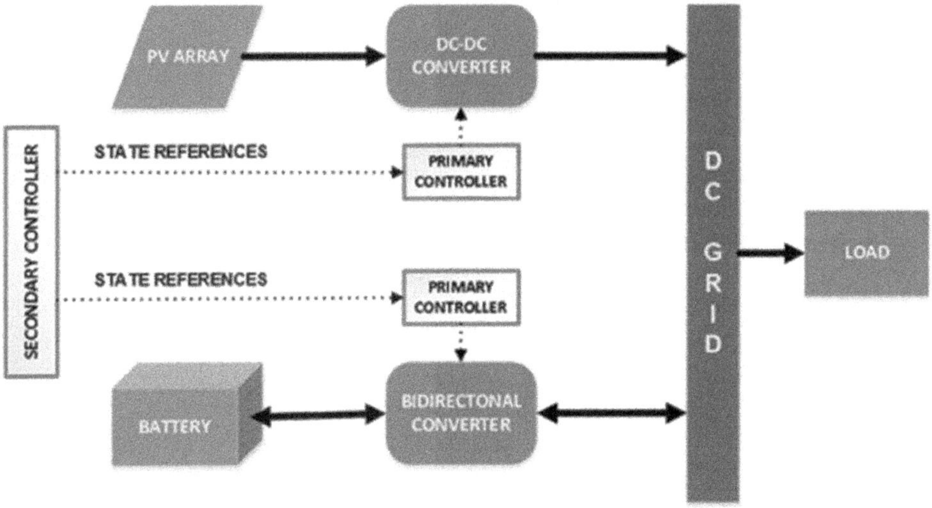

Figure 15.1 Structure of the SPVDG system.

15.2.3 State-Space Model

The power circuit architecture of the SPVDG system is shown in Figure 15.2, which is mathematically modeled in this section. The well-established state-space averaging technique is used for modeling the overall system due to an abundance of power electronic converters in the system.

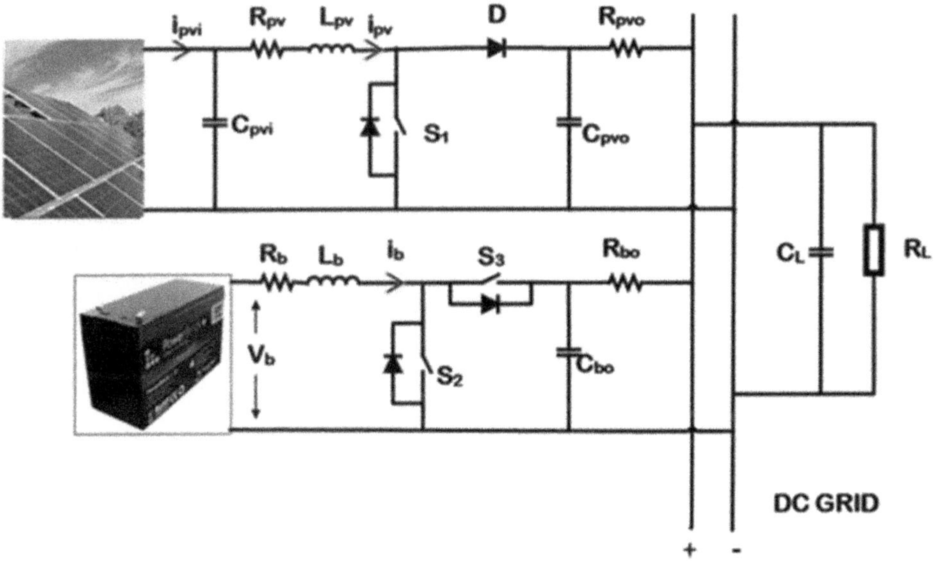

Figure 15.2 Power architecture of the SPVDG system.

The PV system consists of a PV array whose output current equation is given by

$$i_{pv} = n_p I_g - n_p I_s \left(e^{\frac{q(v_{pv} + i_{pv} R_s)}{n_s pKT}} - 1 \right) - \frac{v_{pv} + i_{pv} R_s}{R_{sh}} \tag{15.1}$$

The PV output current i_{pv} varies with a change in temperature and irradiance and needs to be continuously monitored. In the state-space model, this current has been considered as disturbance d_1. The output voltage of the PV panel, which is also the voltage across the capacitor, C_{pvi} has been designated as x_1, the voltage across the capacitor C_{pvo} has been termed x_2, and the current through the inductor L_{pv} has been designated x_3. All these states belong to the PV subsystem.

The battery has been modeled as a voltage source whose voltage is considered as disturbance d_2. Similarly, the current through the inductor L_b has been termed x_4 and the voltage across the capacitor C_{bo} has been designated as x_5. Both these states belong to the battery subsystem. Finally, the DC grid voltage, which is the voltage across C_L and is taken to be x_6 and load admittance $1/R_L$, is modeled as the disturbance d_3.

The state-space model derived for this system is obtained as shown in the following equations:

$$\begin{aligned}
\dot{x}_1 &= \frac{d_1}{C_{pvi}} - \frac{x_3}{C_{pvi}} \\
\dot{x}_2 &= \frac{x_3}{C_{pvo}} - \frac{x_2}{R_{pvo} C_{pvo}} + \frac{x_6}{R_{pvo} C_{pvo}} - \frac{x_3}{C_{pvo}} u_1 \\
\dot{x}_3 &= \frac{x_1}{L_{pv}} - \frac{x_2}{L_{pv}} - \frac{x_3 R_{pv}}{L_{pv}} + \frac{x_2}{L_{pv}} u_1 \\
\dot{x}_4 &= \frac{d_2}{L_b} - \frac{x_4 R_b}{L_b} - \frac{x_5}{L_b} + \frac{x_5}{L_b} u_2 \\
\dot{x}_5 &= \frac{x_4}{C_{bo}} - \frac{x_5}{R_{bo} C_{bo}} + \frac{x_6}{R_{bo} C_{bo}} - \frac{x_4}{C_{bo}} u_2 \\
\dot{x}_6 &= \frac{x_2}{C_L R_{pvo}} + \frac{x_5}{C_L R_{bo}} - \frac{x_6}{C_L} \left[\frac{1}{R_{pvo}} + \frac{1}{R_{bo}} + d_3 \right]
\end{aligned} \tag{15.2}$$

where

$$x = \left[V_{pvi} \; V_{pvo} \; i_{pv} \; i_b \; V_{bo} \; V_L \right]$$

$$d_1 = i_{pvi}, \; d_2 = V_b, \; d_3 = \frac{1}{R_L}$$

15.3 Secondary Level References

This section shows a technique where secondary level references can be generated. These can be generated in a centralized manner or a distributed manner as in [16, 17]. For the SPVDG system discussed in this chapter, the secondary level references are generated by the secondary controller in a centralized fashion and sent to the primary level controllers so that appropriate control action is taken to follow these secondary level references. Successful calculation/estimation of these references holds the key to a power balance of the entire SPVDG system. It also decides whether the bidirectional converter should work in the charging mode or discharging mode.

Since the state-space model of the system is known, equating the state-space equations to zero and solving them is the way forward for generating these references. However, it should be understood that this would not bring the system to work at the desired voltage and current levels. Certain known/desired voltage and current references need to be considered by which other references can also be calculated.

It is assumed that the microgrid needs to operate at a voltage V_{DC} and that the PV output voltage should be maintained at V_{mpp} to ensure maximum power extraction from the panel/array. While V_{DC} is user defined, V_{mpp} is either determined using online MPPT algorithms or model-based algorithms, as clearly described in [18]. These two references serve as the basis for generating the references for all the states in the system. The following set of equations denotes the values of references used for the current SPVDG system:

$$x_{6ref} = V_{DC}\left(known\right)$$

$$x_{1ref} = V_{mpp}\left(known\right)$$

$$x_{3ref} = I_{mpp}\left(known\right)$$

$$a_1 = 1, \quad b_1 = -x_{6ref}$$

$$c_1 = x_{3ref}^2 R_{pvo} R_{pv} - x_{1ref} x_{3ref} R_{pvo}$$

$$x_{2ref} = \left(-b_1 + \sqrt{\frac{(b_1^2 - 4a_1c_1)}{2a_1}}\right)$$

$$x_{5ref} = -\frac{R_{bo}}{R_{pvo}} x_{2ref} + x_{6ref}\left(1 + \frac{R_{bo}}{R_{pvo}} + \frac{R_{bo}}{R_l}\right)$$

$$a_2 = R_{bo}, \quad b_2 = -R_{bo} d_2$$

$$c_2 = x_{5ref}^2 - x_{5ref} x_{6ref}$$

$$x_{4ref} = \left(-b_2 + \sqrt{\frac{(b_2^2 - 4a_2c_2)}{2a_2}}\right) \tag{15.3}$$

Although all the references are generated initially using this technique, in the actual implementation, a combination of references obtained from this section and the virtual control references generated in the upcoming section will be utilized to enforce backstepping control on to the system.

15.4 Backstepping Controller Design

This section delineates the detailed procedure for designing the backstepping-based controller proposed in this chapter. The mathematical derivation of the controllers also serves as the complete stability analysis of the entire design technique. Hence, it serves a two-way purpose.

The entire procedure for designing both the MPPT and DC bus voltage controllers can be summarized in the following set of instructions.

1. Compute the reference values for all states.
2. Compute virtual input α_3 for the PV system to make x_1 follow $x_{1\text{ref}}$.
3. Compute the value of u_1 so that x_3 follows α_3 and x_2 follows $x_{2\text{ref}}$.
4. Compute the value of the virtual input α_5 such that x_6 reaches its reference value $x_{6\text{ref}}$.
5. Compute the value of u_2 such that x_5 follows α_5 and x_4 follows its reference value $x_{4\text{ref}}$.

15.4.1 Controller Design for PV Subsystem

Let us assume a Lyapunov function $V_1 = \frac{C_{\text{pvi}}}{2} e_1^2$, where $e_1 = x_1 - x_{1\text{ref}}$.

Upon differentiation, we get $\dot{V}_1 = e_1 \left(d_1 - x_3\right)$.

If we choose x_3 as

$$\alpha_3 = d_1 + K_1 e_1 \tag{15.4}$$

then $\dot{V}_1 = -K_1 e_1^2$, which means $\dot{V}_1 < 0$.

Then, we consider the Lyapunov function $V_{2,3} = \frac{C_{pvo}}{2} e_2^2 + \frac{L_{pv}}{2} e_3^2$,

where $e_2 = x_2 - x_{2\text{ref}}$ and $e_3 = x_3 - \alpha_3$.

Differentiating, we get

$$\dot{V}_{2,3} = C_{pvo} e_2 \dot{x}_2 + L_{pv} e_3 \left(\dot{x}_3 - \dot{\alpha}_3\right)$$

$$= e_2 \left(x_3 + \frac{x_6 - x_2}{R_{pvo}} \right) + e_3 \left(x_1 - x_2 - R_{pv} x_3 - L_{pv} \dot{\alpha}_3\right) + u_1 \left(-e_2 x_3 + e_3 x_2\right)$$

If u_1 is chosen as

$$u_1 = \frac{\left(-num_{c1} - K_2 e_2^2 - K_3 e_3^2\right)}{e_3 x_2 - e_2 x_3} \tag{15.5}$$

where

$$\text{num}_{c1} = e_2 x_3 - \frac{e_2 x_2}{R_{pvo}} + \frac{e_2 x_6}{R_{pvo}} + e_3 x_1 - e_3 x_2 - e_3 x_3 R_{pv} - e_3 L_{pv} \dot{\alpha}_3 \tag{15.6}$$

$$\dot{\alpha}_3 = K_1 \dot{x}_1 + \dot{d}_1 \tag{15.7}$$

This results in $\dot{V}_{2,3} = -K_2 e_2^2 - K_3 e_3^2$, which means $\dot{V}_{2,3} < 0$. Using this value of u_1, the convergence of states x_1, x_2, and x_3 of the PV array is ensured.

15.4.2 Controller Design for the Battery Subsystem

Now, for finding the controller for the battery subsystem, first we choose a Lyapunov function

$$V_6 = \frac{C_L}{2} e_6^2 \text{ where } e_6 = x_6 - x_{6ref}$$

$$\dot{V}_6 = C_L e_6 \dot{x}_6 = e_6 \left[\frac{x_2}{R_{pvo}} + \frac{x_5}{R_{bo}} - x_6 \left(\frac{1}{R_{pvo}} + \frac{1}{R_{bo}} + d_3 \right) \right]$$

The virtual input α_5 is chosen as shown in Eq. (15.8):

$$\alpha_5 = -\frac{R_{bo}}{R_{pvo}} x_2 + R_{bo} x_6 \left(\frac{1}{R_{pvo}} + \frac{1}{R_{bo}} + d_3 \right) - K_6 e_6 R_{bo} \tag{15.8}$$

This results in $\dot{V}_6 = -K_6 e_6^2$, ensuring stability of x_6.

Following this, $V_{4,5}$ is chosen as $V_{4,5} = \frac{C_{bo}}{2} e_5^2 + \frac{L_b}{2} e_4^2$, where $e_5 = x_5 - \alpha_5$ and $e_4 = x_4 - x_{4ref}$. Upon differentiation,

$$\dot{V}_{4,5} = e_5 \left(x_4 - \frac{x_5}{R_{bo}} + \frac{x_4}{R_{bo}} - x_4 u_2 - C_{bo} \dot{\alpha}_5 \right) + e_4 \left(d_2 - x_5 - R_b x_4 + x_5 u_2 \right)$$

Further, u_2 is chosen as follows to stabilize the system:

$$u_2 = \frac{-\,\text{num}_{c2} - K_4 e_4^2 - K_5 e_5^2}{e_4 x_5 - e_5 x_4} \tag{15.9}$$

where

$$num_{c2} = e_5 \left(x_4 + \frac{x_6 - x_5}{R_{bo}} - C_{bo} \dot{\alpha}_5 \right) + e_4 \left(d_2 - x_5 - R_b x_4 \right) \tag{15.10}$$

$$\dot{\alpha}_5 = -\frac{R_{bo}}{R_{pvo}} \dot{x}_2 + R_{bo} x_6 \dot{d}_3 + \dot{x}_6 \left(-K_6 R_{bo} + \frac{R_{bo}}{R_{pvo}} + 1 + R_{bo} d_3 \right) \tag{15.11}$$

Thus, $\dot{V}_{4,5}$ becomes $\dot{V}_{4,5} = -K_4 e_4^2 - K_5 e_5^2$, which is $\dot{V}_{4,5} < 0$ for $K_4, K_5 > 0$.
The total Lyapunov function,

$$\dot{V} = \dot{V}_1 + \dot{V}_{2,3} + \dot{V}_{4,5} + \dot{V}_6 < 0 \tag{15.12}$$

Thus, adopting the controllers designed in Eqs (15.5) and (15.9) and the virtual controllers designed in Eqs (15.4) and (15.8), which are carefully derived on the basis of the Lyapunov stability theory, the stability of all the system states is guaranteed. It is to be noted that the denominators of the controllers sometimes tend to zero. Under such circumstances it is good practice to limit the duty cycle of the controllers to achieve the best results.

15.5 Results

In this section, different scenarios have been considered where the effects of change in irradiation, temperature, and load were studied when the proposed backstepping control was applied. The DC bus voltage is taken to be 40V. The PV array always operates at a peak power rating of *W*. The battery operates at a constant operating voltage of 24V. The simulation has been carried out in MATLAB. Table 15.1 gives an overall picture of various system and controller parameters used for the simulation.

15.5.1 Case 1: Variation in Irradiance

In this case, the irradiance of the PV panel is changed with time while temperature and load are kept constant at 25°C and 200s. The irradiance changes for every 1.5 seconds from 1500 W/m^2 to 1200 W/m^2 and then to 1000W/m^2, 500W/m^2, and 200W / m^2 at 1.5s°, 3s, 4.5s and 6s, respectively. Figures 15.3 and 15.4 show the evolution of PV and battery states for this case.

When the irradiance changes, both the voltage and current at which maximum power is extracted from the PV panel change. Moreover, the maximum power that can be extracted also reduces with a reduction in the irradiance. This effect is visible in the PV and battery currents shown in Figure 15.3. Since load is constant at 200w, as the irradiance decreases, the PV inductor current reduces and the battery current increases, compensating the reduction

Table 15.1 PVDG parameters

Subsystem	Parameter specification
Battery voltage	24V
PV array at STC	Model: Kyocera Solar KC200GT $V_{OC} = 32.9V, I_{SC} = 8.21\ A$ $V_{MPP} = 26.3V, I_{MPP} = 7.61A$
DC/DC converter	$C_{pvi} = 3mF, C_{pvo} = 3mF, L_{pv} = 10mH,$ $R_{pv} = 0.5\Omega, R_{pvo} = 0.1\Omega$
DC/DC Bidirectional converter	$C_{bo} = 3mF,\ R_b = 0.5\Omega,$ $L_b = 10mH,\ R_{bo} = 0.1\Omega$
Backstepping gains	$K_1 = 10,\ K_2 = 0.04,\ K_3 = 0.04$ $K_4 = 0.04,\ K_5 = 0.04,\ K_6 = 15.0$

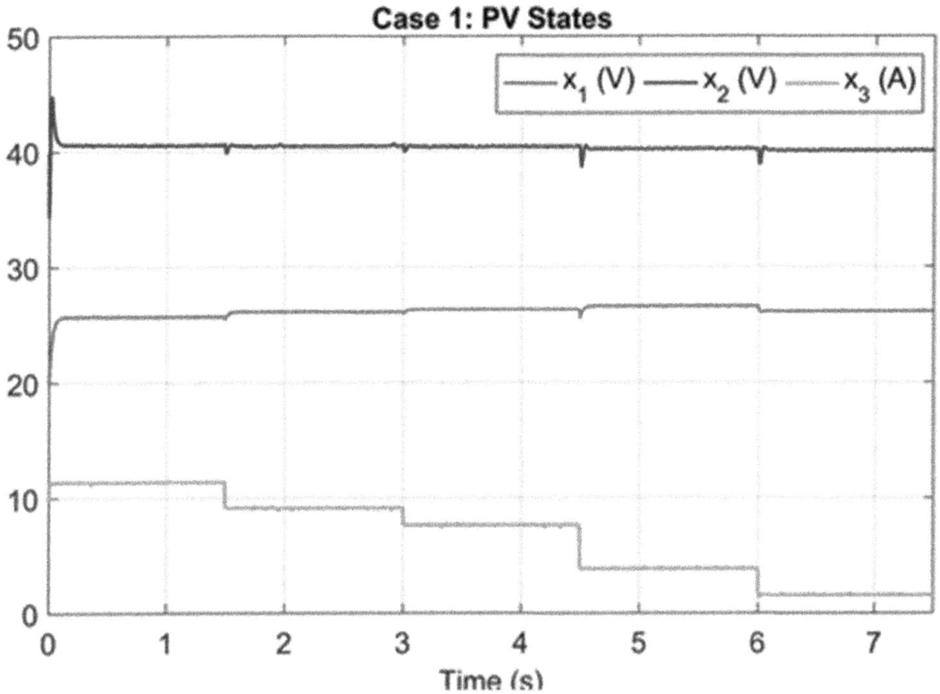

Figure 15.3 Case 1: PV states.

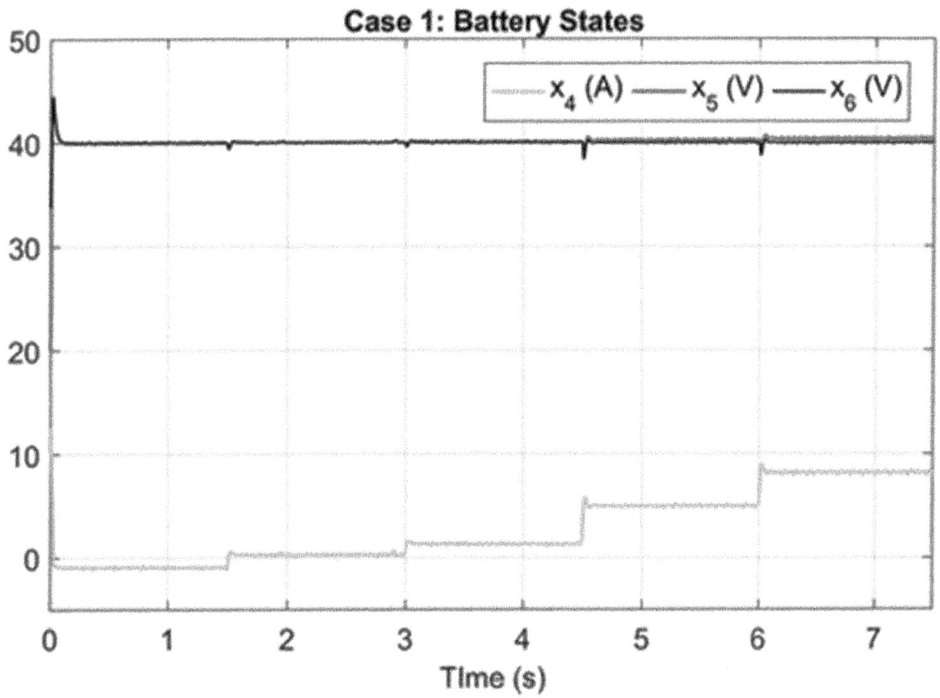

Figure 15.4 Case 1: Battery states.

in the PV current. During changes in irradiance, some transients occur in the DC bus voltage but they die out within 80ms. Therefore, the DC bus voltage remains constant at 40V.

15.5.2 Case 2: Variation in Temperature

With an increase in temperature, the maximum power extracted from the PV panel reduces, and also the values of the MPP current and voltage. In this case, the temperature is varied, keeping the load and irradiation constant at 200W and 1000W/m^2, respectively. The temperature is varied from 75°C to 50°C and to 25°C, 10°C, and 0°C, respectively, at 1.5s, 3s, 4.5s, and 6s, respectively.

Figures 15.5 and 15.6 show the response of the SPVDG system for the change in temperature. However, the variation in MPP current extracted from PV with a change in temperature is less compared to the change in MPP voltage. The grid voltage stays at 40V irrespective of any variation in temperature.

15.5.3 Case 3: Variation in Load

In this case, the PV panel characteristics remain constant with constant temperature and irradiation at 1000W/m^2 and 25°C. However, the load resistance is varied from 5Ω to 7Ω, 9Ω, 11Ω, and then to 8Ω at 1.5s, 3s, 4.5s, and 6s, respectively.

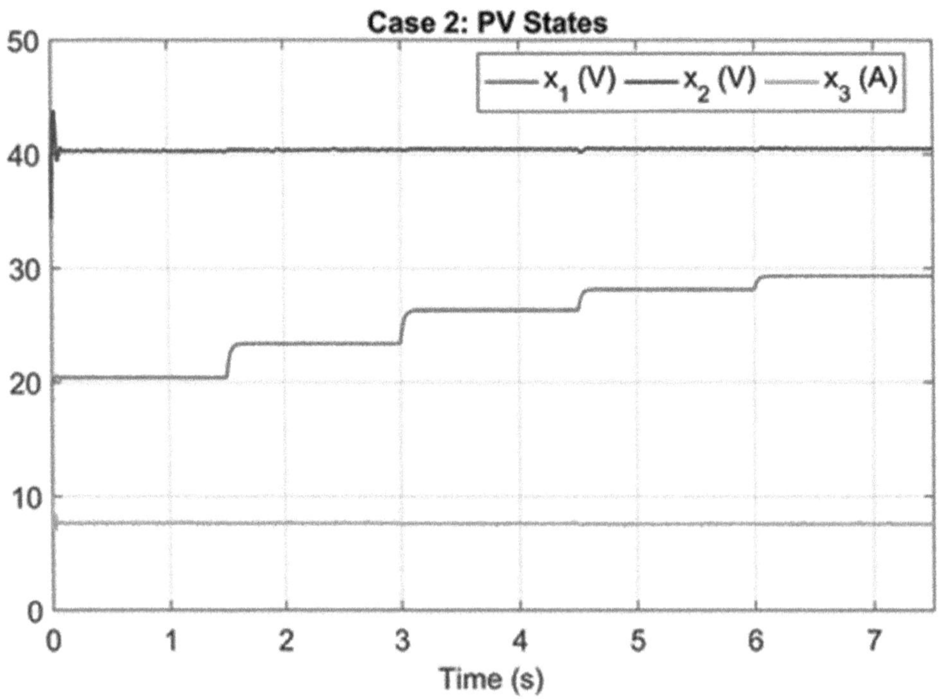

Figure 15.5 Case 2: PV states.

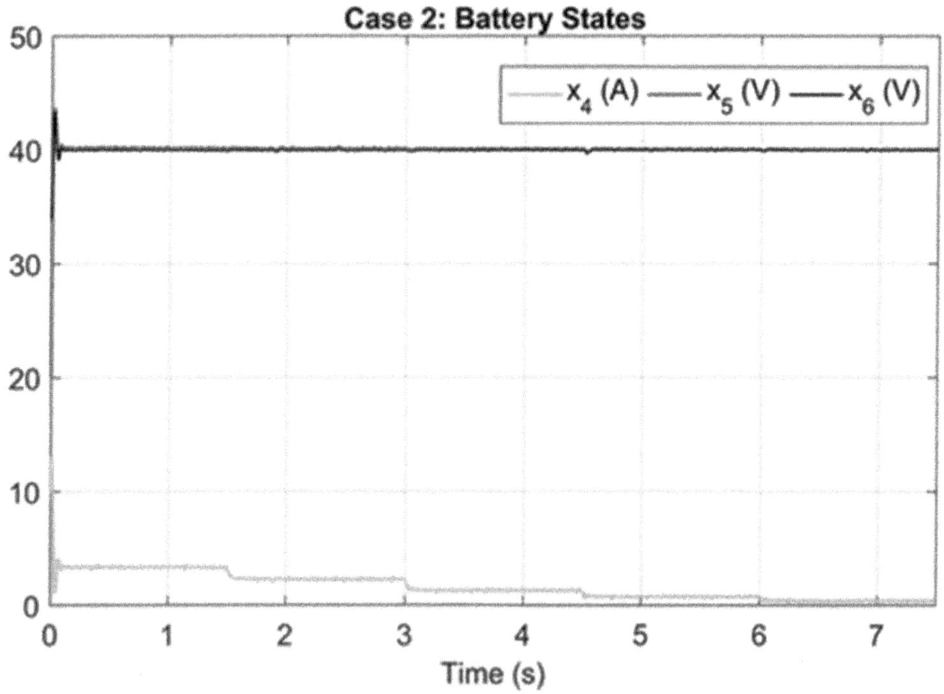

Figure 15.6 Case 2: Battery states.

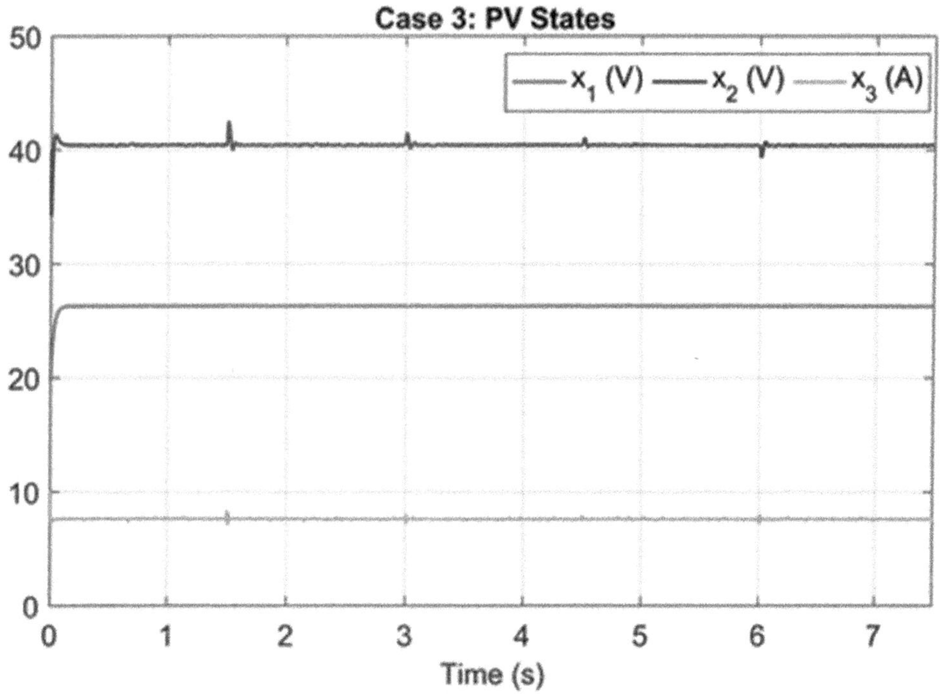

Figure 15.7 Case 3: PV states.

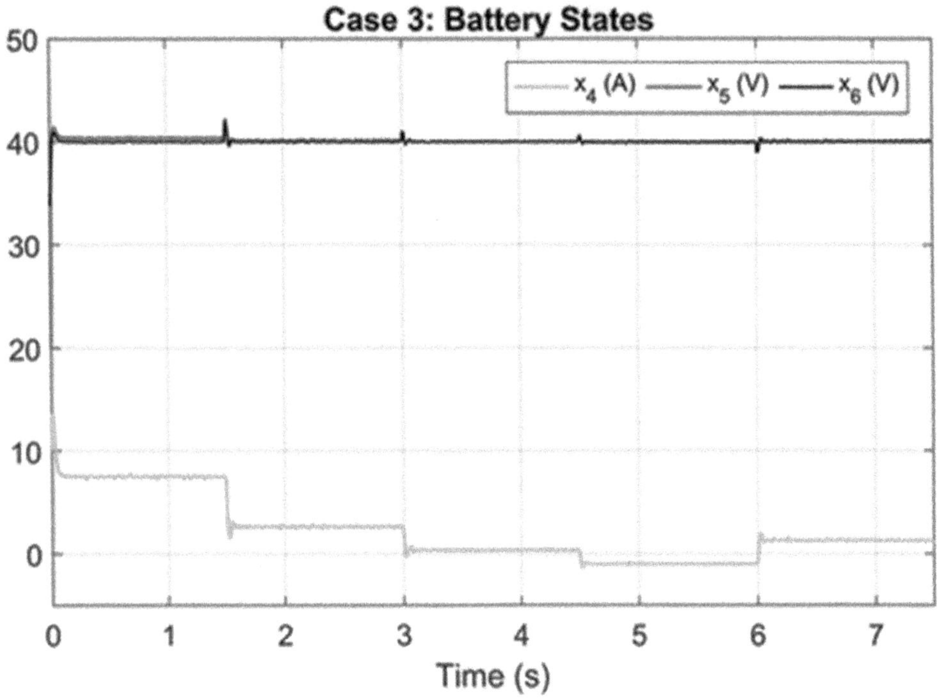

Figure 15.8 Case 3: Battery states.

The change in battery current can easily be observed when PV generation is constant and load is changed, as shown in Figures 15.7 and 15.8. Initially, when the load is very high, the battery also supplies positive current to meet the load requirement and, as the load is reduced, the current supplied by the battery gradually reduces. When the load reduces below the maximum power provided by the PV panel, the battery current becomes negative, showing that the extra power produced by the PV panel is being used to charge the battery. It is worth mentioning that both the MPP voltage K_4 and grid voltage K_5 remain constant.

In all cases, it can be observed that the grid is hardly perturbed without any deviations from the desired value. All the states completely stabilize within 80ms. Hence, the proposed algorithm is fast and very robust to change in the disturbances.

15.6 Conclusion

In this chapter, a nonlinear backstepping control strategy is developed for a standalone PVDG system with battery energy storage in a modular fashion. This technique ensures faster stabilization of all the system states and grid voltage when subjected to large disturbances. It has been verified for three different cases in the presence of large variations in irradiance, temperature, and load. This controller also ensures an appropriate bidirectional power flow depending on the power balance in the standalone system. The simulation results validate the efficacy of the proposed nonlinear control strategy.

References

[1] S.R. Mishra, N.V. Srinath, K.M. Preetam, and L. Behera, "A generalized novel framework for optimal sensor-controller connection design to guarantee a stable cyber physical smart grid," *2015 IEEE 13th International Conference on Industrial Informatics (INDIN)*, pp. 424–429, 2015.

[2] M.P. Korukonda, S.R. Mishra, A. Shukla, and L. Behera, "Handling multi-parametric variations in distributed control of cyber-physical energy systems through optimal communication design," *IET Cyber-Physical Systems: Theory Applications*, vol. 2, no. 2, pp. 90–100, 2017.

[3] J. Neely, S. Pekarek, R. DeCarlo, and N. Vaks, "Real-time hybrid model predictive control of a boost converter with constant power load," *2010 Twenty-Fifth Annual IEEE Applied Power Electronics Conference and Exposition (APEC)*, pp. 480–490, 2010.

[4] A.P.N. Tahim, D.J. Pagano, and E. Ponce, "Nonlinear control of DC–DC bidirectional converters in stand-alone DC microgrids," *2012 IEEE 51st IEEE Conference on Decision and Control (CDC)*, pp. 3068–3073, 2012.

[5] J. Zeng, Z. Zhang, and W. Qiao, "An interconnection and damping assignment passivity-based controller for a DC–DC boost converter with a constant power load," *IEEE Transactions on Industry Applications*, vol. 50, no. 4, pp. 2314–2322, 2014.

[6] B. Khan and P. Singh, "Selecting a meta-heuristic technique for smart micro-grid optimization problem: A comprehensive analysis," *IEEE Access*, vol. 5, no. pp. 13951–13977, 2017.

[7] O.P. Mahela, A.G. Shaik, N. Gupta, M. Khosravy, B. Khan, H.H. Alhelou, and S. Padmanaban, "Recognition of power quality issues associated with grid integrated solar photovoltaic plant in experimental framework," *IEEE Systems Journal*, October 2020. doi: 10.1109/JSYST.2020.3027203.

[8] R.K. Pachauri, O.P. Mahela, A. Sharma, J. Bai, Y.K. Chauhan, B. Khan, and H.H. Alhelou, "Impact of partial shading on various PV array configurations and different modeling approaches: A comprehensive review," *IEEE Access*, vol. 8, pp. 181375–181403, 2020. doi: 10.1109/ACCESS.2020.3028473.

[9] R. Kaushik, O.P. Mahela, P.K. Bhatt, B. Khan, A.R. Garg, H.H. Alhelou, and P. Siano, "Recognition of islanding and operational events in power system with renewable energy penetration using a Stockwell transform-based method," *IEEE Systems Journal*, September 2020. doi: 10.1109/JSYST.2020.3020919.

[10] G.S. Yogee, O.P. Mahela, K.D. Kansal, B. Khan, R. Mahla, H. Haes Alhelou, and P. Siano, "An algorithm for recognition of fault conditions in the utility grid with renewable energy penetration," *Energies*, vol. 13, pp. 2383, 2020.

[11] T. Alnejaili, S. Drid, D. Mehdi, L. Chrifi-Alaoui, R. Belarbi, and A. Hamdouni, "Dynamic control and advanced load management of a stand-alone hybrid renewable power system for remote housing," *Energy Conversion and Management*, vol. 105, pp. 377–392, 2015.

[12] M. Cupelli, M. Mirz, and A. Monti, "Application of backstepping to MVDC ship power systems with constant power loads," *2015 International Conference on Electrical Systems for Aircraft, Railway, Ship Propulsion and Road Vehicles (ESARS)*, pp. 1–6, 2015.
[13] Q. Xu, C. Zhang, C. Wen, and P. Wang, "A novel composite nonlinear controller for stabilization of constant power load in dc microgrid," *IEEE Transactions on Smart Grid*, vol. 10, no. 1, pp. 752–761, 2019.
[14] T.K. Roy and M.A. Mahmud, "Dynamic stability analysis of hybrid islanded dc microgrids using a nonlinear backstepping approach," *IEEE Systems Journal*, vol. 12, no. 4, pp. 3120–3130, 2018.
[15] M.P. Korukonda, S.R. Mishra, K. Rajawat, and L. Behera, "Hybrid adaptive framework for coordinated control of distributed generators in cyber-physical energy systems," *IET Cyber-Physical Systems: Theory Applications*, vol. 3, no. 1, pp. 54–62, 2018.
[16] S.R. Mishra, M.P. Korukonda, L. Behera, and A. Shukla, "Enabling cyber-physical demand response in smart grids via conjoint communication and controller design," *IET Cyber-Physical Systems: Theory Applications*, vol. 4, no. 4, pp. 291–303, 2019.
[17] M.P. Korukonda, S.R. Mishra, A. Shukla, and L. Behera, "Improving microgrid voltage stability through cyber-physical control," *2016 National Power Systems Conference (NPSC)*, pp. 1–6, 2016.
[18] A. Hussain, M.M. Garg, M.P. Korukonda, S. Hasan, and L. Behera, "A parameter estimation based MPPT method for a PV system using Lyapunov control scheme," *IEEE Transactions on Sustainable Energy*, vol. 10, no. 4, pp. 2123–2132, 2019.

Part VIII

Electric Vehicle Technology

16

Vehicle-to-Grid Challenges and Potential Benefits for Smart Microgrids

Mohd Tayyab Saeed

Solution Engineer (Energy & Utilities), Al-Mostajed Technologies Co. LLC, Dubai, UAE

16.1 Introduction

Nowadays, solicitudes towards environmental changes have contributed significantly to technological changes in the urban transportation sector. The sector is going through a transition stage for contributing to a healthy environment by adapting decarbonizing strategies such as replacing fossil-based vehicles with electric or hybrid vehicles. This paradigm is supported by all major automotive brands and government policies. Nevertheless, with the high penetration of new electric vehicle (EV) loads, electrical power grids are foisted by challenges in terms of a dynamic, stable, synergistic, and progressive integration of electric mobility. This new load will require careful and intelligent control to reduce the influence on peak power demand [1–3].

Alliance of the intelligent grid and microgrid with vehicle-to-grid (V2G) can be strategically designed to utilize EVs as a distributed energy resource and thereby allowing EVs to fling power stored in their batteries back to the grid when required and can help to regulate the frequency of the decentralized grid [4, 5]. Consequently, V2G can assist in reducing system operation costs by offering a cost-efficient spinning reserve, frequency and voltage regulation services, peak shaving capability, and others. However, there are many technical, economic, and social challenges that we will discuss in this chapter.

16.2 Microgrid and Role of EV

A microgrid is described by the Department of Energy (DOE) of the United States as a group of interconnected loads and distributed energy resources (DERs) within clearly defined electrical boundaries that act as a single controllable entity concerning the grid. A microgrid can connect and disconnect from the grid to enable it to operate in both grid-connected and island mode [6]. Figure 16.1 presents the microgrid concept redeveloped as per IEEE 1547.4 Standard [7].

Active Electrical Distribution Network: A Smart Approach, First Edition. Edited by Baseem Khan, Josep M. Guerrero, Sanjeevikumar Padmanaban, Hassan Haes Alhelou, Om Prakash Mahela, and Sudeep Tanwar.

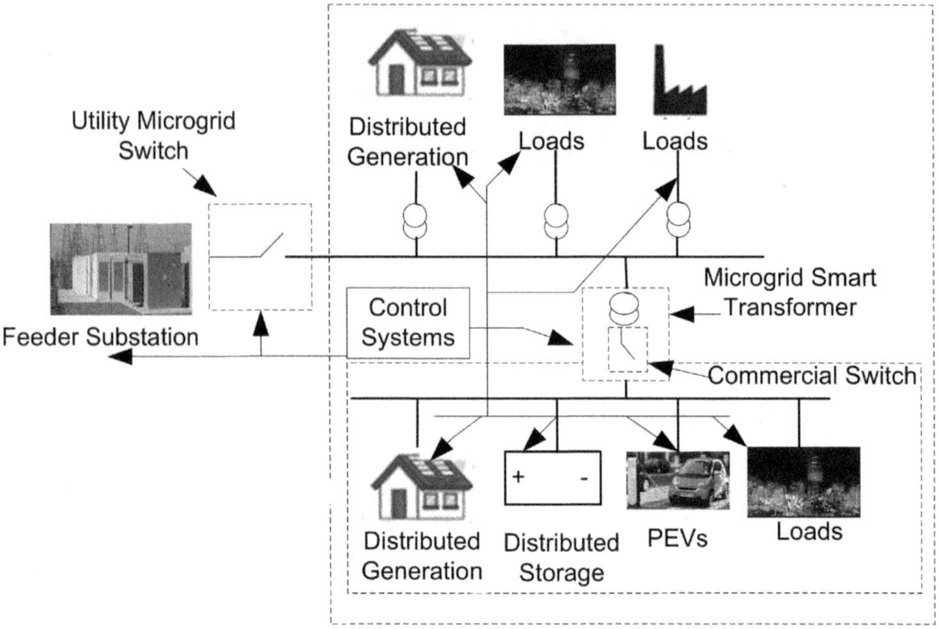

Figure 16.1 Microgrid concept.

Conclusively, microgrids are energy supply arrangements on the low-voltage side made to fulfill the electrical load for a smaller zone, such as a university campus, a market area, or a group of buildings. It can dynamically and autonomously coordinate and manage generations and demands within defined boundaries. In the microgrid paradigm, distributed resources (DRs) such as photovoltaic (PV) plants, microgenerators, and energy storage systems (ESSs) fulfil the energy demand. The majority of the vehicles are idle most of the time. With the V2G framework, vehicles can participate in the energy market by allowing operators to utilize idle batteries as DERs for enhancing microgrid capabilities and making EV owners as prosumers. However, the working of the microgrid involves different modes of operations, such as standard mode, transition mode, island mode, and reconnection mode. It heavily depends on precise control, management services, and reliable communication. IEEE 1547.4 highlights important factors toward planning and managing microgrids [7]. This standard comprehends protection schemes and adjustments; the effect of parameters, such as voltage, frequency, power quality; understanding load demand of the consumer; monitoring; communication and control; addition of a single or multiple points of common coupling (PCC); cold load pickup; identifying the properties of DRs; distinguishing transients and steady-state situations; understanding synergies between devices; load shedding; reserve margins; demand response; device requirements; and extended inverter functionality. This chapter will conceptualize V2G as a DR and discuss the technology and framework requirement for participation in the microgrid.

16.3 V2G Components and Realization Models

Naturally, EVs are connected to the utility grid during the charging of the batteries, though, in the absence of V2G, they cannot contribute energy into the grid. For facilitating the V2G mode of operation, the bidirectional exchange of information and power is required, which needs a variety of adjustments and developments in the EV infrastructure. The control realization for V2Gs can be broadly categorized within two methodologies: centralized and decentralized.

Primarily V2G was proposed by Letendre and Kempton [8] with three critical components: (1) connection to the grid, (2) an information exchange system regulating charging and discharging, and (3) a means to quantify the cooperations delivered to the grid. Concept is depicted in Figure 16.2 [9–11].

Connection of EV to the grid in the V2G paradigm needs a specialized charger, bidirectional communication, and power flow capability. There are many essential attributes of

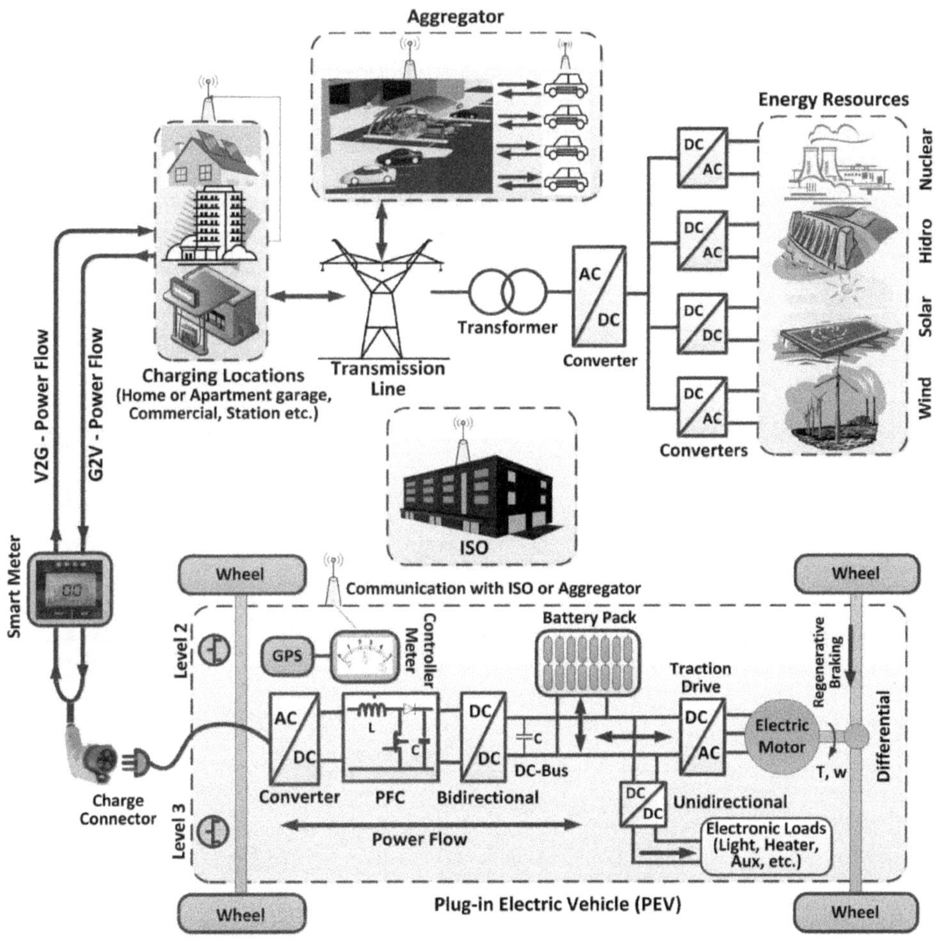

Figure 16.2 Generic V2G structure, ISO: independent system operator.

the V2G charger, such as power capability, characteristics if the charger is of an off-board or an on-board type, and furthermore its connectivity and two-way capabilities [9]. Though V2G is possible at any level, the potential of a single EV is too little in an economic sense to partake in those grid ancillary cooperations. Therefore, there is a requirement of a centralized governing system that can manage a group of EVs to provide a considerable service to the grid. This concept is known as a V2G aggregator [10, 11]. Figure 16.3 presents the participation of a V2G in aggregation.

To participate in the grid, another critical requirement is of a three-tier communication, i.e. between the charger and vehicle, charger and aggregator, aggregator, and grid network. Also, user consent information and the current situation are critical. There are currently numerous communication protocols introduced by different societies to systematize the means of communication between EVSEs and an EV. Some published standards are ISO 15118 [12], SAE J2847 [13], and CHAdeMO [14]. There are likewise various periphery standards for communication between EVSE and the aggregator, such as ISO 61850 and the Open Charge Point Protocol (OCPP) [14]. Figure 16.4 presents a typical communication scenario of a V2G system [15].

Another critical element of the V2G system is an audit system, which is a precise metering system. The agile utility markets demand reliability; therefore, the meter needs to have excellent accuracy and high granularity and should be able to practically provide a real-time report to the aggregator and the power network operator [16, 17].

In the centralized scheduling method, an aggregator is a very vital part of facilitating an EV to participate in the network. Once the EV has sent the initial charge and user-related information, an aggregator will estimate charging and network cooperation maneuvering, which will be validated by the grid operator. Aggregation of EVs makes this method very optimal and practical for providing power backup, frequency/voltage regulation [15, 18],

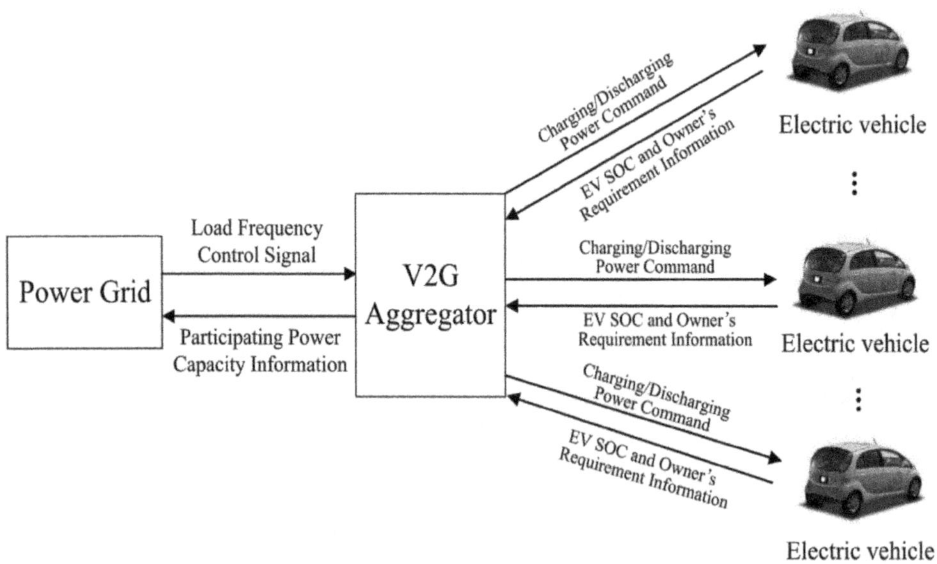

Figure 16.3 Participation of V2G in aggregation.

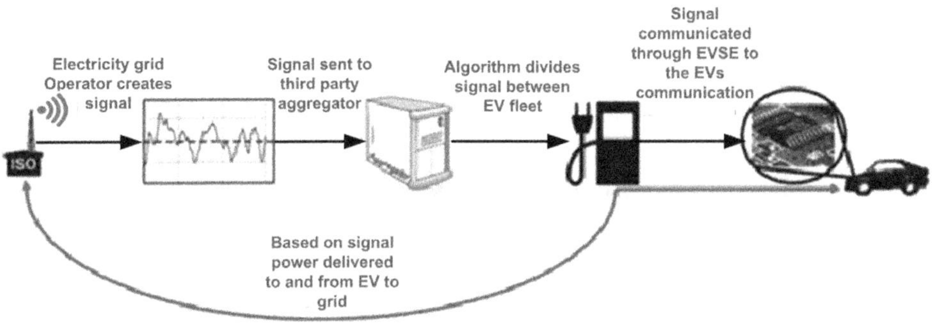

Figure 16.4 Typical communication scenario of a V2G system.

maximizing aggregator margin [19], and ancillary services economically [20]. However, in the centralized approach, the EV user is dependent on the control center and has to surrender the charging control to the aggregator. Another scheduling approach is decentralized, and a single EV can participate directly with the coordination of the principal grid operator. The advantage of such scheduling is a simple architecture and freedom to EV owners on charging and discharging [21]. However, this method allows an arbitrary number of associations of EVs at any given moment, which makes the technique unpredictable and does not provide a guarantee of the availability of backup power and ancillary services [22].

16.4 Applications and Potential Benefits of a Vehicle to Grid

A study says that, on average, a private vehicle spends just 4–5% of the time on the road, while 95–96% of the time the vehicle is idle [8, 23]. If the majority adopted electric cars, they could be utilized for V2G services, benefiting the owners, aggregators, and the grid. The fundamental of vehicle-to-grid is, indeed, the utility grid. Nevertheless, the utility grid is complicated, and to recognize how V2G can provide an advantage, knowledge of its arrangement is vital. The V2G scheme, in general, shall have a high potential at an economical price, have high availability, and should respond promptly. One of the primary purposes of V2G that is drawing attention is that it directly provides benefits to all associated parties, including owners of the EV, the utility, aggregator, and society at large. The advantages of V2G are broad and are still under research and academic exploration. However, we will try to review the technological and social benefits of V2G.

16.4.1 Technological Advantages

16.4.1.1 V2G for Ancillary Services Market, Renewable Integration, and Other Possible Services

Globally the electricity system is significantly diverse, though fundamentals remain the same. Three distinct fundamental markets are peakload, baseload, and ancillary services, as shown in Figure 16.5. The foremost need for an electricity supply system is for the baseload energy market. The continuous large-scale generation of electricity fulfills such demand, typically happening from nuclear, coal, or hydro energy plants that have low generation expenses but offer poor flexibility (though hydro provides baseload with flexibility). V2G is not feasible for such a market because of the continuous demand and need for longer duration participation and very aggressive energy generation per kWh rates [23].

Since demand fluctuates considerably during the day sometimes baseload power is inadequate to satisfy the new electricity need. Also, peaks tend to occur randomly and sporadically daily, weekly, or even less often. For such peak power demands, rather than low-cost and large-scale generation, participants have high production costs. However, they allow increased flexibility and economic capital expenses. Occasionally, such requests are fulfilled with potential markets, wherein operators compensate service providers for having idle capacity available at certain times [24]. Sometimes, these customers are called as prosumers. Prosumer's benefits assure safer investment and filling energy, which is only seldom required. V2G, as an aggregate, can satisfy these demands. However, such needs are intense, and a V2G arrangement usually does not have any economic advantage [25].

As opposed to two former electricity markets, the third kind that adequately harmonizes with the traits of V2G is the spinning reserves and frequency regulation. Often spinning reserves (also sometimes referred to as synchronized reserves) is a market that supports a grid response to unanticipated interruptions or other emergency incidents [23, 24]. The grid operator buys power capacity throughout the day from the participant and the participant should always be available and, when required, the participant must react within a short time (usually within 10–15 minutes) [24]. The potential expenses are profoundly tractable to V2G, as electric vehicles can contribute energy continuously because of their quick response rate and high availability. Although spinning reserves are sporadically

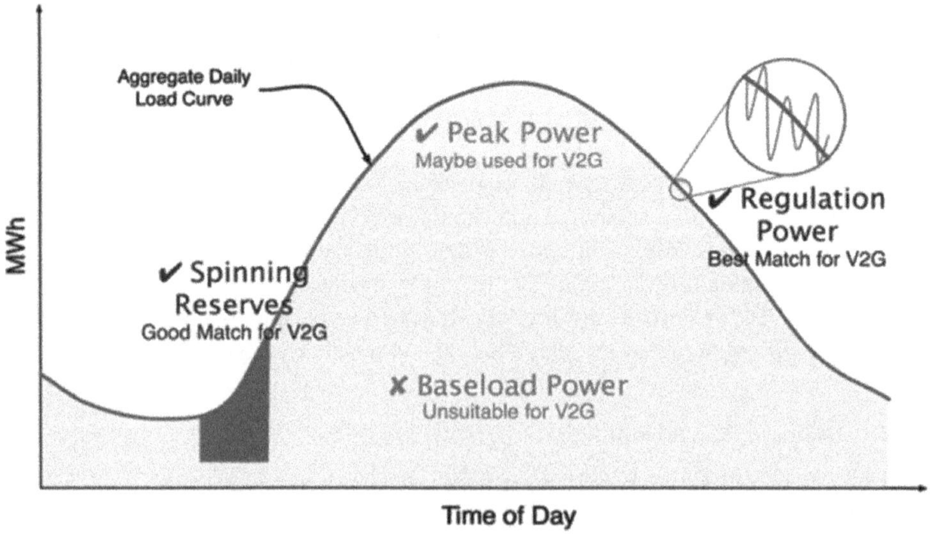

Figure 16.5 Possible energy market and V2G application. *Source*: Willett Kempton

needed, more notably such events (e.g. failed nuclear plant) can drain batteries due to the requirement for a considerable volume of energy. Thus, V2G is good, but not an excellent fit for this market. Another ancillary service is frequency regulation, which is principally a requirement to fine-tune the frequency of the grid to maintain quality and stability continually. It is also known as frequency containment reserves, frequency control, and automatic generation control (AGC). There is always a mismatch between actual power generated and demand. Henceforth, there is a requirement for participants that can provide energy to the grid and can also regulate the frequency around the grid standard of 50 or 60 Hz. A V2G in aggregation (a centralized scheme) is one that is continuously available to respond to charging and discharging quickly [24].

In conclusion, frequency regulation seems to be the essential ancillary service that a V2G can participate in. Since frequency regulation is a 24×7×365 market, the aggregator has to guarantee a capacity to follow a frequency regulation signal by ensuring V2G fleet availability over time. Therefore, the V2G scheme cooperating in a frequency regulation market needs more extensive information from the grid, EVs, and owners [25, 26].

Apart from the ancillary services, a V2G is often discussed with renewable energy integration. The issue with renewable technologies such as solar and wind energy is that they hold a fickle nature and therefore are not governable and create generation-demand balancing issues. As more and more solar and wind energy plants are integrating into the grid, the generation-demand mismatch is increasing, which develops an immediate demand for cooperation such as the V2G to support the power markets described above [24].

Additionally, the V2G can further offer many services to the grid for which markets have not yet been developed, especially on a local level [27]. For instance, the V2G can delay the upgradation cost of transformers, enhance power supply, diminish line congestions and voltage violations, and circumvent critical situations.

16.4.1.2 Economic Energy Storage System

Energy storage is not a new problem and researchers have investigated storage in all possible forms of energy, as shown in Figure 16.6 [28].

All the technologies shown in the figure possess their advantages and disadvantages. Some may offer better specific purposes, such as flywheels proven to be an appropriate choice for short-term storage for ancillary services. In contrast, compressed air energy storage is feasible for long-term energy storage for renewable systems.

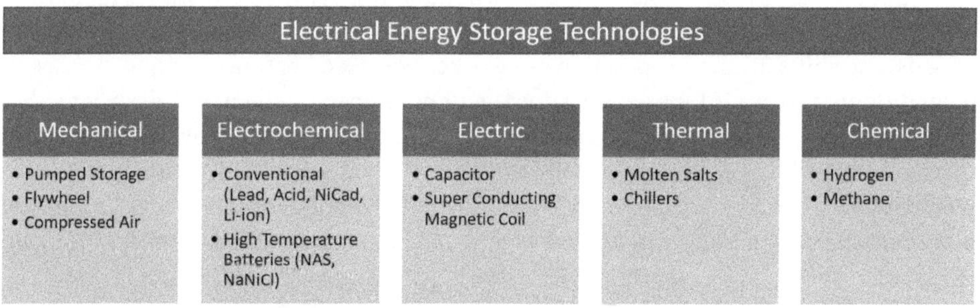

Figure 16.6 Energy storage technologies.

Table 16.1 Comparison between different energy storing technologies [11, 31, 32].

Technology	Cost $/kW	Cost $/kWh	Response	Efficiency	Discharge time	Max cycles or lifetime
Pumped hydro	1400	68	s–min	70–85%	4–16 h	30–60 years
Compressed air	900–1300	40–109	9–12 min	40–70%	2–30 h	20–40 years
Flywheel	870	4800	s	25–45%	s–min	5–30 years
Hydrogen	1500–3200	260–540	s–min	70–95%	min–week	20 000–100 000
V2G	0	0–40	s	70–85%	min	Life of vehicle

In Table 16.1, various grid-scale energy storage technologies are compared and overall the V2G is very beneficial. From the table, it can be inferred that V2G has an extraordinary advantage of providing energy storage at "zero" cost provided EV owners give consent to utilize the currently available storage capability in the electric vehicles for V2G purposes. In the worst case, if we take battery degradation into account, the higher limit of the cost would be US$ 40/kWh. Moreover, in theory, if we consider an average charger of 10 kW and battery of 30 kWh per vehicle, the total fleet's power and energy would be calculated as 2.7 TW and 16.1 TWh, respectively [17, 29]. This potential is significant compared to the US's full-electric grid ability, which is slightly over 1 TW [30]. Precisely, this means that if the complete US EV fleet turned into a VGI (vehicle grid integrated or simply V2G) framework, it would be a precious source for energy storage to the grid. Notably, we can have the same capacity as any technology, but V2G is offering a low cost by utilizing existing EVs. Additionally, there are various technological advantages compared to others, such as V2G offers of comparatively high efficiency and a fast response time economically.

16.4.2 Social Advantages

Besides offering significant technological benefits economically, V2G also provides other societal aids by reducing carbon contribution to the environment and thereby reducing health damage caused by the transport and electricity sectors. Fossil-powered vehicles produce carbon dioxide (CO_2), a greenhouse gas associated with climate change. The Environmental Protection Agency (EPA) of the United States estimates CO_2 emission reduction climate benefits using the social cost of carbon (SCC). The SCC value captures the cost of damage to the environment from carbon emissions, including the damage to agricultural yields, human well-being, resource losses from events such as floods, and shifts in energy operation prices like decreased heating costs and increased air conditioning costs [28]. Based on a document developed by the working society, the social cost of carbon, in 2015, it was calculated as $42.30 per ton [33]. However, researchers have admitted that SCC does not incorporate all commercial, environmental, well-being, and physical losses associated with climate change identified in the literature. Later, researchers at Stanford proposed an integrated assessment model (IAM) to consider slower economic growth rates related to climate change that impacted underdeveloped countries the most. They have reported a $220/ton estimation of carbon's social cost [34].

Choosing an EV (Nissan Leaf) rather than a gasoline-powered vehicle (Honda Civic) results in a reduction of carbon discharges of about 1.85 tons per year [35]. Based on the calculated SCC, the value of reduced carbon discharges over a 10-year period would be roughly $4070. In general, if the life of a vehicle is considered to be 20 years, an EV will save approximately $8140. Further, if the EV has been used as a V2G service participant, it would have displaced other grid stabilizing resources and contributed to the integration of renewable generation sources. Considering such costs to the SCC will provide a better estimate and is under the future plan. VEIC has conducted detailed research in quantifying the social benefits of EV, which are summarized in Figure 16.7. CO_2 estimated in the research is based on EPA and the actual cost is much higher, as discussed above.

Conclusively, EV is enhancing the quality of life by improving well-being by reducing hazardous emissions and unpleasant noise. It enhances the quality of the product by lowering maintenance as it has less mechanical parts and there is no need for an oil change or regular servicing. Further, it reduces the dependency on imported gasoline (National Security) and many more advantages.

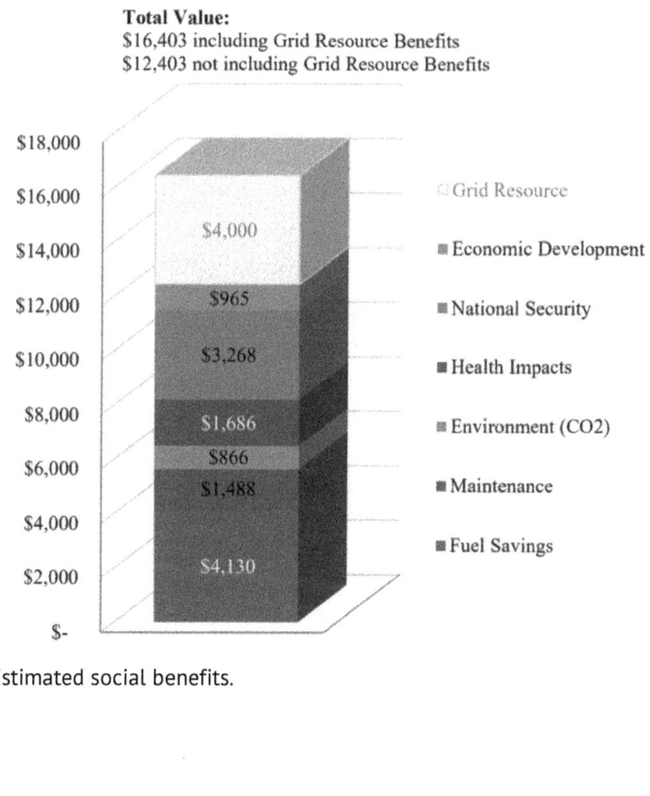

Figure 16.7 Estimated social benefits.

16.5 Challenges

The most significant hurdles to a V2G development are battery technology and its degradation, the high initial costs compared to ICE vehicles, a two-way communication infrastructure, and charger efficiency. An added concern is that the design of the grid was not for bidirectional flow. These issues tend to restrict the service capacities of V2G.

16.5.1 Battery Degradation

The prime challenge is the degradation of the battery as a result of more frequent use. Battery degradation can cause a loss of battery charge-holding capacity over time, impacting an EV range capability, which could affect an EV owner's willingness to cooperate in V2G operations. It will also affect the business case of V2G, as it will necessitate an attractive return to EV owners to compensate for the loss. Estimating battery degradation costs is very challenging because the leading cause is time (calendar aging) and the number of battery cycles (cycle aging). Both of these reasons depend on a multitude of complex factors (such as driving behavior), which are hard to model or quantify. However, the equivalent series resistance (ESR) can help forecast a battery cycle lifetime [36]. A deeper discharge hastens the battery deterioration rate for many battery types, resulting in an increased ESR, although several studies have been conducted to estimate the loss. In [37] and [38] financial damages connected with battery degradation were determined based on data gathered from A123 lithium cell systems for a 16-kWh battery pack. As per the estimate, the maximum cumulative yearly profit is just $10–$120. A study has suggested cost optimization by connecting degraded batteries in developing microgrids. An aggregator can manage these used batteries. Outcomes revealed that used batteries hold a significant monetary charge to contribute to frequent regulation services [31, 39].

16.5.2 Charger Efficiency and Energy Losses

The other scientific hurdle in the V2G path is the gross efficiency of round-trip energy, i.e. from and to the grid, mainly from the charger. As discussed in the benefits section, the system is comparatively efficient, but the V2G aggregator and EV owner pay round-trip energy losses, which reduces their yield. A study performed on an experimental V2G set up by A. Moore and Diaz [40] determined the overall energy losses presented in Table 16.2.

In the practical scenario, transformers are at the utility end and behind the meters. Therefore, transformer losses are under utility costs and are not borne by the aggregator or EV owner. However, other losses are the cost to the EV owners or compensated by an aggregator. Studies show that installing higher rated chargers and designing algorithms that can consider the factors, such as the battery's state of charge, for controlling the current flow can reduce the energy losses during charging or while delivering V2G services (charging or discharging) [40–42].

Table 16.2 Charging and discharging energy losses

Total GIV system percentage losses: building and electric vehicles elements.

Element	AC current (A)	Losses in percentage (%)	
		Charging	Discharging
EV battery	10	0.64	0.64
	40	1.69	1.91
EV PEU	10	6.28	16.67
	40	5.77	19.23
EVSE	10	0.10	1.42
	≈40	0.29	1.39
Breakers	10	0.00	2.80
	≈40	1.30	0.60
Transformer	10	10.20	14.60
	≈40	3.33	6.65
Total	10	17.22	36.13
	40	12.38	29.78

16.5.3 Communication, Data, and Security

With the adaptation of V2G on a country scale, there will be millions of vehicles talking to the aggregator multiple times an hour. It will generate an enormous amount of data each hour in the range of petabytes. The US has 270 million vehicles [37], so it would be challenging to store and process it. Moreover, the critical challenge is security because such data might be susceptible and can have information such as the driver's driving schedule. Apart from privacy concerns, there could be spoofing, eavesdropping, data tampering, unauthorized control, a DoS attack, or a cyber physical attack that needs to be secured reliably. Another challenge in this context is the communication system. The communication system has to be secure, reliable, and standard, so that EVs and chargers from multiple vendors can function seamlessly. Currently, there are many distinct communication protocols introduced by different societies for regularizing the information exchange between the charger and an EV, such as ISO15118 [12], SAE J2847 [13], and CHAdeMO [14]. Also, various communication standards exist for information exchange between aggregators and the charger, like ISO 61850 and Open Charge Point Protocol (OCPP) [43]. The principal discrepancies and relationships between different charger–EV communication standards are shown in Table 16.3.

Until now, no standard has been universally acceptable for V2G due to shortcomings of all of the standards. We anticipate standards will evolve and secure the concept of V2G in the future smart grid.

Table 16.3 Principal discrepancies and relationships between different charger–EV communication standards.

Standard	Description	Response time/ means of communication	Critique
ISO 15118 [7]	ISO 15118 is still in the development stage and has the goal to be accepted globally, though prominently; it is in Western Europe. ISO 15118 defines the technological specifications and information exchange protocols for regulating charging in DC, AC and V2G.	< 60 s Power line communication	Lacking in terms of cybersecurity measures and has low response times
SAE J2847 [8]	Delay in development of ISO 15118 full standard has led to the different standards in various parts of the globe, such as SAE J2847. SAE J2847 is a standard that was developed by the USA for implementation within the USA. It also defines technological specifications and information exchange protocols for regulating charging in DC, AC, and V2G.	< 1 s Power line communication	Aggregation design is dependent on SEP 2.0 based non-vehicle intelligent home technologies.
CHAdeMO [9]	CHAdeMO is a separate standard developed by private organizations. It is designed exclusively for EVs with special plugs and high capacity chargers.	< 1 s Control area network	Only compatible with high-capacity DC CHAdeMO chargers, which are very costly.

16.5.4 High Initial Investment Cost

The economics of the V2G service can directly determine the degree of its spread. It is based on an existing system having a cost and benefits; one such expense is the EV itself, which is often compared to a typical gasoline car. However, EVs also provide several advantages over a standard gasoline vehicle, such as decreased fueling costs (as electricity is more economical) and low maintenance costs, as EVs have fewer mechanical parts. Therefore, EV prices principally consist of high upfront initial costs while electrification benefits accumulate over the vehicle's lifetime. Numerous studies have tried to quantify EV benefits; for example, one study found that an EV is cost-efficient compared to a standard gasoline vehicle only after driving 60 000 miles (or approximately 5–7 years, depending on travel demands) [44]. Another research uncovered the fact that EVs in Sweden were the most economical alternative over the lifetime of a decade [45]. A different study observed that EVs are only more economical than standard gasoline vehicles if a driving need is more than 50 kilometers each day [39]. The point is that EVs are economical depending on the user requirement and require an initial capital investment. The reason is undoubtedly the century of manufacturing cost optimization and mature technology, while EV is relatively new and will need a few years to match production costs. Moreover, battery degradation is also a considerable cost to EV economics. Although less regular maintenance is needed, the battery replacement cost is very high. Considering an EV with a 30 kWh battery pack, the expense of replacement would be about $7000–$9000 (depending on the battery pack) [31, 32]. Another critical element in V2G is the aggregator and the fact that V2G is a novel technology that is under research and experiments within a restricted environment and circumstances. Studies have suggested that the aggregator cost will be high and it is estimated to be 50% of the net revenue, which will leave EV owners with less margin.

16.5.5 Social Challenges

Another challenge is the socioenvironmental challenge and comprises the adversarial externalities connected to V2G systems, particularly those correlated with increased EV connections. Both EVs and a V2G clearly accrue environmental benefits, but a shift from gasoline engines to electric power will increase electricity consumption. This is a hurdle in electrifying transport in regions where water is limited, especially in large urban and dry regions, because electricity is predominantly generated by fossil and nuclear power plants, which demand a considerable water volume to generate steam and cooling processes [46, 47]. Furthermore, the manufacturing of batteries for EVs requires hazardous ingredients that pollute and have environmental costs [48–51]. The solution underlying the batteries is reducing the production cost of lithium ion batteries and reusing the constituents.

16.6 Conclusion

It is clear that V2G has ample benefits to offer society. It can help society by enabling sustainable transitions of the transportation and electrical utility sector in an environmentally friendly and economical manner. V2G enables vehicles to improve the electrical grid's efficiency and profitability, decrease carbon discharges, support renewable energy sources, and realize profits for every stakeholder, including EV owners, aggregator service providers, and TSOs/DSOs. In this chapter, we have discussed crucial V2G infrastructure components and their specifications. In general, V2G will need EV vehicles, a bidirectional charger with a communication system on a standard protocol, an aggregator, and a two-way connection to the power grid. V2G can render ancillary services, efficient energy storage, and improve the quality of life by providing social advantages, including environmental and health benefits. However, the path to transition is not straightforward, and in this context the chapter has elaborated various challenges in three dimensions (technological, economic, and social). The significant challenges are battery degradation, charger efficiency, energy losses, standardization of the communication system, data management, and cybersecurity. Although studies show that battery degradation will not affect the EV's ability to fulfill regular driving demand, EV owners may feel that V2G appends avoidable risk. Similarly, cyberattacks may shift owners completely aside from thinking V2G as an opportunity. These uncertainties necessitate being adequately communicated in a tone focused on the benefits of V2G that can appease and override the concerns of EV owners.

References

[1] B. Khan and P. Singh, “Selecting a meta-heuristic technique for smart micro-grid optimization problem: A comprehensive analysis,” *IEEE Access*, vol. 5, pp. 13951–13977, 2017.

[2] P. Singh and B. Khan, “Smart microgrid energy management using a novel artificial shark optimization,” *Complexity*, vol. 2017, Article ID 2158926, 22 pages, 2017.

[3] B. Khan and P. Singh, “Optimal power flow techniques under characterization of conventional and renewable energy sources: A comprehensive analysis,” *Journal of Engineering*, vol. 2017, Article ID 9539506, 16 pages, 2017.

[4] O.P. Mahela, B. Khan, H.H. Alhelou, and P. Siano, “Power quality assessment and event detection in distribution network with wind energy penetration using Stockwell transform and fuzzy clustering,” *IEEE Transactions on Industrial Informatics*, vol. 16, no. 11, pp. 6922–6932, November 2020. doi:10.1109/TII.2020.2971709.

[5] G.S. Yogee, O.P. Mahela, K.D. Kansal, B. Khan, R. Mahla, H. Haes Alhelou, and P. Siano, “An algorithm for recognition of fault conditions in the utility grid with renewable energy penetration,” *Energies*, vol. 13, p. 2383, 2020.

[6] P. Asmus, “Why microgrids are moving into the mainstream,” *IEEE Electrification Magazine*, March 2014.

[7] IEEE Standards Coordinating Committee 21, “IEEE guide for design, operation, and integration of distributed resource island systems with electric power systems,” IEEE Std 1547.4, 2011.

[8] W. Kempton and S.E. Letendre, “Electric vehicles as a new power source for electric utilities,” *Transportation Research Part D: Transport and Environment*, vol. 2, no. 3, pp. 157175, 1997.

[9] C. Guille and G. Gross, “A conceptual framework for the vehicle-to-grid (V2G) implementation,” *Energy Policy*, vol. 37, no. 11, pp. 4379–4390, 2009.

[10] J. Tomić and W. Kempton, “Using fleets of electric-drive vehicles for grid support,” *Journal of Power Sources*, vol. 168, no. 2, pp. 459–468, 2007.

[11] M. Yilmaz and P.T. Krein, “Review of benefits and challenges of vehicle-to-grid technology,” *2012 IEEE Energy Conversion Congress and Exposition (ECCE)*, Raleigh, NC: IEEE, pp. 3082-3089, 2012.

[12] M. Yilmaz and P.T. Krein, “Review of battery charger topologies, charging power levels, and infrastructure for plug-in electric and hybrid vehicles,” *IEEE Transactions on Power Electronics*, vol. 28, no. 5, pp. 2151–2169, May 2013.

[13] S. Han, S. Han, and K. Sezaki, “Development of an optimal vehicle-to-grid aggregator for frequency regulation,” *IEEE Transactions on Smart Grid*, vol. 1, no. 1, pp. 65–72, June 2010.

[14] P. Chao, Z. Jianxiao, L. Lian, and L. Liying, “An optimal dispatching strategy for V2G aggregator participating in supplementary frequency regulation considering EV driving demand and aggregator’s benefits,” *Applied Energy*, vol. 190, pp. 591–599, 2017.

[15] N. Lance, Z.R. Gerardo, K. Johannes, and K.S. Benjamin, *Vehicle to Grid*, London, UK: Palgrave Macmillan, 2019.

[16] ISO ISO 15118-1:2013, “Vehicle to grid communication interface – Part 1: General information and use-case definitions,” International Standards Organization, 2013.

[17] California Energy Commission, “Transcript of the 12/ 07/16 vehicle–grid integration communications standards workshop,” California Energy Commission, 2016.

[18] J. Kester, L. Noel, X. Lin, G. Zarazua de Rubens, and B.K. Sovacool, "The coproduction of electric mobility: Selectivity, conformity and fragmentation in the sociotechnical acceptance of vehicle-to-grid (V2G) standards," *Journal of Cleaner Production*, vol. 207, pp. 400–410, 2019.

[19] K.S. Ko, S. Han, and D.K. Sung, "Performance-based settlement of frequency regulation for electric vehicle aggregators," *IEEE Transaction: Smart Grid*, vol. 9, pp. 866–875, 2018.

[20] K.R. Reddy and S. Meikandasivam, "Load flattening and voltage regulation using plug-in electric vehicle's storage capacity with vehicle prioritization using ANFIS," *IEEE Transaction: Sustainable Energy*, vol. 11, pp. 260–270, 2020.

[21] F. Ahmad, M.S. Alam, and M. Shahidehpour, "Profit maximization of microgrid aggregator under power market environment," *IEEE Systems Journal*, vol. 13, pp. 3388–3399, 2019.

[22] D.M.D. Van, G.R.C. Mouli, E.G. Morales and L.R. Elizondo, "Erratum to energy management system with pv power forecast to optimally charge evs at the workplace," *IEEE Transactions on Industrial Informatics*, vol. 14, p. 3298, 2018.

[23] K. Zhou and L. Cai, "Randomized PHEV charging under distribution grid constraints," *IEEE Transaction: Smart Grid*, vol. 5, pp. 879–887, 2014.

[24] M. Asaad, Z. Ramon, and L. Tek Tjing, "Integration of electric vehicles in the distribution network: A review of PV based electric vehicle modelling," *Energies*, vol. 13, p. 4541, 2020.

[25] J. Tomic and W. Kempton, "Vehicle-to-grid power fundamentals: Calculating capacity and net revenue," *Journal of Power Sources*, vol. 144, no. 1, pp. 268–279, 2005.

[26] Z. Zhou, T. Levin, and G. Conzelmann, "Survey of U.S. ancillary services markets," Center for Energy, Environmental, and Economic Systems Analysis, Energy Systems Division, Argonne National Laboratory, p. 59, 2016.

[27] B.K. Sovacool, L. Noel, J. Axsen, and W. Kempton, "The neglected social dimensions to a vehicle-to-grid (V2G) transition: A critical and systematic review," *Environmental Research Letters*, vol. 13, no. 1, p. 013001, 2018.

[28] "Clean Energy," [Online]. Available at: https://reneweconomy.com.au/energy-storage-to-reach-cost-holy-grail-mass-adoption-in-5-years-18383 [Accessed 9 September 2020].

[29] T. Markel, A. Meintz, K. Hardy, B. Chen, T. Bohn, and J. Smart, "Multilab EV smart grid integration requirements study," Golden, Colorado: NREL, no. NREL/TP-5400-63963, p. 91, May 2015.

[30] K. Knezovic, M. Marinelli, P. Codani, and Y. Perez, "Distribution grid services and flexibility provision by electric vehicles: A review of options," *IEEE*, pp. 1–6, 2015.

[31] D. Wang, J. Coignard, T. Zeng, C. Zhang, and S. Saxena, "Quantifying electric vehicle battery degradation from driving vs. vehicle-to-grid services," *Journal of Power Sources*, vol. 332, pp. 193–203, 2016.

[32] E. Apostolaki-Iosifidou, P. Codani and W. Kempton, "Measurement of power loss during electric vehicle charging and discharging," *Energy*, vol. 127, pp. 730–742, 2017.

[33] T.E. Baldock and Cooperative Research Centre for Sustainable Tourism. *Renewable Energy Sources for the Australian Tourism Industry*. CRC for Sustainable Tourism, Gold Coast, Qld, 2006.

[34] Fact Sheet: Energy Storage, EESI, 22 February 2019. [Online]. Available at: https://www.eesi.org/papers/view/energy-storage-2019 [Accessed 9 September 2020].

[35] U.S. DOT, "Highway statistics," US Department of Transportation, Federal Highway Administration, June 2018.

[36] EIA, "Electric power annual, Washington, DC: U.S. energy," Information Administration, December 2017.
[37] "EPA," [Online]. Available at: http://www3.epa.gov/climatechange/EPAactivities/economics/scc.html [Accessed September 2020].
[38] U. EPA, "Technical support document: Technical update of the social cost of carbon for regulatory impact analysis – Under executive order 12866 – Interagency working group on the social cost of carbon, United States Government," 2015.
[39] S. Beer, T. Gómez, D. Dallinger, I. Momber, and J. Lai, "An economic analysis of used electric vehicle batteries integrated into commercial building microgrids," *IEEE Transactions on Smart Grid*, vol. 3, no. 1, pp. 517–525, March 2012.
[40] F.C. Moore and D.B. Diaz, "Temperature impacts on economic growth warrant stringent mitigation policy," *Nature Climate Change*, vol. 5, pp. 127–131, 2015.
[41] Available at: https://nyserda.wattplan.com [Online].
[42] I. Malmgren, "Quantifying the societal benefits of electric vehicles," *World Electric Vehicle Journal*, vol. 8, pp. 996–1007, 2016.
[43] J.D. Dogger, B. Roossien, and F.D.J. Nieuwenhou, "Characterization of Li-ion batteries for intelligent management of distributed grid-connected storage," *IEEE Trans. Energy Conversion*, vol. 26, no. 1, pp. 256–263, March 2011.
[44] B. Peterson, J. Whitacre, and J. Apt, "The economics of using plug-in hybrid electric vehicle battery packs for grid storage," *Journal of Power Sources*, vol. 195, no. 8, pp. 2377–2384, 2010.
[45] S. Petersona, J. Apt, and J. Whitacrea, "Lithium-ion battery cell degradation resulting from realistic vehicle and vehicle-to-grid utilization," *Journal of Power Sources*, vol. 195, pp. 2385–2392, November 2010.
[46] D. Roberts, K. Giltman, and J. Sears, "A comparison of electric vehicle Level 1 and Level 2 charging efficiency," *IEEE*, pp. 255–258, 2014.
[47] C. Peng, J. Zou, and L. Lian, "Dispatching strategies of electric vehicles participating in frequency regulation on power grid: A review," *Renewable Sustainable Energy*, vol. 68, pp. 147–152, 2017.
[48] S. Martinenas, S. Vandael, P.B. Andersen, and B. Christensen, "Standards for EV charging and their usability for providing V2G services in the primary reserve market," Montreal, Canada, 2016.
[49] L.K. Mitropoulos, P.D. Prevedouros, and P. Kopelias, "Total cost of ownership and externalities of conventional, hybrid and electric vehicle," *Transportation Research Procedia*, vol. 24, pp. 267–274, 2017.
[50] J. Hagman, S. Ritzén, J.J. Stier, and Y. Susilo, "Total cost of ownership and its potential implications for battery electric vehicle diffusion," *Research in Transportation Business & Management*, vol. 18, pp. 11–17, 2016.
[51] G. Wu, A. Inderbitzin, and C. Bening, "Total cost of ownership of electric vehicles compared to conventional vehicles: A probabilistic analysis and projection across market segments," *Energy Policy*, vol. 80, pp. 196–214, 2015.

Part IX

Reconfiguration of a Smart Distribution Network

17

Reconfiguration of Radial Distribution Systems

Test System

Meisam Mahdavi[1], *Pierluigi Siano*[2], *Hassan Haes Alhelou*[3], *and Baseem Khan*[4]

[1] *Department of Electrical Engineering, São Paulo State University, Ilha Solteira, SP, Brazil*
[2] *Department of Management & Innovation Systems, University of Salerno, Salerno, Italy*
[3] *Department of Electrical Power Engineering, Tishreen University, Lattakia, Syria*
[4] *Department of Electrical and Computer Engineering, Hawassa University, Hawassa, Ethiopia*

17.1 Introduction

The distribution network is an important part of the power system infrastructure that links a high-voltage transmission system [1–8] to end-users of electric energy in order to deliver the electricity produced by the generation system [9–11] (or generation companies (GENCOs) in deregulated environments [12]) to individual customers [13–16]. The purpose of a power system is to provide reliable and safe electricity for its consumers in an economical way [17–22].

Power distribution systems in urban areas are constructed as a meshed structure and are usually operated in a radial topology by opening sectional switches (normally closed) and closing tie lines (normally open), because many operation and control activities (e.g. voltage control and protection) are based on a radial structure. In simple terms, operation of distribution networks in a radial configuration is cheaper because of easier coordination of protective instruments (simple protection and coordination schemes), lower fault and short-circuit currents, and better voltage and power flow control. Tie lines interconnect ends of radial feeders and/or provide connections to alternative supply points. Some of the tie and sectionalizing switches are controlled manually (manual switches) and others may be controlled automatically (remote controlled switches) [23–28].

In modern power distribution networks, a number of remote controlled switches are allocated to provide emergency connections and for reliability improvement [29]. In these networks, switches along feeders are automated and can be controlled using communication systems to improve network efficiency and sustainability [30]. In [31], an assessment and enhancement of distribution system reliability by renewable energy sources and energy storage are presented. In [32], a holistic review of distribution system reliability assessment methods with conventional and renewable energy sources is presented. In [33], a power quality assessment and event detection in a distribution network with wind energy penetration using Stockwell transform and fuzzy clustering are presented. In [34], an

Active Electrical Distribution Network: A Smart Approach, First Edition. Edited by Baseem Khan, Josep M. Guerrero, Sanjeevikumar Padmanaban, Hassan Haes Alhelou, Om Prakash Mahela, and Sudeep Tanwar.

assessment of power quality in the utility grid integrated with wind energy generation is presented. In [35], optimal expansion planning of a distribution system using a grid-based multiobjective harmony search algorithm is presented. However, switching operations can also be used to optimize the steady-state operating point of the system under normal conditions. The main reason for distribution utilities (or distribution companies (DISCOs) in deregulated power systems) to invest in switching devices is to prevent prolonged failures and to reduce the number of customers isolated by faults [36].

Historically, distribution system reconfiguration (DSR) was initiated with the purpose of obtaining the lowest power losses [37]. Nevertheless, today it also includes other objectives such as improvement of a bus voltage profile, an increase in system security [38–43], reliability enhancement (a decrease in the frequency and the duration of power interruptions) [44], congestion management (line loading optimization) [45–48], reduction of line maintenance costs [49, 50], supply capacity expansion, power stability maximization [51], load balancing, an increase in DG [52, 53] penetration, service restoration, quick fault isolation, and implementation of preventive maintenance plans [14, 30, 54–56].

In this process, the configuration of the system changes with the status of the switches while the network has to maintain its radiality and energized loads [57–59]. In general, network reconfiguration is implemented when there is a significant change in loads or a sudden fault occurring on the feeder. DSR is a nonlinear and complex constrained optimization problem because of numerous switching scenarios [60, 61].

Test systems play an important role in showing efficiency of proposed models and methodologies [62]. Nowadays, several test systems available in the specialized literature are used to verify studies regarding distribution network reconfiguration. This chapter presents different test systems, including the load data, network configuration, line characteristics, and nominal voltage. The main objective is to provide all the details and information required to evaluate methods and models developed for operation and reconfiguration of radial distribution systems.

17.2 Test Systems

As mentioned earlier, this section includes several test systems with all their required data and information to implement different reconfiguration models and methods.

17.2.1 System 1: 7-Bus Distribution Network

Figure 17.1 shows this test system with 6 branches (closed switches) and one loop (tie line) before reconfiguration. All data related to this system are given in Tables 17.1 and 17.2 [63].

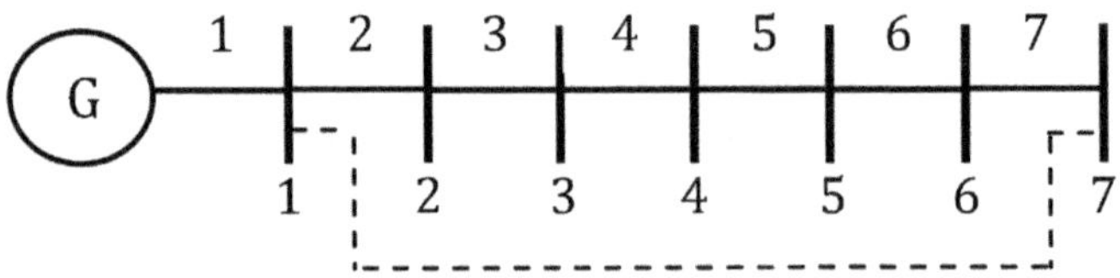

Figure 17.1 Schematic of the 7-bus test system.

Table 17.1 Line characteristics of System 1

Branch (Switch)	From Bus	To us	Resistance (Ω)	Reactance (Ω)	Maximum Current (A)
1	1	2	0.0922	0.0407	500
2	2	3	0.4930	0.2510	500
3	3	4	0.3660	0.1864	500
4	4	5	0.3811	0.1941	500
5	5	6	0.1872	0.6188	500
6	6	7	1.7114	1.2351	500
7 tie line	7	1	1.0300	0.7400	500

Table 17.2 Load data for System 1

Bus Number	Bus Type	Active Power Demand (kW)	Reactive Power Demand (kVAr, or kVA reactive)
1	Generation	0	0
2	Load	100	60
3	Load	90	40
4	Load	90	40
5	Load	120	80
6	Load	60	30
7	Load	60	20

The base power is 1 MVA (megavolt-ampere) and the nominal voltage is 12.66 kV. It should be noted that the voltage magnitude is considered to be 1 per unit (p.u.) at the generation (substation) bus.

17.2.2 System 2: 12-Bus Distribution Network

This actual system is a part of the Baghdad city network in Iraq (AL-Mansoor No.11) and its single line diagram is shown in Figure 17.2. The feeder nominal (base) voltage and apparent power are 11.1 kV and 2250 kVA, which are connected to the AL-Mansoor substation. The system relevant data are listed in Tables 17.3 and 17.4 [64]. Tie lines (tie switches) in this network are branches 7–8 (12) and 7–11 (13).

17.2.3 System 3: 16-Bus Distribution Network

The three-feeder distribution system including 13 sectional switches (branches) and three tie lines has been shown in Figure 17.3. All data of this 23 kV network, such as resistance

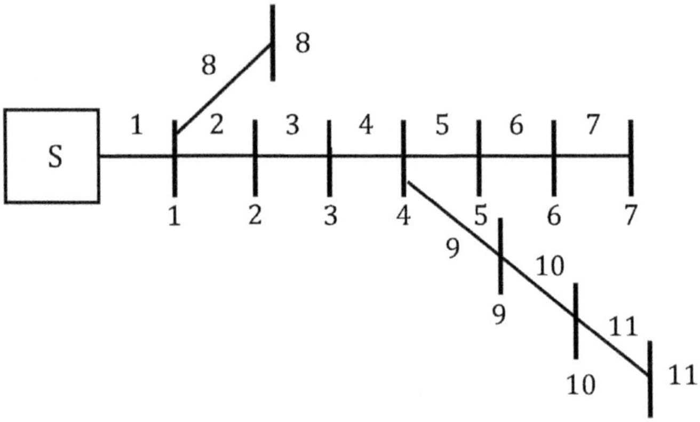

Figure 17.2 AL-Mansoor No.11 network.

and reactance of branches as well as nodal active and reactive demand, have been provided in Tables 17.5 and 17.6 [65, 66].

17.2.4 System 4: 28-Bus Distribution Network

This electrical grid is part of the electrical power distribution system in the city of Koprivnica, Croatia. It consists of 28 nodes, one transformer station of 110/35 kV, two transformer stations of 35/10 kV, 22 loads, and 28 transmission lines. A graph

Table 17.3 Line characteristics of System 2

Branch (Switch)	From Bus	To Bus	Resistance (Ω/km)	Reactance (Ω/km)	Length (km)	Maximum Current (A)
1	0	1	0.124	0.106	0.450	500
2	1	2	0.330	0.350	0.150	500
3	2	3	0.330	0.350	0.200	500
4	3	4	0.330	0.350	0.025	500
5	4	5	0.330	0.350	0.250	500
6	5	6	0.330	0.350	0.100	500
7	6	7	0.330	0.350	0.075	500
8	7	8	0.330	0.350	0.050	500
9	8	9	0.330	0.350	0.100	500
10	9	10	0.330	0.350	0.035	500
11	10	11	0.330	0.350	0.150	500
12 (tie line)	7	8	0.124	0.106	0.450	500
13 (tie line)	7	11	0.330	0.350	0.150	500

Table 17.4 Load data for System 2

Bus Number	Bus Type	Active Power Demand (kW)	Reactive Power Demand (kVAr)
0	Generation	0	0
1	Load	0	0
2	Load	178.711	149.504
3	Load	178.711	149.504
4	Load	0	0
5	Load	178.711	149.504
6	Load	178.711	149.504
7	Load	178.711	149.504
8	Load	178.711	149.504
9	Load	178.711	149.504
10	Load	178.711	149.504
11	Load	178.711	149.504

representation of the Koprivnica distribution grid is shown in Figure 17.4 and its data are represented in Tables 17.7–17.9 [67].

Lines and transformers that connect nodes are represented as edges of the graph. Full lines represent transmission lines that are switched on, while dotted lines represent transmission lines that are switched off in the current topology. Base values for this network are nominal voltages and power.

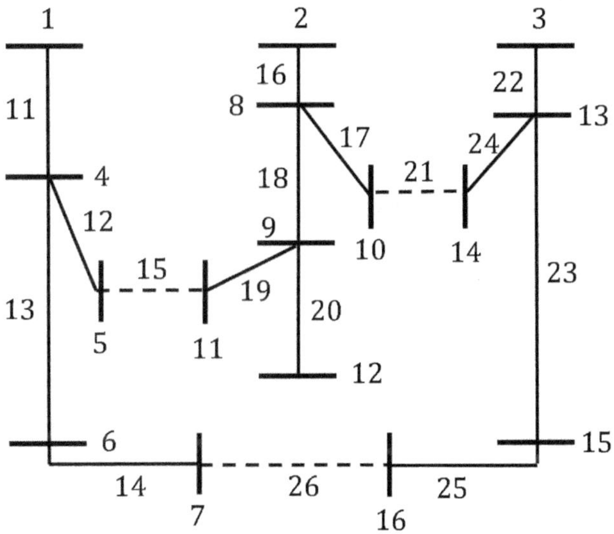

Figure 17.3 Initial distribution network.

Table 17.5 Line characteristics of System 3

Branch (Switch)	From Bus	To Bus	Resistance (Ω)	Reactance (Ω)	Maximum Current (A)
11	1	4	0.39675	0.52900	500
12	4	5	0.42320	0.58190	250
13	4	6	0.47610	0.95220	250
14	6	7	0.21160	0.21160	250
16	2	8	0.58190	0.58190	500
17	8	10	0.58190	0.58190	250
18	8	9	0.42320	0.58190	300
19	9	11	0.58190	0.58190	250
20	9	12	0.42320	0.58190	250
22	3	13	0.58190	0.58190	250
23	13	15	0.42320	0.58190	250
24	13	14	0.47610	0.63480	250
25	15	16	0.21160	0.21160	250
15 (tie line)	5	11	0.21160	0.21160	250
21 (tie line)	10	14	0.21160	0.21160	250
26 (tie line)	7	16	0.63480	0.63480	250

Table 17.6 Load data for System 3

Bus Number	Bus Type	Active Power Demand (kW)	Reactive Power Demand (kVAr)	Fixed Capacitor (kVAr)
1	Generation	0	0	0
2	Generation	0	0	0
3	Generation	0	0	0
4	Load	2000	1600	0
5	Load	3000	1500	1100
6	Load	2000	800	1200
7	Load	1500	1200	0
8	Load	4000	2700	0
9	Load	5000	3000	1200
10	Load	1000	900	0
11	Load	600	100	600
12	Load	4500	2000	3700
13	Load	1000	900	0
14	Load	1000	700	1800
15	Load	1000	900	0
16	Load	2100	1000	1800

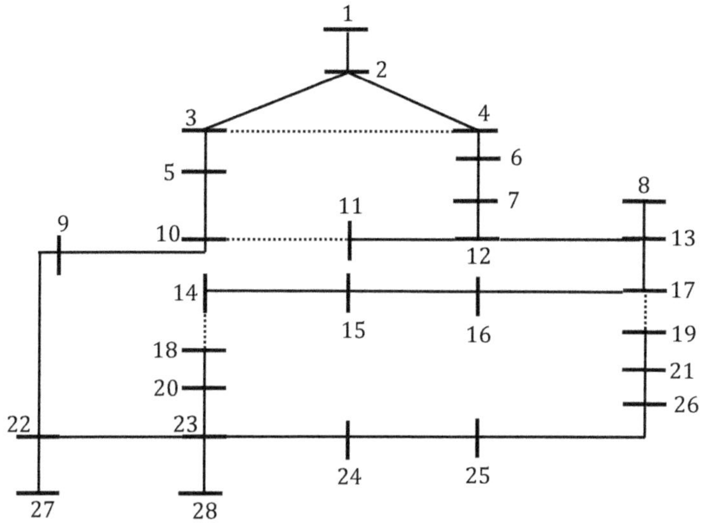

Figure 17.4 Distribution network of Koprivnica.

17.2.5 System 5: 30-Bus Distribution Network

Configuration of this test system with six tie lines and 28 normal branches in the network is shown in Figure 17.5 and its data are listed in Tables 17.10 and 17.11 [68]. The Nominal values are 1 MVA and 18.6 kV.

17.2.6 System 6: 33-Bus Distribution System

The system includes two feeder substations with three 12.66 kV laterals, five tie switches, and 32 normal branches (Figure 17.6). The data of this test system are available in Tables 17.12 and 17.13 [69]. The MVA and kV base are 100 and 12.66, respectively [70, 71]. Also, the voltage magnitude of substation bus (node 0) is adjusted in 1 per unit.

17.2.7 System 7: 49-Bus Distribution System

Figure 17.7 shows the schematic of the real distribution network of Baghdad city, Iraq. The test system is an 11 kV network with 49 buses, five looping branches (tie lines), 48 sectionalizing switches, and six laterals. The system data are listed in Tables 17.14 and 17.15 [72].

17.2.8 System 8: 59-Bus Distribution System

Figure 17.8 shows a 33 kV real distribution network. This system is a region of the distribution network of the city of Ahvaz in the south of Iran. The 59-bus system includes 63 branches and five and 58 tie and sectionalizing switches, respectively. Line and load data for this real distribution network are given in Tables 17.16 and 17.17 [73].

Table 17.7 Line characteristics of System 4

Branch (Switch)	From Bus	To Bus	Resistance (Ω/km)	Reactance (Ω/km)	Length (km)
2	2	3	0.0870	0.1160	0.3224
3	2	4	0.1380	0.1160	0.1360
6	5	10	0.2060	0.1220	1.3500
7	6	7	0.2060	0.1220	0.1420
8	7	12	0.3200	0.1000	0.1610
9	13	8	0.3200	0.1000	0.0870
10	9	10	0.1930	0.1000	0.4000
11	11	12	0.3200	0.1000	0.4550
12	12	13	0.1930	0.1000	0.2870
13	9	22	0.2060	0.1220	0.2850
14	13	17	0.3200	0.1000	0.2020
15	14	15	0.3200	0.1000	0.1400
16	15	16	0.2060	0.1220	0.3000
17	16	17	0.2060	0.1220	0.3000
18	18	20	0.2060	0.1220	0.3880
19	19	21	0.2060	0.1150	0.3700
20	20	23	0.2060	0.1220	0.6310
21	21	26	0.2060	0.1220	0.0880
22	22	23	0.2060	0.1220	0.8690
23	23	24	0.2060	0.1220	0.1150
24	24	25	0.1930	0.1000	0.2460
25	25	26	0.1930	0.1000	0.2050
26	22	27	0.3200	0.1000	0.5130
27	23	28	0.2060	0.1220	0.3110
28 (tie line)	3	4	0.0870	0.1160	0.2658
29 (tie line)	10	11	0.2060	0.1220	0.4150
30 (tie line)	14	18	0.3200	0.1000	0.1600
31 (tie line)	17	19	0.2060	0.1150	0.6320

Table 17.8 Transformer data for System 4

Branch	From Bus	To Bus	Resistance (Ω)	Reactance (Ω)	Primary Voltage (kV)	Secondary Voltage (kV)	Nominal Power (kVA)
1	1	2	1.3000	32.656	110	35	40
4	3	5	1.0311	10.645	35	10	8
5	4	6	1.0311	10.645	35	10	8

Table 17.9 Load data for System 4

Bus Number	Bus Type	Active Power Demand (kW)	Reactive Power Demand (kVAr)	Nominal Voltage (kV)
1	Load	0	0	110
2	Generation	0	0	35
3	Load	0	0	35
4	Load	0	0	35
5	Load	0	0	10
6	Load	0	0	10
7	Load	127.6	37.2	10
8	Load	121.6	36.3	10
9	Load	130.2	38	10
10	Load	75.6	22	10
11	Load	68.3	19.9	10
12	Load	106.4	22.4	10
13	Load	100.7	22.2	10
14	Load	105.4	30.7	10
15	Load	54.9	16	10
16	Load	71.1	21.6	10
17	Load	100.3	27.7	10
18	Load	129.7	37.8	10
19	Load	194.5	56.7	10
20	Load	8.3	2.4	10
21	Load	84.9	24.8	10
22	Load	50.5	20.2	10
23	Load	124.7	36.4	10
24	Load	105.3	30.7	10
25	Load	33.4	9.7	10
26	Load	256.8	74.9	10
27	Load	98.2	28.6	10
28	Load	72.6	21.2	10

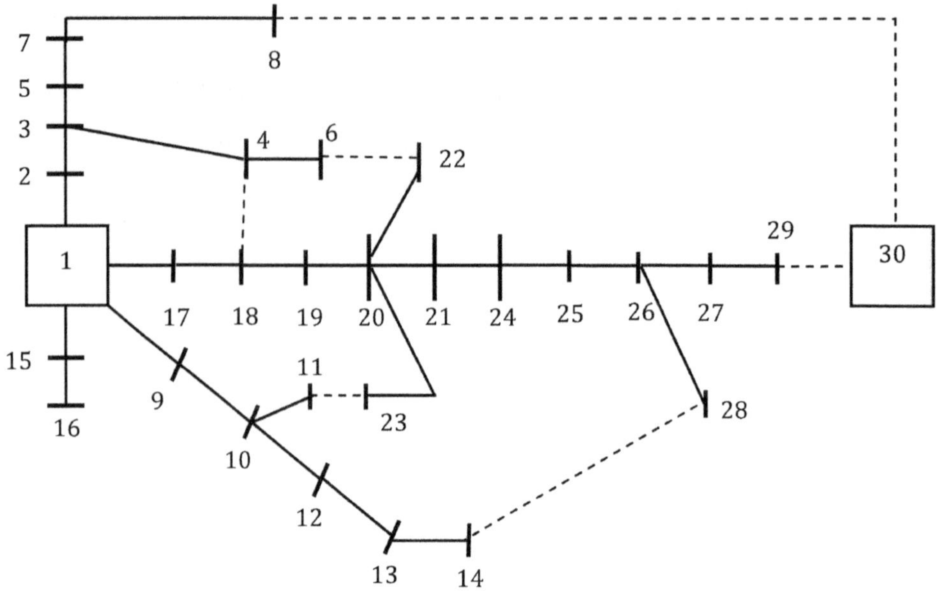

Figure 17.5 The 30-bus distribution system.

Table 17.10 Line characteristics of System 5

Branch (Switch)	From Bus	To Bus	Resistance (Ω/km)	Reactance (Ω/km)	Length (km)	Maximum Current (A)
1	2	1	0.60	0.45	0.140	144
2	3	2	0.60	0.45	2.310	215
3	4	3	0.60	0.45	1.003	215
4	5	3	0.60	0.45	1.370	215
5	6	4	1.52	1.48	3.010	125
6	7	5	0.60	0.45	3.977	215
7	8	7	0.60	0.45	6.440	215
8	9	1	0.19	0.40	0.770	341.7
9	10	9	0.19	0.40	3.137	400
10	11	10	0.60	0.45	3.380	215
11	12	10	0.19	0.40	5.050	400
12	13	12	0.19	0.40	1.720	400
13	14	13	0.19	0.40	1.070	510
14	15	1	0.60	0.45	0.300	144
15	16	15	0.60	0.45	0.300	215
16	17	1	0.30	0.42	1.485	227.8
17	18	17	0.30	0.42	1.485	340

(continued)

Table 17.10 (*continued*)

Branch (Switch)	From Bus	To Bus	Resistance (Ω/km)	Reactance (Ω/km)	Length (km)	Maximum Current (A)
18	19	18	0.30	0.42	1.290	340
19	20	19	0.30	0.42	0.550	340
20	21	20	0.30	0.42	4.518	340
21	22	20	0.60	0.45	0.080	215
22	23	20	0.60	0.45	0.400	215
23	24	21	0.30	0.42	0.845	340
24	25	24	0.30	0.42	1.050	340
25	26	25	0.30	0.42	1.424	340
26	27	26	0.70	0.52	1.420	230
27	28	26	0.70	0.52	1.051	230
28	29	27	0.70	0.52	1.920	230
29 (tie line)	29	30	0.19	0.38	0.568	355.1
30 (tie line)	8	30	0.19	0.38	7.332	355.1
31 (tie line)	4	18	0.30	0.42	0.275	340
32 (tie line)	6	22	0.60	0.45	1.199	215
33 (tie line)	14	28	0.60	0.45	0.069	215
34 (tie line)	11	23	0.60	0.45	0.030	215

Table 17.11 Load data for System 5

Bus Number	Bus Type	Active Power Demand (kW)	Power Factor
1	Generation	0	1
2	Load	0	1
3	Load	452.4	0.86
4	Load	260.4	0.85
5	Load	139.1	0.85
6	Load	776.2	0.85
7	Load	906.1	0.85
8	Load	1091.6	0.85
9	Load	0	1
10	Load	444.5	0.85
11	Load	49.9	0.85
12	Load	231.9	0.85
13	Load	947.4	0.85

(*continued*)

Table 17.11 (*continued*)

Bus Number	Bus Type	Active Power Demand (kW)	Power Factor
14	Load	3051.4	0.85
15	Load	0	1
16	Load	3124.8	0.85
17	Load	0	1
18	Load	71.3	0.85
19	Load	35.7	0.85
20	Load	7.1	0.85
21	Load	3.6	0.85
22	Load	0	1
23	Load	42.8	0.85
24	Load	164.1	0.85
25	Load	873.9	0.85
26	Load	85.6	0.85
27	Load	428.1	0.85
28	Load	2214.5	0.85
29	Load	2238.8	0.85
30	Generation	0	1

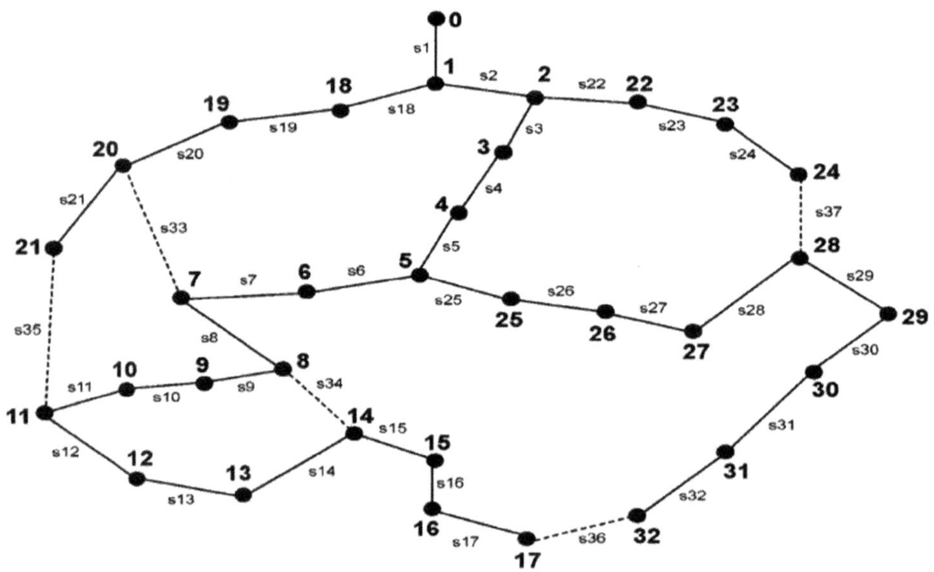

Figure 17.6 Single-diagram of the 33-bus distribution system.

Table 17.12 Line characteristics of System 6

Branch (Switch)	From Bus	To Bus	Resistance (Ω)	Reactance (Ω)	Maximum Current (A)
1	1	2	0.0922	0.0477	500
2	2	3	0.4930	0.2511	500
3	3	4	0.3660	0.1864	500
4	4	5	0.3811	0.1941	500
5	5	6	0.8190	0.7070	500
6	6	7	0.1872	0.6188	500
7	7	8	0.7114	0.2351	500
8	8	9	1.0300	0.7400	500
9	9	10	1.0440	0.7400	500
10	10	11	0.1966	0.0650	500
11	11	12	0.3744	0.1238	500
12	12	13	1.4680	1.1550	500
13	13	14	0.5416	0.7129	500
14	14	15	0.5910	0.5260	500
15	15	16	0.7463	0.5450	500
16	16	17	1.2890	1.7210	500
17	17	18	0.7320	0.5740	500
18	2	19	0.1640	0.1565	500
19	19	20	1.5042	1.3554	500
20	20	21	0.4095	0.4784	500
21	21	22	0.7089	0.9373	500
22	3	23	0.4512	0.3083	500
23	23	24	0.8980	0.7091	500
24	24	25	0.8960	0.7011	500
25	6	26	0.2030	0.1034	500
26	26	27	0.2842	0.1447	500
27	27	28	1.0590	0.9337	500
28	28	29	0.8042	0.7006	500
29	29	30	0.5075	0.2585	500
30	30	31	0.9744	0.9630	500
31	31	32	0.3105	0.3619	500
32	32	33	0.3410	0.5302	500
33 (tie line)	8	21	2.0000	2.0000	500
34 (tie line)	9	15	2.0000	2.0000	500
35 (tie line)	12	22	2.0000	2.0000	500
36 (tie line)	18	33	0.5000	0.5000	500
37 (tie line)	25	29	0.5000	0.5000	500

Table 17.13 Load data for system 6

Bus Number	Bus Type	Active Power Demand (kW)	Reactive Power Demand (kVAr)
1	Generation	0	0
2	Load	100	60
3	Load	90	40
4	Load	120	80
5	Load	60	30
6	Load	60	20
7	Load	200	100
8	Load	200	100
9	Load	60	20
10	Load	60	20
11	Load	45	30
12	Load	60	35
13	Load	60	35
14	Load	120	80
15	Load	60	10
16	Load	60	20
17	Load	60	20
18	Load	90	40
19	Load	90	40
20	Load	90	40
21	Load	90	40
22	Load	90	40
23	Load	90	50
24	Load	420	200
25	Load	420	200
26	Load	60	25
27	Load	60	25
28	Load	60	20
29	Load	120	70
30	Load	200	600
31	Load	150	70
32	Load	210	100
33	Load	60	40

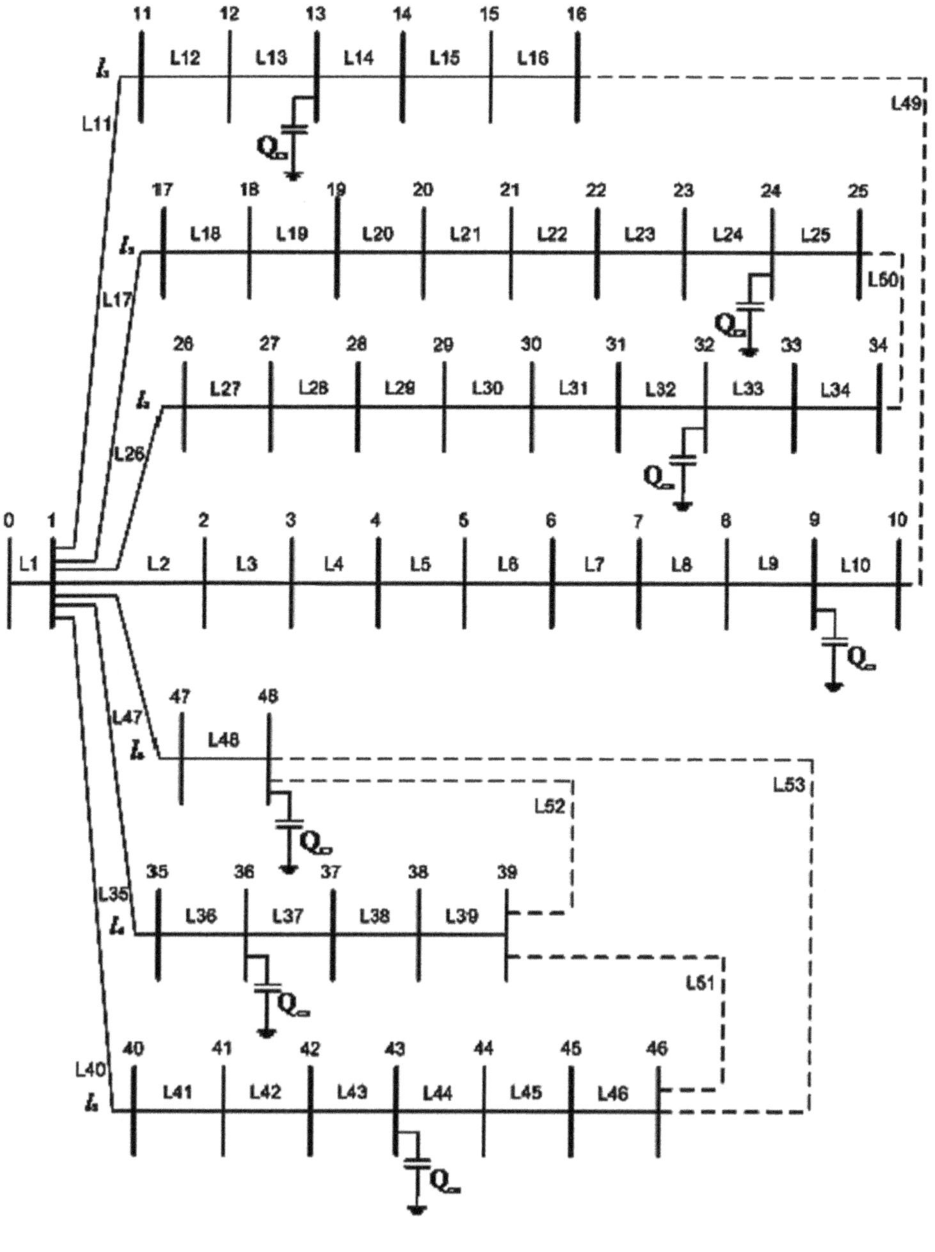

Figure 17.7 Baghdad distribution network.

Table 17.14 Load data for System 7

Bus Number	Bus Type	Active Power Demand (kW)	Reactive Power Demand (kVAr)	Fixed Capacitor (kVAr)
0	Generation	0.00000	0.00000	0
1	Load	0.00000	0.00000	0
2	Load	969.679	469.637	0
3	Load	401.246	194.333	0
4	Load	100.311	48.582	0
5	Load	401.246	194.333	0
6	Load	1003.11	485.832	0
7	Load	936.243	453.470	0
8	Load	735.619	356.284	0
9	Load	869.369	521.054	100
10	Load	1270.61	615.388	0
11	Load	922.058	446.572	0
12	Load	1276.69	618.332	0
13	Load	638.347	384.166	75
14	Load	2908.02	1408.41	0
15	Load	922.058	446.572	0
16	Load	425.565	48.7060	0
17	Load	1791.42	867.626	0
18	Load	1264.53	612.442	0
19	Load	948.402	459.332	0
20	Load	421.512	204.147	0
21	Load	1896.80	918.663	0
22	Load	1686.04	816.590	0
23	Load	316.134	153.110	0
24	Load	1053.78	610.368	100
25	Load	632.267	306.220	0
26	Load	559.313	270.888	0
27	Load	838.970	406.332	0
28	Load	1258.45	609.499	0
29	Load	1258.45	609.499	0
30	Load	1817.77	880.386	0
31	Load	2237.25	1083.55	0
32	Load	1817.77	1080.38	200
33	Load	1957.59	948.108	0
34	Load	1957.59	948.108	0
35	Load	607.950	294.444	0

(continued)

Table 17.14 (*continued*)

Bus Number	Bus Type	Active Power Demand (kW)	Reactive Power Demand (kVAr)	Fixed Capacitor (kVAr)
36	Load	1925.17	1132.40	200
37	Load	303.975	147.221	0
38	Load	3343.72	1619.44	0
39	Load	222.915	107.962	0
40	Load	719.407	348.425	0
41	Load	1726.57	836.220	0
42	Load	1582.69	766.535	0
43	Load	3597.03	2092.12	350
44	Load	2877.63	1393.69	0
45	Load	3884.79	1881.49	0
46	Load	4677.46	2265.39	0
47	Load	1491.50	722.367	0
48	Load	1750.89	1047.99	200

Table 17.15 Line characteristics of System 7

Branch (Switch)	From Bus	To Bus	Resistance (Ω)	Reactance (Ω)	Maximum Current (A)
1	0	1	0.0310	0.0220	500
2	1	2	0.0310	0.0220	500
3	2	3	0.7440	0.5470	500
4	3	4	0.8680	0.6380	500
5	4	5	0.4340	0.3190	500
6	5	6	0.1628	0.1200	500
7	6	7	0.6200	0.4600	500
8	7	8	0.5890	0.4330	500
9	8	9	0.3570	0.2620	500
10	9	10	1.0186	1.0137	500
11	1	11	0.0434	0.0319	500
12	11	12	1.0171	1.0125	500
13	12	13	0.4340	0.3190	500
14	13	14	1.0403	1.0296	500
15	14	15	0.3660	0.2690	500
16	15	16	0.1860	0.1370	500
17	1	17	0.0233	0.0171	500

(*continued*)

Table 17.15 (*continued*)

Branch (Switch)	From Bus	To Bus	Resistance (Ω)	Reactance (Ω)	Maximum Current (A)
18	17	18	0.3880	0.2850	500
19	18	19	0.2790	0.2050	500
20	19	20	0.5520	0.4050	500
21	20	21	0.3970	0.2920	500
22	21	22	1.0233	1.0171	500
23	22	23	0.0780	0.0570	500
24	23	24	0.9900	0.7300	500
25	24	25	0.9900	0.7300	500
26	1	26	0.0160	0.0120	500
27	26	27	0.3720	0.2730	500
28	27	28	0.4400	0.3230	500
29	28	29	0.3660	0.2690	500
30	29	30	0.4500	0.3300	500
31	30	31	0.2950	0.2160	500
32	31	32	0.2850	0.2100	500
33	32	33	0.4590	0.3370	500
34	33	34	1.0217	1.0160	500
35	1	35	0.0372	0.0273	500
36	35	36	0.2880	0.2120	500
37	36	37	0.4190	0.3080	500
38	37	38	0.4030	0.2960	500
39	38	39	0.8680	0.6380	500
40	1	40	0.0620	0.0456	500
41	40	41	0.2110	0.1550	500
42	41	42	0.2170	0.1600	500
43	42	43	0.4190	0.3080	500
44	43	44	1.0388	1.0285	500
45	44	45	0.6360	0.4670	500
46	45	46	1.0140	0.6105	500
47	1	47	0.1023	0.0752	500
48	47	48	0.0285	0.0210	500
49 (tie line)	10	16	0.388	0.285	500
50 (tie line)	25	34	0.118	0.087	500
51 (tie line)	39	46	0.481	0.353	500
52 (tie line)	39	48	0.837	0.615	500
53 (tie line)	46	48	0.357	0.262	500

17.2.9 System 9: 69-Bus Distribution System

This 12.66 kV radial distribution system has 69 nodes, 68 normal switches, and five tie lines, as shown in Figure 17.9. Data for this system are given in Tables 17.18 and 17.19 [74, 75].

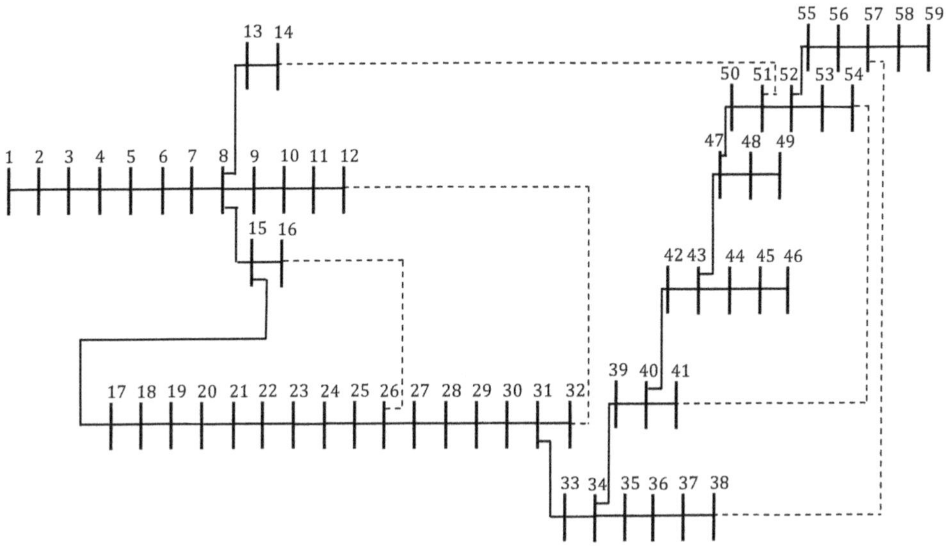

Figure 17.8 The real 59-bus distribution system.

Table 17.16 Line characteristics of System 8

Branch (Switch)	From Bus	To Bus	Resistance (Ω)	Reactance (Ω)	Maximum Current (A)
1	1	2	0.081	0.061	500
2	2	3	0.135	0.101	500
3	3	4	0.027	0.020	500
4	4	5	0.027	0.020	500
5	5	6	0.027	0.020	500
6	6	7	0.027	0.020	500
7	7	8	0.540	0.405	500
8	8	9	0.540	0.405	500
9	9	10	0.270	0.203	500
10	10	11	0.540	0.405	500
11	11	12	0.540	0.405	500

(continued)

Table 17.16 (*continued*)

Branch (Switch)	From Bus	To Bus	Resistance (Ω)	Reactance (Ω)	Maximum Current (A)
12	8	13	0.027	0.020	500
13	13	14	0.027	0.020	500
14	8	15	0.027	0.020	500
15	15	16	0.027	0.020	500
16	15	17	0.027	0.020	500
17	17	18	0.027	0.020	500
18	18	19	0.027	0.020	500
19	19	20	0.027	0.020	500
20	20	21	0.027	0.020	500
21	21	22	0.270	0.203	500
22	22	23	0.027	0.020	500
23	23	24	0.027	0.020	500
24	24	25	0.027	0.020	500
25	25	26	0.270	0.203	500
26	26	27	0.027	0.020	500
27	27	28	0.027	0.020	500
28	28	29	0.027	0.020	500
29	29	30	0.027	0.020	500
30	30	31	0.027	0.020	500
31	31	32	0.027	0.020	500
32	31	33	0.135	0.101	500
33	33	34	0.027	0.020	500
34	34	35	0.054	0.041	500
35	35	36	0.027	0.020	500
36	36	37	0.027	0.020	500
37	37	38	0.027	0.020	500
38	34	39	0.027	0.020	500
39	39	40	0.027	0.020	500
40	40	41	0.054	0.041	500
41	40	42	0.054	0.041	500
42	42	43	0.027	0.020	500
43	43	44	0.027	0.020	500
44	44	45	0.027	0.020	500
45	45	46	0.027	0.020	500
46	43	47	0.270	0.203	500

(continued)

Table 17.16 (continued)

Branch (Switch)	From Bus	To Bus	Resistance (Ω)	Reactance (Ω)	Maximum Current (A)
47	47	48	0.027	0.020	500
48	48	49	0.270	0.203	500
49	47	50	0.027	0.020	500
50	50	51	0.027	0.020	500
51	51	52	0.027	0.020	500
52	52	53	0.027	0.020	500
53	53	54	0.081	0.061	500
54	52	55	0.540	0.405	500
55	55	56	0.135	0.101	500
56	56	57	0.054	0.041	500
57	57	58	0.027	0.020	500
58	58	59	0.054	0.041	500
59 (tie line)	16	26	0.500	0.500	500
60 (tie line)	14	50	0.500	0.500	500
61 (tie line)	12	32	0.500	0.500	500
62(tie line)	38	57	0.500	0.500	500
63 (tie line)	41	54	0.500	0.500	500

Table 17.17 Load data for System 8

Bus Number	Bus Type	Active Power Demand (kW)	Reactive Power Demand (kVAr)
1	Generation	0	0
2	Load	72	54
3	Load	51	38.25
4	Load	50	37.5
5	Load	50	37.5
6	Load	85	63.75
7	Load	108	81
8	Load	0	0
9	Load	80	60
10	Load	320	240
11	Load	288	216
12	Load	125	93.75
13	Load	108	81
14	Load	200	150

(continued)

Table 17.17 (*continued*)

Bus Number	Bus Type	Active Power Demand (kW)	Reactive Power Demand (kVAr)
15	Load	0	0
16	Load	95	71.25
17	Load	200	150
18	Load	225	168.75
19	Load	100	75
20	Load	231	173.25
21	Load	410	307.5
22	Load	450	337.5
23	Load	230	172.5
24	Load	262	196.5
25	Load	250	187.5
26	Load	400	310.43
27	Load	203	152.25
28	Load	200	150
29	Load	320	240
30	Load	200	150
31	Load	260	199.5
32	Load	240	186.25
33	Load	150	112.5
34	Load	0	0
35	Load	158	118.5
36	Load	40	30
37	Load	190	142.5
38	Load	87	65.25
39	Load	450	337.5
40	Load	0	0
41	Load	420	315
42	Load	100	75
43	Load	0	0
44	Load	199	149.25
45	Load	80	60
46	Load	420	315
47	Load	0	0
48	Load	141	105.75
49	Load	165	123.75
50	Load	0	0

(*continued*)

Table 17.17 (continued)

Bus Number	Bus Type	Active Power Demand (kW)	Reactive Power Demand (kVAr)
51	Load	300	225
52	Load	0	0
53	Load	100	80.23
54	Load	178	133.5
55	Load	0	0
56	Load	235	176.25
57	Load	154	123.55
58	Load	0	0
59	Load	310	248.71

17.2.10 System 10: 70-Bus Distribution System

The tested system is an 11 kV radial distribution system with two substations, four feeders, 70 nodes, and 76 branches (including tie lines), as shown in Figure 17.10. Data for this system are provided in Tables 17.20 and 17.21 [76].

17.2.11 System 11: Distribution System of Taiwan Power Company (TPC)

As shown in Figure 17.11, this accrual 11.4 kV network consists of two substations, 11 feeders, 83 normal switches, and 13 tie lines, in which its data are available in Tables 17.22 and 17.23 [77]. The substation voltage is set at 1.0 p.u. and each line current-carrying capacity is 410 A [78].

17.2.12 System 12: 119-Bus Distribution System

This test system, as shown in Figure 17.12, is an 11 kV distribution network with three feeders, 118 lines, and 15 tie switches. The parameters and related data of this network can be found in Tables 17.24 and 17.25 [79].

17.2.13 System 13: 136-Bus Real Distribution System

A shown in Figure 17.13, it is a part of the Tres Lagoas distribution network in Brazil. Its data are listed in Tables 17.26 and 17.27 [80].

17.2.14 System 14: 203-Bus Real Test System

The initial configuration of this real 13.8 kV distribution network is shown in Figure 17.14 and its data are shown in Tables 17.28 and 17.29 [81].

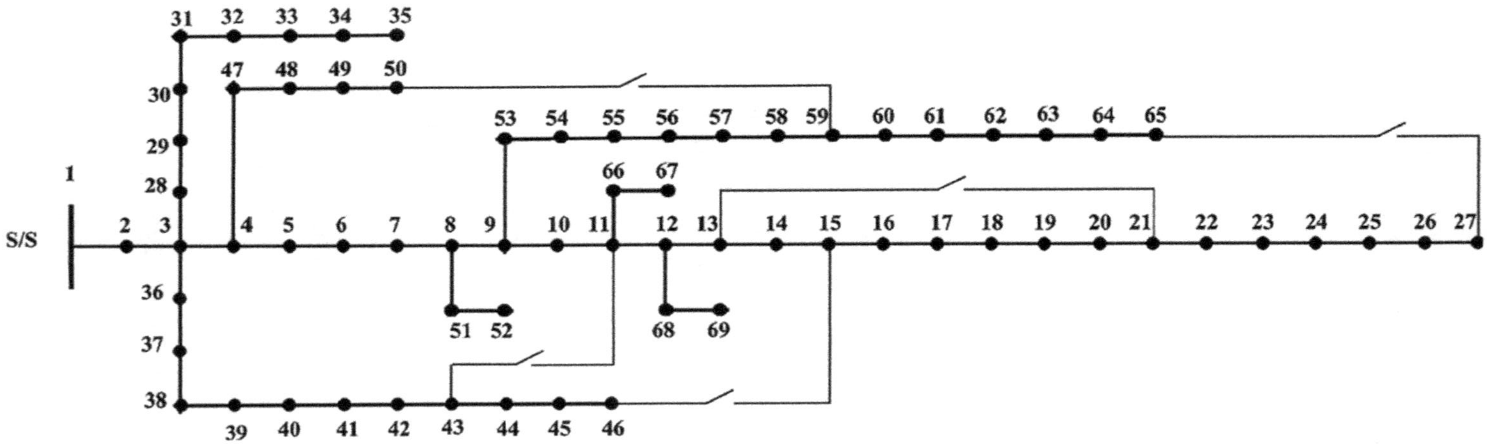

Figure 17.9 Schematic diagram of the 69-bus distribution system.

Table 17.18 Line characteristics of System 9

Branch (Switch)	From Bus	To Bus	Resistance (Ω)	Reactance (Ω)	Maximum Current (A)
1	1	2	0.0005	0.0012	400
2	2	3	0.0005	0.0012	400
3	3	4	0.0015	0.0036	400
4	4	5	0.0251	0.0294	400
5	5	6	0.3660	0.1864	400
6	6	7	0.381	0.1941	400
7	7	8	0.0922	0.0470	400
8	8	9	0.0493	0.0251	400
9	9	10	0.8190	0.2707	400
10	10	11	0.1872	0.0691	200
11	11	12	0.7114	0.2351	200
12	12	13	1.0300	0.3400	200
13	13	14	1.0440	0.3450	200
14	14	15	1.0580	0.3496	200
15	15	16	0.1966	0.0650	200
16	16	17	0.3744	0.1238	200
17	17	18	0.0047	0.0016	200
18	18	19	0.3276	0.1083	200
19	19	20	0.2106	0.0690	200
20	20	21	0.3416	0.1129	200
21	21	22	0.0140	0.0046	200
22	22	23	0.1591	0.0526	200
23	23	24	0.3463	0.1145	200
24	24	25	0.7488	0.2475	200
25	25	26	0.3089	0.1021	200
26	26	27	0.1732	0.0572	200
27	3	28	0.0044	0.0108	200
28	28	29	0.0640	0.1565	200
29	29	30	0.3978	0.1315	200
30	30	31	0.0702	0.0232	200
31	31	32	0.3510	0.1160	200
32	32	33	0.8390	0.2816	200
33	33	34	1.7080	0.5646	200
34	34	35	1.4740	0.4873	200

(continued)

Table 17.18 (*continued*)

Branch (Switch)	From Bus	To Bus	Resistance (Ω)	Reactance (Ω)	Maximum Current (A)
35	3	36	0.0044	0.0108	200
36	36	37	0.0640	0.1565	200
37	37	38	0.1053	0.1230	200
38	38	39	0.0304	0.0355	200
39	39	40	0.0018	0.0021	200
40	40	41	0.7283	0.8509	200
41	41	42	0.3100	0.3623	200
42	42	43	0.0410	0.0478	200
43	43	44	0.0092	0.0116	200
44	44	45	0.1089	0.1373	200
45	45	46	0.0009	0.0012	200
46	4	47	0.0034	0.0084	300
47	47	48	0.0851	0.2083	300
48	48	49	0.2898	0.7091	300
49	49	50	0.0822	0.2011	300
50	8	51	0.0928	0.0473	200
51	51	52	0.3319	0.1114	200
52	9	53	0.1740	0.0886	300
53	53	54	0.2030	0.1034	300
54	54	55	0.2842	0.1447	300
55	55	56	0.2813	0.1433	300
56	56	57	1.5900	0.5337	300
57	57	58	0.7837	0.2630	300
58	58	59	0.3042	0.1006	300
59	59	60	0.3861	0.1172	300
60	60	61	0.5075	0.2585	300
61	61	62	0.0974	0.0496	300
62	62	63	0.1450	0.0738	300
63	63	64	0.7105	0.3619	300
64	64	65	1.0410	0.5302	300
65	11	66	0.2012	0.0611	200
66	66	67	0.0047	0.0014	200
67	12	68	0.7394	0.2444	200
68	68	69	0.0047	0.0016	200
69 (tie line)	11	43	0.5000	0.5000	200

(*continued*)

Table 17.18 (*continued*)

Branch (Switch)	From Bus	To Bus	Resistance (Ω)	Reactance (Ω)	Maximum Current (A)
70 (tie line)	13	21	0.5000	0.5000	200
71 (tie line)	15	46	1.0000	0.5000	200
72 (tie line)	50	59	2.0000	1.0000	200
73 (tie line)	27	65	1.0000	0.5000	200

Table 17.19 Load data for System 9

Bus Number	Bus Type	Active Power Demand (kW)	Reactive Power Demand (kVAr)
1	Generation	0	0
2	Load	0	0
3	Load	0	0
4	Load	0	0
5	Load	0	0
6	Load	2.6	2.2
7	Load	40.4	30
8	Load	75	54
9	Load	30	22
10	Load	28	19
11	Load	145	104
12	Load	145	104
13	Load	8	5
14	Load	8	5.5
15	Load	0	0
16	Load	45.5	30
17	Load	60	35
18	Load	60	35
19	Load	0	0
20	Load	1	0.6
21	Load	114	81
22	Load	5	3.5
23	Load	0	0
24	Load	28	20
25	Load	0	0

(*continued*)

Table 17.19 (*continued*)

Bus Number	Bus Type	Active Power Demand (kW)	Reactive Power Demand (kVAr)
26	Load	14	10
27	Load	14	10
28	Load	26	18.6
29	Load	26	18.6
30	Load	0	0
31	Load	0	0
32	Load	0	0
33	Load	14	10
34	Load	19.5	14
35	Load	6	4
36	Load	26	18.55
37	Load	26	18.55
38	Load	0	0
39	Load	24	17
40	Load	24	17
41	Load	1.2	1
42	Load	0	0
43	Load	6	4.3
44	Load	0	0
45	Load	39.22	26.3
46	Load	39.22	26.3
47	Load	0	0
48	Load	79	56.4
49	Load	384.7	274.5
50	Load	384.7	274.5
51	Load	40.5	28.3
52	Load	3.6	2.7
53	Load	4.35	3.5
54	Load	26.4	19
55	Load	24	17.2
56	Load	0	0
57	Load	0	0
58	Load	0	0
59	Load	100	72
60	Load	0	0
61	Load	1244	888

(*continued*)

Table 17.19 *(continued)*

Bus Number	Bus Type	Active Power Demand (kW)	Reactive Power Demand (kVAr)
62	Load	32	23
63	Load	0	0
64	Load	227	162
65	Load	59	42
66	Load	18	13
67	Load	18	13
68	Load	28	20
69	Load	28	20

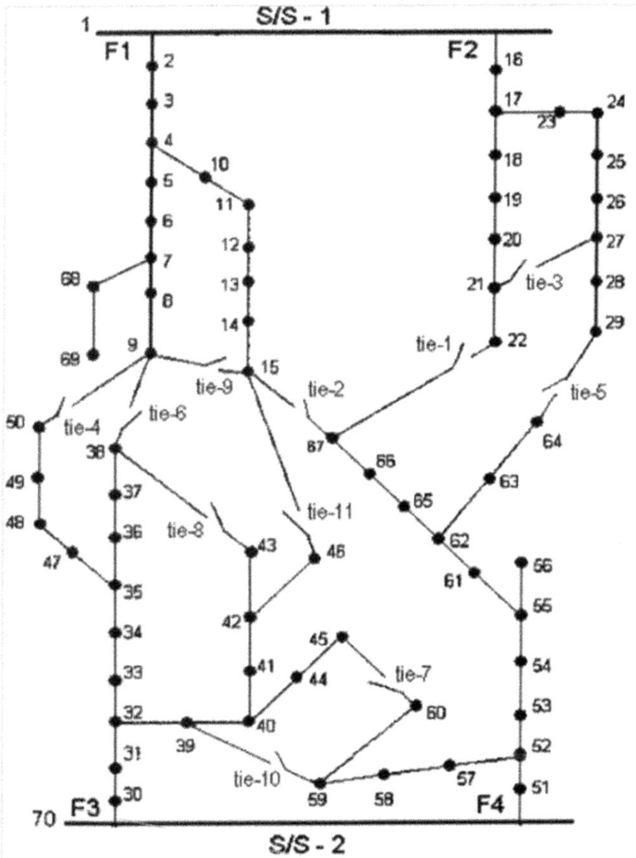

Figure 17.10 The 70-bus radial distribution system.

Table 17.20 Line characteristics of System 10

Branch	From Bus	To Bus	Resistance (Ω)	Reactance (Ω)	Maximum Current (A)
1	1	2	1.097	1.074	270
2	2	3	1.463	1.432	270
3	3	4	0.731	0.716	270
4	4	5	0.366	0.358	270
5	5	6	1.828	1.790	270
6	6	7	1.097	1.074	270
7	7	8	0.731	0.716	270
8	8	9	0.731	0.716	270
9	4	10	1.080	0.734	208
10	10	11	1.620	1.101	208
11	11	12	1.080	0.734	208
12	12	13	1.350	0.917	208
13	13	14	0.810	0.550	208
14	14	15	1.944	1.321	208
15	7	68	1.080	0.734	208
16	68	69	1.620	1.101	208
17	1	16	1.097	1.074	270
18	16	17	0.366	0.358	270
19	17	18	1.463	1.432	270
20	18	19	0.914	0.895	270
21	19	20	0.804	0.787	270
22	20	21	1.133	1.110	270
23	21	22	0.475	0.465	270
24	17	23	2.214	1.505	208
25	23	24	1.620	1.110	208
26	24	25	1.080	0.734	208
27	25	26	0.540	0.367	208
28	26	27	0.540	0.367	208
29	27	28	1.080	0.734	208
30	28	29	1.080	0.734	208
31	70	30	0.366	0.358	270
32	30	31	0.731	0.716	270
33	31	32	0.731	0.716	270
34	32	33	0.804	0.787	270
35	33	34	1.170	1.145	270
36	34	35	0.768	0.752	270

(continued)

Table 17.20 (continued)

Branch	From Bus	To Bus	Resistance (Ω)	Reactance (Ω)	Maximum Current (A)
37	35	36	0.731	0.716	270
38	36	37	1.097	1.074	270
39	37	38	1.463	1.432	270
40	32	39	1.080	0.734	208
41	39	40	0.540	0.367	208
42	40	41	1.080	0.734	208
43	41	42	1.836	1.248	208
44	42	43	1.296	0.881	208
45	40	44	1.188	0.807	208
46	44	45	0.540	0.367	208
47	42	46	1.080	0.734	208
48	35	47	0.540	0.367	208
49	47	48	1.080	0.734	208
50	48	49	1.080	0.734	208
51	49	50	1.080	0.734	208
52	70	51	0.366	0.358	270
53	51	52	1.463	1.432	270
54	52	53	1.463	1.432	270
55	53	54	0.914	0.895	270
56	54	55	1.097	1.074	270
57	55	56	1.097	1.074	270
58	52	57	0.270	0.183	208
59	57	58	0.270	0.183	208
60	58	59	0.810	0.550	208
61	59	60	1.296	0.881	208
62	55	61	1.188	0.807	208
63	61	62	1.188	0.807	208
64	62	63	0.810	0.550	208
65	63	64	1.620	1.101	208
66	62	65	1.080	0.734	208
67	65	66	0.540	0.367	208
68	66	67	1.080	0.734	208
69 (tie line)	9	50	0.908	0.726	234
70 (tie line)	9	38	0.381	0.244	234
71 (tie line)	15	46	0.681	0.544	234
72 (tie line)	22	67	0.254	0.203	234

(continued)

Table 17.20 (*continued*)

Branch	From Bus	To Bus	Resistance (Ω)	Reactance (Ω)	Maximum Current (A)
73 (tie line)	29	64	0.254	0.203	234
74 (tie line)	45	60	0.254	0.203	234
75 (tie line)	43	38	0.454	0.363	234
76 (tie line)	39	59	0.454	0.363	234
77 (tie line)	21	27	0.454	0.363	234
78 (tie line)	15	9	0.681	0.544	234
79 (tie line)	67	15	0.454	0.363	234

Table 17.21 Load data for System 10

Bus Number	Bus Type	Active Power Demand (kW)	Reactive Power Demand (kVAr)
1	Generation	0	0
2	Load	100	90
3	Load	60	40
4	Load	150	130
5	Load	75	50
6	Load	15	9
7	Load	18	14
8	Load	13	10
9	Load	16	11
10	Load	20	10
11	Load	16	9
12	Load	50	40
13	Load	105	90
14	Load	25	15
15	Load	40	25
16	Load	60	30
17	Load	40	25
18	Load	15	9
19	Load	13	7
20	Load	30	20
21	Load	90	50
22	Load	50	30
23	Load	60	40
24	Load	100	80

(*continued*)

Table 17.21 *(continued)*

Bus Number	Bus Type	Active Power Demand (kW)	Reactive Power Demand (kVAr)
25	Load	80	65
26	Load	100	60
27	Load	100	55
28	Load	120	70
29	Load	105	70
30	Load	80	50
31	Load	60	40
32	Load	13	8
33	Load	16	9
34	Load	50	30
35	Load	40	28
36	Load	60	40
37	Load	40	30
38	Load	30	25
39	Load	150	100
40	Load	60	35
41	Load	120	70
42	Load	90	60
43	Load	18	10
44	Load	16	10
45	Load	100	50
46	Load	60	40
47	Load	90	70
48	Load	85	55
49	Load	100	70
50	Load	140	90
51	Load	60	40
52	Load	20	11
53	Load	40	30
54	Load	36	24
55	Load	30	20
56	Load	43	30
57	Load	80	50
58	Load	240	120
59	Load	125	110
60	Load	25	10

(continued)

Table 17.21 (*continued*)

Bus Number	Bus Type	Active Power Demand (kW)	Reactive Power Demand (kVAr)
61	Load	10	5
62	Load	150	130
63	Load	50	30
64	Load	30	20
65	Load	130	120
66	Load	150	130
67	Load	25	15
68	Load	100	60
69	Load	40	30
70	Generation	0	0

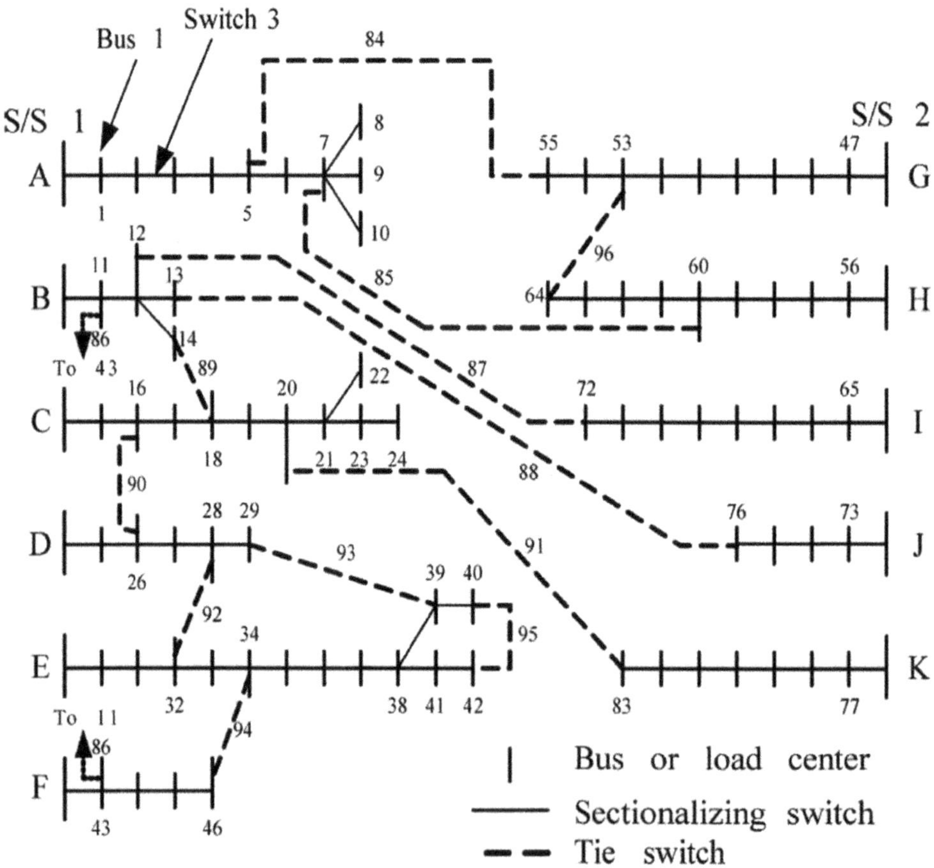

Figure 17.11 Distribution system of TPC.

Table 17.22 Line characteristics of System 11

Branch	From Bus	To Bus	Resistance (Ω)	Reactance (Ω)	Maximum Current (A)
1	A	1	0.1944	0.6624	410
2	1	2	0.2096	0.4304	410
3	2	3	0.2358	0.4842	410
4	3	4	0.0917	0.1883	410
5	4	5	0.2096	0.4304	410
6	5	6	0.0393	0.0807	410
7	6	7	0.0405	0.1380	410
8	7	8	0.1048	0.2152	410
9	7	9	0.2358	0.4842	410
10	7	10	0.1048	0.2152	410
11	B	11	0.0786	0.1614	410
12	11	12	0.3406	0.6944	410
13	12	13	0.0262	0.0538	410
14	12	14	0.0786	0.1614	410
15	C	15	0.1134	0.3864	410
16	15	16	0.0524	0.1076	410
17	16	17	0.0524	0.1076	410
18	17	18	0.1572	0.3228	410
19	18	19	0.0393	0.0807	410
20	19	20	0.1703	0.3497	410
21	20	21	0.2358	0.4842	410
22	21	22	0.1572	0.3228	410
23	21	23	0.1965	0.4035	410
24	23	24	0.1310	0.2690	410
25	D	25	0.0567	0.1932	410
26	25	26	0.1048	0.2152	410
27	26	27	0.2489	0.5111	410
28	27	28	0.0486	0.1656	410
29	28	29	0.1310	0.2690	410
30	E	30	0.1965	0.3960	410
31	30	31	0.1310	0.2690	410
32	31	32	0.1310	0.2690	410
33	32	33	0.0262	0.0538	410
34	33	34	0.1703	0.3497	410
35	34	35	0.0524	0.1076	410
36	35	36	0.4978	1.0222	410

(continued)

Table 17.22 (*continued*)

Branch	From Bus	To Bus	Resistance (Ω)	Reactance (Ω)	Maximum Current (A)
37	36	37	0.0393	0.0807	410
38	37	38	0.0393	0.0807	410
39	38	39	0.0786	0.1614	410
40	39	40	0.2096	0.4304	410
41	38	41	0.1965	0.4035	410
42	41	42	0.2096	0.4304	410
43	F	43	0.0486	0.1656	410
44	43	44	0.0393	0.0807	410
45	44	45	0.1310	0.2690	410
46	45	46	0.2358	0.4842	410
47	G	47	0.2430	0.8280	410
48	47	48	0.0655	0.1345	410
49	48	49	0.0655	0.1345	410
50	49	50	0.0393	0.0807	410
51	50	51	0.0786	0.1614	410
52	51	52	0.0393	0.0807	410
53	52	53	0.0786	0.1614	410
54	53	54	0.0524	0.1076	410
55	54	55	0.1310	0.2690	410
56	H	56	0.2268	0.7728	410
57	56	57	0.5371	1.1029	410
58	57	58	0.0524	0.1076	410
59	58	59	0.0405	0.1380	410
60	59	60	0.0393	0.0807	410
61	60	61	0.0262	0.0538	410
62	61	62	0.1048	0.2152	410
63	62	63	0.2358	0.4842	410
64	63	64	0.0243	0.0828	410
65	I	65	0.0486	0.1656	410
66	65	66	0.1703	0.3497	410
67	66	67	0.1215	0.4140	410
68	67	68	0.2187	0.7452	410
69	68	69	0.0486	0.1656	410
70	69	70	0.0729	0.2484	410
71	70	71	0.0567	0.1932	410
72	71	72	0.0262	0.0528	410

(continued)

Table 17.22 (*continued*)

Branch	From Bus	To Bus	Resistance (Ω)	Reactance (Ω)	Maximum Current (A)
73	J	73	0.3240	1.1040	410
74	73	74	0.0324	0.1104	410
75	74	75	0.0567	0.1932	410
76	75	76	0.0486	0.1656	410
77	K	77	0.2511	0.8556	410
78	77	78	0.1296	0.4416	410
79	78	79	0.0486	0.1656	410
80	79	80	0.1310	0.2640	410
81	80	81	0.1310	0.2640	410
82	81	82	0.0917	0.1883	410
83	82	83	0.3144	0.6456	410
84 (tie line)	5	55	0.1310	0.2690	410
85 (tie line)	7	60	0.1310	0.2690	410
86 (tie line)	11	43	0.1310	0.2690	410
87 (tie line)	12	72	0.3406	0.6994	410
88 (tie line)	13	76	0.4585	0.9415	410
89 (tie line)	14	18	0.5371	1.0824	410
90 (tie line)	16	26	0.0917	0.1883	410
91 (tie line)	20	83	0.0786	0.1614	410
92 (tie line)	28	32	0.0524	0.1076	410
93 (tie line)	29	39	0.0786	0.1614	410
94 (tie line)	34	46	0.0262	0.0538	410
95 (tie line)	40	42	0.1965	0.4035	410
96 (tie line)	53	64	0.0393	0.0807	410

Table 17.23 Load data for System 11

Bus Number	Bus Type	Active Power Demand (kW)	Reactive Power Demand (kVAr)
A, B, C, D, E, F	Generation	0	0
G, H, I, J, K	Generation	0	0
1	Load	0	0
2	Load	100	50
3	Load	300	200

(*continued*)

Table 17.23 (*continued*)

Bus Number	Bus Type	Active Power Demand (kW)	Reactive Power Demand (kVAr)
4	Load	350	250
5	Load	220	100
6	Load	1100	800
7	Load	400	320
8	Load	300	200
9	Load	300	230
10	Load	300	260
11	Load	0	0
12	Load	1200	800
13	Load	800	600
14	Load	700	500
15	Load	0	0
16	Load	300	150
17	Load	500	350
18	Load	700	400
19	Load	1200	1000
20	Load	300	300
21	Load	400	350
22	Load	50	20
23	Load	50	20
24	Load	50	10
25	Load	50	30
26	Load	100	60
27	Load	100	70
28	Load	1800	1300
29	Load	200	120
30	Load	0	0
31	Load	1800	1600
32	Load	200	150
33	Load	200	100
34	Load	800	600
35	Load	100	60
36	Load	100	60
37	Load	20	10

(*continued*)

Table 17.23 (*continued*)

Bus Number	Bus Type	Active Power Demand (kW)	Reactive Power Demand (kVAr)
38	Load	20	10
39	Load	20	10
40	Load	20	10
41	Load	200	160
42	Load	50	30
43	Load	0	0
44	Load	30	20
45	Load	800	700
46	Load	200	150
47	Load	0	0
48	Load	0	0
49	Load	0	0
50	Load	200	160
51	Load	800	600
52	Load	500	300
53	Load	500	350
54	Load	500	300
55	Load	200	80
56	Load	0	0
57	Load	30	20
58	Load	600	420
59	Load	0	0
60	Load	20	10
61	Load	20	10
62	Load	200	130
63	Load	300	240
64	Load	300	200
65	Load	0	0
66	Load	50	30
67	Load	0	0
68	Load	400	360
69	Load	0	0
70	Load	0	0
71	Load	2000	1500

(*continued*)

Table 17.23 (*continued*)

Bus Number	Bus Type	Active Power Demand (kW)	Reactive Power Demand (kVAr)
72	Load	200	150
73	Load	0	0
74	Load	0	0
75	Load	1200	950
76	Load	300	180
77	Load	0	0
78	Load	400	360
79	Load	2000	1300
80	Load	200	140
81	Load	500	360
82	Load	100	30
83	Load	400	360

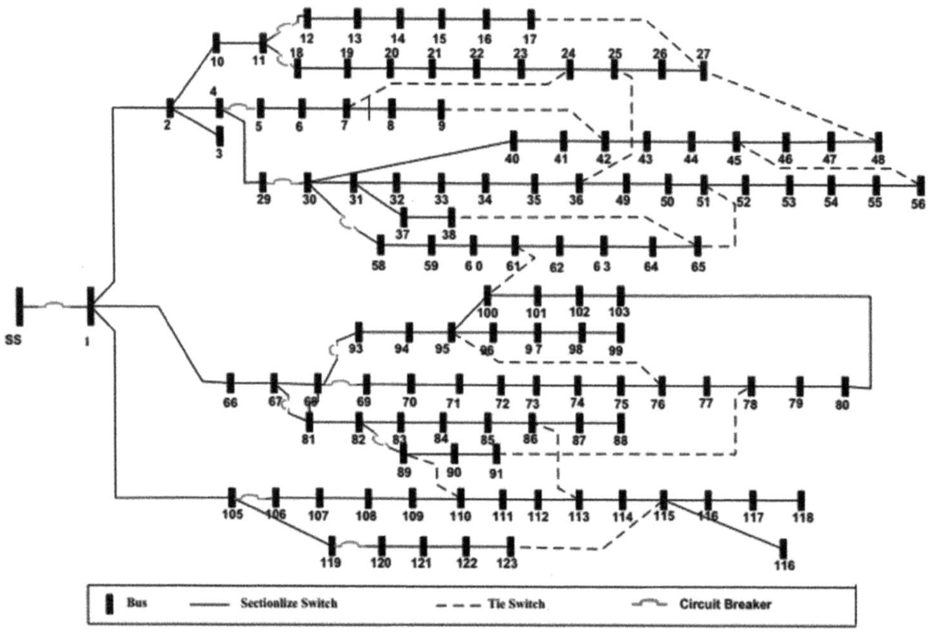

Figure 17.12 The 119-bus test system.

Table 17.24 Line characteristics of System 12

Branch	From Bus	To Bus	Resistance (Ω)	Reactance (Ω)
1	0	1	0	0
2	1	2	0.036	0.01296
3	2	3	0.033	0.01188
4	2	4	0.045	0.0162
5	4	5	0.015	0.054
6	5	6	0.015	0.054
7	6	7	0.015	0.0125
8	7	8	0.018	0.014
9	8	9	0.021	0.063
10	2	10	0.166	0.1344
11	10	11	0.112	0.0789
12	11	12	0.187	0.313
13	12	13	0.142	0.1512
14	13	14	0.18	0.118
15	14	15	0.15	0.045
16	15	16	0.16	0.18
17	16	17	0.157	0.171
18	11	18	0.218	0.285
19	18	19	0.118	0.185
20	19	20	0.16	0.196
21	20	21	0.12	0.189
22	21	22	0.12	0.0789
23	22	23	1.41	0.723
24	23	24	0.293	0.1348
25	24	25	0.133	0.104
26	25	26	0.178	0.134
27	26	27	0.178	0.134
28	4	29	0.015	0.0296
29	29	30	0.012	0.0276
30	30	31	0.12	0.2766
31	31	32	0.21	0.243
32	32	33	0.12	0.054
33	33	34	0.178	0.234
34	34	35	0.178	0.234

(continued)

Table 17.24 (*continued*)

Branch	From Bus	To Bus	Resistance (Ω)	Reactance (Ω)
35	35	36	0.154	0.162
36	31	37	0.187	0.261
37	37	38	0.133	0.099
38	30	40	0.33	0.194
39	40	41	0.31	0.194
40	41	42	0.13	0.194
41	42	43	0.28	0.15
42	43	44	1.18	0.85
43	44	45	0.42	0.2436
44	45	46	0.27	0.0972
45	46	47	0.339	0.1221
46	47	48	0.27	0.1779
47	36	49	0.21	0.1383
48	49	50	0.12	0.0789
49	50	51	0.15	0.0987
50	51	52	0.15	0.0987
51	52	53	0.24	0.1581
52	53	54	0.12	0.0789
53	54	55	0.405	0.1458
54	55	56	0.405	0.1458
55	30	58	0.391	0.141
56	58	59	0.406	0.1461
57	59	60	0.406	0.1461
58	60	61	0.706	0.5461
59	61	62	0.338	0.1218
60	62	63	0.338	0.1218
61	63	64	0.207	0.0747
62	64	65	0.247	0.8922
63	1	66	0.028	0.0418
64	66	67	0.117	0.2016
65	67	68	0.255	0.0918
66	68	69	0.21	0.0759
67	69	70	0.383	0.138
68	70	71	0.504	0.3303

(*continued*)

Table 17.24 (*continued*)

Branch	From Bus	To Bus	Resistance (Ω)	Reactance (Ω)
69	71	72	0.406	0.1461
70	72	73	0.962	0.761
71	73	74	0.165	0.06
72	74	75	0.303	0.1092
73	75	76	0.303	0.1092
74	76	77	0.206	0.144
75	77	78	0.233	0.084
76	78	79	0.591	0.1773
77	79	80	0.126	0.0453
78	67	81	0.559	0.3687
79	81	82	0.186	0.1227
80	82	83	0.186	0.1227
81	83	84	0.26	0.139
82	84	85	0.154	0.148
83	85	86	0.23	0.128
84	86	87	0.252	0.106
85	87	88	0.18	0.148
86	82	89	0.16	0.182
87	89	90	0.2	0.23
88	90	91	0.16	0.393
89	68	93	0.669	0.2412
90	93	94	0.266	0.1227
91	94	95	0.266	0.1227
92	95	96	0.266	0.1227
93	96	97	0.266	0.1227
94	97	98	0.233	0.115
95	98	99	0.496	0.138
96	95	100	0.196	0.18
97	100	101	0.196	0.18
98	101	102	0.1866	0.122
99	102	103	0.0746	0.318
100	1	105	0.0625	0.0265
101	105	106	0.1501	0.234
102	106	107	0.1347	0.0888

(*continued*)

Table 17.24 (*continued*)

Branch	From Bus	To Bus	Resistance (Ω)	Reactance (Ω)
103	107	108	0.2307	0.1203
104	108	109	0.447	0.1608
105	109	110	0.1632	0.0588
106	110	1111	0.33	0.099
107	111	112	0.156	0.0561
108	112	113	0.3819	0.1374
109	113	114	0.1626	0.0585
110	114	115	0.3819	0.1374
111	115	116	0.2445	0.0879
112	115	117	0.2088	0.0753
113	117	118	0.2301	0.0828
114	105	119	0.6102	0.2196
115	119	120	0.1866	0.127
116	120	121	0.3732	0.246
117	121	122	0.405	0.367
118	122	123	0.489	0.438
119 (tie line)	48	27	0.5258	0.2925
120 (tie line)	17	27	0.5258	0.2916
121 (tie line)	8	24	0.4272	0.1539
122 (tie line)	56	45	0.48	0.1728
123 (tie line)	65	51	0.36	0.1296
124 (tie line)	38	65	0.57	0.572
125 (tie line)	9	42	0.53	0.3348
126 (tie line)	61	100	0.3957	0.1425
127 (tie line)	76	95	0.68	0.648
128 (tie line)	91	78	0.4062	0.1464
129 (tie line)	103	80	0.4626	0.1674
130 (tie line)	113	86	0.651	0.234
131 (tie line)	110	89	0.8125	0.2925
132 (tie line)	115	123	0.7089	0.2553
133 (tie line)	25	36	0.5	0.5

Table 17.25 Load data for System 12

Bus Number	Bus Type	Active Power Demand (kW)	Reactive Power Demand (kVAr)
0	Generation	0	0
1	Load	0	0
2	Load	133.84	101.14
3	Load	16.214	11.292
4	Load	34.315	21.845
5	Load	73.016	63.602
6	Load	144.2	68.604
7	Load	104.47	61.725
8	Load	28.547	11.503
9	Load	87.56	51.073
10	Load	198.2	106.77
11	Load	146.8	75.995
12	Load	26.04	18.687
13	Load	52.1	23.22
14	Load	141.9	117.5
15	Load	21.87	28.79
16	Load	33.37	26.45
17	Load	32.43	25.23
18	Load	20.234	11.906
19	Load	156.94	78.523
20	Load	546.29	351.4
21	Load	180.31	164.2
22	Load	93.167	54.594
23	Load	85.18	39.65
24	Load	168.1	95.178
25	Load	125.11	150.22
26	Load	16.03	24.62
27	Load	26.03	24.62
29	Load	594.56	522.62
30	Load	120.62	59.117
31	Load	102.38	99.554
32	Load	513.4	318.5
33	Load	475.25	456.14
34	Load	151.43	136.79

(continued)

Table 17.25 (*continued*)

Bus Number	Bus Type	Active Power Demand (kW)	Reactive Power Demand (kVAr)
35	Load	205.38	83.302
36	Load	131.6	93.082
37	Load	448.4	369.79
38	Load	440.52	321.64
40	Load	112.54	55.134
41	Load	53.963	38.998
42	Load	393.05	342.6
43	Load	326.74	278.56
44	Load	536.26	240.24
45	Load	76.247	66.562
46	Load	53.52	39.76
47	Load	40.328	31.964
48	Load	39.653	20.758
49	Load	66.195	42.361
50	Load	73.904	51.653
51	Load	114.77	57.965
52	Load	918.37	1205.1
53	Load	210.3	146.66
54	Load	66.68	56.608
55	Load	42.207	40.184
56	Load	433.74	283.41
58	Load	62.1	26.86
59	Load	92.46	88.38
60	Load	85.188	55.436
61	Load	345.3	332.4
62	Load	22.5	16.83
63	Load	80.551	49.156
64	Load	95.86	90.758
65	Load	62.92	47.7
66	Load	478.8	463.74
67	Load	120.94	52.006
68	Load	139.11	100.34
69	Load	391.78	193.5
70	Load	27.741	26.713
71	Load	52.814	25.257

(*continued*)

Table 17.25 (continued)

Bus Number	Bus Type	Active Power Demand (kW)	Reactive Power Demand (kVAr)
72	Load	66.89	38.713
73	Load	467.5	395.14
74	Load	594.85	239.74
75	Load	132.5	84.363
76	Load	52.699	22.482
77	Load	869.79	614.775
78	Load	31.349	29.817
79	Load	192.39	122.43
80	Load	65.75	45.37
81	Load	238.15	223.22
82	Load	294.55	162.47
83	Load	485.57	437.92
84	Load	243.53	183.03
85	Load	243.53	183.03
86	Load	134.25	119.29
87	Load	22.71	27.96
88	Load	49.513	26.515
89	Load	383.78	257.16
90	Load	49.64	20.6
91	Load	22.473	11.806
93	Load	62.93	42.96
94	Load	30.67	34.93
95	Load	62.53	66.79
96	Load	114.57	81.748
97	Load	81.292	66.526
98	Load	31.733	15.96
99	Load	33.32	60.48
100	Load	531.28	224.85
101	Load	507.03	367.42
102	Load	26.39	11.7
103	Load	45.99	30.392
105	Load	100.66	47.572
106	Load	456.48	350.3
107	Load	522.56	449.29
108	Load	408.43	168.46

(continued)

Table 17.25 (*continued*)

Bus Number	Bus Type	Active Power Demand (kW)	Reactive Power Demand (kVAr)
109	Load	141.48	134.25
110	Load	104.43	66.024
111	Load	96.793	83.647
112	Load	493.92	419.34
113	Load	225.38	135.88
114	Load	509.21	387.21
115	Load	188.5	173.46
116	Load	918.03	898.55
117	Load	305.08	215.37
118	Load	54.38	40.97
119	Load	211.14	192.9
120	Load	67.009	53.336
121	Load	162.07	90.321
122	Load	48.785	29.156
123	Load	33.9	18.98

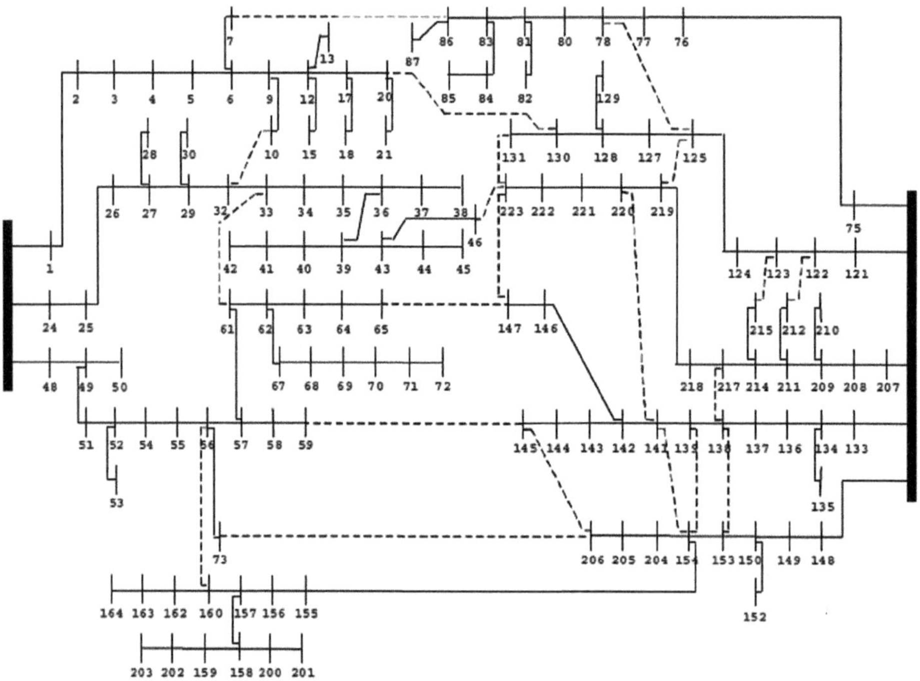

Figure 17.13 Diagram of the 136-bus distribution system.

Table 17.26 Line characteristics of System 13

Branch	From Bus	To Bus	Resistance (Ω)	Reactance (Ω)	Maximum Current (A)
1	0	1	0.33205	0.76653	200
2	1	2	0.00188	0.00433	200
3	2	3	0.22324	0.51535	200
4	3	4	0.09943	0.22953	200
5	4	5	0.15571	0.35945	200
6	5	6	0.16321	0.37677	200
7	6	7	0.11444	0.26417	200
8	6	9	0.05675	0.05666	200
9	9	10	0.52124	0.27418	200
10	9	12	0.10877	0.10860	200
11	12	13	0.39803	0.20937	200
12	12	15	0.91744	0.31469	200
13	12	17	0.11823	0.11805	200
14	17	18	0.50228	0.26421	200
15	17	20	0.05675	0.05666	200
16	20	21	0.29379	0.15454	200
17	0	24	0.33205	0.76653	200
18	24	25	0.00188	0.00433	200
19	25	26	0.22324	0.51535	200
20	26	27	0.10881	0.25118	200
21	27	28	0.71078	0.37388	200
22	27	29	0.18197	0.42008	200
23	29	30	0.30326	0.15952	200
24	29	32	0.02439	0.05630	200
25	32	33	0.04502	0.10394	200
26	33	34	0.01876	0.04331	200
27	34	35	0.11823	0.11805	200
28	35	36	0.02365	0.02361	200
29	36	37	0.18954	0.09970	200
30	37	38	0.39803	0.20937	200
31	36	39	0.05675	0.05666	200
32	39	40	0.09477	0.04985	200
33	40	41	0.41699	0.21934	200
34	41	42	0.11372	0.05982	200
35	39	43	0.07566	0.07555	200

(continued)

Table 17.26 (*continued*)

Branch	From Bus	To Bus	Resistance (Ω)	Reactance (Ω)	Maximum Current (A)
36	43	44	0.36960	0.19442	200
37	44	45	0.26536	0.13958	200
38	43	46	0.05675	0.05666	200
39	0	48	0.33205	0.76653	200
40	48	49	0.11819	0.27283	200
41	49	50	2.96288	1.01628	200
42	49	51	0.00188	0.00433	200
43	51	52	0.06941	0.16024	200
44	52	53	0.81502	0.42872	200
45	52	54	0.06378	0.14724	200
46	54	55	0.13132	0.30315	200
47	55	56	0.06191	0.14291	200
48	56	57	0.11444	0.26417	200
49	57	58	0.28374	0.28331	200
50	58	59	0.28374	0.28331	200
51	57	61	0.04502	0.10394	200
52	61	62	0.02626	0.06063	200
53	62	63	0.06003	0.13858	200
54	63	64	0.03002	0.06929	200
55	64	65	0.02064	0.04764	200
56	62	67	0.10881	0.25118	200
57	67	68	0.25588	0.13460	200
58	68	69	0.41699	0.21934	200
59	69	70	0.50228	0.26421	200
60	70	71	0.33170	0.17448	200
61	71	72	0.20849	0.10967	200
62	56	73	0.13882	0.32047	200
63	0	75	0.00750	0.01732	200
64	75	76	0.27014	0.62362	200
65	76	77	0.38270	0.88346	200
66	77	78	0.33018	0.76220	200
67	78	80	0.32830	0.75787	200
68	80	81	0.17072	0.39409	200

(*continued*)

Table 17.26 (*continued*)

Branch	From Bus	To Bus	Resistance (Ω)	Reactance (Ω)	Maximum Current (A)
69	81	82	0.55914	0.29412	200
70	81	83	0.05816	0.13425	200
71	83	84	0.70130	0.36890	200
72	84	85	1.02352	0.53839	200
73	83	86	0.06754	0.15591	200
74	86	87	1.32352	0.45397	200
75	0	121	0.01126	0.02598	200
76	121	122	0.72976	1.68464	200
77	122	123	0.22512	0.51968	200
78	123	124	0.20824	0.48071	200
79	124	125	0.04690	0.10827	200
80	125	127	0.61950	0.61857	200
81	127	128	0.34049	0.33998	200
82	128	129	0.56862	0.29911	200
83	128	130	0.10877	0.10860	200
84	130	131	0.56862	0.29911	200
85	0	133	0.01126	0.02598	200
86	133	134	0.41835	0.96575	200
87	134	135	0.10499	0.13641	200
88	134	136	0.43898	1.01338	200
89	136	137	0.07520	0.02579	200
90	137	138	0.07692	0.17756	200
91	138	139	0.33205	0.76653	200
92	139	141	0.08442	0.19488	200
93	141	142	0.13320	0.30748	200
94	142	143	0.29320	0.29276	200
95	143	144	0.21753	0.21721	200
96	144	145	0.26482	0.26443	200
97	142	146	0.10318	0.23819	200
98	146	147	0.13507	0.31181	200
99	0	148	0.00938	0.02165	200
100	148	149	0.16884	0.38976	200
101	149	150	0.11819	0.27283	200

(*continued*)

Table 17.26 (*continued*)

Branch	From Bus	To Bus	Resistance (Ω)	Reactance (Ω)	Maximum Current (A)
102	150	152	2.28608	0.78414	200
103	150	153	0.45587	1.05236	200
104	153	154	0.69600	1.60669	200
105	154	155	0.45774	1.05669	200
106	155	156	0.20298	0.26373	200
107	156	157	0.21348	0.27737	200
108	157	158	0.54967	0.28914	200
109	158	159	0.54019	0.28415	200
110	157	160	0.04550	0.05911	200
111	160	162	0.47385	0.24926	200
112	162	163	0.86241	0.45364	200
113	163	164	0.56862	0.29911	200
114	158	200	0.77711	0.40878	200
115	200	201	1.08038	0.56830	200
116	159	202	1.09933	0.57827	200
117	202	203	0.47385	0.24926	200
118	154	204	0.32267	0.74488	200
119	204	205	0.14633	0.33779	200
120	205	206	0.12382	0.28583	200
121	0	207	0.01126	0.02598	200
122	207	208	0.64910	1.49842	200
123	208	209	0.04502	0.10394	200
124	209	210	0.52640	0.18056	200
125	209	211	0.02064	0.04764	200
126	211	212	0.53071	0.27917	200
127	211	214	0.09755	0.22520	200
128	214	215	0.11819	0.27283	200
129	214	217	0.13882	0.32047	200
130	217	218	0.04315	0.09961	200
131	218	219	0.09192	0.21220	200
132	219	220	0.16134	0.37244	200
133	220	221	0.37832	0.37775	200
134	221	222	0.39724	0.39664	200
135	222	223	0.29320	0.29276	200

(*continued*)

Table 17.26 (*continued*)

Branch	From Bus	To Bus	Resistance (Ω)	Reactance (Ω)	Maximum Current (A)
136 (tie line)	7	86	0.13132	0.30315	200
137 (tie line)	10	32	0.26536	0.13958	200
138 (tie line)	20	130	0.14187	0.14166	200
139 (tie line)	46	223	0.08512	0.08499	200
140 (tie line)	33	61	0.04502	0.10394	200
141 (tie line)	59	145	0.14187	0.14166	200
142 (tie line)	65	147	0.14187	0.14166	200
143 (tie line)	73	206	0.03940	0.09094	200
144 (tie line)	78	125	0.12944	0.29882	200
145 (tie line)	125	219	0.01688	0.03898	200
146 (tie line)	131	223	0.33170	0.17448	200
147 (tie line)	139	154	0.14187	0.14166	200
148 (tie line)	138	217	0.07692	0.17756	200
149 (tie line)	138	153	0.07692	0.17756	200
150 (tie line)	141	154	0.07692	0.17756	200
151 (tie line)	141	220	0.07692	0.17756	200
152 (tie line)	145	206	0.26482	0.26443	200
153 (tie line)	160	56	0.49696	0.64567	200
154 (tie line)	212	122	0.17059	0.08973	200
155 (tie line)	215	123	0.05253	0.12126	200
156 (tie line)	223	147	0.29320	0.29276	200

Table 17.27 Load data for System 13

Bus Number	Bus Type	Active Power Demand (kW)	Reactive Power Demand (kVAr)
0	Generation	0	0
1	Load	0	0
2	Load	47.78	19.009
3	Load	42.551	16.929
4	Load	87.022	34.622
5	Load	311.31	123.855
6	Load	148.869	59.228

(*continued*)

Table 17.27 (*continued*)

Bus Number	Bus Type	Active Power Demand (kW)	Reactive Power Demand (kVAr)
7	Load	238.672	94.956
9	Load	62.299	24.786
10	Load	124.598	49.571
12	Load	140.175	55.768
13	Load	116.813	46.474
15	Load	249.203	99.145
17	Load	291.447	115.952
18	Load	303.72	120.835
20	Load	215.396	85.695
21	Load	198.586	79.007
24	Load	0	0
25	Load	0	0
26	Load	0	0
27	Load	30.127	14.729
28	Load	230.972	112.92
29	Load	60.256	29.458
30	Load	230.972	112.92
32	Load	120.507	58.915
33	Load	0	0
34	Load	56.981	27.857
35	Load	364.665	178.281
36	Load	0	0
37	Load	124.647	60.939
38	Load	56.981	27.857
39	Load	0	0
40	Load	85.473	41.787
41	Load	0	0
42	Load	396.735	193.96
43	Load	0	0
44	Load	181.152	88.563
45	Load	242.172	118.395
46	Load	75.316	36.821
48	Load	0	0
49	Load	1.254	0.531

(*continued*)

Table 17.27 (*continued*)

Bus Number	Bus Type	Active Power Demand (kW)	Reactive Power Demand (kVAr)
50	Load	6.274	2.66
51	Load	0	0
52	Load	117.88	49.971
53	Load	62.668	26.566
54	Load	172.285	73.034
55	Load	458.556	194.388
56	Load	262.962	111.473
57	Load	235.761	99.942
58	Load	0	0
59	Load	109.215	46.298
61	Load	0	0
62	Load	72.809	30.865
63	Load	258.473	109.57
64	Load	69.169	29.322
65	Load	21.843	9.26
67	Load	0	0
68	Load	20.527	8.702
69	Load	150.548	63.819
70	Load	220.687	93.552
71	Load	92.384	39.163
72	Load	0	0
73	Load	226.693	96.098
75	Load	0	
76	Load	294.016	116.974
77	Load	83.015	33.028
78	Load	83.015	33.028
80	Load	103.77	41.285
81	Load	176.408	70.184
82	Load	83.015	33.028
83	Load	217.917	86.698
84	Load	23.294	9.267
85	Load	5.075	2.019
86	Load	72.638	28.899
87	Load	405.99	161.523
121	Load	0	0

(*continued*)

Table 17.27 (*continued*)

Bus Number	Bus Type	Active Power Demand (kW)	Reactive Power Demand (kVAr)
122	Load	100.182	42.468
123	Load	142.523	60.417
124	Load	96.042	40.713
125	Load	300.454	127.366
127	Load	141.238	59.873
128	Load	279.847	118.631
129	Load	87.312	37.013
130	Load	243.849	103.371
131	Load	247.75	105.025
133	Load	0	0
134	Load	89.878	38.101
135	Load	1137.28	482.108
136	Load	458.339	194.296
137	Load	385.197	163.29
138	Load	0	0
139	Load	79.608	33.747
141	Load	87.312	37.013
142	Load	0	0
143	Load	74.001	31.37
144	Load	232.05	98.369
145	Load	141.819	60.119
146	Load	0	0
147	Load	76.449	32.408
148	Load	0	
149	Load	51.322	21.756
150	Load	59.874	25.381
152	Load	9.065	3.843
153	Load	2.092	0.887
154	Load	16.735	7.094
155	Load	1506.522	638.634
156	Load	313.023	132.694
157	Load	79.831	33.842
158	Load	51.322	21.756

(*continued*)

Table 17.27 (continued)

Bus Number	Bus Type	Active Power Demand (kW)	Reactive Power Demand (kVAr)
159	Load	0	0
160	Load	202.435	85.815
162	Load	60.823	25.784
163	Load	45.618	19.338
164	Load	0	
200	Load	157.07	66.584
201	Load	0	
202	Load	250.148	106.041
203	Load	0	
204	Load	69.809	29.593
205	Load	32.072	13.596
206	Load	61.084	25.894
207	Load	0	
208	Load	94.622	46.26
209	Load	49.858	24.375
210	Load	123.164	60.214
211	Load	78.35	38.304
212	Load	145.475	71.121
214	Load	21.369	10.447
215	Load	74.789	36.564
217	Load	227.926	111.431
218	Load	35.614	17.411
219	Load	249.295	121.877
220	Load	316.722	154.842
221	Load	333.817	163.199
222	Load	249.295	121.877
223	Load	0	0

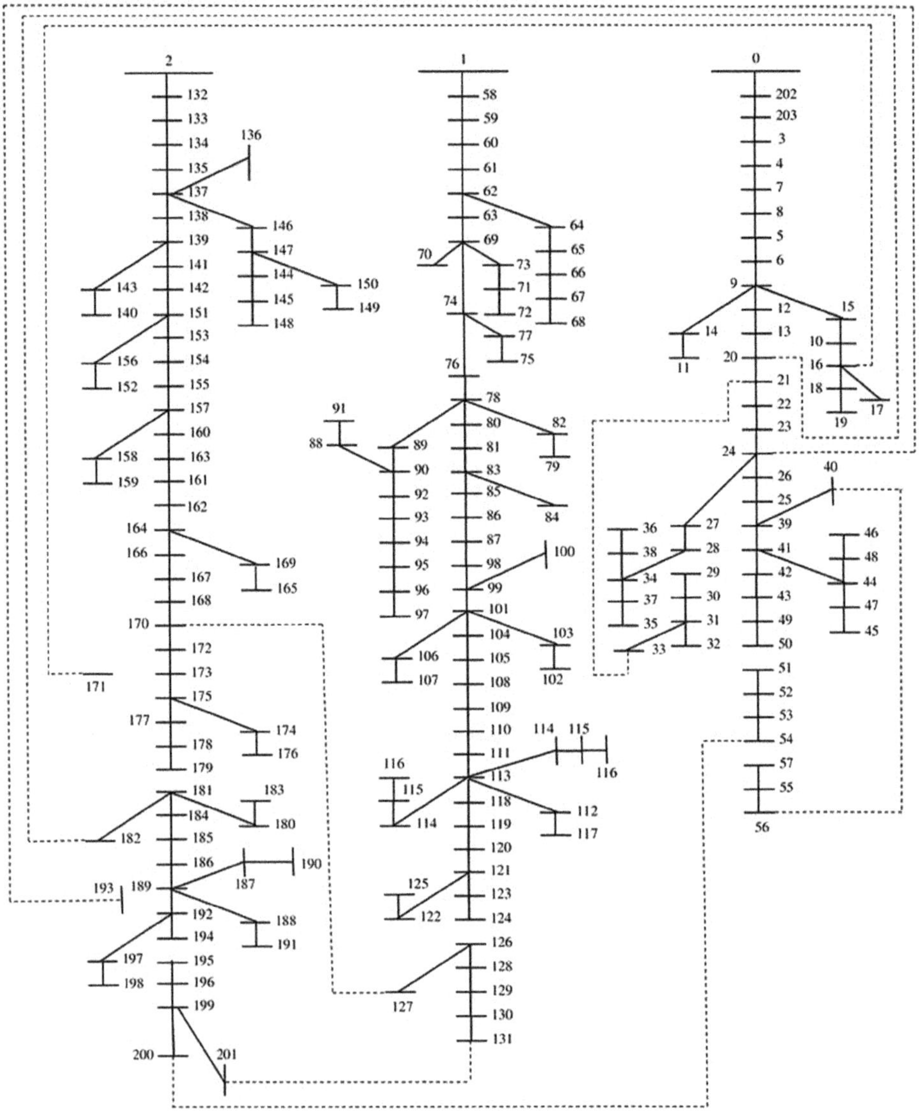

Figure 17.14 Schematic diagram of the 203-bus distribution system.

Table 17.28 Line characteristics of System 14

Branch	From Bus	To Bus	Resistance (Ω)	Reactance (Ω)
1	202	203	0.01883	0.04232
2	203	3	0.065905	0.14812
3	3	4	0.03766	0.08464
4	5	6	0.016947	0.038088
5	6	9	0.01883	0.04232
6	16	17	0.074728	0.03912
7	16	18	0.18682	0.0978
8	18	19	0.074728	0.03912
9	20	21	0.041426	0.093104
10	22	23	0.0065905	0.014812
11	23	24	0.0065905	0.014812
12	28	29	0.046705	0.02445
13	29	30	0.046705	0.02445
14	30	31	0.046705	0.02445
15	31	32	0.059416	0.02026
16	31	33	0.089124	0.03039
17	28	34	0.09341	0.0489
18	39	40	0.037135	0.0126625
19	25	39	0.009415	0.02116
20	39	41	0.009415	0.02116
21	41	44	0.059416	0.02026
22	41	42	0.009415	0.02116
23	42	43	0.009415	0.02116
24	43	49	0.0141225	0.03174
25	49	50	0.013181	0.029624
26	50	51	0.011298	0.025392
27	51	52	0.011298	0.025392
28	52	53	0.011298	0.025392
29	53	54	0.011298	0.025392
30	4	7	0.01883	0.04232
31	8	5	0.01883	0.04232
32	9	12	0.01883	0.04232
33	13	20	0.03766	0.08464
34	26	25	0.024479	0.055016
35	14	11	0.05935	0.04654
36	10	16	0.09341	0.0489

(continued)

Table 17.28 *(continued)*

Branch	From Bus	To Bus	Resistance (Ω)	Reactance (Ω)
37	15	10	0.046705	0.02445
38	27	28	0.065387	0.03423
39	37	35	0.037364	0.01956
40	38	36	0.037364	0.01956
41	47	45	0.037135	0.0126625
42	48	46	0.037135	0.0126625
43	54	57	0.07427	0.025325
44	55	56	0.103978	0.035455
45	58	59	0.01597	0.04095
46	61	62	0.041522	0.10647
47	62	64	0.019164	0.04914
48	69	70	0.111405	0.0379875
49	71	72	0.081697	0.0278575
50	69	74	0.020713	0.046552
51	74	76	0.011298	0.025392
52	76	78	0.007532	0.016928
53	92	93	0.056046	0.027924
54	93	94	0.028023	0.013962
55	94	95	0.028023	0.013962
56	95	96	0.028023	0.013962
57	96	97	0.0513755	0.025597
58	78	80	0.009415	0.02116
59	83	84	0.013181	0.029624
60	83	85	0.0047075	0.01058
61	85	86	0.005649	0.012696
62	86	87	0.009415	0.02116
63	87	98	0.020713	0.046552
64	99	101	0.005649	0.012696
65	99	100	0.0141225	0.03174
66	101	106	0.0065905	0.014812
67	106	107	0.0103565	0.023276
68	101	104	0.0047075	0.01058
69	104	105	0.005649	0.012696
70	105	108	0.011298	0.025392
71	109	110	0.007532	0.016928
72	110	111	0.007532	0.016928

(continued)

Table 17.28 (continued)

Branch	From Bus	To Bus	Resistance (Ω)	Reactance (Ω)
73	111	113	0.007532	0.016928
74	113	114	0.005649	0.012696
75	114	115	0.007532	0.016928
76	115	116	0.009415	0.02116
77	113	118	0.007532	0.016928
78	118	119	0.007532	0.016928
79	119	120	0.0065905	0.014812
80	120	121	0.005649	0.012696
81	123	124	0.015064	0.033856
82	124	126	0.009415	0.02116
83	126	127	0.009415	0.02116
84	126	128	0.015064	0.033856
85	128	129	0.013181	0.029624
86	129	130	0.013181	0.029624
87	59	60	0.099014	0.25389
88	63	69	0.01883	0.04232
89	80	81	0.011298	0.025392
90	122	125	0.065387	0.03003
91	65	66	0.01597	0.04095
92	67	68	0.133686	0.045585
93	73	71	0.133686	0.045585
94	77	75	0.163394	0.055715
95	82	79	0.126259	0.0430525
96	89	90	0.178248	0.06078
97	88	91	0.111405	0.0379875
98	103	102	0.07427	0.025325
99	112	117	0.051989	0.0177275
100	130	131	0.016947	0.038088
101	134	135	0.015064	0.033856
102	135	137	0.060256	0.135424
103	138	139	0.020713	0.046552
104	139	141	0.009415	0.02116
105	141	142	0.005649	0.012696
106	142	151	0.005649	0.012696
107	151	153	0.005649	0.012696
108	153	154	0.005649	0.012696

(continued)

Table 17.28 (*continued*)

Branch	From Bus	To Bus	Resistance (Ω)	Reactance (Ω)
109	154	155	0.005649	0.012696
110	155	157	0.005649	0.012696
111	157	158	0.028023	0.01467
112	158	159	0.046705	0.02445
113	157	160	0.009415	0.02116
114	162	164	0.01883	0.04232
115	164	166	0.005649	0.012696
116	166	167	0.005649	0.012696
117	167	168	0.007532	0.016928
118	170	171	0.07122	0.055848
119	170	172	0.009415	0.02116
120	172	173	0.009415	0.02116
121	173	175	0.009415	0.02116
122	175	177	0.003766	0.008464
123	177	178	0.003766	0.008464
124	178	179	0.005649	0.012696
125	181	182	0.046705	0.02445
126	181	184	0.005649	0.012696
127	184	185	0.005649	0.012696
128	185	186	0.005649	0.012696
129	186	189	0.005649	0.012696
130	189	192	0.022596	0.050784
131	192	193	0.065387	0.03423
132	192	194	0.046705	0.02445
133	194	195	0.046705	0.02445
134	195	196	0.046705	0.02445
135	196	199	0.046705	0.02445
136	199	200	0.112092	0.05868
137	199	201	0.09341	0.0489
138	192	197	0.009415	0.02116
139	197	198	0.009415	0.02116
140	137	136	0.009415	0.02116
141	132	133	0.069671	0.156584
142	160	163	0.011298	0.025392
143	146	147	0.074728	0.03912
144	147	144	0.056046	0.02934

(*continued*)

Table 17.28 (continued)

Branch	From Bus	To Bus	Resistance (Ω)	Reactance (Ω)
145	150	149	0.046705	0.02445
146	143	140	0.028023	0.01467
147	156	152	0.037364	0.01956
148	169	165	0.065387	0.03423
149	174	176	0.074728	0.03912
150	180	183	0.0607165	0.031785
151	188	191	0.0504475	0.039559
152	187	190	0.0385775	0.030251
153	145	148	0.037364	0.01956
154	0	202	0.0000001	0.00000001
155	2	132	0.0000001	0.0000001
156	133	134	0.0000001	0.0000001
157	137	146	0.0000001	0.0000001
158	144	145	0.0000001	0.0000001
159	147	150	0.0000001	0.0000001
160	137	138	0.0000001	0.0000001
161	139	143	0.0000001	0.0000001
162	151	156	0.0000001	0.0000001
163	163	161	0.0000001	0.0000001
164	161	162	0.0000001	0.0000001
165	164	169	0.0000001	0.0000001
166	168	170	0.0000001	0.0000001
167	175	174	0.0000001	0.0000001
168	179	181	0.0000001	0.0000001
169	181	180	0.0000001	0.0000001
170	189	187	0.0000001	0.0000001
171	189	188	0.0000001	0.0000001
172	7	8	0.0000001	0.00000001
173	9	15	0.0000001	0.00000001
174	9	14	0.0000001	0.00000001
175	12	13	0.0000001	0.00000001
176	21	22	0.0000001	0.00000001
177	24	27	0.0000001	0.00000001
178	34	37	0.0000001	0.00000001
179	34	38	0.0000001	0.00000001
180	24	26	0.0000001	0.00000001

(continued)

Table 17.28 (*continued*)

Branch	From Bus	To Bus	Resistance (Ω)	Reactance (Ω)
181	44	47	0.0000001	0.00000001
182	44	48	0.0000001	0.00000001
183	57	55	0.0000001	0.00000001
184	1	58	0.0000001	0.0000001
185	60	61	0.0000001	0.0000001
186	64	65	0.0000001	0.0000001
187	66	67	0.0000001	0.0000001
188	62	63	0.0000001	0.0000001
189	69	73	0.0000001	0.0000001
190	74	77	0.0000001	0.0000001
191	78	82	0.0000001	0.0000001
192	78	89	0.0000001	0.0000001
193	90	88	0.0000001	0.0000001
194	90	92	0.0000001	0.0000001
195	98	99	0.0000001	0.0000001
196	101	103	0.0000001	0.0000001
197	108	109	0.0000001	0.0000001
198	113	112	0.0000001	0.0000001
199	121	122	0.0000001	0.0000001
200	121	123	0.0000001	0.0000001
201	81	83	0.0000001	0.0000001
202 (tie line)	3	135	0.0000001	0.0000001
203 (tie line)	9	159	0.0000001	0.0000001
204 (tie line)	16	171	0.0000001	0.0000001
205 (tie line)	20	182	0.0000001	0.0000001
206 (tie line)	24	193	0.0000001	0.0000001
207 (tie line)	54	200	0.0000001	0.0000001
208 (tie line)	202	58	0.0000001	0.0000001
209 (tie line)	33	21	0.0000001	0.0000001
210 (tie line)	56	40	0.0000001	0.0000001
211 (tie line)	61	136	0.0000001	0.0000001
212 (tie line)	107	161	0.0000001	0.0000001
213 (tie line)	116	161	0.0000001	0.0000001
214 (tie line)	164	125	0.0000001	0.0000001
215 (tie line)	127	170	0.0000001	0.0000001
216 (tie line)	131	201	0.0000001	0.0000001

Table 17.29 Load data for System 14

Bus Number	Bus Type	Active Power Demand (kW)	Reactive Power Demand (kVAr)
0	Generation	0	0
1	Generation	0	0
2	Generation	0	0
3	Load	0	0
4	Load	63.75	39.5
5	Load	38.25	23.7
6	Load	95.2	58.99
7	Load	0	0
8	Load	0	0
9	Load	63.75	39.5
10	Load	63.75	39.5
11	Load	510	316.02
12	Load	0	0
13	Load	0	0
14	Load	0	0
15	Load	0	0
16	Load	127.5	79.01
17	Load	63.75	39.5
18	Load	63.75	39.5
19	Load	255	158.01
20	Load	0	0
21	Load	0	0
22	Load	255	158.01
23	Load	255	158.01
24	Load	382.5	237.02
25	Load	191.25	118.51
26	Load	0	0
27	Load	0	0
28	Load	51	31.6
29	Load	95.63	59.25
30	Load	95.63	59.25
31	Load	0	0
32	Load	350.63	217.26
33	Load	350.63	217.26
34	Load	0	0
35	Load	95.63	59.25
36	Load	63.75	39.5

(continued)

Table 17.29 (*continued*)

Bus Number	Bus Type	Active Power Demand (kW)	Reactive Power Demand (kVAr)
37	Load	0	0
38	Load	0	0
39	Load	0	0
40	Load	255	158.01
41	Load	0	0
42	Load	191.25	118.51
43	Load	95.63	59.25
44	Load	0	0
45	Load	605.63	375.27
46	Load	63.75	39.5
47	Load	0	0
48	Load	0	0
49	Load	573.75	355.52
50	Load	191.25	118.51
51	Load	255	158.01
52	Load	765	474.03
53	Load	255	158.01
54	Load	0	0
55	Load	318.75	197.51
56	Load	318.75	197.51
57	Load	0	0
58	Load	0	0
59	Load	25.5	15.8
60	Load	0	0
61	Load	0	0
62	Load	0	0
63	Load	0	0
64	Load	95.63	59.25
65	Load	0	0
66	Load	38.25	23.7
67	Load	0	0
68	Load	102	63.2
69	Load	0	0
70	Load	89.25	55.3
71	Load	0	0
72	Load	38.25	23.7
73	Load	0	0

(*continued*)

Table 17.29 (continued)

Bus Number	Bus Type	Active Power Demand (kW)	Reactive Power Demand (kVAr)
74	Load	0	0
75	Load	510	316.02
76	Load	595	368.69
77	Load	0	0
78	Load	0	0
79	Load	350.63	217.26
80	Load	382.5	237.02
81	Load	0	0
82	Load	0	0
83	Load	0	0
84	Load	38.25	23.7
85	Load	255	158.01
86	Load	446.25	276.52
87	Load	382.5	237.02
88	Load	0	0
89	Load	0	0
90	Load	0	0
91	Load	95.63	59.25
92	Load	191.25	118.51
93	Load	216.75	134.31
94	Load	95.63	59.25
95	Load	191.25	118.51
96	Load	63.75	39.5
97	Load	382.5	237.02
98	Load	0	0
99	Load	0	0
100	Load	63.75	39.5
101	Load	0	0
102	Load	63.75	39.5
103	Load	0	0
104	Load	127.5	79.01
105	Load	95.63	59.25
106	Load	106.25	65.84
107	Load	127.5	79.01
108	Load	95.63	59.25
109	Load	0	0
110	Load	127.5	79.01

(continued)

Table 17.29 (*continued*)

Bus Number	Bus Type	Active Power Demand (kW)	Reactive Power Demand (kVAr)
111	Load	127.5	79.01
112	Load	0	0
113	Load	0	0
114	Load	318.75	197.51
115	Load	350.63	217.26
116	Load	446.25	276.52
117	Load	63.75	39.5
118	Load	127.5	79.01
119	Load	63.75	39.5
120	Load	382.5	237.02
121	Load	0	0
122	Load	76.5	47.4
123	Load	255	158.01
124	Load	191.25	118.51
125	Load	255	158.01
126	Load	233.75	144.84
127	Load	573.75	355.52
128	Load	276.25	171.18
129	Load	255	158.01
130	Load	393.13	243.6
131	Load	510	316.02
132	Load	0	0
133	Load	0	0
134	Load	0	0
135	Load	0	0
136	Load	0	0
137	Load	0	0
138	Load	63.75	39.5
139	Load	0	0
140	Load	63.75	39.5
141	Load	38.25	23.7
142	Load	38.25	23.7
143	Load	0	0
144	Load	0	0
145	Load	0	0
146	Load	0	0
147	Load	63.75	39.5

(*continued*)

Table 17.29 (continued)

Bus Number	Bus Type	Active Power Demand (kW)	Reactive Power Demand (kVAr)
148	Load	95.63	59.25
149	Load	63.75	39.5
150	Load	0	0
151	Load	0	0
152	Load	63.75	39.5
153	Load	63.75	39.5
154	Load	212.5	131.68
155	Load	63.75	39.5
156	Load	0	0
157	Load	0	0
158	Load	63.75	39.5
159	Load	38.25	23.7
160	Load	63.75	39.5
161	Load	0	0
162	Load	95.63	59.25
163	Load	0	0
164	Load	0	0
165	Load	605.63	375.27
166	Load	446.25	276.52
167	Load	191.25	118.51
168	Load	255	158.01
169	Load	0	0
170	Load	63.75	39.5
171	Load	127.5	79.01
172	Load	63.75	39.5
173	Load	191.68	118.77
174	Load	0	0
175	Load	0	0
176	Load	208.25	129.04
177	Load	191.25	118.51
178	Load	127.5	79.01
179	Load	127.5	79.01
180	Load	0	0
181	Load	0	0
182	Load	255	158.01
183	Load	127.5	79.01
184	Load	850	526.7

(continued)

Table 17.29 (*continued*)

Bus Number	Bus Type	Active Power Demand (kW)	Reactive Power Demand (kVAr)
185	Load	127.5	79.01
186	Load	510	316.02
187	Load	0	0
188	Load	0	0
189	Load	0	0
190	Load	127.5	79.01
191	Load	448.38	277.83
192	Load	255	158.01
193	Load	510	316.02
194	Load	382.5	237.02
195	Load	255	158.01
196	Load	191.25	118.51
197	Load	446.25	276.52
198	Load	133.88	82.96
199	Load	95.63	59.25
200	Load	382.5	237.02
201	Load	1030.2	638.36
202	Load	0	0
203	Load	38.25	23.7

17.2.15 System 15: 417-Bus Real Distribution System

Data of this real distribution system with 417 buses and nominal voltage of 13.8 kV is provided in Tables 17.30 and 17.31 [82].

Table 17.30 Line characteristics of System 15

Branch	From Bus	To Bus	Resistance (Ω)	Reactance (Ω)
1	384	68	0.0310675	0.021420
2	89	67	0.1078225	0.07434
3	225	224	0.0694450	0.04788
4	21	72	0.193715	0.13356
5	66	384	0.0347225	0.02394
6	3	20	0.610385	0.42084
7	224	222	0.069445	0.04788
8	72	74	0.02193	0.01512
9	70	382	0.0310675	0.02142

(*continued*)

Table 17.30 (continued)

Branch	From Bus	To Bus	Resistance (Ω)	Reactance (Ω)
10	20	51	0.08772	0.06048
11	222	221	0.08772	0.06048
12	76	72	0.054825	0.0378
13	382	66	0.032895	0.02268
14	51	52	0.069445	0.04788
15	221	220	0.084065	0.05796
16	75	81	0.084065	0.05796
17	59	70	0.0537285	0.037044
18	52	56	0.11696	0.08064
19	1	273	0.0142545	0.009828
20	81	76	0.0529975	0.03654
21	68	85	0.02924	0.02016
22	56	27	0.09503	0.06552
23	1	274	0.115498	0.079632
24	74	75	0.105995	0.07308
25	68	93	0.084065	0.05796
26	27	31	0.10234	0.07056
27	274	275	0.0836995	0.057708
28	72	73	0.098685	0.06804
29	60	58	0.010965	0.00756
30	31	40	0.105995	0.07308
31	275	209	0.139621	0.096264
32	73	80	0.0731	0.0504
33	65	71	0.0712725	0.04914
34	40	47	0.091375	0.063
35	275	276	0.1158635	0.079884
36	73	77	0.076755	0.05292
37	69	65	0.0310675	0.02142
38	51	54	0.054825	0.0378
39	276	264	0.1151325	0.07938
40	77	78	0.05117	0.03528
41	59	69	0.021199	0.014616
42	54	24	0.0731	0.0504
43	263	264	0.068714	0.047376
44	79	83	0.24123	0.16632
45	63	61	0.10965	0.0756

(continued)

Table 17.30 (*continued*)

Branch	From Bus	To Bus	Resistance (Ω)	Reactance (Ω)
46	24	29	0.09503	0.06552
47	262	263	0.0630936	0.0532656
48	82	80	0.0731	0.0504
49	71	63	0.022661	0.015624
50	29	34	0.1151325	0.07938
51	261	262	0.037973	0.032058
52	78	79	0.02193	0.01512
53	61	60	0.04386	0.03024
54	34	47	0.1736125	0.1197
55	259	261	0.0511175	0.043155
56	83	90	0.1209805	0.083412
57	58	62	0.129387	0.089208
58	47	21	0.084065	0.05796
59	257	259	0.068714	0.047376
60	77	59	0.38743	0.26712
61	64	67	0.0449565	0.030996
62	20	44	0.01462	0.01008
63	372	239	0.0265811	0.0224406
64	123	1	0.013524	0.015953
65	67	68	0.086258	0.059472
66	44	25	0.127925	0.0882
67	264	265	0.072369	0.049896
68	120	94	0.2538284	0.2584552
69	62	64	0.0603075	0.04158
70	25	35	0.127925	0.0882
71	369	347	0.13158	0.09072
72	120	123	0.294875	0.30025
73	58	20	0.59211	0.40824
74	35	38	0.08041	0.05544
75	369	370	0.208335	0.14364
76	114	115	0.047515	0.03276
77	1	92	0.388892	0.268128
78	38	41	0.1041675	0.07182
79	370	348	0.1187875	0.0819
80	115	112	0.040205	0.02772
81	92	2	0.3410115	0.235116

(*continued*)

Table 17.30 (*continued*)

Branch	From Bus	To Bus	Resistance (Ω)	Reactance (Ω)
82	41	43	0.11696	0.08064
83	383	251	0.010965	0.00756
84	94	113	0.032895	0.02268
85	92	87	0.0128524	0.0108504
86	43	48	0.08041	0.05544
87	383	370	0.018275	0.0126
88	113	114	0.076755	0.05292
89	2	17	0.08772	0.06048
90	48	50	0.0731	0.0504
91	208	277	0.067252	0.046368
92	106	95	0.084065	0.05796
93	6	7	0.040205	0.02772
94	20	33	0.06579	0.04536
95	277	278	0.032895	0.02268
96	95	112	0.032895	0.02268
97	7	3	0.051901	0.035784
98	84	33	0.0420325	0.02898
99	277	279	0.048246	0.033264
100	94	111	0.06579	0.04536
101	18	5	0.1224425	0.08442
102	28	36	0.091375	0.063
103	279	280	0.034357	0.023688
104	111	106	0.06579	0.04536
105	5	6	0.0493425	0.03402
106	36	45	0.13158	0.09072
107	280	281	0.062135	0.04284
108	119	121	0.0102235	0.008631
109	17	18	0.11696	0.08064
110	49	45	0.04386	0.03024
111	281	235	0.0493425	0.03402
112	94	119	0.056028	0.066091
113	2	14	0.0164475	0.01134
114	55	23	0.076755	0.05292
115	281	282	0.0537285	0.037044
116	118	84	0.040205	0.02772
117	8	9	0.09503	0.06552

(*continued*)

Table 17.30 (continued)

Branch	From Bus	To Bus	Resistance (Ω)	Reactance (Ω)
118	23	28	0.069445	0.04788
119	235	234	0.083334	0.057456
120	119	122	0.05117	0.03528
121	9	3	0.01462	0.01008
122	33	55	0.04386	0.03024
123	235	236	0.0420325	0.02898
124	122	118	0.201025	0.1386
125	16	19	0.157165	0.10836
126	21	50	0.03655	0.0252
127	208	236	0.076755	0.05292
128	104	105	0.049657	0.041922
129	19	8	0.02193	0.01512
130	32	44	0.25585	0.1764
131	232	349	0.075946	0.064116
132	84	104	0.0731	0.0504
133	14	16	0.098685	0.06804
134	37	32	0.105995	0.07308
135	349	207	0.032131	0.027126
136	105	108	0.06579	0.04536
137	2	13	0.032131	0.027126
138	42	37	0.0676175	0.04662
139	380	229	0.0411861	0.0347706
140	109	108	0.049657	0.041922
141	4	12	0.067183	0.056718
142	46	42	0.025585	0.01764
143	229	228	0.0630936	0.0532656
144	84	109	0.047515	0.03276
145	12	15	0.105156	0.088776
146	46	50	0.03655	0.0252
147	232	231	0.0968575	0.06678
148	110	84	0.04386	0.03024
149	15	11	0.1329055	0.112203
150	21	49	0.135235	0.09324
151	232	293	0.0749275	0.05166
152	103	110	0.076755	0.05292
153	13	4	0.0452755	0.038223

(continued)

Table 17.30 (*continued*)

Branch	From Bus	To Bus	Resistance (Ω)	Reactance (Ω)
154	39	30	0.11696	0.08064
155	293	236	0.069445	0.04788
156	107	103	0.054825	0.0378
157	11	86	0.035052	0.029592
158	49	39	0.08772	0.06048
159	221	344	0.02924	0.02016
160	84	107	0.06579	0.04536
161	86	91	0.09503	0.06552
162	53	22	0.040205	0.02772
163	304	306	0.069445	0.04788
164	97	116	0.04386	0.03024
165	10	11	0.046736	0.039456
166	30	26	0.054825	0.0378
167	378	377	0.1114775	0.07686
168	84	99	0.13158	0.09072
169	3	10	0.1078225	0.07434
170	26	53	0.1005125	0.0693
171	304	378	0.0588455	0.040572
172	101	96	0.105995	0.07308
173	10	88	0.1095375	0.092475
174	22	33	0.04386	0.03024
175	309	307	0.098685	0.06804
176	100	101	0.05848	0.04032
177	88	89	0.08041	0.05544
178	21	57	0.215645	0.14868
179	314	317	0.0815065	0.056196
180	102	100	0.054825	0.0378
181	99	102	0.05117	0.03528
182	187	167	0.084709	0.071514
183	314	318	0.072369	0.049896
184	138	194	0.0127925	0.00882
185	96	117	0.0785825	0.05418
186	167	175	0.090551	0.076446
187	301	305	0.04386	0.03024
188	155	151	0.069445	0.04788
189	117	97	0.040205	0.02772

(*continued*)

Table 17.30 (*continued*)

Branch	From Bus	To Bus	Resistance (Ω)	Reactance (Ω)
190	175	181	0.084709	0.071514
191	310	315	0.061404	0.042336
192	126	124	0.02924	0.02016
193	98	94	0.040205	0.02772
194	160	180	0.04386	0.03024
195	310	305	0.082603	0.056952
196	1	351	0.22661	0.15624
197	116	98	0.062135	0.04284
198	180	177	0.054825	0.0378
199	129	346	0.261698	0.180432
200	1	350	0.22661	0.15624
201	57	147	0.157165	0.10836
202	169	162	0.076755	0.05292
203	90	370	0.157896	0.108864
204	350	203	0.023392	0.016128
205	153	141	0.073025	0.06165
206	162	183	0.084065	0.05796
207	382	383	0.1502205	0.103572
208	351	379	0.023392	0.016128
209	147	142	0.105995	0.07308
210	183	182	0.032895	0.02268
211	66	252	0.192984	0.133056
212	203	204	0.0511175	0.043155
213	142	136	0.08041	0.05544
214	160	181	0.091375	0.063
215	384	243	0.227341	0.156744
216	204	205	0.062135	0.04284
217	132	153	0.08772	0.06048
218	181	178	0.05848	0.04032
219	85	244	0.1399865	0.096516
220	203	206	0.1034034	0.0872964
221	136	132	0.149855	0.10332
222	171	163	0.064262	0.054252
223	352	414	0.0705415	0.048636
224	379	380	0.1034034	0.0872964
225	57	149	0.069445	0.04788

(*continued*)

Table 17.30 (continued)

Branch	From Bus	To Bus	Resistance (Ω)	Reactance (Ω)
226	163	184	0.075946	0.064116
227	414	385	0.0720035	0.049644
228	206	207	0.0598805	0.050553
229	57	125	0.091375	0.063
230	160	129	0.227838	0.192348
231	414	401	0.0727345	0.050148
232	1	375	0.010965	0.00756
233	127	124	0.02924	0.02016
234	129	141	0.032131	0.027126
235	401	386	0.059211	0.040824
236	208	209	0.0383775	0.02646
237	128	127	0.27047	0.18648
238	131	157	0.03655	0.0252
239	401	402	0.070907	0.048888
240	209	210	0.11696	0.08064
241	125	128	0.02193	0.01512
242	130	131	0.054825	0.0378
243	402	387	0.0566525	0.03906
244	209	211	0.0559215	0.038556
245	200	119	0.076755	0.05292
246	141	130	0.046736	0.039456
247	402	403	0.0595765	0.041076
248	211	1	0.056287	0.038808
249	200	201	0.00731	0.00504
250	157	133	0.113305	0.07812
251	403	388	0.083334	0.057456
252	375	202	0.149855	0.10332
253	158	200	0.02193	0.01512
254	133	140	0.105995	0.07308
255	403	404	0.0683485	0.047124
256	202	1	0.48246	0.33264
257	124	158	0.10234	0.07056
258	140	147	0.15351	0.10584
259	368	345	0.1151325	0.07938
260	373	352	0.1041675	0.07182
261	199	192	0.01462	0.01008

(continued)

Table 17.30 (*continued*)

Branch	From Bus	To Bus	Resistance (Ω)	Reactance (Ω)
262	129	154	0.035052	0.029592
263	345	346	0.05848	0.04032
264	352	212	0.1136705	0.078372
265	199	197	0.069445	0.04788
266	137	144	0.09503	0.06552
267	345	369	0.0529975	0.03654
268	352	353	0.2138175	0.14742
269	159	199	0.27047	0.18648
270	144	148	0.09503	0.06552
271	400	404	0.047515	0.03276
272	353	213	0.201025	0.1386
273	197	191	0.010965	0.00756
274	154	156	0.032131	0.027126
275	404	405	0.040205	0.02772
276	353	214	0.0500735	0.034524
277	159	190	0.2359	0.2402
278	156	137	0.084065	0.05796
279	405	389	0.0588455	0.040572
280	1	215	0.3475905	0.239652
281	159	161	0.084065	0.05796
282	143	150	0.098685	0.06804
283	405	406	0.071638	0.049392
284	215	216	0.105995	0.07308
285	161	172	0.08772	0.06048
286	150	148	0.02193	0.01512
287	406	390	0.045322	0.031248
288	216	217	0.099314	0.083844
289	172	188	0.091375	0.063
290	134	139	0.119761	0.101106
291	406	391	0.0731	0.0504
292	216	218	0.06579	0.04536
293	188	168	0.091375	0.063
294	139	143	0.062135	0.04284
295	404	407	0.083334	0.057456
296	218	219	0.002924	0.002016
297	176	179	0.06579	0.04536

(*continued*)

Table 17.30 (continued)

Branch	From Bus	To Bus	Resistance (Ω)	Reactance (Ω)
298	154	134	0.078867	0.066582
299	407	392	0.0785825	0.05418
300	219	220	0.08041	0.05544
301	168	176	0.113305	0.07812
302	148	57	0.1462	0.1008
303	407	408	0.048977	0.033768
304	220	223	0.0709803	0.0599238
305	161	185	0.13158	0.09072
306	129	151	0.032895	0.02268
307	408	409	0.0668865	0.046116
308	223	237	0.081788	0.069048
309	185	164	0.091375	0.063
310	145	146	0.032895	0.02268
311	409	393	0.073831	0.050904
312	237	238	0.0622173	0.0525258
313	164	165	0.032895	0.02268
314	146	149	0.054825	0.0378
315	409	394	0.0690795	0.047628
316	237	372	0.0537464	0.0453744
317	165	173	0.13889	0.09576
318	152	135	0.076755	0.05292
319	408	410	0.0551905	0.038052
320	239	240	0.005842	0.004932
321	173	179	0.113305	0.07812
322	135	145	0.142545	0.09828
323	410	415	0.06579	0.04536
324	239	243	0.0835406	0.0705276
325	179	160	0.069445	0.04788
326	135	195	0.0091375	0.0063
327	410	411	0.064328	0.044352
328	243	355	0.0054825	0.00378
329	159	182	0.054096	0.063812
330	151	152	0.054825	0.0378
331	411	395	0.0595765	0.041076
332	355	354	0.127925	0.0882
333	182	186	0.0483	0.056975

(continued)

Table 17.30 (continued)

Branch	From Bus	To Bus	Resistance (Ω)	Reactance (Ω)
334	1	190	0.14154	0.14412
335	411	412	0.0676175	0.04662
336	354	242	0.018275	0.0126
337	186	166	0.037744	0.038432
338	190	189	0.0731	0.0504
339	412	399	0.0566525	0.03906
340	354	241	0.0731	0.0504
341	166	170	0.037744	0.038432
342	189	198	0.10965	0.0756
343	412	398	0.071638	0.049392
344	243	356	0.0091375	0.0063
345	170	174	0.02921	0.02466
346	198	196	0.03655	0.0252
347	412	413	0.0873545	0.060228
348	356	357	0.05117	0.03528
349	174	180	0.063693	0.064854
350	189	193	0.023392	0.016128
351	413	397	0.0705415	0.048636
352	357	244	0.0127925	0.00882
353	159	184	0.22661	0.15624
354	149	138	0.127925	0.0882
355	413	396	0.069445	0.04788
356	357	245	0.0054825	0.00378
357	184	187	0.035052	0.029592
358	138	155	0.105995	0.07308
359	290	289	0.062866	0.043344
360	243	246	0.0365125	0.030825
361	246	247	0.052632	0.036288
362	265	266	0.1078225	0.07434
363	289	288	0.046784	0.032256
364	306	310	0.126463	0.087192
365	247	248	0.0727345	0.050148
366	266	267	0.099416	0.068544
367	288	300	0.1867705	0.128772
368	310	320	0.0127925	0.00882
369	243	249	0.0487807	0.0411822

(continued)

Table 17.30 (continued)

Branch	From Bus	To Bus	Resistance (Ω)	Reactance (Ω)
370	267	268	0.156434	0.107856
371	300	286	0.10965	0.0756
372	310	314	0.072369	0.049896
373	249	358	0.043129	0.029736
374	268	269	0.0549647	0.0559666
375	1	373	0.027778	0.019152
376	314	309	0.0698105	0.048132
377	358	359	0.0054825	0.00378
378	269	365	0.07077	0.07206
379	300	285	0.081141	0.055944
380	309	313	0.0785825	0.05418
381	359	250	0.0201025	0.01386
382	365	366	0.025585	0.01764
383	300	301	0.0932025	0.06426
384	313	317	0.0683485	0.047124
385	359	251	0.1407175	0.09702
386	366	270	0.1005125	0.0693
387	301	302	0.055556	0.038304
388	317	319	0.123539	0.085176
389	251	252	0.1882325	0.12978
390	366	271	0.0164475	0.01134
391	302	303	0.045322	0.031248
392	319	318	0.066521	0.045864
393	251	360	0.098685	0.06804
394	365	363	0.0483595	0.049241
395	303	304	0.195177	0.134568
396	318	315	0.069445	0.04788
397	360	253	0.049708	0.034272
398	363	362	0.018872	0.019216
399	304	305	0.208335	0.14364
400	315	311	0.083334	0.057456
401	253	254	0.0458597	0.0387162
402	363	381	0.26316	0.18144
403	305	308	0.050439	0.034776
404	311	304	0.105264	0.072576
405	254	361	0.0588455	0.040572

(continued)

Table 17.30 (*continued*)

Branch	From Bus	To Bus	Resistance (Ω)	Reactance (Ω)
406	362	272	0.0164475	0.01134
407	308	316	0.2218585	0.152964
408	296	260	0.0478805	0.033012
409	361	256	0.040205	0.02772
410	362	1	0.02415	0.0284875
411	316	312	0.105264	0.072576
412	260	287	0.035819	0.024696
413	254	255	0.0274125	0.0189
414	1	364	0.28509	0.19656
415 (tie line)	312	307	0.072369	0.049896
416 (tie line)	260	324	0.1078225	0.07434
417 (tie line)	254	257	0.078867	0.066582
418 (tie line)	364	284	0.1041675	0.07182
419 (tie line)	307	300	0.078217	0.053928
420 (tie line)	324	326	0.0054825	0.00378
421 (tie line)	257	258	0.105995	0.07308
422 (tie line)	364	285	0.05848	0.04032
423 (tie line)	334	336	0.0712725	0.04914
424 (tie line)	324	266	0.103802	0.071568
425 (tie line)	257	283	0.0537464	0.0453744
426 (tie line)	381	268	0.3095785	0.213444
427 (tie line)	336	323	0.054825	0.0378
428 (tie line)	324	325	0.048246	0.033264
429 (tie line)	283	282	0.0663067	0.0559782
430 (tie line)	268	327	0.0383775	0.02646
431 (tie line)	336	337	0.0646935	0.044604
432 (tie line)	324	323	0.08772	0.06048
433 (tie line)	282	234	0.055499	0.046854
434 (tie line)	327	328	0.054825	0.0378
435 (tie line)	337	338	0.02924	0.02016
436 (tie line)	323	322	0.0676175	0.04662
437 (tie line)	234	233	0.0277495	0.023427
438 (tie line)	268	297	0.050439	0.034776
439 (tie line)	338	339	0.0420325	0.02898
440 (tie line)	322	321	0.09503	0.06552
441 (tie line)	233	376	0.011684	0.009864

(*continued*)

Table 17.30 (*continued*)

Branch	From Bus	To Bus	Resistance (Ω)	Reactance (Ω)
442 (tie line)	297	296	0.050439	0.034776
443 (tie line)	339	265	0.067983	0.046872
444 (tie line)	321	299	0.070176	0.048384
445 (tie line)	230	231	0.0455676	0.0384696
446 (tie line)	296	295	0.0741965	0.051156
447 (tie line)	265	367	0.008772	0.006048
448 (tie line)	299	296	0.038012	0.026208
449 (tie line)	230	228	0.026289	0.022194
450 (tie line)	296	298	0.0376465	0.025956
451 (tie line)	367	340	0.0010965	0.000756
452 (tie line)	322	329	0.031433	0.021672
453 (tie line)	228	371	0.046736	0.039456
454 (tie line)	295	294	0.022661	0.015624
455 (tie line)	340	341	0.0274125	0.0189
456 (tie line)	329	330	0.0266815	0.018396
457 (tie line)	371	374	0.0394335	0.033291
458 (tie line)	294	292	0.0917405	0.063252
459 (tie line)	341	342	0.113305	0.07812
460 (tie line)	330	331	0.041667	0.028728
461 (tie line)	374	227	0.003655	0.00252
462 (tie line)	268	292	0.098685	0.06804
463 (tie line)	334	335	0.0559215	0.038556
464 (tie line)	331	332	0.0274125	0.0189
465 (tie line)	374	226	0.0023368	0.0019728
466 (tie line)	292	290	0.016813	0.011592
467 (tie line)	341	368	0.179095	0.12348
468 (tie line)	329	333	0.05848	0.04032
469 (tie line)	226	225	0.0274125	0.0189
470 (tie line)	290	291	0.0054825	0.00378
471 (tie line)	376	230	0.0277495	0.023427
472 (tie line)	333	334	0.041667	0.028728
473 (tie line)	368	343	0.0091375	0.0063

Table 17.31 Load data for System 15

Bus Number	Bus Type	Active Power Demand (kW)	Reactive Power Demand (kVAr)
1	Generation	0.1	0
2	Load	0.1	0
3	Load	122	59
4	Load	0.1	0
5	Load	0.1	0
6	Load	114	55
7	Load	51	25
8	Load	83	40
9	Load	0.1	0
10	Load	122	59
11	Load	122	59
12	Load	77	37
13	Load	0.1	0
14	Load	68	33
15	Load	71	34
16	Load	0.1	0
17	Load	0.1	0
18	Load	77	37
19	Load	30	14
20	Load	13	6
21	Load	0.1	0
22	Load	194	94
23	Load	77	37
24	Load	77	37
25	Load	77	37
26	Load	122	59
27	Load	104	51
28	Load	0.1	0
29	Load	77	37
30	Load	243	118
31	Load	77	37
32	Load	77	37
33	Load	0.1	0
34	Load	121	58
35	Load	194	94
36	Load	122	59

(continued)

Table 17.31 (*continued*)

Bus Number	Bus Type	Active Power Demand (kW)	Reactive Power Demand (kVAr)
37	Load	77	37
38	Load	0.1	0
39	Load	0.1	0
40	Load	96	47
41	Load	122	59
42	Load	70	34
43	Load	58	28
44	Load	0.1	0
45	Load	122	59
46	Load	122	59
47	Load	122	59
48	Load	86	42
49	Load	77	37
50	Load	0.1	0
51	Load	122	59
52	Load	122	59
53	Load	122	59
54	Load	0.1	0
55	Load	42	20
56	Load	122	59
57	Load	13	6
58	Load	77	37
59	Load	50	24
60	Load	0.1	0
61	Load	60	29
62	Load	71	34
63	Load	122	59
64	Load	77	37
65	Load	95	46
66	Load	121	58
67	Load	77	37
68	Load	103	50
69	Load	52	25
70	Load	57	27
71	Load	171	83
72	Load	121	58

(*continued*)

Table 17.31 (*continued*)

Bus Number	Bus Type	Active Power Demand (kW)	Reactive Power Demand (kVAr)
73	Load	39	19
74	Load	119	58
75	Load	49	24
76	Load	0.1	0
77	Load	50	24
78	Load	44	21
79	Load	82	40
80	Load	132	64
81	Load	194	94
82	Load	0.1	0
83	Load	122	59
84	Load	3	1
85	Load	52	25
86	Load	103	50
87	Load	21	10
88	Load	21	10
89	Load	21	10
90	Load	5	3
91	Load	0.1	0
92	Load	0.1	0
93	Load	21	10
94	Load	18	9
95	Load	122	59
96	Load	116	56
97	Load	7	3
98	Load	64	31
99	Load	49	24
100	Load	16	8
101	Load	77	37
102	Load	119	58
103	Load	71	34
104	Load	77	37
105	Load	194	94
106	Load	122	59
107	Load	55	27
108	Load	38	18
109	Load	22	10

(*continued*)

Table 17.31 (*continued*)

Bus Number	Bus Type	Active Power Demand (kW)	Reactive Power Demand (kVAr)
110	Load	0.1	0
111	Load	77	37
112	Load	122	59
113	Load	38	18
114	Load	122	59
115	Load	27	13
116	Load	64	31
117	Load	0.1	0
118	Load	0.1	0
119	Load	0.1	0
120	Load	0.1	0
121	Load	79	38
122	Load	247	119
123	Load	0.1	0
124	Load	0.1	0
125	Load	0.1	0
126	Load	0.1	0
127	Load	0.1	0
128	Load	0.1	0
129	Load	77	37
130	Load	122	59
131	Load	62	30
132	Load	105	51
133	Load	122	59
134	Load	77	37
135	Load	77	37
136	Load	0.1	0
137	Load	85	41
138	Load	8	4
139	Load	122	59
140	Load	35	17
141	Load	55	27
142	Load	153	74
143	Load	316	153
144	Load	105	51
145	Load	20	10
146	Load	95	46

(*continued*)

Table 17.31 (*continued*)

Bus Number	Bus Type	Active Power Demand (kW)	Reactive Power Demand (kVAr)
147	Load	38	18
148	Load	83	40
149	Load	50	24
150	Load	77	37
151	Load	122	59
152	Load	102	49
153	Load	60	29
154	Load	77	37
155	Load	194	94
156	Load	56	27
157	Load	122	59
158	Load	0.1	0
159	Load	3	1
160	Load	0.1	0
161	Load	52	25
162	Load	194	94
163	Load	10	5
164	Load	128	62
165	Load	16	8
166	Load	37	18
167	Load	154	75
168	Load	0.1	0
169	Load	0.1	0
170	Load	78	38
171	Load	0.1	0
172	Load	27	13
173	Load	103	50
174	Load	122	59
175	Load	72	35
176	Load	0.1	0
177	Load	122	59
178	Load	194	94
179	Load	0.1	0
180	Load	77	37
181	Load	82	40
182	Load	53	26
183	Load	122	59

(*continued*)

Table 17.31 (*continued*)

Bus Number	Bus Type	Active Power Demand (kW)	Reactive Power Demand (kVAr)
184	Load	64	31
185	Load	86	42
186	Load	82	40
187	Load	0.1	0
188	Load	167	81
189	Load	0.1	0
190	Load	0.1	0
191	Load	131	64
192	Load	21	10
193	Load	21	10
194	Load	21	10
195	Load	21	10
196	Load	0.1	0
197	Load	0.1	0
198	Load	0.1	0
199	Load	0.1	0
200	Load	0.1	0
201	Load	52	25
202	Load	202	98
203	Load	202	98
204	Load	73	35
205	Load	81	39
206	Load	115	56
207	Load	118	57
208	Load	318	154
209	Load	161	78
210	Load	126	61
211	Load	59	29
212	Load	24	12
213	Load	24	12
214	Load	5	2
215	Load	126	61
216	Load	56	27
217	Load	62	30
218	Load	0.1	0
219	Load	32	15
220	Load	158	77

(*continued*)

Table 17.31 (*continued*)

Bus Number	Bus Type	Active Power Demand (kW)	Reactive Power Demand (kVAr)
221	Load	126	61
222	Load	202	98
223	Load	41	20
224	Load	81	39
225	Load	126	61
226	Load	202	98
227	Load	18	9
228	Load	202	98
229	Load	106	51
230	Load	175	85
231	Load	174	84
232	Load	113	55
233	Load	202	98
234	Load	0.1	0
235	Load	202	98
236	Load	403	195
237	Load	202	98
238	Load	126	61
239	Load	0.1	0
240	Load	29	14
241	Load	9	4
242	Load	5	2
243	Load	0.1	0
244	Load	18	9
245	Load	45	22
246	Load	32	16
247	Load	61	30
248	Load	81	39
249	Load	126	61
250	Load	23	11
251	Load	14	7
252	Load	18	9
253	Load	126	61
254	Load	5	3
255	Load	26	13
256	Load	18	9
257	Load	126	61

(*continued*)

Table 17.31 (*continued*)

Bus Number	Bus Type	Active Power Demand (kW)	Reactive Power Demand (kVAr)
258	Load	81	39
259	Load	0.1	0
260	Load	126	61
261	Load	41	20
262	Load	0.1	0
263	Load	28	14
264	Load	30	14
265	Load	81	39
266	Load	202	98
267	Load	318	154
268	Load	176	85
269	Load	126	61
270	Load	6	3
271	Load	31	15
272	Load	60	29
273	Load	56	27
274	Load	126	61
275	Load	118	57
276	Load	126	61
277	Load	202	98
278	Load	10	5
279	Load	126	61
280	Load	202	98
281	Load	318	154
282	Load	82	40
283	Load	126	61
284	Load	22	10
285	Load	13	6
286	Load	71	34
287	Load	101	49
288	Load	113	54
289	Load	126	61
290	Load	0.1	0
291	Load	34	17
292	Load	83	40
293	Load	59	29
294	Load	126	61

(*continued*)

Table 17.31 *(continued)*

Bus Number	Bus Type	Active Power Demand (kW)	Reactive Power Demand (kVAr)
295	Load	145	70
296	Load	202	98
297	Load	50	24
298	Load	41	20
299	Load	202	98
300	Load	77	37
301	Load	161	78
302	Load	477	231
303	Load	68	33
304	Load	14	7
305	Load	4	2
306	Load	0.1	0
307	Load	66	32
308	Load	0.1	0
309	Load	60	29
310	Load	95	46
311	Load	0.1	0
312	Load	19	9
313	Load	67	32
314	Load	18	9
315	Load	33	16
316	Load	21	10
317	Load	33	16
318	Load	42	20
319	Load	52	25
320	Load	43	21
321	Load	126	61
322	Load	107	52
323	Load	202	98
324	Load	126	61
325	Load	24	12
326	Load	101	49
327	Load	81	39
328	Load	155	75
329	Load	23	11
330	Load	31	15
331	Load	126	61

(continued)

Table 17.31 (*continued*)

Bus Number	Bus Type	Active Power Demand (kW)	Reactive Power Demand (kVAr)
332	Load	13	6
333	Load	56	27
334	Load	120	58
335	Load	126	61
336	Load	403	195
337	Load	81	39
338	Load	177	86
339	Load	151	73
340	Load	9	4
341	Load	29	14
342	Load	14	7
343	Load	23	11
344	Load	45	22
345	Load	0.1	0
346	Load	36	17
347	Load	5	2
348	Load	9	4
349	Load	80	39
350	Load	0.1	0
351	Load	0.1	0
352	Load	0.1	0
353	Load	0.1	0
354	Load	0.1	0
355	Load	0.1	0
356	Load	0.1	0
357	Load	0.1	0
358	Load	0.1	0
359	Load	0.1	0
360	Load	0.1	0
361	Load	0.1	0
362	Load	0.1	0
363	Load	0.1	0
364	Load	0.1	0
365	Load	0.1	0
366	Load	0.1	0
367	Load	0.1	0
368	Load	0.1	0

(*continued*)

Table 17.31 (*continued*)

Bus Number	Bus Type	Active Power Demand (kW)	Reactive Power Demand (kVAr)
369	Load	0.1	0
370	Load	0.1	0
371	Load	0.1	0
372	Load	0.1	0
373	Load	0.1	0
374	Load	0.1	0
375	Load	0.1	0
376	Load	0.1	0
377	Load	86	41
378	Load	86	41
379	Load	0.1	0
380	Load	0.1	0
381	Load	0.1	0
382	Load	0.1	0
383	Load	0.1	0
384	Load	0.1	0
385	Load	68	33
386	Load	15	7
387	Load	56	27
388	Load	12	6
389	Load	50	24
390	Load	115	56
391	Load	14	7
392	Load	5	2
393	Load	19	9
394	Load	34	17
395	Load	10	5
396	Load	40	19
397	Load	28	14
398	Load	14	7
399	Load	17	8
400	Load	41	20
401	Load	0.1	0
402	Load	0.1	0
403	Load	0.1	0
404	Load	0.1	0

(continued)

Table 17.31 (continued)

Bus Number	Bus Type	Active Power Demand (kW)	Reactive Power Demand (kVAr)
405	Load	0.1	0
406	Load	0.1	0
407	Load	0.1	0
408	Load	0.1	0
409	Load	0.1	0
410	Load	0.1	0
411	Load	0.1	0
412	Load	0.1	0
413	Load	0.1	0
414	Load	0.1	0
415	Load	19	9

17.3 Conclusion

In this chapter, properties of most of the required test systems have been described and their important data and information, such as the network configuration, load arrangement, and lines characteristics, were provided. These data are very useful to evaluate the different reconfiguration methods and models proposed by the specialized literature. This chapter provides important information for readers and researchers who are trying to solve the DSR problem more efficiently. In the next chapter, the results of implementation of several mathematical models and approaches are described using different examples. In that chapter, the proposed reconfiguration methods are applied to these test systems and the results are presented.

References

[1] M. Mahdavi, C.S. Antunez, M. Ajalli, and R. Romero, "Transmission expansion planning: Literature review and classification," *IEEE Systems Journal*, vol. 3, no. 3, pp. 3129–3140, 2019.

[2] M. Mahdavi and H. Monsef, "Review of static transmission expansion planning," *Journal of Electrical and Control Engineering*, vol. 1, no. 1, pp. 11–18, 2011.

[3] M. Mahdavi, C.S. Antunez, A. Bagheri, and R. Romero, "Line maintenance within transmission expansion planning: A multistage framework," *IET Generation, Transmission and Distribution*, vol. 14, no. 13, pp. 3057–3065, 2019.

[4] H. Shayeghi, M. Mahdavi, A. Kazemi, and H.A. Shayanfar, "Studying effect of bundle lines on TNEP considering network losses using decimal codification genetic algorithm," *Energy Conversion and Management*, vol. 51, no. 12, pp. 2685–2691, 2010.

[5] H. Shayeghi and M. Mahdavi, "Genetic algorithm based studying of bundle lines effect on network losses in transmission network expansion planning," *Journal of Electrical Engineering*, vol. 60, no. 5, pp. 237–245, 2009.
[6] A. Kazemi, S. Jalilzadeh, M. Mahdavi, and H. Haddadian, "Genetic algorithm-based investigation of load growth factor effect on the network loss in TNEP," *3rd IEEE Conference on Industrial Electronics and Applications*, Singapore, vol. 1, pp. 764–769, 2008.
[7] M. Mahdavi and E. Mahdavi, "Evaluating the effect of load growth on annual network losses in TNEP considering bundle lines using DCGA," *International Journal on Technical and Physical Problems of Engineering*, vol. 3, no. 4, pp. 1–9, 2011.
[8] S. Jalilzadeh, A. Kazemi, M. Mahdavi, and H. Haddadian, "TNEP considering voltage level, network losses and number of bundle lines using GA," *Third International Conference on Electric Utility Deregulation and Restructuring and Power Technologies*, China, pp. 1580–1585, 2008.
[9] H. Shayeghi, H. Hosseini, A. Shabani, and M. Mahdavi, "GEP considering purchase prices, profits of IPPs and reliability criteria using hybrid GA and PSO," *International Journal of Electrical and Computer Engineering*, vol. 2, no. 8, pp. 1619–1625, 2008.
[10] M. Mahdavi, L.H. Macedo, and R. Romero, "Transmission and generation expansion planning considering system reliability and line maintenance," *26th Iranian Conference on Electrical Engineering*, Mashhad, Iran, pp. 1005–1010, 2018.
[11] M. Mahdavi, L.H. Macedo, and R. Romero, "Considering line maintenance and repair in GTEP," *4th Conference on Modern Technologies in Electrical, Telecommunication and Mechatronics Engineering*, Iran, pp. 1–6, 2018.
[12] H. Hosseini, S. Jalilzadeh, V. Nabaei, G.R.Z. Govar, and M. Mahdavi, "Enhancing deregulated distribution network reliability for minimizing penalty cost based on reconfiguration using BPSO," *2nd IEEE International Conference on Power and Energy*, Malaysia, pp. 983–987, 2008.
[13] O. Badran, S. Mekhilef, H. Mokhlis, and W. Dahalan, "Optimal reconfiguration of distribution system connected with distributed generations: A review of different methodologies," *Renewable and Sustainable Energy Reviews*, vol. 73, pp. 854–867, June 2017.
[14] N.G. Paterakis, A. Mazza, S.F. Santos, O. Erdinç, G. Chicco, A.G. Bakirtzis, and J.P.S. Catalão, "Multi-objective reconfiguration of radial distribution systems using reliability indices," *IEEE Transactions on Power Systems*, vol. 31, no. 2, pp. 1048–1062, March 2016.
[15] M. Khodayari, M. Mahdavi, and H. Monsef, "Simultaneous scheduling of energy & spinning reserve considering customer and supplier choice on reliability," *19th Iranian Conference on Electrical Engineering*, Tehran, Iran, pp. 491–496, 2011.
[16] M. Khodayari, H. Monsef, and M. Mahdavi, "Customer reliability based energy & SR scheduling at hierarchical level II," *27th International Power System Conference*, Tehran, Iran, pp. 1–11, 2012.
[17] S.S. Duttagupta, "A reliability assessment methodology for distribution systems with distributed generation," MSc Thesis, Texas A&M University, Electrical Engineering Department, pp. 1–90, May 2006.

[18] M. Mahdavi and R. Romero, "Transmission system expansion planning based on LCC criterion considering uncertainty in VOLL using DCGA," *32nd International Power System Conference*, Iran, pp. 1–7, 2017.

[19] M. Mahdavi, A.R. Kheirkhah, L.H. Macedo, and R. Romero, "A genetic algorithm for transmission network expansion planning considering line maintenance," *IEEE Congress on Evolutionary Computation (CEC)*, Glasgow, United Kingdom, pp. 1–6, 2020.

[20] M. Mahdavi and R. Romero, "A new model for dynamic transmission expansion planning," *4th National Conference on Technology in Electrical and Computer Engineering*, Iran, pp. 1–7, 2018.

[21] M. Mahdavi, A.R. Kheirkhah, and R. Romero, "Transmission expansion planning considering line failures and maintenance," *33rd International Power System Conference*, Iran, pp. 1–7, 2018.

[22] A. Bagheri, M. Mahdavi, and H. Monsef, "Suggestion of an economical plan for electrical power transmission," *The 9th International Energy Conference*, Iran, pp. 1–10, 2013.

[23] Y.Y. Fu and H.D. Chiang, "Toward optimal multiperiod network reconfiguration for increasing the hosting capacity of distribution networks," *IEEE Transactions on Power Delivery*, vol. 33, no. 5, pp. 2294–2304, October 2018.

[24] S. Lei, Y. Hou, F. Qiu, and J. Yan, "Identification of critical switches for integrating renewable distributed generation by dynamic network reconfiguration," *IEEE Transactions on Sustainable Energy*, vol. 9, no. 1, pp. 420–432, January 2018.

[25] H. Xing and X. Sun, "Distributed generation locating and sizing in active distribution network considering network reconfiguration," *IEEE Access*, vol. 5, pp. 14768–14774, August 2017.

[26] A. Azizivahed, H. Narimani, M. Fathi, E. Naderi, H.R. Safarpour, and M.R. Narimani, "Multi-objective dynamic distribution feeder reconfiguration in automated distribution systems," *Energy*, vol. 147, pp. 896–914, March 2018.

[27] J. Shukla, B. Das, and V. Pant, "Stability constrained optimal distribution system reconfiguration considering uncertainties in correlated loads and distributed generations," *International Journal of Electrical Power and Energy Systems*, vol. 99, pp. 121–133, 2018.

[28] K. Chen, W. Wu, B. Zhang, S. Djokic, and G.P. Harrison, "A method to evaluate total supply capability of distribution systems considering network reconfiguration and daily load curves," *IEEE Transactions on Power Systems*, vol. 31, no. 3, pp. 2096–2104, May 2016.

[29] N.C. Koutsoukis, D.O. Siagkas, P.S. Georgilakis, and N.D. Hatziargyriou, "Online reconfiguration of active distribution networks for maximum integration of distributed generation," *IEEE Transactions on Automation Science and Engineering*, vol. 14, no. 2, pp. 437–448, April 2017.

[30] A. Asrari, S. Lotfifard, and M. Ansari, "Reconfiguration of smart distribution systems with time varying loads using parallel computing," *IEEE Transactions on Smart Grid*, vol. 7, no. 6, pp. 2713–2723, November 2016.

[31] Y. Kifle, B. Khan, and P. Singh, "Assessment and enhancement of distribution system reliability by renewable energy sources and energy storage," *Journal of Green Engineering*, vol. 8, no. 3 (2), pp. 219–262, July 2018.

[32] B. Khan, H.H. Alhelou, and F. Mebrahtu, "A holistic analysis of distribution system reliability assessment methods with conventional and renewable energy sources," *AIMS Energy*, vol. 7, no. 4, pp. 413–429, 2019.

[33] O.P. Mahela, B. Khan, H.H. Alhelou, and P. Siano, "Power quality assessment and event detection in distribution network with wind energy penetration using Stockwell transform and fuzzy clustering," *IEEE Transactions on Industrial Informatics*, vol. 16, no. 11, pp. 6922–6932, November 2020, doi: 10.1109/TII.2020.2971709.

[34] O.P. Mahela, B. Khan, H. Haes Alhelou, and S. Tanwar, "Assessment of power quality in the utility grid integrated with wind energy generation," *IET Power Electronics*, vol. 13, no. 13, pp. 2917–2925, 14 October 2020, doi: 10.1049/iet-pel.2019.1351.

[35] T.F. Agajie, B. Khan, H.H. Alhelou and O.P. Mahela, "Optimal expansion planning of distribution system using grid-based multi-objective harmony search algorithm," *Computers & Electrical Engineering*, vol. 87, p. 106823, 2020.

[36] J.C. López, M. Lavorato, and M.J. Rider, "Optimal reconfiguration of electrical distribution systems considering reliability indices improvement," *International Journal of Electrical Power and Energy Systems*, vol. 78, pp. 837–845, June 2016.

[37] M. Mahdavi and A. Bagheri, "BPSO applied to TNEP considering adequacy criterion," *American Journal of Neural Networks and Applications*, vol. 4, no. 1, pp. 1–7, 2018.

[38] H. Shayeghi, M. Mahdavi, and A. Kazemi, "Discrete particle swarm optimization algorithm used for TNEP considering network adequacy restriction," *International Journal of Electrical and Computer Engineering*, vol. 3, no. 3, pp. 521–528, 2009.

[39] H. Shayeghi, M. Mahdavi, and H. Haddadian, "DCGA based-transmission network expansion planning considering network adequacy," *International Journal of Electrical and Computer Engineering*, vol. 2, no. 12, pp. 2875–2880, 2008.

[40] A. Kazemi, S. Jalilzadeh, H. Haddadian, and M. Mahdavi, "Transmission expansion planning considering network adequacy and losses using DCGA," *IEEE International Conference on Sustainable Energy Technologies*, Singapore, pp. 145–150, 2008.

[41] A. Kazemi, H. Haddadian, and M. Mahdavi, "Transmission network adequacy optimization using genetic algorithm," *5th International Conference on Electrical and Electronics Engineering*, Turkey, pp. 1–5, 2007.

[42] M. Mahdavi, A. Bagheri, and E. Mahdavi, "Comparing efficiency of PSO with GA in transmission expansion planning considering network adequacy," *WSEAS Transactions on Power Systems*, vol. 7, no. 1, pp. 34–43, 2012.

[43] M. Mahdavi and E. Mahdavi, "Transmission expansion planning considering network adequacy and investment cost limitation using genetic algorithm," *International Journal of Electrical and Computer Engineering*, vol. 5, no. 8, pp. 1–5, 2011.

[44] A. Kimiyaghalam, M. Mahdavi, A. Ashouri, and M. Bagherivand, "Optimal placement of PMUs for reliable observability of network under probabilistic events using BABC algorithm," *22nd International Conference on Electricity Distribution*, Stockholm, Sweden, pp. 1–4, 2013.

[45] M. Mahdavi, H. Monsef, and A. Bagheri, "Transmission lines loading enhancement using ADPSO approach," *International Journal of Electrical and Computer Engineering*, vol. 4, no. 3, pp. 556–561, 2010.

[46] M. Mahdavi, H. Monsef, and A. Bagheri, "Lines loading optimization in transmission expansion planning based on binary PSO algorithm," *I-manager's Journal on Information Technology*, vol. 1, no. 1, pp. 24–32, 2012.

[47] H. Shayeghi and M. Mahdavi, "Optimization of transmission lines loading in TNEP using decimal codification based GA," *International Journal of Electrical and Computer Engineering*, vol. 2, no. 5, pp. 942–947, 2008.

[48] H. Shayeghi, M. Mahdavi, and A. Bagheri, "Discrete PSO algorithm based optimization of transmission lines loading in TNEP problem," *Energy Conversion and Management*, vol. 51, no. 1, pp. 112–121, 2010.

[49] M. Mahdavi, H. Monsef, and R. Romero, "Reliability effects of maintenance on TNEP considering preventive and corrective repairs," *IEEE Transactions on Power Systems*, vol. 32, no. 5, pp. 3768–3781, 2017.

[50] M. Mahdavi, H. Monsef, and R. Romero, "Reliability and economic effects of maintenance on TNEP considering line loading and repair," *IEEE Transactions on Power Systems*, vol. 31, no. 5, pp. 3381–3393, 2016.

[51] M. Mahdavi, A. Nazari, V. Hosseinnezhad, and A. Safari, "A PSO-based static synchronous compensator controller for power system stability enhancement," *Journal of Artificial Intelligence in Electrical Engineering*, vol. 1, no. 1, pp. 18–25, 2012.

[52] M. Mahdavi, H. Monsef, A. Bagheri, and A. Kimiyaghalam, "Dynamic transmission network expansion planning considering network losses, DG sources and operational costs – Part 1: Review and problem formulation," *WULFENIA*, vol. 19, no. 10, pp. 242–257, 2012.

[53] M. Mahdavi, H. Monsef, A. Bagheri, and A. Kimiyaghalam, "Dynamic transmission network expansion planning considering network losses, DG sources and operational costs – Part 2: Solution method and numerical results," *WULFENIA*, vol. 19, no. 10, pp. 258–273, 2012.

[54] M. Rahmani-Andebili and M. Fotuhi Firuzabad, "An adaptive approach for PEVs charging management and reconfiguration of electrical distribution system penetrated by renewables," *IEEE Transactions on Industrial Informatics*, vol. 14, no. 5, pp. 2001–2010, May 2018.

[55] Z. Li, S. Jazebi, and F. de León, "Determination of the optimal switching frequency for distribution system reconfiguration," *IEEE Transactions on Power Delivery*, vol. 32, no. 4, pp. 2060–2069, August 2017.

[56] S.S.F. Souza, R. Romero, J. Pereira, and J.T. Saraiva, "Artificial immune algorithm applied to distribution system reconfiguration with variable demand," *International Journal of Electrical Power and Energy Systems*, vol. 82, pp. 561–568, November 2016.

[57] L H. Macedo, J.F. Franco, M. Mahdavi, and R. Romero, "A contribution to the optimization of the reconfiguration problem in radial distribution systems," *Journal of Control, Automation and Electrical Systems*, vol. 29, no. 6, pp. 756–768, 2018.

[58] M. Mahdavi, L.H. Macedo, and R. Romero, "A mathematical formulation for distribution network reconfiguration," *7th Regional Conference on Electricity Distribution*, Iran, pp. 1–5, 2019.

[59] M. Mahdavi and R. Romero, "Reconfiguration of radial distribution systems: An efficient mathematical model," *IEEE Latin America Transactions*, vol. 100, no. 1e (2020): Early Access, 2020.

[60] M.A. Muhammad, H. Mokhlis, K. Naidu, J.F. Franco, H.A. Illias, and L. Wang, "Integrated database approach in multiobjective network reconfiguration for distribution system using discrete optimisation techniques," *IET Generation, Transmission & Distribution*, vol. 12, no. 4, pp. 976–986, 2018.

[61] F. Ding and K.A. Loparo, "Feeder reconfiguration for unbalanced distribution systems with distributed generation: A hierarchical decentralized approach," *IEEE Transactions on Power Systems*, vol. 31, no. 2, pp. 1633–1642, March 2016.
[62] M. Mahdavi, C. Sabillón, M. Ajalli, H. Monsef, and R. Romero, "A real test system for power system planning, operation, and reliability," *Journal of Control, Automation and Electrical Systems*, vol. 29, no. 2, pp. 192–208, 2018.
[63] K. Masteri and B. Venkatesh, "A new optimization method for distribution system reconfiguration", *IEEE 27th Canadian Conference on Electrical and Computer Engineering*, Canada, May 2014, pp. 1–4.
[64] J.S.R.K. AL-Sujada and O.Y.M. Al-Rawi, "Novel load flow algorithm for multi-phase balanced/unbalanced radial distribution systems," *International Journal of Advanced Research in Electrical, Electronics and Instrumentation Engineering*, vol. 3, no. 7, pp. 10928–10942, July 2014.
[65] J.P. Chiou, C.F. Chang, and C.T. Su, "Variable scaling hybrid differential evolution for solving network reconfiguration of distribution systems," *IEEE Transactions on Power Systems*, vol. 20, no. 2, pp. 668–674, May 2005.
[66] A.M. Eldurssi and R.M. O'Connell, "A fast nondominated sorting guided genetic algorithm for multi-objective power distribution system reconfiguration problem," *IEEE Transactions on Power Systems*, vol. 30, no. 2, pp. 593–601, March 2015.
[67] B. Novoselnik, M. Bolfek, M. Bošković, and M. Baotić, "Electrical power distribution system reconfiguration: Case study of a real-life grid in Croatia," *IFAC PapersOnLine*, vol. 50, no. 1, pp. 61–66, 2017.
[68] N. Kagan, "Configuração de redes de distribuição através de algoritmos genéticos e tomada de decisão Fuzzy," PhD Thesis, University of São Paulo, São Paulo, Brazil, pp. 1–187, 1999.
[69] M.E. Baran and F.F. Wu, "Network reconfiguration in distribution systems for loss reduction and load balancing," *IEEE Transactions on Power Delivery*, vol. 4, no. 2, pp. 1401–1407, April 1989.
[70] S. Goswami and S. Basu, "A new algorithm for the reconfiguration of distribution feeders for loss minimization," *IEEE Transactions on Power Delivery*, vol. 7, no. 3, pp. 1484–1491, July 1992.
[71] R. Rajaram, K. Sathish Kumar, and N. Rajasekar, "Power system reconfiguration in a radial distribution network for reducing losses and to improve voltage profile using modified plant growth simulation algorithm with distributed generation (DG)," *Energy Reports*, vol. 1, pp. 116–122, November 2015.
[72] R.A. Abttan, "Electric distribution system reconfiguration for loss reduction using genetic algorithm," *Journal of Baghdad College of Economic Sciences University*, vol. 0, no. 43, pp. 457–478, 2015.
[73] A. Lotfipour and H. Afrakhte, "A discrete teaching–learning-based optimization algorithm to solve distribution system reconfiguration in presence of distributed generation," *International Journal of Electrical Power and Energy Systems*, vol. 82, pp. 264–273, November 2016.
[74] N.C. Sahoo and K. Prasad, "A fuzzy genetic approach for network reconfiguration to enhance voltage stability in radial distribution systems," *Energy Conversion and Management*, vol. 47, no. 18–19, pp. 3288–3306, November 2006.

[75] R. Parasher, "Load flow analysis of radial distribution network using linear data structure", MSc Dissertation, Department of Computer Science and Engineering, Yagyavalkya Institute of Technology, Jaipur, pp. 1–38, October 2013.
[76] D. Das, "A fuzzy multiobjective approach for network reconfiguration of distribution systems," *IEEE Transactions on Power Delivery*, vol. 21, no. 1, pp. 202–209, January 2006.
[77] C.T. Su and C.S. Lee, "Network reconfiguration of distribution systems using improved mixed-integer hybrid differential evolution," *IEEE Transactions on Power Delivery*, vol. 18, no. 3, pp. 1022–1027, July 2003.
[78] R.A. Jabr, R. Singh, and B.C. Pal, "Minimum loss network reconfiguration using mixed-integer convex programming," *IEEE Transactions on Power Systems*, vol. 27, no. 2, pp. 1106–1115, May 2012.
[79] D. Zhang, Z. Fu, and L. Zhang, "An improved TS algorithm for loss-minimum reconfiguration in large-scale distribution systems," *Electric Power Systems Research*, vol. 77, no. 5–6, pp. 685–694, April 2007.
[80] J.R.S. Mantovani, F. Casari, and R.A. Romero, "Reconfiguração de sistemas de distribuição radiais utilizando o critério de queda de tensão," *SBA Controle & Automação*, vol. 11, no. 3, pp. 150–159, 2000.
[81] J.A. Taylor and F.S. Hover, "Convex models of distribution system reconfiguration," *IEEE Transactions on Power Systems*, vol. 27, no. 3, pp. 1407–1413, August 2012.
[82] I.J. Ramírez-Rosado and J.L. Bernal-Augustín, "Genetic algorithms applied to the design of large power distribution systems," *IEEE Transactions on Power Systems*, vol. 13, no. 2, pp. 696–703, May 1998.

18

Distribution System Reconfiguration: Case Studies

Meisam Mahdavi[1], *Pierluigi Siano*[2], *Hassan Haes Alhelou*[3], *and Josep M. Guerrero*[4]

[1]*Bioenergy Research Institute, São Paulo State University, Ilha Solteira, SP, Brazil*
[2]*Department of Management & Innovation Systems, University of Salerno, Salerno, Italy*
[3]*Department of Electrical Power Engineering, Tishreen University, Lattakia, Syria*
[4]*Department of Energy Technology, Faculty of Engineering and Science, Aalborg University, Aalborg, Denmark*

18.1 Introduction

The distribution network is an important part of the power system infrastructure that links a high-voltage transmission system [1–13] to end-users of electric energy [14–18]. It is constructed as a meshed structure and is usually operated in a radial topology by opening sectional switches and closing tie lines (distribution system reconfiguration) [19–22].

Distribution losses are significant compared to generation and transmission losses [23–37] because of a higher ratio of current to voltage in distribution systems that can affect the operational costs and the voltage profile, especially in heavily loaded power systems [14, 38]. Therefore, active power loss minimization is important to enhance the distribution system efficiency. Three common methods to reduce the power loss of distribution systems are capacitor placement, distributed generation (DG) allocation [39–44], and distribution system reconfiguration (DSR) [45, 46]. In DSR, the configuration of the system changes with the status of the switches while the network has to maintain its radiality [38, 47].

DSR can be classified into static and dynamic reconfigurations. In a static reconfiguration, the distribution system topology is considered to be fixed for a yearly, seasonal, monthly, or shorter period. In simple terms, load is considered to be constant during these specific timeframes and switching operations are not executed frequently, while in a dynamic reconfiguration, the topology permanently changes with the real-time operational conditions and requirements using automatic (smart) switches [19, 48, 49].

Although dynamic DSR causes the power losses to reduce more than in a conventional reconfiguration (static one) because of considering load fluctuations and generation variability, it is more complex than the static problem [50, 51]. Also, frequent switching may be costly and may lead to outages and transient disturbances that could reduce system reliability [52–59] and create undesirable overvoltages [60] in a dynamic reconfiguration. In addition, it is not economically feasible to replace all manual switches by smart switches for dynamic applications [19, 61]. In order to simplify the dynamic reconfiguration problem, most studies determine the operation time first and then solve the problem for each

Active Electrical Distribution Network: A Smart Approach, First Edition. Edited by Baseem Khan, Josep M. Guerrero, Sanjeevikumar Padmanaban, Hassan Haes Alhelou, Om Prakash Mahela, and Sudeep Tanwar.

subperiod (multiperiod reconfiguration). In this manner, network topologies remain unchanged for a given time period [19].

Although in a radial topology each consumer has a single source of supply, distribution networks are usually the main cause of outages in the power system (80% of customer interruptions happen in the distribution system) [62]. According to the North American Electric Reliability Corporation [63], almost one-third of these power outages are related to technical causes like equipment failures, operational errors, supply shortages, and system overloading [64–70]. Considering that there may be a 37% increase in the global energy demand by 2040 and an annual growth rate [71–73] of 0.5% and 0.8% for electricity consumption in residential and commercial sectors, respectively (from 2013 to 2040), overloading of the distribution feeders plays a more important role nowadays than a few years ago in power deficiency [62].

In this chapter, first a mathematical modeling method of the conventional DSR problem, aiming at a loss minimization, is described. Then, results and solutions obtained by other researchers and literature in the field of network reconfiguration are presented. Also, all of the results and methods are compared with each other from the loss reduction amount, computational time, and voltage improvement point of view in order to show advantages and disadvantages of the proposed methods for distribution system reconfiguration.

18.2 Mathematical Modeling of the DSR Problem

Figure 18.1 illustrates a radial distribution system with two lines (branches) and three buses (nodes).

According to this figure, the following set of equations can be used to calculate the power flow [74–77]:

$$P_k^S = P_{ki} + R_{ki}\left|I_{ki}\right|^2 \tag{18.1}$$

$$Q_k^S = Q_{ki} + X_{ki}\left|I_{ki}\right|^2 \tag{18.2}$$

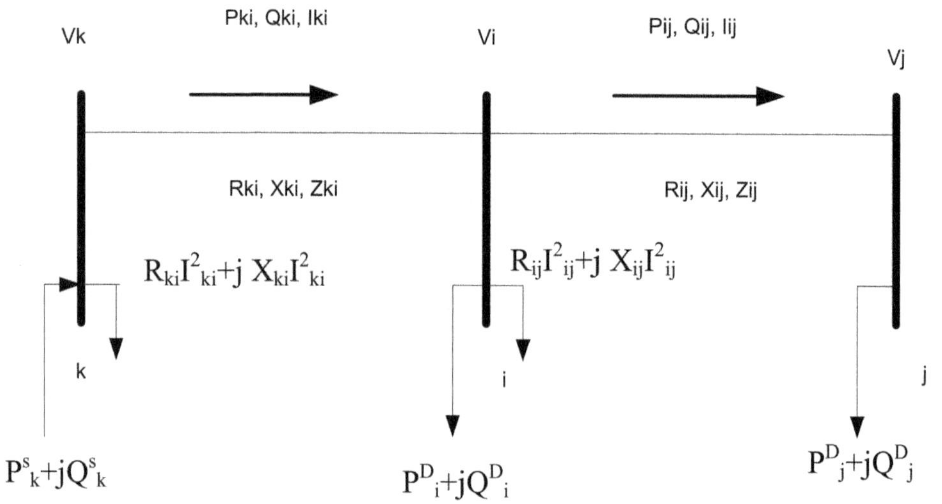

Figure 18.1 Single-line diagram of a three-node system.

$$P_{ki} = P_{ij} + P_i^D + R_{ij}\left|I_{ij}\right|^2 \tag{18.3}$$

$$Q_{ki} = Q_{ij} + Q_i^D + X_{ij}\left|I_{ij}\right|^2 \tag{18.4}$$

$$\left|V_k\right|^2 = \left|V_i\right|^2 - 2\left(R_{ki}P_{ki} + X_{ki}Q_{ki}\right) + \left|Z_{ij}\right|^2\left|I_{ki}\right|^2 \tag{18.5}$$

$$\left|V_i\right|^2 = \left|V_j\right|^2 - 2\left(R_{ij}P_{ij} + X_{ij}Q_{ij}\right) + \left|Z_{ij}\right|^2\left|I_{ij}\right|^2 \tag{18.6}$$

$$P_{ij} = P_i^D \tag{18.7}$$

$$Q_{ij} = Q_j^D \tag{18.8}$$

where

$$\left|I_{ij}\right| = \frac{\left|S_{ij}\right|}{\left|V_j\right|} \tag{18.9}$$

$$\left|S_{ij}\right| = \sqrt{P_{ij}^2 + Q_{ij}^2} \tag{18.10}$$

$$\left|Z_{ij}\right| = \sqrt{R_{ij}^2 + X_{ij}^2} \tag{18.11}$$

P_k^S, Q_k^S: Active and reactive power provided by substation at bus *k*.
P_{ij}, Q_{ij}: Active and reactive power flow from bus *i* to *j* (branch *i–j*).
$\left|S_{ij}\right|$, $\left|Z_{ij}\right|$: Amplitudes of apparent (complex) power and impedance of branch *i–j*.
P_i^D, Q_i^D: Active and reactive power demand at bus *i*.
R_{ij}, X_{ij}: Resistance and reactance of branch *i–j*.
$\left|I_{ij}\right|$: Amplitude of current flow from bus *i* to *j*.
$\left|V_j\right|$: Voltage magnitude of bus *j*.

The DSR problem with the objective of loss (P_{Loss}) reduction considering equality and inequality constraints can be formulated as follows. The goal is to reconfigure the distribution network in order to obtain a minimal power loss considering operational and technical restrictions.

18.2.1 Objective Function

The objective function for loss minimization can be represented by [78–81]

$$\min P_{Loss} = \sum_{ij \in \Omega^l} R_{ij}\left|I_{ij}\right|^2 \tag{18.12}$$

where Ω^l is a set of all branches (including normal and tie lines).

18.2.2 Constraints

The following limits need to be considered during optimization of Eq. (18.12) without any violation.

- *Nodal active power balance*

For any electrical network, Kirchhoff's first law needs to be established. For real power, this condition can be described by general representation of

$$P_i^S + \sum_{k=1}^{NB} P_{ki} = \sum_{j=1}^{NB} P_{ij} + \sum_{j=1}^{NB} R_{ij} \left| I_{ij} \right|^2 + P_i^D, \quad \forall ij, ki \in \Omega^l, i = 1,2,\ldots,NB \tag{18.13}$$

where *NB* is the total number of buses.

- *Nodal reactive power balance*

Kirchhoff's first law for reactive power can be written by general representation of

$$Q_i^S + \sum_{k=1}^{NB} Q_{ki} = \sum_{j=1}^{NB} Q_{ij} + \sum_{j=1}^{NB} X_{ij} \left| I_{ij} \right|^2 + Q_i^D, \quad \forall ij, ki \in \Omega^l, i = 1,2,\ldots,NB \tag{18.14}$$

- *Voltage magnitude limit*

A change in voltage magnitude of buses is not desirable in distribution systems because of stability and technical concerns. Therefore, the voltage magnitude at each bus must be maintained within its limits (acceptable operating bounds) during the optimization process in the following manner [82–90]:

$$V_{min} \le \left| V_i \right| \le V_{max}, \quad \forall i = 1,2,\ldots,NB \tag{18.15}$$

where V_{min} and V_{max} are the lower and upper bounds of bus voltage limits, respectively.

- *Voltage drop constraint*

Generally, the voltage drop of each branch based on Kirchhoff's second law (Kirchhoff's voltage law) can be exhibited by Eq. (18.6). Kirchhoff's voltage law (KVL) indicates that the net sum of voltage drops of all branches (links) in a planar loop has to be equal to zero. Therefore, this equation does not include open branches while in a reconfiguration problem some of the branches need to be opened and others (tie switches) must be closed. To resolve this problem, Eq. (18.6) needs to be revised as follows [91]:

$$\left| V_i \right|^2 = \left| V_j \right|^2 - 2\left(R_{ij} P_{ij} + X_{ij} Q_{ij} \right) + \left| Z_{ij} \right|^2 \left| I_{ij} \right|^2 + b_{ij}, \quad \forall ij \in \Omega^l, i = 1,\ldots,NB \tag{18.16}$$

In Eq. (18.16), b_{ij} is a slack variable that needs to be zero when branch i–j is closed (i.e. when Eq. (18.6) is satisfied) and is a real number when the corresponding branch is open. Let us consider what does happen when a switch becomes open. In this situation, $\left|I_{ij}\right| = P_{ij} = Q_{ij} = 0$ and therefore $b_{ij} = \left|V_i\right|^2 - \left|V_j\right|^2$. According to Eq. (18.15), the maximum value of this continuous variable (b_{ij}) will be $V_{max}^2 - V_{min}^2$ and its minimum value is $V_{min}^2 - V_{max}^2$. Therefore, the following constraint can be considered for the slack variable to provide $b_{ij} = 0$ when the branch is being operated and $b_{ij} \neq 0$ when the branch is out of service.

- *Slack variable limit*

As mentioned earlier, the slack variable constraint needs to be defined as follows [92]:

$$\left|b_{ij}\right| \leq \left(V_{max}^2 - V_{min}^2\right)\left(1 - y_{ij}\right), \quad \forall ij \in \Omega^l \tag{18.17}$$

In this equation, $\left|b_{ij}\right|$ is an absolute value of b_{ij}. Also, y_{ij} is a binary variable that represents the status of each switch (0 and 1 for open and closed switches, respectively). It can be observed that $b_{ij} = 0$ for closed branches ($y_{ij} = 1$) and for open ones it will be $V_{min}^2 - V_{max}^2 \leq b_{ij} \leq V_{max}^2 - V_{min}^2$.

- *Permissible line current capacity*

The current magnitude in each branch must satisfy the branch's capacity, because of thermal and security problems. This limitation can be presented as follows [61, 79, 81, 89, 93]:

$$\left|I_{ij}\right| \leq I_{ij}^{max}, \quad \forall ij \in \Omega^l \tag{18.18}$$

where I_{ij}^{max} is the current rating limit of branch i–j. This rating value is determined according to the thermal rating limit and permissible operating capacity. The thermal limit that is defined by the line (branch) manufacturer shows the maximum current capacity of a branch without any line damage. The permissible operating capacity is a margin that a line can carry current without any security problems (part of a line capacity should be allocated for maintaining stability and reliability of the system in contingency states, e.g. line outages).

- *Radiality constraint*

This constraint indicates that the weekly meshed distribution network has to be operated in a radial structure because of advantages mentioned in the Introduction. Distribution systems may be operated as meshed structures during the switching actions because of operation of some tie lines. This is an important constraint, because a non-radial structure found after reconfiguration is not practicable. In addition, a large part of the computational time and effort is related to finding feasible radial topologies. In addition to radial solutions, the connectivity of the network has to be maintained during the switching operation (i.e. none of the buses must be separated from the network except for transfer nodes). An appropriate definition of this condition can help us to find more accurate solutions in a shorter time. Most of the specialized literature was used in the following equation to find feasible solutions in distribution networks with one substation:

$$\sum_{ij\in\Omega_l} y_{ij} = NB - 1 \tag{18.19}$$

This means that the sum of branches under operation (closed switches) has to be one unit less than total nodes. This condition is based on spanning trees in graph theory and its implementation in the reconfiguration problem is easy. However, there are several substation nodes in real distribution systems. Therefore, Eq. (18.19) cannot be employed to find the radial configurations. Equation (18.20) is an extension of this condition to include multisubstation distribution networks [94]:

$$\sum_{ij\in\Omega_l} y_{ij} = NB - NB_s \tag{18.20}$$

where NB_s is the total number of substation buses. This constraint can be applied for reconfiguration of both single- and multisubstation distribution networks. Nevertheless, Eq. (18.20) cannot guarantee network connectivity in distribution systems with transfer nodes. Power balance, given in Eqs (18.13) and (18.14), can meet connectivity conditions if there are no transfer buses in the network. Thus, a way to resolve this shortcoming is to consider small demand values for transferring nodes. However, these small values may reduce the accuracy of solutions. Another way to solve this problem is to define the following constraint in addition to Eq. (18.20):

$$\sum_{k=1}^{NB} y_{ki} + \sum_{j=1}^{NB} y_{ij} \geq 1, \quad \forall\, ij, ki \in \Omega^l \tag{18.21}$$

This inequality constraint represents the condition that at least one branch has to be connected to every node. In this way, no isolated transfer buses appeared in the proposed configurations. Nevertheless, considering Eq. (18.21) in the problem constraints leads to an increase in the computing time. Although these two mentioned schemes can help the constraint in Eq. (18.20) to find feasible radial topologies in distribution networks with several substations and transfer nodes, these modifications cannot guarantee a radial configuration when size or complexity of the network increases. Therefore, the following minimal spanning tree problem can be used instead of Eq. (18.20) [95]:

$$\min \sum_{ij\in\Omega_l} w_{ij}\beta_{ij} \tag{18.22}$$

subject to

$$\sum_{j\in N_i} \beta_{ij} = 1, \quad \forall\, i = 1,2,\ldots, NB-1 \tag{18.23}$$

$$\sum_{l\in N_m} x_{ml,e} = 1 \quad \forall m = 1,2,\ldots,f-1 \tag{18.24}$$

$$\beta_{ij} + \beta_{ji} + x_{ml,e} + x_{lm,e} = 1, \quad \forall ij \in \Omega^l \tag{18.25}$$

where

w_{ij}: Weight of the branch i–j.

N_i: Set of buses directly connected to the bus i in the meshed network.

N_m: Set of nodes directly connected to node m in the dual graph of the meshed network.

f: Total number of nodes in the dual graph.

β_{ij}: Orientation variable of the branch i–j.

$x_{ml,e}$: Status of the branch connecting node m to l in the dual graph. Index e is used to distinguish between multiple branches connecting the same two nodes.

Although solving Eqs (18.22) to (18.25) simultaneously can provide the radial configuration, solution of the two optimization problems (DSR and minimum spanning tree) increases the computational burden. Moreover, solving the minimal spanning tree problem for large distribution systems is not easy because determination of a dual graph and finding its branches that connect different nodes is boring and time consuming. In simple words, we need to draw the dual graph correctly and find the relationship between its nodes and the main graph nodes for each case study. Also, this method can only be applied to meshed distribution systems with a planar graph. In addition, determination of appropriate values for w_{ij} is important and needs additional calculations. Therefore, the set of Eqs (18.26) to (18.31) can be defined to satisfy the radiality constraint instead of Eqs (18.22) to (18.25) without checking the network loops and solving an additional optimization problem [96]:

$$\beta_{ij} + \beta_{ji} = y_{ij}, \quad \forall ij \in \Omega^l \tag{18.26}$$

$$y \ = y_{ji}, \forall ij \in \Omega^l \tag{18.27}$$

$$\sum_{j\in N_i} \beta_{ij} = 1, \quad \forall i = 2,3,\ldots,NB \tag{18.28}$$

$$\beta_{1j} = 0, \quad \forall j \in N_1 \tag{18.29}$$

$$0 \le y_{ij} \le 1, \quad \forall ij \subset \Omega^l \tag{18.30}$$

$$\beta_{ij} \in \{0,1\}, \forall i = 2,3,\ldots,NB, j \in N_i \tag{18.31}$$

Constraints in Eqs (18.26) to (18.31) can guarantee the radial structure of the distribution network with or without transfer nodes. However, these constraints were defined for distribution networks with one substation bus. Also, Eq. (18.27) is not necessary and just increases the computation time. Therefore, the set of Eq. (18.26) and Eqs (18.30) to (18.33) can be considered in order to meet the radiality condition for multisubstation distribution networks [97]:

$$\beta_{ij} + \beta_{ji} = y_{ij}, \quad \forall ij \in \Omega^l \tag{18.26}$$

$$\sum_{j \in N_i} \beta_{ij} \leq 1, \quad \forall i \in \Omega^b \tag{18.32}$$

$$\sum_{j \in N_i} \beta_{ij} = 0, \forall i \in \Omega^s \tag{18.33}$$

$$0 \leq y_{ij} \leq 1, \quad \forall ij \in \Omega^l \tag{18.30}$$

$$\beta_{ij} \in \{0,1\}, \forall i \in \Omega^b, j \in N_i \tag{18.31}$$

where Ω^b is a set of buses without a substation and Ω^s is a set of buses including a substation The following equations define the above-mentioned radiality constraints in easier terms [20]:

$$\beta_{ij} + \beta_{ji} = y_{ij}, \quad \forall ij \in \Omega^l \tag{18.26}$$

$$\sum_{j \in N_i} \beta_{ij} = 1, \quad \forall i \in \Omega^b \tag{18.34}$$

$$\beta_{ij} = 0, \quad \forall i \in \Omega^s, j \in N_i \tag{18.35}$$

$$0 \leq y_{ij} \leq 1, \quad \forall ij \in \Omega^l \tag{18.30}$$

$$\beta_{ij} \in \{0,1\}, \forall i \in \Omega^b, j \in N_i \tag{18.31}$$

The constraint in Eq. (18.30) is not mandatory because y_{ij} is a binary variable and this form of definition increases the computing time of the optimization algorithm. In addition, determination of sets N_i for each distribution network takes time and needs more calculations. For these reasons the following equations can be presented [98]:

$$\beta_{ij} + \beta_{ji} = y_{ij}, \quad \forall ij \in \Omega^l \tag{18.26}$$

$$\sum_{j=1}^{NB} \beta_{ji} = 1 \quad \forall ij \in \Omega^l, i \in \Omega^b \tag{18.36}$$

$$\beta_{ij} = 0, \quad \forall j \in \Omega^s, ij \in \Omega^l \tag{18.37}$$

$$\beta_{ij} \geq 0, \quad \forall ij \in \Omega^l \tag{18.38}$$

These equations can meet the radiality constraint for distribution networks with different substations and transfer nodes. In order to increase the convergence speed of the solution method and easier implementation, these constraints can be written as Eqs (18.26), (18.39)–(18.42):

$$\beta_{ij} + \beta_{ji} = y_{ij}, \quad \forall ij \in \Omega^l \tag{18.26}$$

$$\sum_{j=1}^{NB} \beta_{ij} = 1, \quad \forall ij \,||\, ji \in \Omega^l, i \in \Omega^b \tag{18.39}$$

$$\beta_{ij} = 0, \quad \forall i \in \Omega^s, ij \in \Omega^l \tag{18.40}$$

$$\beta_{ji} = 0, \forall j \in \Omega^s, ij \in \Omega^l \tag{18.41}$$

$$\beta_{ij} \in \{0,1\}, \quad \forall ij \in \Omega^l \tag{18.42}$$

- *Operational constraint*

According to Eq. (18.10), there is an equality relationship between apparent power of each branch and its real and imaginary components [99]:

$$|S_{ij}|^2 = P_{ij}^2 + Q_{ij}^2, \quad \forall ij \in \Omega^l \tag{18.43}$$

Regarding Eq. (18.9), we have

$$|S_{ij}|^2 = |V_j|^2 |I_{ij}|^2, \quad \forall ij \in \Omega^l, j = 1,2,\ldots,NB \tag{18.44}$$

- *Active power flow limit*

This constraint declared that the real power flow of each branch should not violate its operational limits [100]:

$$|P_{ij}| \leq P_{ij}^{max} y_{ij}, \quad \forall ij \in \Omega^l \tag{18.45}$$

where

$$P_{ij}^{max} = V_{max} I_{ij}^{max} \tag{18.46}$$

In Eq. (18.45), P_{ij}^{max} is the real power flow capacity of branch *i*–*j* and $|P_{ij}|$ is the absolute value of P_{ij}. This constraint is not necessary because Eqs (18.15) and (18.18) can meet this condition. Nevertheless, the constraint in Eq. (18.45) beside Eqs (18.15) and (18.18) decreases the computing time of the solution method significantly for large distribution systems.

- *Reactive power flow limit*

Similar to the active power limit, the reactive one can be exhibited as follows [98]:

$$|Q_{ij}| \leq Q_{ij}^{max} y_{ij}, \quad \forall ij \in \Omega^l \tag{18.47}$$

where

$$Q_{ij}^{max} = V_{max} I_{ij}^{max} \tag{18.48}$$

In the above equations, Q_{ij}^{max} is the reactive power flow capacity of branch *i*–*j* and $|Q_{ij}|$ is the absolute value of Q_{ij}. This constraint is not essential for the same reasons as those mentioned for the active power flow limit, but embedding it in the problem constraints reduces the CPU time considerably for large-scale distribution networks.

18.3 Case Studies

In this section, DSR models and formulations proposed by the specialized literature are applied to the test systems described in Chapter 17 and the results are presented. It should be mentioned that the voltage magnitude of the substation nodes has been fixed on 1 per unit.

18.3.1 Case 1

The results of implementation of the DSR problem on System 1 (see Section 2 in Chapter 17 for more information about System 1) are listed in Table 18.1. It should be noted that the best voltage magnitudes before and after reconfiguration is 1 p.u. at substation bus 1. Buses 5 and 7 have the worst voltage magnitudes in the original and reconfigured networks, respectively.

Table 18.1 Numerical results for Case 1

Items		Power loss (kW)		Worst voltage (p.u.)		
Models	Open switches	Before	After	Before	After	Run time (s[1])
[102]	4	1.44	1.2	0.995	0.99	–

[1]Seconds.

18.3.2 Case 2

The DSR problem was applied to System 2 (refer to Section 2 of Chapter 17 for more information) and results are provided in Table 18.2. It should be noted that the best voltage magnitudes before and after reconfiguration is 1 p.u. at bus 1. Bus 10 has the worst voltage magnitude before and after network reconfiguration.

18.3.3 Case 3

The proposed approaches in the literature were applied to System 3 (see Section 2 in Chapter 17) and results are provided in Table 18.3. In this table, methods were arranged according to their publication dates.

Furthermore, the solution methods presented for solving the DSR problem were ranked based on the power loss reduction, worst voltage increase, and computing time in Tables 18.4, 18.5, and 18.6, respectively.

Table 18.2 Numerical results for Case 2

Items		Power loss (kW)		Worst voltage (p.u.)		
Models	Open switches	Before	After	Before	After	Run time (s)
[103]	12, 13	3.55	3.55	0.9958	0.996	0.4

Table 18.3 Numerical results of proposed models for Case 3

Items		Power loss (kW)		Worst voltage (p.u.)		
Models	Open switches	Before	After	Before	After	Run time (s)
[104]	15, 17, 26	514	488.46	0.969	–	–
[105]	15, 17, 26	514	485.9	0.968	–	–
[106]	17, 19, 26	511.4	466.1	0.9693	0.9716	6
[107]	15, 17, 26	514	483.88	0.969	–	–
[108]	17, 19, 26	511.1	466	0.9693	0.9716	7.5
[109]	17, 19, 26	511.4	466.1	0.9693	–	–
[74]	17, 19, 26	511.5	466.1	0.9693	0.9716	4.5
[110]	17, 19, 26	511.43	466.1	0.9693	0.9694	–
[111]	17, 19, 26	511.43	466.1	0.9693	0.9694	–
[112]	17, 19, 26	511.4	466.13	–	–	–
[113]	17, 19, 26	511.4	466.1	0.9693	0.9716	> [56]

(continued)

Table 18.3 (*continued*)

Items		Power loss (kW)		Worst voltage (p.u.)		
Models	Open switches	Before	After	Before	After	Run time (s)
SA[2] [75, 76]	17, 19, 26	511.4	466.1	0.9693	0.9716	2.07
GA[3] [75, 76]	17, 19, 26	511.4	466.1	0.9693	0.9716	2.32
ACO[4] [75, 76]	17, 19, 26	511.4	466.1	0.9693	0.9716	1.81
[114]	17, 19, 26	511.44	466.12	0.9693	0.97158	2.0269
[115]	17, 19, 26	511.4	466.13	–	–	<[79]
[116]	17, 19, 26	511.4	466.13	–	–	<[81]
[117]	17, 19, 26	511.5	466.1	0.9693	0.9716	1.35
[118]	17, 19, 26	511.44	466.12	0.9693	0.97158	0.156
[119]	17, 19, 26	511.5	466.1	0.9693	0.9716	1.7
[120]	17, 19, 26	511.43	466.1	0.9693	0.9694	–
[121]	17, 19, 26	511	465.8	–	–	0.212
[122]	17, 19, 26	511.44	466.13	0.969	0.972	0.45
[123]	17, 19, 26	511.4	466.14	–	–	–
[95]	17, 19, 26	514	468.3	–	–	0.16
[124]	17, 19, 26	511.44	466.13	0.969	0.972	0.27
MIQCP[5] [125]	17, 19, 26	514	468.3	0.969	–	–
MILP[6] [125]	17, 19, 26	514	468.3	0.969	–	–

[2]Simulated annealing.
[3]Genetic algorithm.
[4]Ant colony optimization.
[5]Mixed-integer quadratic conic programming.
[6]Mixed-integer linear programming.

Table 18.4 Ranking of models based on loss reduction for Case 3

Models	Power loss (kW)
[104]	488.46
[105]	485.9
[107]	483.88
[95, 125]	468.3
[123]	466.14
[112, 115, 116, 122, 124]	466.13
[114, 118]	466.12
[74–76, 106, 109–112, 114, 117, 119, 120]	466.1
[108]	466
[121]	465.8

Table 18.5 Ranking of models regarding voltage improvement for Case 3

Models	Worst voltage (p.u.)
[110, 111, 120]	0.9694
[114, 118]	0.97158
[74–76, 106, 108, 109, 113, 117, 119]	0.9716
[122, 124]	0.972

Table 18.6 Ranking of models based on execution time for Case 3

Models	Run time (s)
[104, 105, 107, 112, 113]	Long
[115]	<[51]
[116]	>[54]
[108]	7.5
[106]	6
[74]	4.5
GA [75, 76]	2.32
SA [75, 76]	2.07
[119]	2.027
ACO [75, 76]	1.81
[114]	1.7
[117]	1.35
[122]	0.45
[124]	0.27
[121]	0.212
[95]	0.16
[118]	0.156

18.3.4 Case 4

The DSR problem was implemented on System 4 (refer to Chapter 17) and numerical results are illustrated in Table 18.7.

18.3.5 Case 5

The proposed technique of [101] was applied to System 5 (Chapter 17) and results are provided in Table 18.8. The best voltage magnitudes before and after reconfiguration are 1 p.u. at substation bus 1, while the worst voltage before and after reconfiguration belongs to buses 29 and 14, respectively.

18.3.6 Case 6

The proposed formulations were applied to System 6 and the results are listed in Table 18.9 chronologically. Also, the voltage magnitude of the substation bus (node 0) is adjusted by 1 per unit.

The solution methods are arranged in Tables 18.10–18.12 from the viewpoints of convergence speed, voltage stability, and accuracy, respectively.

Table 18.7 Numerical results for Case 4

Items		Power loss (kW)		Worst voltage (p.u.)		
Models	Open switches	Before	After	Before	After	Run time (s)
[126]	11, 20, 21, 28	47	40	0.9	0.92	–

Table 18.8 Numerical results for Case 5

Items		Power loss (kW)		Worst voltage (p.u.)		
Models	Open switches	Before	After	Before	After	Run time (s)
[101]	3, 7, 10, 25, 32, 33	1240	368	0.8251	0.92	–

Table 18.9 Numerical results of proposed models for Case 6

Items		Power loss (kW)		Worst voltage (p.u.)		
Models	Open switches	Before	After	Before	After	Run time (s)
[127]	11, 28, 31, 33, 34	205.81	148.858	0.88	0.92	Long
[128]	7, 10, 14, 32, 37	205.81	141.541	0.913	0.94	0.14
[105]	7, 9, 14, 32, 37	202.67	139.532	0.9131	–	<48
[129]	7, 9, 14, 32, 37	205.81	140.815	0.91253	0.9472	0.87
[130]	7, 9, 14, 32, 37	202.71	141.6	0.9131	0.929	19.1
[106]	7, 9, 14, 32, 37	203.2	139.7	0.913	0.9378	48
[131]	7, 9, 14, 32, 37	202.67	139.55	0.913	0.94749	1.99
[132]	6, 9, 14, 32, 37	204.14	142.83	0.91	0.9378	–
[133]	7, 9, 14, 32, 37	202.67	139.55	0.913	0.94749	647.03
[109]	7, 9, 14, 32, 37	202.7	139.55	0.9131	0.9378	7.41
[134]	7, 9, 14, 32, 37	211.22	139.83	0.9038	0.9378	55.04
[110]	7, 9, 14, 32, 37	203	139.8	0.9129	0.9378	6.852

Table 18.9 (*continued*)

Items		Power loss (kW)		Worst voltage (p.u.)		
Models	**Open switches**	**Before**	**After**	**Before**	**After**	**Run time (s)**
[135]	9, 28, 33, 34, 36	205.81	140.6	0.842	0.95	–
[112]	7, 9, 14, 32, 37	202.67	139.55	0.913	0.94749	1.66
[136]	7, 9, 14, 32, 37	202.68	139.5	–	–	6.3
[137]	7, 9, 14, 36, 37	202.71	139.28	0.9131	0.921	8.1
[116]	7, 9, 14, 32, 37	202.68	139.55	–	–	–
[77]	7, 9, 14, 32, 37	202.68	139.55	–	–	<1.66
[138]	7, 9, 14, 32, 37	202.68	139.55	0.9131	0.9378	0.41
GA [117]	7, 10, 14, 36, 37	202.68	142.68	–	–	160
GAMT[7] [117]	7, 9, 14, 32, 37	202.68	139.55	0.9131	0.9378	7.2
[139]	7, 9, 14, 32, 37	202.7	139.5	–	–	0.5
[140]	7, 9, 14, 32, 37	202.68	139.55	0.9131	0.9378	6
Discrete PSO[8] [141]	7, 9, 14, 32, 37	202.67	139.53	0.913	0.9378	8
[142]	7, 9, 14, 32, 37	202.67	139.53	0.913	0.9378	8
[143, 144]	7, 9, 14, 32, 37	202.68	139.55	0.913	0.94749	0.11
[145]	7, 9, 14, 32, 37	202.6	139.5	0.9131	0.9378	5.693
[120]	7, 9, 14, 32, 37	202.6	139.5	0.9131	0.9378	< 5.693
[100]	7, 9, 14, 32, 37	202.68	139.55	0.9131	0.9378	8
[80]	7, 10, 14, 32, 37	202.71	140.26	–	–	–
[146]	7, 10, 14, 36, 37	202.71	142.68	0.9131	0.9342	7.2
EICPSO[9] [141]	7, 9, 14, 28, 32	203.33	139.98	0.9131	0.9378	6.343
[147]	7, 9, 14, 32, 37	202.68	139.55	0.9131	0.9378	0.3
[94]	7, 9, 14, 32, 37	202.68	139.55	0.9131	0.9378	19
[148]	7, 9, 14, 32, 37	203	139.8	0.9129	0.9378	–
[149]	7, 9, 14, 32, 37	202.68	139.55	0.9131	0.9378	–
QP[10] [98]	7, 9, 14, 32, 37	202.7	139.5	0.913	0.9412	0.21
QCP[11] [98]	7, 9, 14, 32, 37	202.7	139.5	0.913	0.9412	1.43
SOCP[12] [98]	7, 9, 14, 32, 37	202.7	139.5	0.913	0.9412	12.8
[121]	7, 9, 14, 32, 37	202.68	139.55	–	–	5.704
[123]	7, 9, 14, 32, 37	202.68	139.55	–	–	5.28
[150]	7, 9, 14, 32, 37	202.67	139.53	0.9131	0.9412	–
MIQCP [125]	7, 9, 14, 32, 37	202.68	139.5	0.9131	0.9378	35.5
MILP [125]	7, 9, 14, 32, 37	202.68	139.5	0.9131	0.9378	0.15

(*continued*)

Table 18.9 (*continued*)

Items		Power loss (kW)		Worst voltage (p.u.)		
Models	**Open switches**	**Before**	**After**	**Before**	**After**	**Run time (s)**
[95]	7, 9, 14, 32, 37	202.7	139.6	–	–	3
[151]	7, 9, 14, 32, 37	202.68	139.5	0.9131	0.9378	3.2
[152]	7, 9, 14, 32, 37	202.7	139.5	0.9131	0.9378	–
[153]	7, 9, 14, 32, 37	202.67	139.5	0.9052	0.9343	0.16
ACO [154]	7, 9, 14, 32, 37	202.68	139.551	0.9131	0.9378	6.439
HSA[13] [154]	7, 10, 14, 36, 37	202.68	139.551	0.9131	0.9378	3.274
GA [154]	7, 10, 14, 36, 37	202.68	139.551	0.9131	0.9378	10.83
[155]	7, 9, 14, 32, 37	202.68	139.55	0.9131	0.9378	3.7

[7] GA based on Matroid theory.
[8] Particle swarm optimization.
[9] Enhanced integer coded PSO.
[10] Quadratic programming.
[11] Quadratically constrained programming.
[12] Second-order cone programming.
[13] Harmony search algorithm.

Table 18.10 Models based on computational time in Case 6

Models	Run time (s)
[133]	647.03
GA [117]	160
[134]	55.04
[106]	48
MIQCP [125]	35.5
[130]	19.1
[94]	19
SOCP [98]	12.8
GA [154]	10.83
[137]	8.1
[100, 142]	8
[109]	7.41
GAMT [117, 146]	7.2
[110]	6.852
ACO [154]	6.439
[141]	6.343

(continued)

Table 18.10 (*continued*)

Models	Run time (s)
[136]	6.3
[140]	6
[145]	5.693
[121]	5.7
[123]	5.28
[155]	3.7
HSA [154]	3.274
[151]	3.2
[95]	3
[131]	1.99
[112]	1.66
QCP [98]	1.43
[129]	0.87
[139]	0.5
[138]	0.41
[147]	0.3
QP [98]	0.21
[153]	0.16
MILP [125]	0.15
[128]	0.14
[144]	0.11

Table 18.11 Models according to voltage enhancement in Case 6

Models	Minimum voltage (p.u.)
[127, 137]	0.92
[130]	0.929
[146, 153]	0.934
[100, 106, 109, 110, 117, 120, 125, 132, 134, 138, 140–142, 145, 147–149, 151, 152, 154, 155]	0.9378
[98, 128, 150]	0.94
[129]	0.9472
[112, 131, 133, 143, 144]	0.9475
[135]	0.95

Table 18.12 Arrangement of models according to loss reduction for Case 6

Models	Power loss (kW)
[127]	148.858
[132]	142.83
GA [117]	142.68
[130]	141.6
[128]	141.54
[129]	140.815
[135]	140.6
[80]	140.26
[141]	139.98
[134]	139.83
[110, 148]	139.8
[106]	139.7
[95]	139.6
[77, 94, 100, 109, 112, 116], GAMT [117, 121, 123, 131, 133, 138, 140, 143, 144, 149, 154, 155]	139.55
[141, 142, 150]	139.53
[98, 120, 125, 136, 139, 145, 147, 151–153]	139.5
[137, 146]	139.28

18.3.7 Case 7

The results are obtained according to Table 18.13 by implementing the DSR problem on System 7. Data of this system with power demand information are available in Chapter 17.

18.3.8 Case 8

Table 18.14 shows the simulation results after applying the DSR problem to System 8. All data of System 8 have been provided in Chapter 17.

Table 18.13 Numerical results for Case 7

Items		Power loss (kW)		Worst voltage (p.u.)		
Models	Open switches	Before	After	Before	After	Run time (s)
[156]	15, 34, 38, 45, 52	10.593	8.976	0.9957	0.9970	–

Table 18.14 Numerical results for Case 8

Items		Power loss (kW)		Worst voltage (p.u.)		
Models	Open switches	Before	After	Before	After	Run time (s)
[46]	25, 31, 39, 40, 62	178.66	131.69	0.9767	0.9847	–

18.3.9 Case 9

The proposed models were applied to System 9 and the results were provided in Tables 18.15–18.18. Data for this system have been given in Chapter 17. Base values are 12.66 kV and 100 MVA.

Table 18.15 Numerical results for Case 9

Items		Power loss (kW)		Worst voltage (p.u.)		
Models	Open switches	Before	After	Before	After	Run time (s)
[106]	14, 55, 61, 69, 70	224.95	99.62	0.9092	0.9428	165
[157]	14, 58, 61, 69, 70	225.05	99.62	0.9092	0.9428	–
SA [158]	14, 57, 61, 69, 70	225	99.62	0.9092	0.9428	250
Efficient SA [158]	14, 57, 61, 69, 70	225	99.62	0.9092	0.9428	20.57
GA [159]	12, 20, 58, 64, 69	226.13	113.8	0.9092	0.9328	25
Fuzzy GA [159]	12, 20, 58, 64, 69	226.13	113.8	0.9092	0.9328	50
[160]	14, 56, 61, 69, 70	224.95	99.59	0.9092	0.9428	–
[116]	14, 56, 62, 70, 71	225	99.72	0.9092	0.9428	–
[138]	14, 55, 61, 69, 70	224.95	99.62	0.9092	0.9428	–
[161]	14, 58, 61, 69, 70	225	99.62	0.9092	0.9310	–
[145]	14, 55, 61, 69, 70	224.95	99.62	0.9092	0.9428	8
[120]	14, 55, 61, 69, 70	224.95	99.62	0.9092	0.9428	150
[162]	14, 23, 51, 60, 72	226.92	136.87	0.9092	0.9428	37
[80]	12, 57, 61, 69, 70	225	99.82	0.9092	0.9428	–
[163]	15, 56, 62, 70, 71	225	99.62	0.9092	0.9428	–
GA [164]	14, 53, 61, 69, 70	225	103.29	0.9092	0.9411	–
RGA[14] [164]	13, 17, 55, 61, 69	225	100.28	0.9092	0.9428	–
NCUP[15] [123]	14, 59, 62, 70, 71	225	99.72	0.9092	0.9428	–
Heuristic [123]	11, 14, 21, 56, 62	225	106.67	0.9092	0.9428	–
[165]	14, 58, 61, 69, 70	225.05	99.62	0.9092	0.9428	1181
[124]	14, 57, 61, 69, 70	224.93	99.62	0.9092	0.9430	20.2

(continued)

Table 18.15 (*continued*)

Items		Power loss (kW)		Worst voltage (p.u.)		
Models	Open switches	Before	After	Before	After	Run time (s)
[166]	14, 58, 61, 69, 70	225.05	99.62	0.9092	0.9428	20.69
[46]	14, 57, 61, 69, 70	224.9	99.62	0.9092	0.9428	–
[167]	14, 57, 61, 69, 70	225.04	99.62	0.9092	0.9428	7
[168]	14, 58, 61, 69, 70	225	99.62	0.9092	0.9428	12.5

[14]Refined GA.
[15]Neighbor-chain updating process.

Table 18.16 Ranking of models based on loss reduction for Case 9

Models	Power loss (kW)
[162]	136.87
[159]	113.8
Heuristic [123]	106.67
GA [164]	103.29
RGA [164]	100.28
[80]	99.82
NCUP [123] and [116]	99.72
[46, 106, 120, 124, 138, 145, 157, 158, 163, 165–168]	99.62

Table 18.17 Ranking of models regarding voltage improvement for Case 9

Models	Worst voltage (p.u.)
[161]	0.9310
[159]	0.9328
GA [164]	0.9411
[46, 80, 106, 116, 120, 123, 138, 145, 157, 158, 160, 162, 163], RGA [164–168]	0.9428
[124]	0.9430

Table 18.18 Ranking of models based on execution time for Case 9

Models	Run time (s)
[165]	1181
SA [158]	250
[106]	165
[120]	150
Fuzzy GA [159]	50
[162]	37
GA [159]	25
[166]	20.69
Efficient SA [158]	20.57
[124]	20.2
[168]	12.5
[145]	8
[167]	7

18.3.10 Case 10

Tables 18.19–18.22 show the simulation results of application of the models proposed by other literature to System 10. Data of this system are available in Chapter 17.

Table 18.19 Numerical results for Case 10

Items		Power loss (kW)		Worst voltage (p.u.)		
Models	Open switches	Before	After	Before	After	Run time (s)
[169]	14, 28, 39, 46, 51, 67, 70, 71, 73, 76, 79	227.53	205.32	0.9052	0.9268	3
GA [121]	13, 28, 45, 51, 67, 70, 73, 75, 76, 78, 79	227.346	203.67	0.9052	0.9268	4.64
Brute force [121]	13, 28, 45, 51, 67, 70, 73, 75, 76, 78, 79	227.346	203.67	0.9052	0.9268	7633.35
MIQCP [125]	14, 30, 45, 51, 66, 70, 75–79	227.53	201.4	0.9052	0.9268	485
MILP [125]	14, 30, 45, 51, 66, 70, 75–79	227.53	201.4	0.9052	0.9268	0.91
AACO[16] [147]	13, 30, 45, 51, 66, 70, 75–79	227.53	201.39	0.9052	0.9312	91.13

(*continued*)

Table 18.19 (*continued*)

Items		Power loss (kW)		Worst voltage (p.u.)		
Models	**Open switches**	**Before**	**After**	**Before**	**After**	**Run time (s)**
Proposed AACO [147]	13, 30, 45, 51, 66, 70, 75–79	227.53	201.39	0.9052	0.9312	19.72
[95]	14, 30, 45, 51, 66, 70, 75–79	227.5	201.4	0.9052	0.9268	5.7
GAMT [117]	14, 30, 38, 46, 51, 66, 70, 71, 76, 77, 79	227.52	203.17	0.9052	0.9297	160
GA [117]	13, 30, 45, 51, 66, 70, 75–79	227.52	201.4	0.9052	0.9297	1900
[170]	30, 46, 51, 66, 70, 71, 75–79	227.53	202.18	0.9052	0.9316	–
[171, 172]	14, 28, 39, 46, 67, 51, 70, 71, 73, 76, 79	227.53	205.32	0.9052	0.9268	5.4
[173]	14, 28, 39, 46, 67, 51, 70, 71, 73, 76, 79	227.53	205.32	0.9052	0.9268	8

[16]Adaptive ACO.

Table 18.20 Ranking of models based on loss reduction for Case 10

Models	Power loss (kW)
[168, 170–172]	205.32
[121]	203.67
[117]	203.17
[95, 117, 125]	201.4
[147]	201.39

Table 18.21 Ranking of models regarding voltage improvement for Case 10

Models	Worst voltage (p.u.)
[95, 121, 125, 168, 170–172]	0.9268
[117]	0.9297
[147]	0.9312
[169]	0.9316

Table 18.22 Ranking of models based on execution time for Case 10

Models	Run time (s)
Brute force [121]	7633.35
GA [117]	1900
MIQCP [125]	485
GAMT [117]	160
AACO [147]	91.13
Proposed AACO [147]	19.72
[172]	8
[95]	5.7
[170, 171]	5.4
SOReco [121]	4.64
[168]	3
MILP [125]	0.91

18.3.11 Case 11

Tables 18.23–18.26 show the results of implementation of models proposed in specialized literature on System 11. All data of this system are available in Chapter 17. The substation voltage is set at 1.0 p.u. and voltage base is 11.4 kV.

Table 18.23 Numerical results of proposed models for Case 11

Items		Power loss (kW)		Worst voltage (p.u.)		
Models	Open switches	Before	After	Before	After	Run time (s)
[128]	7, 13, 34, 39, 41, 55, 62, 72, 83, 86, 89, 90, 92	531.99	471.44	0.9285	0.9532	–
SA [74, 96]	7, 13, 34, 39, 41, 55, 62, 72, 83, 86, 89, 90, 92,	531.99	471.08	0.9285	0.9532	195.21
MIHDE[17] [74]	7, 13, 34, 39, 41, 55, 62, 72, 83, 86, 89, 90, 92	531.99	471.08	0.9285	0.9532	36.15
HDE[18] [113]	7, 13, 34, 39, 41, 55, 62, 72, 83, 86, 89, 90, 92	531.99	471.08	0.9285	0.9532	318

(continued)

Table 18.23 (*continued*)

Items		Power loss (kW)		Worst voltage (p.u.)		
Models	**Open switches**	**Before**	**After**	**Before**	**After**	**Run time (s)**
VSHDE[19] [96, 113]	7, 13, 34, 39, 41, 55, 62, 72, 83, 86, 89, 90, 92	531.99	471.08	0.9285	0.9532	585
GA [75, 76, 147]	7, 13, 34, 39, 41, 55, 62, 72, 83, 86, 89, 90, 92	531.99	471.08	0.9285	0.9532	303.66
ACO [75, 76, 147]	7, 13, 34, 39, 41, 55, 62, 72, 83, 86, 89, 90, 92	531.99	471.08	0.9285	0.9532	241.51
[112]	7, 13, 34, 39, 41, 55, 62, 72, 83, 86, 89, 90, 92	531.99	470.88	0.9285	0.9532	–
AIS[20]–ACO [147, 174]	7, 13, 34, 39, 42, 55, 62, 72, 86, 89, 90, 91, 92	531.99	472.32	0.9285	0.9479	15
[116]	7, 13, 34, 39, 42, 55, 62, 72, 83, 86, 89, 90, 92	531.99	469.878	0.9285	0.9532	3.92
[77]	7, 13, 34, 39, 41, 55, 62, 72, 83, 86, 89, 90, 92	531.99	471.08	0.9285	0.9532	113.25
[138]	7, 13, 34, 39, 42, 55, 62, 72, 83, 86, 89, 90, 92	531.99	469.878	0.9285	0.9532	–
Efficient GA [175]	7, 13, 34, 39, 42, 55, 62, 72, 83, 86, 89, 90, 92	531.99	469.878	0.9285	0.9532	450
GA [176]	7, 33, 55, 61, 72, 83, 86, 88, 89, 90, 92, 93, 95	531.99	471.51	0.9285	0.9532	–
ACO [176]	7, 13, 34, 39, 41, 55, 62, 72, 83, 86, 89, 90, 92	531.99	471.078	0.9285	0.9517	–
HBMO[21] [100, 147]	7, 14, 34, 39, 42, 55, 62, 72, 83, 86, 88, 90, 92	531.99	482.14	0.9285	0.9529	13
AACO [147]	7, 13, 34, 39, 42, 55, 62, 72, 83, 86, 89, 90, 92	531.99	469.878	0.9285	0.9532	95.88
MILP [96]	7, 13, 34, 39, 42, 55, 62, 72, 83, 86, 89, 90, 92	531.99	469.88	0.9285	0.9532	207.7

(*continued*)

Table 18.23 (*continued*)

Items		Power loss (kW)		Worst voltage (p.u.)		
Models	**Open switches**	**Before**	**After**	**Before**	**After**	**Run time (s)**
MICP[22] [96]	7, 13, 34, 39, 42, 55, 62, 72, 83, 86, 89, 90, 92	531.99	469.88	0.9285	0.9532	245.4
[149]	7, 13, 34, 39, 42, 55, 62, 72, 83, 86, 89, 90, 92	531.99	469.88	0.9285	0.9532	–
[121]	7, 13, 34, 39, 42, 55, 62, 72, 83, 86, 89, 90, 92	531.99	469.878	0.9285	0.9532	7
[123]	7, 13, 34, 39, 42, 55, 62, 72, 83, 86, 89, 90, 92	531.99	469.878	0.9285	0.9532	5.28
[150]	7, 13, 34, 39, 42, 55, 62, 72, 83, 86, 89, 90, 92	531.99	469.88	0.93	0.95	160
[95]	7, 13, 34, 39, 42, 62, 72, 83, 84, 86, 89, 90, 92	532	469.95	0.9285	0.9532	7.8
SA [154]	7, 13, 34, 39, 42, 62, 72, 83, 84, 86, 89, 90, 92	531.99	470.513	0.9285	0.9532	22.061
GA [154]	7, 13, 34, 39, 63, 72, 83, 84, 86, 89, 90, 92, 95	531.99	470.214	0.9285	0.9517	30.526
ACO [154]	7, 13, 34, 39, 42, 55, 62, 72, 83, 86, 89, 90, 92	531.99	469.878	0.9285	0.9532	19.736
HBMO [154]	7, 13, 34, 39, 42, 55, 62, 72, 83, 86, 89, 90, 92	531.99	469.878	0.9285	0.9532	17.104
RGA [154]	7, 13, 34, 39, 42, 63, 72, 83, 84, 86, 89, 90, 92	531.99	470.084	0.9285	0.9517	170
HSA [154]	7, 13, 34, 39, 42, 55, 62, 72, 83, 86, 89, 90, 92	531.99	469.88	0.9285	0.9532	9.871
GA + BE[23] [154]	7, 13, 34, 39, 42, 55, 62, 72, 83, 86, 89, 90, 92	531.99	469.878	0.9285	0.9532	37.983

(*continued*)

Table 18.23 (*continued*)

Items		Power loss (kW)		Worst voltage (p.u.)		
Models	**Open switches**	**Before**	**After**	**Before**	**After**	**Run time (s)**
[177]	7,13, 34, 39, 42, 55, 62, 72, 83, 86, 89, 90, 92	531.99	469.89	0.9285	0.9532	7.75
[155]	7, 13, 34, 39, 42, 63, 72, 83, 84, 86, 89, 90, 92	531.99	470.01	0.9285	0.9518	69

[17]Mixed-integer hybrid differential evolution.
[18]Hybrid differential evolution.
[19]Variable scaling HDE.
[20]Artificial immune system.
[21]Honey bee mating optimization.
[22]Mixed-integer conic programming.
[23]Branch exchange.

Table 18.24 List of models regarding loss minimization in Case 11

Models	Losses (kW)
[100]	482.14
[174]	472.32
GA [176]	471.51
[128]	471.44
[74–77], VSHDE [96, 113], ACO [147], ACO [176]	471.08
[112]	470.88
SA [154]	470.513
GA [154]	470.214
RGA [154]	470.084
[155]	470.01
[95]	469.95
[177]	469.89
[96, 149, 150], HSA [154]	469.88
[116, 121, 123, 138], Efficient GA [147], AACO [147], ACO [154], HBMO [154], GA + BE [154, 175]	469.878

Table 18.25 Models based on nodal voltage enhancement of Case 11

Models	Minimum voltage (p.u.)
AIS-ACO [147, 174]	0.9479
[150]	0.95
GA [154], RGA [154], ACO [176]	0.9517
[155]	0.9518
[100], HBMO [147]	0.9529
[74–77, 95, 96, 112, 113, 116, 121, 123, 128, 138], GA [147], Eficiente GA [147], ACO [147], AACO [147, 149], SA [154], ACO [154], HBMO [154], GA + BE [154], HSA [154, 175], GA [176, 177]	0.9532

Table 18.26 Classification of models according to running time in Case 11

Models	Computing time (s)
VSHDE [113]	585
[175]	450
HDE [113]	318
GA [75]	303.66
MICP [96]	245.4
ACO [75]	241.51
MILP [96]	207.7
SA [74]	195.21
RGA [154]	170
[150]	160
[77]	113.25
AACO [147]	95.88
[155]	69
GA + BE [154]	37.983
MIHDE [74]	36.15
GA [154]	30.526
SA [154]	22.061
ACO [154]	19.736
HBMO [154]	17.104
AIS-ACO [147, 174]	15
[100], HBMO [147]	13
HSA [154]	9.871
[95]	7.8
[177]	7.75
[121]	7
[123]	5.28
[116]	3.92

18.3.12 Case 12

Tables 18.27–18.30 show relevant results obtained for System 12. The parameters and related data of this network can be found in Chapter 17.

Table 18.27 Numerical results for Case 12

Items		Power loss (kW)		Worst voltage (p.u.)		
Models	Open switches	Before	After	Before	After	Run time (s)
[137]	24, 27, 35, 40, 43, 52, 59, 72, 75, 96, 98, 110, 123, 130, 131	1294.3	865.86	0.9321	0.9323	9.38
[138]	24, 27, 35, 40, 43, 52, 59, 72, 75, 96, 99, 110, 123, 130, 131	1298.09	870.35	0.8783	0.9321	–
MTS[24] [120]	24, 27, 35, 40, 43, 52, 59, 72, 75, 96, 98, 110, 123, 130, 131	1294.3	865.86	0.866	0.9323	18 000
TS[25] [120]	23, 27, 33, 40, 43, 49, 52, 62, 72, 74, 77, 83, 110, 126, 131	1294.3	884.163	0.866	0.9321	–
HSA [146]	24, 27, 35, 40, 43, 52, 59, 72, 75, 96, 98, 110, 123, 130, 131	1298.09	865.86	0.8783	0.9323	8.61
GA [146]	24, 31, 35, 40, 43, 49, 51, 62, 73, 74, 77, 83,110, 120, 126	1298.09	885.56	0.8783	0.9321	24.45
RGA [146]	23, 27, 33, 40, 43, 49, 52, 62, 73, 74, 77, 83, 110, 126, 131	1298.09	883.13	0.8783	0.9321	17.53
AACO [147]	24, 27, 35, 40, 43, 52, 59, 72, 75, 96, 98, 110, 123, 130, 131	1298.1	865.86	0.8688	0.9323	430.74
[150]	24, 26, 35, 40, 43, 51, 62, 72, 74, 77, 83, 110, 122, 126, 131	1296.6	870.91	0.87	0.93	704.1
[95]	24, 27, 35, 40, 43, 52, 59, 72, 75, 96, 98, 110, 123, 130, 131	1298.1	869.7	0.8783	0.9321	39.4
MIQCP [125]	24, 27, 35, 40, 43, 52, 59, 72, 75, 96, 98, 110, 123, 130, 131	1298.1	869.7	0.8783	0.9321	1009
MILP [125]	24, 27, 35, 40, 43, 52, 59, 72, 75, 96, 98, 110, 123, 130, 131	1298.1	869.7	0.8783	0.9321	2.8

(continued)

Table 18.27 (*continued*)

Items		Power loss (kW)		Worst voltage (p.u.)		
Models	**Open switches**	**Before**	**After**	**Before**	**After**	**Run time (s)**
GA [178]	5, 27, 34, 43, 47, 54, 61, 71, 74, 83, 96, 99, 110, 125, 131	1298.09	980.54	0.8687	0.9321	–
GA [168]	24, 27, 35, 40, 43, 52, 59, 72, 75, 96, 98, 110, 123, 130, 131	1298.1	865.86	0.8688	0.9323	42.13
[179]	24, 27, 35, 40, 43, 52, 59, 72, 75, 96, 98, 110, 123, 130, 131	1301.9	865.86	0.8688	0.9323	7.64

[24]Modified tabu search.
[25]Tabu search.

Table 18.28 Ranking of models based on loss reduction for Case 12

Models	Power loss (kW)
[178]	980.54
GA [146]	885.56
TS [120]	884.163
RGA [146]	883.13
[150]	870.91
[138]	870.35
[95, 125]	869.7
MTS [120, 137], HSA [146, 147, 168, 179]	865.86

Table 18.29 Ranking of models regarding voltage improvement for Case 12

Models	Worst voltage (p.u.)
[150]	0.93
[95], TS [120, 125, 138], GA [146], RGA [146, 178]	0.9321
MTS [120, 137], HSA [146, 147, 168, 179]	0.9323

Table 18.30 Ranking of models based on execution time for Case 12

Models	Run time (s)
MTS [120]	18 000
MIQCP [125]	1009
[150]	704.1
[147]	430.74
[168]	42.13
[95]	39.4
GA [146]	24.45
RGA [146]	17.53
[137]	9.38
HSA [146]	8.61
[179]	7.64
MILP [125]	2.8

18.3.13 Case 13

The proposed models were applied to System 13 and the results are listed in Tables 18.31–18.33. Data of this actual system can be found in Chapter 17.

It should be mentioned that the minimum voltage of all models after reconfiguration is almost 0.96 p.u. For this, the arrangement of methods based on voltage improvement was ignored.

Table 18.31 Numerical results of proposed models for Case 13

Items		Power loss (kW)		Worst voltage (p.u.)		
Methods	Open switches	Before	After	Before	After	Run time (s)
[180]	51, 106, 136–139, 141–152, 154–156	320.17	285.5	0.9307	0.9638	–
[116]	7, 9, 38, 51, 55, 90, 92, 95, 104, 106, 120, 126, 128, 135, 138, 141,144–146, 148, 150	320.36	286	0.9307	0.963	8.96
Decimal GA (DGA) [147, 181]	7, 51, 53, 84, 90, 96, 106, 118, 126, 128, 137–139, 141, 144, 145, 147, 148,150, 151, 156	320.3	280.22	0.9307	0.9605	3600
[182]	7, 38, 51, 53, 90, 96, 106, 118, 126, 128, 137, 138, 141, 144–148, 150, 151, 156	320.3	280.4	0.9307	0.9605	227

(continued)

Table 18.31 (*continued*)

Items		Power loss (kW)		Worst voltage (p.u.)		
Methods	Open switches	Before	After	Before	After	Run time (s)
[183]	51, 53, 90, 96, 106, 118, 136–139, 141, 144–148, 150, 151, 154–156	320.3	281.7	0.9307	0.9605	~ 40
AACO [147]	7, 35, 51, 90, 95, 106, 118,126, 135, 137, 138, 141, 142,144–148, 150, 151, 155	320.3	280.19	0.9307	0.9602	894.2
[94]	7, 35, 51, 90, 96, 106, 118, 126, 135, 137, 138, 141, 142, 144–148, 150, 151, 155	320.36	280.19	0.9307	0.9589	4473
[96]	7, 35, 51, 90, 96, 106, 118, 126, 135, 137, 138, 141, 142, 144–148, 150, 151, 155	320.36	280.19	0.9307	0.9589	1800
[184]	7, 51, 53, 84, 90, 96, 106, 118, 126, 128, 137–139, 141, 144, 145, 147, 148, 150, 151, 156	320.36	280.22	0.9307	0.9605	5120
[185]	7, 35, 51, 90, 96, 106, 118, 126, 135, 137, 138, 141, 142, 144–148, 150, 151, 155	320.36	280.19	0.93065	0.95892	–
[123]	7, 38, 51, 55, 90, 97, 106, 118, 126, 137, 138, 141, 144–148, 150–152, 155	320.36	282.77	0.9307	0.963	23.93
[95]	7, 38, 51, 54, 84, 90, 96, 106, 118, 126, 128, 135, 137, 138, 141, 144, 145, 147, 148, 150, 151	320.4	280.38	0.9307	0.9599	132.5
[151]	7, 38, 51, 54, 84, 90, 96, 106, 118, 126, 128, 135, 137, 138, 141, 144, 145, 147, 148, 150, 151	320.4	280.38	0.9307	0.9599	188.4
MIQCP [125]	7, 35, 51, 90, 96, 106, 118, 126, 135, 137, 138, 141, 142, 144–148, 150, 151, 155	320.36	280.19	0.9307	0.9589	1785
MILP [125]	7, 35, 51, 90, 96, 106, 118, 126, 135, 137, 138, 141, 142, 144–148, 150, 151, 155	320.36	280.19	0.9307	0.9589	5.6
[155]	7, 38, 51, 54, 84, 90, 96, 106, 119, 126, 135, 137, 138, 141, 144, 145, 147, 148, 150, 151, 155	320.35	281.02	0.9307	0.9588	530

Table 18.32 Arrangement of the results based on power loss reduction in Case 13

Models	Power loss (kW)
[116]	286
[180]	285.5
[123]	282.77
[183]	281.7
[155]	281.02
[182]	280.4
[95, 151]	280.38
[181, 184]	280.22
[94, 96, 125, 147, 185]	280.19

Table 18.33 Computation times for Case 13

Models	Time (s)
[184]	5120
[94]	4473
[181]	3600
[96]	1800
MIQCP [125]	1785
[147]	894.2
[155]	530
[182]	227
[151]	188.4
[95]	132.5
[183]	~ 40
[123]	23.93
[116]	8.96
MILP [125]	5.6

18.3.14 Case 14

The DSR problem was applied to System 14 and the results are listed in Table 18.34. Related data and information for this real large-scale distribution system have been provided in Chapter 17. It should be noted that base voltage in this system is 13.8 kV.

Table 18.34 Numerical results compared to other methods for Case 14

Items		Power loss (kW)		Worst voltage (p.u.)		
Methods	Open switches	Before	After	Before	After	Run time (s)
TS [186]	29, 66, 74, 83, 111, 118, 125, 131, 135, 136, 140, 177, 199, 202, 208	548.89	511.5	0.9574	0.9611	49.98

18.4 Conclusion

In this chapter, the formulation method of the conventional DSR problem was detailed. The proposed reconfiguration models were applied to 14 test systems including small-, medium-, and large-sized distribution networks using different methods and the results are compared to other reconfiguration techniques reported in specialized literature. Evaluation of simulation results shows that most of these models solve the DSR problem efficiently from the convergence speed and accuracy point of view. This chapter provides valuable information for readers to know which method and model is more effective for reconfiguration of distribution systems.

References

[1] M. Mahdavi, C.S. Antunez, A. Bagheri, and R. Romero, "Line maintenance within transmission expansion planning: A multistage framework," *IET Generation, Transmission and Distribution*, vol. 14, no. 13, pp. 3057–3065, 2019.

[2] M. Mahdavi, C. Sabillón, M. Ajalli, H. Monsef, and R. Romero, "A real test system for power system planning, operation, and reliability," *Journal of Control, Automation and Electrical Systems*, vol. 29, no. 2, pp. 192–208, 2018.

[3] M. Mahdavi, C.S. Antunez, M. Ajalli, and R. Romero, "Transmission expansion planning: Literature review and classification," *IEEE Systems Journal*, vol. 3, no. 3, pp. 3129–3140, 2019.

[4 S. Jalilzadeh, A. Kazemi, H. Shayeghi, and M. Mahdavi, "Technical and economic evaluation of voltage level in transmission network expansion planning using GA," *Energy Conversion and Management*, vol. 49, no. 5, pp. 1119–1125, 2008.

[5] H. Shayeghi, M. Mahdavi, and A. Bagheri, "Discrete PSO algorithm based optimization of transmission lines loading in TNEP problem," *Energy Conversion and Management*, vol. 51, no. 1, pp. 112–121, 2010.

[6] S. Jalilzadeh, A. Kazemi, M. Mahdavi, and H. Haddadian, "Technical and economic evaluation of voltage level effect on TNEP using genetic algorithm," *The 8th International Power Engineering Conference*, Singapore, pp. 1034–1039, 2007.

[7] M. Mahdavi, H. Shayeghi, and A. Kazemi, "DCGA based evaluating role of bundle lines in TNEP considering expansion of substations from voltage level point of view," *Energy Conversion and Management*, vol. 50, no. 8, pp. 2067–2073, 2009.

[8] H. Shayeghi, M. Mahdavi, and A. Bagheri, "An improved DPSO with mutation based on similarity algorithm for optimization of transmission lines loading," *Energy Conversion and Management*, vol. 51, no. 12, pp. 2715–2723, 2010.

[9] H. Shayeghi, M. Mahdavi, and H. Haddadian, "DCGA based-transmission network expansion planning considering network adequacy," *International Journal of Electrical and Computer Engineering*, vol. 2, no. 12, pp. 2875–2880, 2008.

[10] A. Kazemi, S. Jalilzadeh, H. Haddadian, and M. Mahdavi, "Transmission network expansion planning considering network adequacy using genetic algorithm," *The 8th International Power Engineering Conference*, Singapore, pp. 1040–1044, 2007.

[11] H. Shayeghi, H. A. Shayanfar, M. Mahdavi, and A. Bagheri, "Application of binary particle swarm optimization for transmission expansion planning considering lines loading," *The 2009 International Conference on Artificial Intelligence*, Las Vegas, Nevada, USA, pp. 653–659, 2009.

[12] A. Bagheri, M. Mahdavi, and H. Monsef, "Suggestion of an economical plan for electrical power transmission," *The 9th International Energy Conference*, Tehran, Iran, pp. 1–10, 2013.

[13] A. Kazemi, H. Haddadian, and M. Mahdavi, "Transmission network adequacy optimization using genetic algorithm," *5th International Conference on Electrical and Electronics Engineering*, Bursa, Turkey, pp. 1–5, 2007.

[14] M. Khodayari, M. Mahdavi, and H. Monsef, "Simultaneous scheduling of energy & spinning reserve considering customer and supplier choice on reliability," *19th Iranian Conference on Electrical Engineering*, Iran, pp. 491–496, 2011.

[15] M. Khodayari, H. Monsef, and M. Mahdavi, "Customer reliability based energy & SR scheduling at hierarchical level II," *27th International Power System Conference*, Iran, pp. 1–11, 2012.

[16] O. Badran, S. Mekhilef, H. Mokhlis, and W. Dahalan, "Optimal reconfiguration of distribution system connected with distributed generations: A review of different methodologies," *Renewable and Sustainable Energy Reviews*, vol. 73, pp. 854–867, June 2017.

[17] N.G. Paterakis, A. Mazza, S.F. Santos, O. Erdinç, G. Chicco, A.G. Bakirtzis, and J.P.S. Catalão, "Multi-objective reconfiguration of radial distribution systems using reliability indices," *IEEE Transactions on Power Systems*, vol. 31, no. 2, pp. 1048–1062, March 2016.

[18] M. Mahdavi and R. Romero, "Transmission system expansion planning based on LCC criterion considering uncertainty in VOLL using DCGA," *32nd International Power System Conference*, Iran, pp. 1–7, 2017.

[19] Y.Y. Fu and H.D. Chiang, "Toward optimal multiperiod network reconfiguration for increasing the hosting capacity of distribution networks," *IEEE Transactions on Power Delivery*, vol. 33, no. 5, pp. 2294–2304, October 2018.

[20] S. Lei, Y. Hou, F. Qiu, and J. Yan, "Identification of critical switches for integrating renewable distributed generation by dynamic network reconfiguration," *IEEE Transactions on Sustainable Energy*, vol. 9, no. 1, pp. 420–432, January 2018.

[21] H. Hosseini, S. Jalilzadeh, V. Nabaei, G.R.Z. Govar, and M. Mahdavi, "Enhancing deregulated distribution network reliability for minimizing penalty cost based on reconfiguration using BPSO," *2nd IEEE International Conference on Power and Energy*, Malaysia, pp. 983–987, 2008.

[22] M. Mahdavi, L.H. Macedo, and R. Romero, "A mathematical formulation for distribution network reconfiguration," *7th Regional Conference on Electricity Distribution*, Iran, pp. 1–5, 2019.
[23] H. Shayeghi, M. Mahdavi, A. Kazemi, and H.A. Shayanfar, "Studying effect of bundle lines on TNEP considering network losses using decimal codification genetic algorithm," *Energy Conversion and Management*, vol. 51, no. 12, pp. 2685–2691, 2010.
[24] H. Shayeghi and M. Mahdavi, "Genetic algorithm based studying of bundle lines effect on network losses in transmission network expansion planning," *Journal of Electrical Engineering*, vol. 60, no. 5, pp. 237–245, 2009.
[25] S. Jalilzadeh, H. Shayeghi, M. Mahdavi, and H. Haddadian, "A GA based transmission network expansion planning considering voltage level, network losses and number of bundle lines," *American Journal of Applied Sciences*, vol. 6, no. 5, pp. 987–994, 2009.
[26] H. Shayeghi, S. Jalilzadeh, M. Mahdavi, and H. Haddadian, "Role of voltage level and network losses in TNEP solution," *Journal of Electrical Systems*, vol. 5, no. 3, pp. 1–18, 2009.
[27] M. Mahdavi and H. Haddadian, "Evaluation of GA performance in TNEP considering voltage level, network losses and number of bundle lines," *International Journal of Discrete Mathematic*, vol. 3, no. 1, pp. 1–10, 2018.
[28] M. Mahdavi and E. Mahdavi, "Evaluating the effect of load growth on annual network losses in TNEP considering bundle lines using DCGA," *International Journal on Technical and Physical Problems of Engineering*, vol. 3, no. 4, pp. 1–9, 2011.
[29] H. Shayeghi and M. Mahdavi, "Studying the effect of losses coefficient on transmission expansion planning using decimal codification based GA," *International Journal on Technical and Physical Problems of Engineering*, vol. 1, no. 1, pp. 58–64, 2009.
[30] H. Shayeghi, S. Jalilzadeh, M. Mahdavi, and H. Haddadian, "Studying influence of two effective parameters on network losses in transmission expansion planning using DCGA," *Energy Conversion and Management*, vol. 49, no. 11, pp. 3017–3024, 2008.
[31] S. Jalilzadeh, A. Kazemi, M. Mahdavi, and H. Haddadian, "TNEP considering voltage level, network losses and number of bundle lines using GA," *Third International Conference on Electric Utility Deregulation and Restructuring and Power Technologies*, Nanjing, China, pp. 1580–1585, 2008.
[32] S. Jalilzadeh, A. Kazemi, H. Haddadian, and M. Mahdavi, "Transmission expansion planning considering network adequacy and losses using DCGA," *IEEE International Conference on Sustainable Energy Technologies*, Singapore, pp. 145–150, 2008.
[33] S. Jalilzadeh, A. Kazemi, M. Mahdavi, and H. Haddadian, "Studying of voltage level role in TNEP considering the losses for years after expansion using GA with different trails," *43rd International Universities Power Engineering Conference*, Padova, Italy, pp. 1–6, 2008.
[34] A. Kazemi, S. Jalilzadeh, M. Mahdavi, and H. Haddadian, "DCGA-based evaluation of inflation rate effect on the network loss in transmission expansion planning," *43rd International Universities Power Engineering Conference*, Padova, Italy, pp. 1–5, 2008.
[35] A. Kazemi, S. Jalilzadeh, M. Mahdavi, and H. Haddadian, "Evaluation of network losses role in TNEP considering voltage level and number of bundle lines using GA," *3rd IEEE Conference on Industrial Electronics and Applications*, Singapore, pp. 1–6, 2008.
[36] S. Jalilzadeh, A. Kazemi, M. Mahdavi, and H. Haddadian, "Transmission expansion planning considering voltage level and network loss using genetic algorithm," *5th International Conference on Electrical and Electronics Engineering*, Bursa, Turkey, pp. 1–5, 2007.

[37] A. Kimiyaghalam, M. Mahdavi, A. Ashouri, and H. Soheil, “Transmission expansion planning considering uncertainty in fuel price using DABC,” *The 9th International Energy Conference*, Tehran, Iran, pp. 1–10, 2013.

[38] M. Rahmani-Andebili and M. Fotuhi Firuzabad, “An adaptive approach for PEVs charging management and reconfiguration of electrical distribution system penetrated by renewables,” *IEEE Transactions on Industrial Informatics*, vol. 14, no. 5, pp. 2001–2010, May 2018.

[39] M. Mahdavi, H. Monsef, A. Bagheri, and A. Kimiyaghalam, “Dynamic transmission network expansion planning considering network losses, DG sources and operational costs –Part 1: Review and problem formulation,” *Wulfenia Journal*, vol. 19, no. 10, pp. 242–257, 2012.

[40] M. Mahdavi, H. Monsef, A. Bagheri, and A. Kimiyaghalam, “Dynamic transmission network expansion planning considering network losses, DG sources and operational costs – Part 2: Solution method and numerical results,” *Wulfenia Journal*, vol. 19, no. 10, pp. 258–273, 2012.

[41] M. Mahdavi and R.A.V. Ramos, “Optimal allocation of bioenergy distributed generators in electrical energy systems,” *IV Bioenergy Workshop*, Bioenergy Research Institute, Brazil, 2020.

[42] M. Mahdavi and L.A. Okamura, “Biofuels production from petrochemical industry wastewater,” Presentation, *Advanced Topics in Biorefinery*, Bioenergy Research Institute, Brazil, 2020.

[43] M. Mahdavi, “Temperature effects on the methanogenesis and sulfidogenesis in the wastewater treatment,” Presentation, *Advanced Topics in Biorefinery*, Bioenergy Research Institute, Brazil, 2020.

[44] M. Mahdavi, “Iran energy network overview: Renewable energies in Iran,” Presentation, *Advanced Topics in Bioenergy*, Bioenergy Research Institute, Brazil, 2020.

[45] M. Mahdavi and R. Romero, “New models and optimization techniques applied to the problem of optimal reconfiguration of radial distribution systems,” Report, São Paulo Research Foundation, no. 16/12190-7, Brazil, 2019.

[46] A. Lotfipour and H. Afrakhte, “A discrete teaching-learning-based optimization algorithm to solve distribution system reconfiguration in presence of distributed generation,” *International Journal of Electrical Power and Energy Systems*, vol. 82, pp. 264–273, November 2016.

[47] M.A. Muhammad, H. Mokhlis, K. Naidu, J.F. Franco, H.A. Illias, and L. Wang, “Integrated database approach in multiobjective network reconfiguration for distribution system using discrete optimisation techniques,” *IET Generation, Transmission & Distribution*, vol. 12, no. 4, pp. 976–986, 2018.

[48] N.C. Koutsoukis, D.O. Siagkas, P.S. Georgilakis, and N.D. Hatziargyriou, “Online reconfiguration of active distribution networks for maximum integration of distributed generation,” *IEEE Transactions on Automation Science and Engineering*, vol. 14, no. 2, pp. 437–448, April 2017.

[49] R.A. Jabr, I. Džafić, and I. Huseinagić, “Real time optimal reconfiguration of multiphase active distribution networks,” *IEEE Transactions on Smart Grid*, vol. 9, no. 6, pp. 6829–6839, November 2018.

[50] M. Mahdavi and R. Romero, "A new model for dynamic transmission expansion planning," *4th National Conference on Technology in Electrical and Computer Engineering*, Iran, pp. 1–7, 2018.

[51] M. Mahdavi and H. Monsef, "Review of static transmission expansion planning," *Journal of Electrical and Control Engineering*, vol. 1, no. 1, pp. 11–18, 2011.

[52] M. Mahdavi, H. Monsef, and R. Romero, "Reliability effects of maintenance on TNEP considering preventive and corrective repairs," *IEEE Transactions on Power Systems*, vol. 32, no. 5, pp. 3768–3781, 2017.

[53] M. Mahdavi, H. Monsef, and R. Romero, "Reliability and economic effects of maintenance on TNEP considering line loading and repair," *IEEE Transactions on Power Systems*, vol. 31, no. 5, pp. 3381–3393, 2016.

[54] M. Mahdavi, A.R. Kheirkhah, L.H. Macedo, and R. Romero, "A genetic algorithm for transmission network expansion planning considering line maintenance," *IEEE Congress on Evolutionary Computation*, Glasgow, United Kingdom, pp. 1–6, 2020.

[55] M. Mahdavi, L.H. Macedo, and R. Romero, "Considering line maintenance and repair in GTEP," *4th Conference on Modern Technologies in Electrical, Telecommunication and Mechatronics Engineering*, Iran, pp. 1–6, 2018.

[56] M. Mahdavi, A.R. Kheirkhah, and R. Romero, "Transmission expansion planning considering line failures and maintenance," *33rd International Power System Conference*, Iran, pp. 1–7, 2018.

[57] M. Mahdavi, L.H. Macedo, and R. Romero, "Transmission and generation expansion planning considering system reliability and line maintenance," *26th Iranian Conference on Electrical Engineering*, Iran, pp. 1005–1010, 2018.

[58] A. Kimiyaghalam, M. Mahdavi, A. Ashouri, and M. Bagherivand, "Optimal placement of PMUs for reliable observability of network under probabilistic events using BABC algorithm," *22nd International Conference on Electricity Distribution*, Sweden, pp. 1–4, 2013.

[59] H. Shayeghi, H. Hosseini, A. Shabani, and M. Mahdavi, "GEP considering purchase prices, profits of IPPs and reliability criteria using hybrid GA and PSO," *International Journal of Electrical and Computer Engineering*, vol. 2, no. 8, pp. 1619–1625, 2008.

[60] M. Mahdavi, A. Nazari, V. Hosseinnezhad, and A. Safari, "A PSO-based static synchronous compensator controller for power system stability enhancement," *Journal of Artificial Intelligence in Electrical Engineering*, vol. 1, no. 1, pp. 18–25, 2012.

[61] Z. Li, S. Jazebi, and F. de León, "Determination of the optimal switching frequency for distribution system reconfiguration," *IEEE Transactions on Power Delivery*, vol. 32, no. 4, pp. 2060–2069, August 2017.

[62] B. Sultana, M.W. Mustafa, U. Sultana, and A.R. Bhatti, "Review on reliability improvement and power loss reduction in distribution system via network reconfiguration," *Renewable and Sustainable Energy Reviews*, vol. 66, pp. 297–310, 2016.

[63] P. Hines, K. Balasubramaniam, and E.C. Sasnchez, "Cascading failures in power grids," *IEEE Potentials*, pp. 24–30, September/October 2009.

[64] M. Mahdavi and A. Bagheri, "BPSO applied to TNEP considering adequacy criterion," *American Journal of Neural Networks and Applications*, vol. 4, no. 1, pp. 1–7, 2018.

[65] M. Mahdavi, A. Bagheri, and E. Mahdavi, "Comparing efficiency of PSO with GA in transmission expansion planning considering network adequacy," *WSEAS Transactions on Power Systems*, vol. 7, no. 1, pp. 34–43, 2012.
[66] M. Mahdavi, H. Monsef, and A. Bagheri, "Lines loading optimization in transmission expansion planning based on binary PSO algorithm," *I-manager's Journal on Information Technology*, vol. 1, no. 1, pp. 24–32, 2012.
[67] M. Mahdavi and E. Mahdavi, "Transmission expansion planning considering network adequacy and investment cost limitation using genetic algorithm," *International Journal of Electrical and Computer Engineering*, vol. 5, no. 8, pp. 1–5, 2011.
[68] M. Mahdavi, H. Monsef, and A. Bagheri, "Transmission lines loading enhancement using ADPSO approach," *International Journal of Electrical and Computer Engineering*, vol. 4, no. 3, pp. 556–561, 2010.
[69] H. Shayeghi, M. Mahdavi, and A. Kazemi, "Discrete particle swarm optimization algorithm used for TNEP considering network adequacy restriction," *International Journal of Electrical and Computer Engineering*, vol. 3, no. 3, pp. 521–528, 2009.
[70] H. Shayeghi and M. Mahdavi, "Optimization of transmission lines loading in TNEP using decimal codification based GA," *International Journal of Electrical and Computer Engineering*, vol. 2, no. 5, pp. 942–947, 2008.
[71] A. Kazemi, S. Jalilzadeh, M. Mahdavi, and H. Haddadian, "Genetic algorithm-based investigation of load growth factor effect on the network loss in TNEP," *3rd IEEE Conference on Industrial Electronics and Applications*, Singapore, vol. 1, pp. 764–769, 2008.
[72] S. Jalilzadeh, A. Kimiyaghalam, M. Mahdavi, and A. Ashouri, "STNEP considering voltage level and uncertainty in demand using IABPSO," *International Review of Electrical Engineering*, vol. 7, no. 4, pp. 5186–5195, 2012.
[73] A. Kimiyaghalam, M. Mahdavi, S. Jalilzadeh, and A. Ashouri, "Improved binary particle swarm optimization based TNEP considering network losses, voltage level, and uncertainty in demand," *Journal of Artificial Intelligence in Electrical Engineering*, vol. 1, no. 2, pp. 29–42, 2012.
[74] C.T. Su and C.S. Lee, "Network reconfiguration of distribution systems using improved mixed-integer hybrid differential evolution," *IEEE Transactions on Power Delivery*, vol. 18, no. 3, pp. 1022–1027, July 2003.
[75] C.T. Su, C.F. Chang, and J.P. Chiou, "Distribution network reconfiguration for loss reduction by ant colony search algorithm," *Electric Power Systems Research*, vol. 75, no. 2–3, pp. 190–199, August 2005.
[76] C.F. Chang, "Reconfiguration and capacitor placement for loss reduction of distribution systems by ant colony search algorithm," *IEEE Transactions on Power Systems*, vol. 23, no. 4, pp. 1747–1755, November 2008.
[77] C. Wang and H.Z. Cheng, "Optimization of network configuration in large distribution systems using plant growth simulation algorithm," *IEEE Transaction on Power Systems*, vol. 23, no. 1, pp. 119–126, February 2008.
[78] A. Azizivahed, H. Narimani, M. Fathi, E. Naderi, H.R. Safarpour, and M.R. Narimani, "Multi-objective dynamic distribution feeder reconfiguration in automated distribution systems," *Energy*, vol. 147, pp. 896–914, March 2018.

[79] E. Carpaneto and G. Chicco, "Distribution system minimum loss reconfiguration in the hyper-cube ant colony optimization framework," *Electric Power Systems Research*, vol. 78, no. 12, pp. 2037–2045, December 2008.

[80] A. Saffar, R. Hooshmand, and A. Khodabakhshian, "A new fuzzy optimal reconfiguration of distribution systems for loss reduction and load balancing using ant colony search-based algorithm," *Applied Soft Computing*, vol. 11, no. 5, pp. 4021–4028, July 2011.

[81] A.R. Malekpour, T. Niknam, A. Pahwa, and A.K. Fard, "Multi-objective stochastic distribution feeder reconfiguration in systems with wind power generators and fuel cells using the point estimate method," *IEEE Transactions on Power Systems*, vol. 28, no. 2, pp. 1483–1492, May 2013.

[82] A. Asrari, S. Lotfifard, and M. Ansari, "Reconfiguration of smart distribution systems with time varying loads using parallel computing," *IEEE Transactions on Smart Grid*, vol. 7, no. 6, pp. 2713–2723, November 2016.

[83] J.C. López, M. Lavorato, and M.J. Rider, "Optimal reconfiguration of electrical distribution systems considering reliability indices improvement," *International Journal of Electrical Power and Energy Systems*, vol. 78, pp. 837–845, June 2016.

[84] S.S.F. Souza, R. Romero, J. Pereira, and J.T. Saraiva, "Artificial immune algorithm applied to distribution system reconfiguration with variable demand," *International Journal of Electrical Power and Energy Systems*, vol. 82, pp. 561–568, November 2016.

[85] B. Venkatesh, R. Ranjan, and H.B. Gooi, "Optimal reconfiguration of radial distribution systems to maximize loadability," *IEEE Transactions on Power Systems*, vol. 19, no. 1, pp. 260–266, February 2004.

[86] H. Salazar, R. Gallego, and R. Romero, "Artificial neural networks and clustering techniques applied in the reconfiguration of distribution systems," *IEEE Transactions on Power Delivery*, vol. 21, no. 1, pp. 1735–1742, July 2006.

[87] D. Singh and R.K. Misra, "Load type impact on distribution system reconfiguration," *International Journal of Electrical Power and Energy Systems*, vol. 42, pp. 582–592, November 2012.

[88] A. Zidan and E.F. El-Saadany, "Distribution system reconfiguration for energy loss reduction considering the variability of load and local renewable generation," *Energy*, vol. 59, pp. 698–707, September 2013.

[89] A. Tyagil, A. Vermal, and P.R. Bijwe, "Reconfiguration for loadability limit enhancement of distribution systems," *IET Generation, Transmission and Distribution*, vol. 12, no. 1, pp. 88–93, February 2018.

[90] J. Shukla, B. Das, and V. Pant, "Consideration of small signal stability in multi-objective DS reconfiguration in the presence of distributed generation," *IET Generation, Transmission & Distribution*, vol. 11, no. 1, pp. 236–245, January 2017.

[91] L.H. Macedo, J.F. Franco, M. Mahdavi, and R. Romero, "A contribution to the optimization of the reconfiguration problem in radial distribution systems," *Journal of Control, Automation and Electrical Systems*, vol. 29, no. 6, pp. 756–768, 2018.

[92] M. Mahdavi and R. Romero, "Reconfiguration of radial distribution systems: An efficient mathematical model," *IEEE Latin America Transactions*, vol. 100, no. 1e, pp. 1–10, 2020.

[93] M. Sedighizadeh, M. Esmaili, and M. Esmaeili, "Application of the hybrid Big Bang–Big Crunch algorithm to optimal reconfiguration and distributed generation power allocation in distribution systems," *Energy*, vol. 76, pp. 920–930, November 2014.

[94] M. Lavorato, J.F. Franco, M.J. Rider, and R. Romero, "Imposing radiality constraints in distribution system optimization problems," *IEEE Transactions on Power Systems*, vol. 27, no. 1, pp. 172–179, February 2012.
[95] H. Ahmadi and J.R. Martí, "Mathematical representation of radiality constraint in distribution system reconfiguration problem," *International Journal of Electrical Power and Energy Systems*, vol. 64, pp. 293–299, January 2015.
[96] R.A. Jabr, R. Singh, and B.C. Pal, "Minimum loss network reconfiguration using mixed-integer convex programming," *IEEE Transactions on Power Systems*, vol. 27, no. 2, pp. 1106–1115, May 2012.
[97] Y. Wang, C. Liu, M. Shahidehpour, and C. Guo, "Critical components for maintenance outage scheduling considering weather conditions and common mode outages in reconfigurable distribution systems," *IEEE Transactions on Smart Grid*, vol. 7, no. 6, pp. 2807–2816, November 2016.
[98] J.A. Taylor and F.S. Hover, "Convex models of distribution system reconfiguration," *IEEE Transactions on Power Systems*, vol. 27, no. 3, pp. 1407–1413, August 2012.
[99] H. Hijazi and S. Thiébaux, "Optimal distribution systems reconfiguration for radial and meshed grids," *International Journal of Electrical Power and Energy Systems*, vol. 72, pp. 136–143, November 2015.
[100] T. Niknam, "An efficient multi-objective HBMO algorithm for distribution feeder reconfiguration," *Expert Systems with Applications*, vol. 38, no. 3, pp. 2878–2887, March 2011.
[101] N. Kagan, "Configuração de redes de distribuição através de algoritmos genéticos e tomada de decisão Fuzzy," PhD Thesis, University of São Paulo, São Paulo, Brazil, pp. 1–187, 1999.
[102] K. Masteri and B. Venkatesh, "A new optimization method for distribution system reconfiguration," *IEEE 27th Canadian Conference on Electrical and Computer Engineering*, Canada, pp. 1–4, May 2014.
[103] J.S.R.K. AL-Sujada and O.Y.M. Al-Rawi, "Novel load flow algorithm for multi-phase balanced/unbalanced radial distribution systems," *International Journal of Advanced Research in Electrical, Electronics and Instrumentation Engineering*, vol. 3, no. 7, pp. 10928–10942, July 2014.
[104] S. Civanlar, J.J. Grainger, H. Yin, and S.S.H. Lee, "Distribution feeder reconfiguration for loss reduction," *IEEE Transactions on Power Delivery*, vol. 4, no. 3, pp. 1217–1223, July 1988.
[105] C.A. Castro, J.R. Watanabe, and A.A. Watanabe, "An efficient reconfiguration algorithm for loss reduction of distribution systems," *Electric Power Systems Research*, vol. 19, no. 2, pp. 137–144, August 1990.
[106] H.C. Chang and C.C. Kuo, "Network reconfiguration in distribution systems using simulated annealing," *Electric Power Systems Research*, vol. 29, no. 3, pp. 227–238, May 1994.
[107] R.J. Sárf, M.M.A. Salama, and A.Y. Chikhan, "Distribution system reconfiguration for loss reduction: An algorithm based on network partitioning theory," *IEEE Transactions on Power Systems*, vol. 11, no. 1, pp. 504–510, February 1996.
[108] C.T. Su and C.S. Lee, "Feeder reconfiguration and capacitor setting for loss reduction of distribution systems," *Electric Power Systems Research*, vol. 58, no. 2, pp. 97–102, June 2001.

[109] J.Z. Zhu, "Optimal reconfiguration of electrical distribution network using the refined genetic algorithm," *Electric Power Systems Research*, vol. 62, no. 1, pp. 37–42, May 2002.

[110] Y.J. Jeon and J.C. Kim, "Application of simulated annealing and tabu search for loss minimization in distribution systems," *International Journal of Electrical Power and Energy Systems*, vol. 26, no. 1, pp. 9–18, January 2004.

[111] Y. Mishima, K. Nara, T. Satoh, T. Ito, and H. Kaneda, "Method for minimum-loss reconfiguration of distribution system by tabu search," *Electrical Engineering in Japan*, vol. 152, no. 2, pp. 18–25, May 2005.

[112] F.V. Gomes, S. Carneiro, Jr., J.L.R. Pereira, M.P. Vinagre, P.A.N. Garcia, and L.R. Araujo, "A new heuristic reconfiguration algorithm for large distribution systems," *IEEE Transactions on Power Systems*, vol. 20, no. 3, pp. 1373–1378, August 2005.

[113] J.P. Chiou, C.F. Chang, and C.T. Su, "Variable scaling hybrid differential evolution for solving network reconfiguration of distribution systems," *IEEE Transactions on Power Systems*, vol. 20, no. 2, pp. 668–674, May 2005.

[114] S. Sivanagaraju, Y. Sreekanth, and E.J. Babu, "An efficient genetic algorithm for loss minimum distribution system reconfiguration," *Electric Power Components and Systems*, vol. 34, no. 3, pp. 249–258, March 2006.

[115] F.V. Gomes, S. Carneiro, Jr., J.L.R. Pereira, M.P. Vinagre, P.A.N. Garcia, and L.R. Araujo, "A new distribution system reconfiguration approach using optimum power flow and sensitivity analysis for loss reduction," *IEEE Transactions on Power Systems*, vol. 21, no. 4, pp. 1616–1623, November 2006.

[116] G.K.V. Raju and P.R. Bijwe, "Efficient reconfiguration of balanced and unbalanced distribution systems for loss minimisation," *IET Generation, Transmission and Distribution*, vol. 2, no. 1, pp. 7–12, January 2008.

[117] B. Enacheanu, B. Raison, R. Caire, O. Devaux, W. Bienia, and N. HadjSaid, "Radial network reconfiguration using genetic algorithm based on the Matroid theory," *IEEE Transactions on Power Systems*, vol. 23, no. 1, pp. 186–195, August 2008.

[118] S. Sivanagaraju, J.V. Rao, and P.S. Raju, "Discrete particle swarm optimization to network reconfiguration for loss reduction and load balancing," *Electric Power Components and Systems*, vol. 36, no. 5, pp. 513–524, May 2008.

[119] N.T. Linh and N.Q. Anh, "Application of artificial bee colony algorithm (ABC) for reconfiguring distribution network," *Second International Conference on Computer Modeling and Simulation*, China, pp. 102–106, January 2010.

[120] A.Y. Abdelaziz, F.M. Mohammed, S.F. Mekhamer, and M.A.L. Badr, "Distribution system reconfiguration using a modified tabu search algorithm," *Electric Power Systems Research*, vol. 80, no. 8, pp. 943–953, August 2010.

[121] B. Tomoiagă, M. Chindris, A. Sumper, R. Villafafila-Robles, and A. Sudria-Andreu, "Distribution system reconfiguration using genetic algorithm based on connected graphs," *Electric Power Systems Research*, vol. 104, pp. 216–225, November 2013.

[122] C. Wang and Y. Gao, "Determination of power distribution network configuration using non-revisiting genetic algorithm," *IEEE Transactions on Power Systems*, vol. 28, no. 4, pp. 3638–3648, November 2013.

[123] A.K. Ferdavani, A.A.M. Zin, A. Khairuddin, and M.M. Naeini, "Reconfiguration of distribution system through two minimum-current neighbour-chain updating methods," *IET Generation, Transmission and Distribution*, vol. 7, no. 12, pp. 1492–1497, December 2013.

[124] A.M. Eldurssi and R.M. O'Connell, "A fast nondominated sorting guided genetic algorithm for multi-objective power distribution system reconfiguration problem," *IEEE Transactions on Power Systems*, vol. 30, no. 2, pp. 593–601, March 2015.

[125] H. Ahmadi and J.R. Martí, "Linear current flow equations with application to distribution systems reconfiguration," *IEEE Transactions on Power Systems*, vol. 30, no. 4, pp. 2073–2080, March 2015.

[126] B. Novoselnik, M. Bolfek, M. Bošković, and M. Baotić, "Electrical power distribution system reconfiguration: Case study of a real-life grid in Croatia," *IFAC PapersOnLine*, vol. 50, no. 1, pp. 61–66, 2017.

[127] M.E. Baran and F.F. Wu, "Network reconfiguration in distribution systems for loss reduction and load balancing," *IEEE Transactions on Power Delivery*, vol. 4, no. 2, pp. 1401–1407, April 1989.

[128] D. Shirmohammadi and W.H. Hong, "Reconfiguration of electric distribution networks for resistive line loss reduction," *IEEE Transactions on Power Delivery*, vol. 4, no. 2, pp. 1492–1498, April 1989.

[129] S. Goswami and S. Basu, "A new algorithm for the reconfiguration of distribution feeders for loss minimization," *IEEE Transactions on Power Delivery*, vol. 7, no. 3, pp. 1484–1491, July 1992.

[130] K. Nara, A. Shiose, M. Kitagawa, and T. Ishihara, "Implementation of genetic algorithm for distribution system loss minimum reconfiguration," *IEEE Transactions on Power Systems*, vol. 7, no. 3, pp. 1044–1051, August 1992.

[131] T.E. McDermott, I. Drezga, and R.P. Broadwater, "A heuristic nonlinear constructive method for distribution system reconfiguration," *IEEE Transactions on Power Systems*, vol. 14, no. 2, pp. 478–483, May 1999.

[132] M.A. Kashem, G.B. Jasmon, and V. Ganapathy, "A new approach of distribution system reconfiguration for loss minimization," *International Journal of Electrical Power and Energy Systems*, vol. 22, no. 4, pp. 269–276, May 2000.

[133] A.B. Morton and I.M.Y. Mareels, "An efficient brute-force solution to the network reconfiguration problem," *IEEE Transactions on Power Delivery*, vol. 15, no. 3, pp. 996–1000, July 2000.

[134] B. Venkatesh and R. Ranjan, "Optimal radial distribution system reconfiguration using fuzzy adaptation of evolutionary programming," *International Journal of Electrical Power and Energy Systems*, vol. 25, no. 10, pp. 775–780, December 2003.

[135] Y.Y. Hong and S.Y. Ho, "Determination of network configuration considering multiobjective in distribution systems using genetic algorithms," *IEEE Transactions on Power Systems*, vol. 20, no. 2, pp. 1062–1069, May 2005.

[136] J. Mendoza, R. Lopez, D. Morales, E. Lopez, P. Dessante, and R. Moaaga:, "Minimal loss reconfiguration using genetic algorithms with restricted population and addressed operators: Real application," *IEEE Transaction on Power Systems*, vol. 21, no. 2, pp. 948–954, May 2006.

[137] D. Zhang, Z. Fu, and L. Zhang, "An improved TS algorithm for loss-minimum reconfiguration in large-scale distribution systems," *Electric Power Systems Research*, vol. 77, no. 5–6, pp. 685–694, April 2007.

[138] G.K.V. Raju and P.R. Bijwe, "An efficient algorithm for minimum loss reconfiguration of distribution systems based on sensitivity and heuristics," *IEEE Transactions on Power Systems*, vol. 23, no. 3, pp. 1280–1287, August 2008.

[139] J.A. Martín and A.J. Gil, "A new heuristic approach for distribution systems loss reduction," *Electric Power Systems Research*, vol. 78, no. 11, pp. 1953–1958, November 2008.

[140] T. Niknam, "Application of honey bee mating optimization on distribution state estimation including distributed generators," *Journal of Zhejiang University Science A*, vol. 9, no. 12, pp. 1753–1764, December 2008.

[141] W.C. Wu and M.S. Tsai, "Application of enhanced integer coded particle swarm optimization for distribution system feeder reconfiguration," *IEEE Transactions on Power Systems*, vol. 26, no. 3, pp. 1591–1599, August 2011.

[142] T. Niknam, "A new hybrid algorithm for multiobjective distribution feeder reconfiguration," *Energy Conversion and Management*, vol. 50, no. 8, pp. 2074–2082, August 2009.

[143] H.M. Khodr, J.M. Crespo, M.A. Matos, and J. Pereira, "Distribution systems reconfiguration based on OPF using Benders decomposition," *IEEE Transactions on Power Systems*, vol. 24, no. 4, pp. 2166–2176, October 2009.

[144] H.M. Khodr and J.J.M. Crespo, "Integral methodology for distribution systems reconfiguration based on optimal power flow using Benders decomposition technique," *IET Generation, Transmission & Distribution*, vol. 3, no. 1, pp. 521–534, June 2009.

[145] A.Y. Abdelaziz, F.M. Mohammed, S.F. Mekhamer, and M.A.L. Badr, "Distribution systems reconfiguration using a modified particle swarm optimization algorithm," *Electric Power Systems Research*, vol. 79, no. 11, pp. 1521–1530, November 2009.

[146] R.S. Rao, S.V.L. Narasimham, M.R. Raju, and A.S. Rao, "Optimal network reconfiguration of large-scale distribution system using harmony search algorithm," *IEEE Transactions on Power Systems*, vol. 26, no. 3, pp. 1080–1088, August 2011.

[147] A. Swarnkar, N. Gupta, and K.R. Niazi, "Adapted ant colony optimization for efficient reconfiguration of balanced and unbalanced distribution systems for loss minimization," *Swarm and Evolutionary Computation*, vol. 1, no. 3, pp. 129–137, September 2011.

[148] A.Y. Abdelaziz, R.A. Osama, and S.M. El-Khodary, "Reconfiguration of distribution systems for loss reduction using the hyper-cube ant colony optimisation algorithm," *IET Generation, Transmission and Distribution*, vol. 12, no. 4, pp. 976–986, February 2012.

[149] F. Llorens-Iborra, J. Riquelme-Santos, and E. Romero-Ramos, "Mixed-integer linear programming model for solving reconfiguration problems in large-scale distribution systems," *Electric Power Systems Research*, vol. 88, pp. 137–145, July 2012.

[150] L.W. de Oliveira, E.J. de Oliveira, F.V. Gomes, I.C. Silva, Jr., A.L.M. Marcato, and P.V.C. Resende, "Artificial immune systems applied to the reconfiguration of electrical power distribution networks for energy loss minimization," *International Journal of Electrical Power and Energy Systems*, vol. 56, pp. 64–74, March 2014.

[151] H. Ahmadi and J.R. Martí, "Distribution system optimization based on a linear power flow formulation," *IEEE Transactions on Power Delivery*, vol. 30, no. 1, pp. 25–33, February 2015.

[152] I. Ali, M.S. Thomas, and P. Kumar, "Energy efficient reconfiguration for practical load combinations in distribution systems," *IET Generation, Transmission & Distribution*, vol. 9, no. 1, pp. 1051–1060, August 2015.
[153] R. Rajaram, K. Sathish Kumar, and N. Rajasekar, "Power system reconfiguration in a radial distribution network for reducing losses and to improve voltage profile using modified plant growth simulation algorithm with distributed generation (DG)," *Energy Reports*, vol. 1, pp. 116–122, November 2015.
[154] H.R. Esmaeilian and R. Fadaeinedjad, "Energy loss minimization in distribution systems utilizing an enhanced reconfiguration method integrating distributed generation," *IEEE Systems Journal*, vol. 9, no. 4, December 2015.
[155] K. Jasthi and D. Das, "Simultaneous distribution system reconfiguration and DG sizing algorithm without load flow solution," *IET Generation, Transmission and Distribution*, vol. 12, no. 6, pp. 1303–1313, March 2018.
[156] R.A. Abttan, "Electric distribution system reconfiguration for loss reduction using genetic algorithm," *Journal of Baghdad College of Economic Sciences University*, vol. 2015, no. 43, pp. 457–478, 2015.
[157] M.A. Kashem, V. Ganapathy, and G.B. Jasmon, "A geometrical approach for network reconfiguration based loss minimization in distribution systems," *International Journal of Electrical Power and Energy Systems*, vol. 21, no. 4, pp. 295–304, May 2001.
[158] Y.J. Jeon, J.C. Kim, J.O. Kim, J.R. Shin, and K.Y. Lee, "An efficient simulated annealing algorithm for network reconfiguration in large-scale distribution systems," *IEEE Transactions on Power Delivery*, vol. 17, no. 4, pp. 1070–1078, October 2002.
[159] N.C. Sahoo and K. Prasad, "A fuzzy genetic approach for network reconfiguration to enhance voltage stability in radial distribution systems," *Energy Conversion and Management*, vol. 47, no. 18–19, pp. 3288–3306, November 2006.
[160] J.S. Savier and D. Das, "Impact of network reconfiguration on loss allocation of radial distribution systems," *IEEE Transactions on Power Delivery*, vol. 22, no. 4, pp. 2473–2480, October 2007.
[161] M. Arun and P. Aravindhababu, "A new reconfiguration scheme for voltage stability enhancement of radial distribution systems," *Energy Conversion and Management*, vol. 50, no. 9, pp. 2148–2151, September 2009.
[162] S. Chandramohan, N. Atturulu, R.P.K. Devi, and B. Venkatesh, "Operating cost minimization of a radial distribution system in a deregulated electricity market through reconfiguration using NSGA method," *International Journal of Electrical Power and Energy Systems*, vol. 32, no. 2, pp. 126–132, February 2010.
[163] A.A.M. Zin, A.K. Ferdavani, A.B. Khairuddin, and M.M. Naeini, "Reconfiguration of radial electrical distribution network through minimum-current circular-updating-mechanism method," *IEEE Transactions on Power Systems*, vol. 27, no. 2, pp. 968–974, 2012.
[164] R.S. Rao, K. Ravindra, K. Satish, and S.V.L. Narasimham, "Power loss minimization in distribution system using network reconfiguration in the presence of distributed generation," *IEEE Transactions on Power Systems*, vol. 28, no. 1, pp. 317–325, February 2013.
[165] S.A. Taher and M.H. Karimi, "Optimal reconfiguration and DG allocation in balanced and unbalanced distribution systems," *Ain Shams Engineering Journal*, vol. 5, no. 3, pp. 735–749, September 2014.

[166] S. Naveen, K.S. Kumar, and K. Rajalakshmi, "Distribution system reconfiguration for loss minimization using modified bacterial foraging optimization algorithm," *International Journal of Electrical Power and Energy Systems*, vol. 69, pp. 90–97, July 2015.

[167] A.V.S. Reddy and M.D. Reddy, "Optimization of distribution network reconfiguration using dragonfly algorithm," *Journal of Electrical Engineering*, vol. 16, no. 4, pp. 1–10, 2016.

[168] B. Khorshid-Ghazani, H. Seyedi, B. Mohammadi-ivatloo, K. Zare, and S. Shargh, "Reconfiguration of distribution networks considering coordination of the protective devices," *IET Generation, Transmission & Distribution*, vol. 11, no. 1, pp. 82–92, January 2017.

[169] D. Das, "A fuzzy multiobjective approach for network reconfiguration of distribution systems," *IEEE Transactions on Power Delivery*, vol. 21, no. 1, pp. 202–209, January 2006.

[170] T. Niknam and E.A. Farsani, "A hybrid self-adaptive particle swarm optimization and modified shuffled frog leaping algorithm for distribution feeder reconfiguration," *Engineering Applications of Artificial Intelligence*, vol. 23, no. 8, pp. 1340–1349, December 2010.

[171] T. Niknam, "An efficient hybrid evolutionary algorithm based on PSO and ACO for distribution feeder reconfiguration," *European Transactions on Electrical Power*, vol. 20, no. 5, pp. 575–590, July 2010.

[172] T. Niknam, R. Khorshidi, and B. Bahmani Firouzi, "A hybrid evolutionary algorithm for distribution feeder reconfiguration," *Sadhana*, vol. 35, no. 2, pp. 139–162, April 2010.

[173] T. Niknam, "An efficient hybrid evolutionary algorithm based on PSO and HBMO algorithms for multi-objective distribution feeder reconfiguration," *Energy Conversion and Management*, vol. 50, no. 8, pp. 2074–2082, August 2009.

[174] A. Ahuja, S. Das, and A. Pahwa, "An AIS-ACO hybrid approach for multi-objective distribution system reconfiguration," *IEEE Transactions on Power Systems*, vol. 22, no. 3, pp. 1101–1111, August 2007.

[175] E.M. Carreno, R. Romero, and A. Padilha-Feltrin, "An efficient codification to solve distribution network reconfiguration for loss reduction problem," *IEEE Transactions on Power Systems*, vol. 23, no. 4, pp. 1542–1551, November 2008.

[176] Y.K. Wu, C.Y. Lee, L.C. Liu, and S.H. Tsai, "Study of reconfiguration for the distribution system with distributed generators," *IEEE Transaction on Power Delivery*, vol. 5, no. 3, pp. 1678–1685, July 2010.

[177] P. Kumar, I. Ali, M.S. Thomas, and S. Singh, "Imposing voltage security and network radiality for reconfiguration of distribution systems using efficient heuristic and metaheuristic approach," *IET Generation, Transmission & Distribution*, vol. 11, no. 10, pp. 2457–2467, July 2017.

[178] N. G. Paterakis, A. Mazza, S. F. Santos, O. Erdinç, G. Chicco, A. G. Bakirtzis, and J. P. S. Catalão, "Multi-Objective Reconfiguration of Radial Distribution Systems Using Reliability Indices," in *IEEE Transactions on Power Systems*, vol. 31, no. 2, pp. 1048–1062, March 2016.

[179] S. Ghasemi, "Balanced and unbalanced distribution networks reconfiguration considering reliability indices," *Ain Shams Engineering Journal*, vol. 9, no. 4, pp. 1567–1579, December 2018.

[180] J.R.S. Mantovani, F. Casari, and R.A. Romero, "Reconfiguração de sistemas de distribuição radiais utilizando o critério de queda de tensão," *SBA Controle & Automação*, vol. 11, no. 3, pp. 150–159, 2000.
[181] J.C. Cebrian and N. Kagan, "Reconfiguration of distribution networks to minimize loss and disruption costs using genetic algorithms," *Electric Power Systems Research*, vol. 80, no. 1, pp. 53–62, January 2010.
[182] M.A.N. Guimarães, C.A. Castro, and R. Romero, "Distribution systems operation optimisation through reconfiguration and capacitor allocation by a dedicated genetic algorithm," *IET Generation, Transmission, and Distribution*, vol. 4, no. 11, pp. 1213–1222, April 2010.
[183] N. Gupta, A. Swarnkar, K.R. Niazi, and R.C. Bansal, "Multi-objective reconfiguration of distribution systems using adaptive genetic algorithm in fuzzy framework," *IET Generation, Transmission and Distribution*, vol. 4, no. 12, pp. 1288–1298, December 2010.
[184] A. Ahuja, A. Pahwa, B.K. Panigrahi, and S. Das, "Pheromone-based crossover operator applied to distribution system reconfiguration," *IEEE Transactions on Power Systems*, vol. 28, no. 4, pp. 4144–4151, November 2013.
[185] J.F. Franco, M.J. Rider, M. Lavorato, and R. Romero, "A mixed-integer LP model for the reconfiguration of radial electric distribution systems considering distributed generation," *Electric Power Systems Research*, vol. 97, pp. 51–60, April 2013.
[186] M.N.A. Guimarães and C.A. Castro, "Reconfiguration of distribution systems for loss reduction using tabu search," *The 15th Power Systems Computation Conference*, Belgium, pp. 1–6, August 2005.

19

Genetic Algorithm Application in Distribution System Reconfiguration

Meisam Mahdavi[1], *Pierluigi Siano*[2], *Hassan Haes Alhelou*[3], *and Sanjeevikumar Padmanaban*[4]

[1] *Bioenergy Research Institute, São Paulo State University, Ilha Solteira, SP, Brazil*
[2] *Department of Management & Innovation Systems, University of Salerno, Salerno, Italy*
[3] *Department of Electrical Power Engineering, Tishreen University, Lattakia, Syria*
[4] *CTIF Global Capsule (CGC) Laboratory, Department of Business Development and Technology, Aarhus University, Herning, Denmark*

19.1 Introduction

Power is lost in the form of heat during power distribution. Power loss of a distribution network is more than generation and transmission systems [1–15]. Active power loss minimization is important because it affects the operational costs and the voltage profile [16, 17]. Capacitor placement, distributed generation (DG) allocation [18–24], and distribution system reconfiguration (DSR) [25] are common ways to decrease the power distribution loss [26–28].

Although the main purpose of DSR is power loss minimization, it also includes other objectives such as voltage profile improvement, network reliability enhancement [29–33], line loading optimization [34–39], network adequacy [40–46], supply capacity expansion, power quality maximization, load balancing, DG penetration increase, service restoration, and quick fault isolation [17, 47–51]. In this process, the configuration of the system changes by opening sectional branches and closing tie switches in order to maintain radiality of the distribution network [17, 52, 53].

DSR is a nonlinear complex optimization problem that can be solved by mathematical, heuristic, and meta-heuristic methods [54–63]. One of the most important meta-heuristic approaches for solving nonlinear equations and complex optimization problems is use of a genetic algorithm (GA) [64]. GA can be applied to several optimization problems such as DSR [26, 65], transmission expansion planning (TEP) [66–69], generation expansion planning (GEP) [70], and generation and transmission expansion planning (GTEP) [71, 72]. This algorithm can find the optimal solution randomly based on the individual selection. A good initial estimation is not necessary for solving the problem. In simple terms, a complicated problem can be solved by a weak initial estimation of the solution and then this solution is corrected during a fitness evolutionary method. Generally, GA includes three basic operators (reproduction, crossover, and mutation) that conduct chromosomes into the best fitness condition [73–75].

Active Electrical Distribution Network: A Smart Approach, First Edition. Edited by Baseem Khan, Josep M. Guerrero, Sanjeevikumar Padmanaban, Hassan Haes Alhelou, Om Prakash Mahela, and Sudeep Tanwar.

A selection operator chooses population chromosomes for reproduction. A chromosome with a higher probability will be selected. Therefore, the selection operator chooses more appropriate individuals by tournament selection. After selecting parent chromosomes, the crossover operator is implemented on each parent pairs. This operator exchanges values of bits between two parent chromosomes. Then, in order to create the new generation, each child obtained from the crossover is subjected to the mutation. The mutation operator improves GA ability in order to approach an approximate solution by maintaining an adequate genetic variety. In this operation, it is necessary that the whole solution space is used to search for the most optimal solution. In other terms, it prevents losing information (values) of genes. Usually, the mutation operator is applied randomly to individuals of a population with a small probability.

In this chapter, GA is described in detail and then several examples are presented to show its efficiency and effectiveness in solving the problem of distribution network reconfiguration.

19.2 DSR Mathematical Model

DSR is a complex and combinatorial optimization problem including integer and continuous variables with nonlinear and linear constraints. An efficient solution for this problem requires the selection of the most appropriate topology among all feasible configurations [2, 76–78].

19.2.1 Basic Model

As mentioned in Section 2 of Chapter 18, the DSR problem with an objective of loss (P_{Loss}) minimization can be formulated as follows [26, 79]:

$$\min P_{Loss} = \sum_{ij \in \Omega^l} R_{ij} \left| I_{ij} \right|^2 \tag{19.1}$$

Subject to

$$P_i^S + \sum_{k=1}^{NB} P_{ki} = \sum_{j=1}^{NB} P_{ij} + \sum_{j=1}^{NB} R_{ij} \left| I_{ij} \right|^2 + P_i^D, \qquad \forall ij, ki \in \Omega^l, i = 1,2,\ldots,NB \tag{19.2}$$

$$Q_i^S + \sum_{k=1}^{NB} Q_{ki} = \sum_{j=1}^{NB} Q_{ij} + \sum_{j=1}^{NB} X_{ij} \left| I_{ij} \right|^2 + Q_i^D, \qquad \forall ij, ki \in \Omega^l, i = 1,2,\ldots,NB \tag{19.3}$$

$$V_{min} \leq \left| V_i \right| \leq V_{max}, \qquad \forall i = 1,2,\ldots,NB \tag{19.4}$$

$$\left| V_i \right|^2 = \left| V_j \right|^2 - 2\left(R_{ij} P_{ij} + X_{ij} Q_{ij} \right) + \left| Z_{ij} \right|^2 \left| I_{ij} \right|^2, \qquad \forall ij \in \Omega^l, i = 1,\ldots,NB \tag{19.5}$$

$$0 \le \left|I_{ij}\right| \le I_{ij}^{max}, \qquad \forall ij \in \Omega^l \tag{19.6}$$

$$\sum_{ij \in \Omega_l} y_{ij} = NB - 1 \tag{19.7}$$

$$\left|S_{ij}\right|^2 = P_{ij}^2 + Q_{ij}^2, \qquad \forall ij \in \Omega^l \tag{19.8}$$

$$\left|S_{ij}\right|^2 = \left|V_j\right|^2 \left|I_{ij}\right|^2, \qquad \forall ij \in \Omega^l, j = 1,2,\ldots,NB \tag{19.9}$$

$$\left|P_{ij}\right| \le P_{ij}^{max} y_{ij}, \qquad \forall ij \in \Omega^l \tag{19.10}$$

$$P_{ij}^{max} = V_{max} I_{ij}^{max} \tag{19.11}$$

$$\left|Q_{ij}\right| \le Q_{ij}^{max} y_{ij}, \qquad \forall ij \in \Omega^l \tag{19.12}$$

$$Q_{ij}^{max} = V_{max} I_{ij}^{max} \tag{19.13}$$

In the equation above we have
Ω^l: Set of all branches (including normal and tie lines).

NB: Total number of buses.

R_{ij}, X_{ij}: Resistance and reactance of branch *ij*.

$\left|I_{ij}\right|$: Amplitude of current flow from bus *i* to *j*.

P_i^S, Q_i^S: Active and reactive power provided by the substation at bus *i*.

P_i^D, Q_i^D: Active and reactive power demand at bus *i*.

P_{ij}, Q_{ij}: Active and reactive power flow from bus *i* to *j* (branch *ij*).

$\left|V_i\right|$: Voltage magnitude of bus *i*.

V_{min}, V_{max}: Lower and upper bounds of bus voltage limits.

$\left|S_{ij}\right|$, $\left|Z_{ij}\right|$: Amplitudes of apparent power and impedance of branch *ij*.

I_{ij}^{max}: Current rating limit of branch *ij*.

y_{ij}: A binary variable that represents the status of each switch (0 and 1 for open and closed switches, respectively).

P_{ij}^{max}: Real power flow capacity of branch *ij*.

$\left|P_{ij}\right|$: Absolute value of P_{ij}.

Q_{ij}^{max}: Reactive power flow capacity of branch *ij*.

$\left|Q_{ij}\right|$: Absolute value of Q_{ij}.

19.2.2 GA-Based Model

In order to solve the reconfiguration problem using a genetic algorithm, the DSR problem is formulated as follows:

$$\min F = P_{Loss} + \lambda V \tag{19.14}$$

s.t.

$$S_i^S + \sum_{k=1}^{NB} S_{ki} = \sum_{j=1}^{NB} S_{ij} + \sum_{j=1}^{NB} S_{ij}^L + S_i^D, \qquad \forall ij, ki \in \Omega^l, i = 1,2,\ldots,NB \tag{19.15}$$

$$S_i^S = P_i^S + jQ_i^S, \qquad i = 1,2,\ldots,NB \tag{19.16}$$

$$S_i^D = P_i^D + jQ_i^D, \qquad i = 1,2,\ldots,NB \tag{19.17}$$

$$S_{ij} = P_{ij} + jQ_{ij}, \qquad \forall ij \in \Omega^l \tag{19.18}$$

$$S_{ij}^L = R_{ij}\left|I_{ij}\right|^2 + jX_{ij}\left|I_{ij}\right|^2, \qquad \forall ij \in \Omega^l \tag{19.19}$$

$$V_j = V_i - I_{ij}Z_{ij}, \qquad \forall ij \in \Omega^l, i = 1,\ldots,NB \tag{19.20}$$

$$Z_{ij} = R_{ij} + jX_{ij}, \qquad \forall ij \in \Omega^l \tag{19.21}$$

$$\text{rank}(A) = NB - 1 \tag{19.22}$$

$$V_{min} \le \left|V_j\right| \le V_{max}, \qquad \forall j = 1,2,\ldots,NB \tag{19.23}$$

$$\left|I_{ij}\right| \le I_{ij}^{max}, \qquad \forall ij \in \Omega^l \tag{19.24}$$

$$S_{ij} = V_j I_{ij}^*, \qquad \forall ij \in \Omega^l, j = 1,2,\ldots,NB \tag{19.25}$$

where

λ: Voltage penalty constant (its value will be 0 if the associated voltage constraint in Eq. (19.23) is not violated and will be a large amount if the associated constraint is violated).

V: Squared sum of the violated voltage constraints.

S_i^S: Apparent power provided by substation at substation bus *i*.

S_i^D: Apparent power demand at bus *i*.

S_{ij}: Apparent power of branch *i*–*j*.

S_{ij}^L: Apparent power losses of branch *i*–*j*.

Z_{ij}: Impedance of branch *i*–*j*.

A: Bus incidence matrix.

rank (*A*): Number of linearly independent rows or columns of matrix *A*.

I_{ij}^*: Conjugate value of I_{ij}.

Equation (19.22) states the radiality constraint. The DSR model can be easily solved by the GA.

19.3 Genetic Algorithm

In recent decades, meta-heuristics were popular optimization methods to solve the reconfiguration problem. In meta-heuristics based on biological or natural phenomena like genetic algorithms, the implementation method of recombination and mutation operators has to be defined. The method of representation of a solution and adding or removing bit values from the under construction proposal have to be determined [65]. The genetic algorithm is the part of evolutionary algorithms that is based on natural biological evolution. In 1960, it was introduced by John Holland at the University of Michigan and then in 1970 it was promoted by him and his colleagues [54]. GA is a method that can solve nonlinear optimization problems like DSR and nonlinear equations randomly. Individual selection is one of the bases of the genetic algorithm. A good initial estimation is not necessary for solving the problem. In simple terms, a problem can be solved by a weak initial estimation and then is corrected during fitness evolution. In order to code the distribution branches in GA, two following methods exist:

1. Binary codification of a branch (BCGA).
2. Decimal codification of a branch (DCGA) [80, 81].

19.3.1 Binary Codification GA (BCGA)

This type of genetic algorithm employs binary chromosomes (individual) that are probable solutions of the problem. A binary string with bit values of 0 and 1 is used to exhibit an individual. The number of bits in each chromosome depends on parameters of a problem. In an optimization problem, a search space including all possible solutions exists. In each evolutionary stage (generation), the individuals of a population are evaluated by their fitness function values. In the reproduction process, each population individual collaborates in optimization according to its fitness function. In simple terms, higher fitness chromosomes have more of a chance for selection as parents while chromosomes with a lower probability have less chance to be selected [54]. During reproduction, the mutation operator is used randomly. The most optimal chromosome is chosen as the problem solution in an iterative process. Generally, GA includes three basic operators (selection or reproduction, crossover, and mutation) that conduct chromosomes into the best fitness. These operators are explained as follows [41, 82]:

- **Selection**

This operator chooses population individuals for reproduction. More appropriate individuals (higher probability chromosomes) are chosen. The main idea of selection is to choose better chromosomes in a competitive tournament, in which each chromosome tries to remain in the population. The roulette wheel technique is a popular method to use when choosing chromosomes for mating, where a section is allocated to each individual in a roulette wheel where its area is equal to the chromosome's fitness value. This process increases the probability of selection of more appropriate chromosomes for the next generation. Then crossover is employed to create children from the parent chromosomes [56].

- **Crossover**

The crossover exchanges values of genes between two parents. A number of individuals are selected randomly using crossover probability. Afterwards, a position is selected randomly for each chromosome pair. Then each individual pair exchanges their bit values. There are two types of crossover: single and uniform crossovers [1].

- **Single crossover**

According to Figure 19.1, in a single crossover a random position (a point) in the chromosomes is selected and sub-sequences are swapped before this point and after that in order to make a new chromosome from the parent [54].

- **Uniform crossover**

This kind of crossover selects several random points and swaps sub-sequences [54]. Figure 19.2 shows a three points crossover.

- **Mutation**

In order to create a new generation, the mutation operator is implemented on each offspring obtained from the crossover. The mutation operator improves the GA ability in order to approach an approximate solution by maintaining an adequate genetic variety. In this operation, it is necessary that the whole solution space is used in the search for the most optimal solution. The values of genes (bites) are randomly changed with a small mutation probability (e.g. from 0.01 to 0.1). The value of each bit is changed from 0 to 1 or inversely. Mutation is categorized into single and multiple, as shown in Figure 19.3 [39, 43, 80, 83, 84].

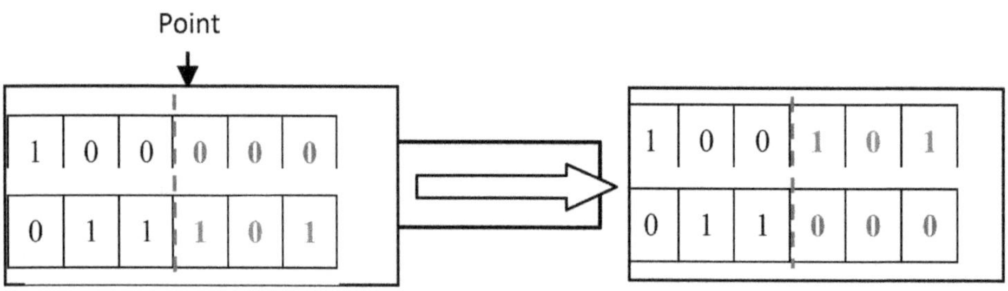

Figure 19.1 Single point.

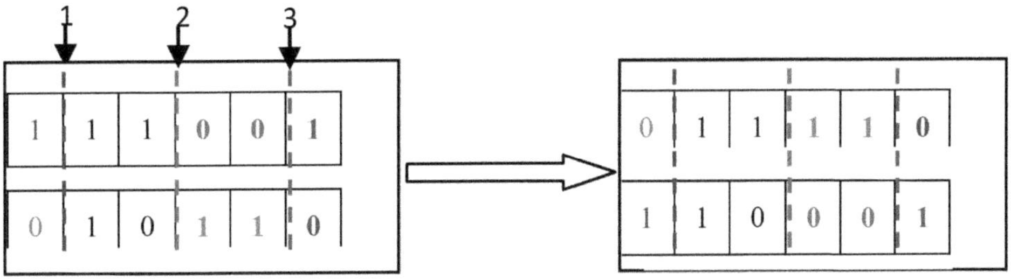

Figure 19.2 Multiple points.

19.3.2 Decimal Codification GA (DCGA)

The second codification method (DCGA) is more efficient than BCGA for network reconfiguration due to the following reasons [3, 85]:

1. Removal problems, happening during codification and decodification.
2. Decrease of algorithm divergence probability because of creation of very different children from a parent.

In this chapter, the decimal codification genetic algorithm (DCGA) technique was applied to solve Eqs (19.14) to (19.25). In the first step, a population with d chromosomes ($d = 20$) is randomly created:

$$Y = \begin{bmatrix} Y_1 \\ Y_2 \\ \vdots \\ Y_d \end{bmatrix} \tag{19.26}$$

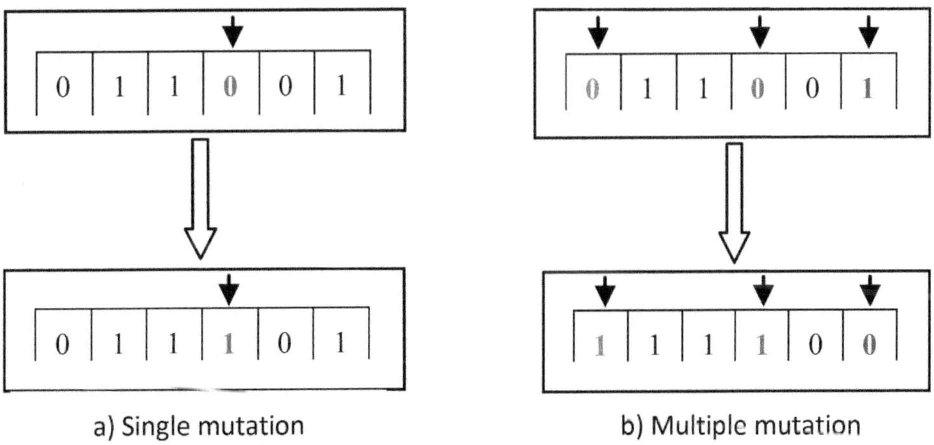

Figure 19.3 Mutation operator.

In Eq. (19.26), Yd is the dth chromosome in population Y. Vector Y includes integer number bites. These bites show variables of the problem (branch numbers):

$$Y_d = \left[n_1, n_2, \ldots, n_i, \ldots, n_{|\Omega^t|}\right] \tag{19.27}$$

where ni and Ω^t indicate the branch number of line i and set of tie lines, respectively:

$$n_i = \left[1, 2, \ldots, |\Omega^l|\right] \tag{19.28}$$

$$|\Omega^l| = |\Omega^t| + NB - 1 \tag{19.29}$$

In these equations, $|\Omega^l|$ and $|\Omega^t|$ are the numbers of all branches and tie lines. Equation (19.30) describes a typical population with three chromosomes (d = 3) for the example system shown in Figure 9.4. This seven-bus system includes six normal lines (sectionalized branches) and three tie lines (normally open branches):

$$\begin{aligned} Y_1 &= [2, 6, 8] \\ Y_2 &= [1, 7, 6] \\ Y_3 &= [1, 6, 8] \end{aligned} \tag{19.30}$$

where Y_1 proposes that branches 2 (normal line), 6 (tie line), and 8 (normal line) have to be open and others (branches 1, 3, 4, 5, and 7) have to be closed.

The branch current and nodal voltage magnitude of radial topologies (chromosomes that satisfy the radiality constraint (19.22)) are computed by forward–backward sweep load flow (radial power flow). Then, power losses (PLoss) and voltage and current penalties (λV and μI) are calculated, and consequently the objective function (19.14) is determined for feasible chromosomes (radial topologies). Then the selection chooses more appropriate chromosomes for reproduction. The reproduction regenerates chromosomes according to their objective function values (Eq. (19.14)). Accordingly, the selection probability of chromosomes with appropriate objective functions is higher than other chromosomes.

The crossover is implemented on each selected pair. In this way, random places between two variables are chosen for crossover regarding crossover probability (PC). Equations

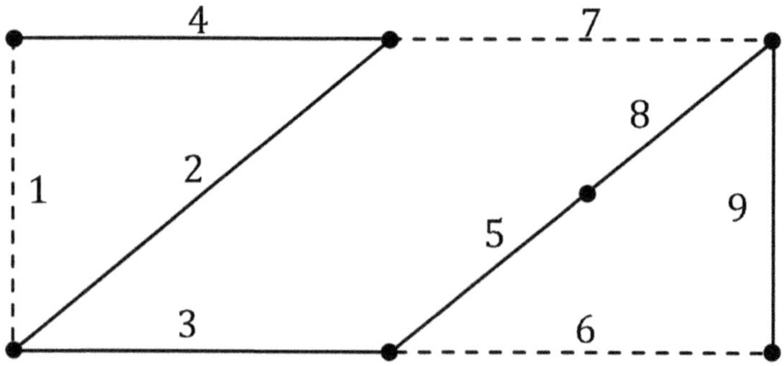

Figure 19.4 Example system.

(19.31) and (19.32) show a pair of the selected chromosomes Y_1 and Y_2 before and after the crossover, respectively:

$$\begin{aligned} Y_1 &= [2,6,8] \\ &\quad \updownarrow \updownarrow \\ Y_2 &= [1,7,6] \end{aligned} \tag{19.31}$$

$$\begin{aligned} Y_1' &= [2,7,6] \\ Y_2' &= [1,6,8] \end{aligned} \tag{19.32}$$

Then the mutation operator is applied to these chromosomes with the mutation probability (PM). It should be noted that constraint (19.33) has to be met, so that the branch number does not violate its limits:

$$1 \le n_i \le |\Omega^l| \tag{19.33}$$

Equations (19.34) and (19.35) show Y_2', before and after mutation, respectively:

$$Y_2' = [1,6,8] \tag{19.34}$$

$$\widehat{Y_2} = [\overset{1}{\hat{4}},6,\overset{8}{\hat{5}}] \tag{19.35}$$

Finally, weak chromosomes are replaced by elite chromosomes (elitist strategy). After replacement, the process is repeated by evaluation of the objective function (19.14) and is ended by defining maximum iterations. The DCGA flowchart can be seen in Figure 19.5.

19.3.3 Improved DCGA (IDCGA)

In order to enhance the convergence speed of the DCGA algorithm, an improved DCGA (IDCGA) is proposed by inserting $\lambda V = \mu I = 0$ in the objective function (19.14). Moreover, the mutation probability is defined as follows instead of a constant value ($PM = 0.1$):

$$P_M(t) = P_M^0 - \left(P_M^0 - P_M^{max}\right) / t_{max} t \tag{19.36}$$

where

t: The index for algorithm iteration (generation).

$P_M(t)$: The mutation probability (P_M) at iteration t.

P_M^0: The start point of $P_M(t)$ (P_M in the DCGA method).

P_M^{max}: The end point of $P_M(t)$.

t_{max}: The maximum number of iterations.

19.3.4 Efficient DCGA (EDCGA)

For improving both accuracy and computational time of the DCGA method, an efficient DCGA (EDCGA) can be presented for the DSR problem. The EDCGA employed in this chapter has better characteristics than the conventional genetic algorithms. EDCGA is efficient for solving large-scale multiconstraint DSR problems, in which many unfeasible solutions (non-radial topologies) are generated in the optimization process.

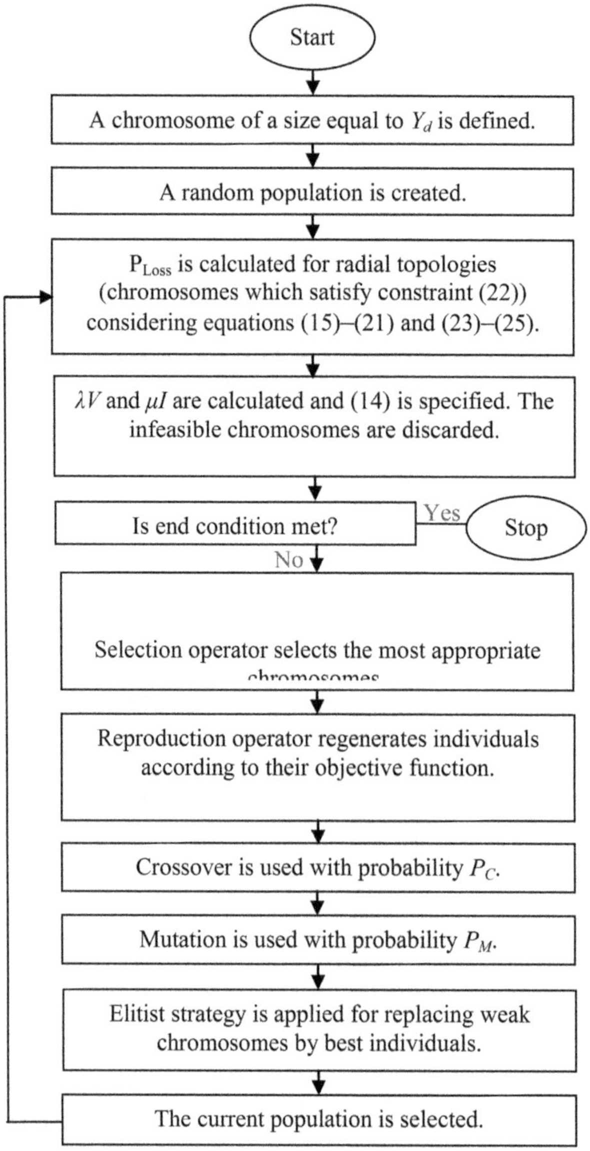

Figure 19.5 Flowchart of the DCGA method.

In the first step ($t = 0$), a d-sized population ($d = 20$) including feasible chromosomes (radial topologies) is constructed randomly in which constraint (19.22) must be satisfied:

$$Y = \begin{bmatrix} Y_1 \\ \vdots \\ Y_i \\ \vdots \\ Y_d \end{bmatrix} \tag{19.37}$$

The representation form (codification) of each chromosome (solution proposal) of the initial population (Eq. (19.37)) is very important. This representation (codification) should present high-quality solutions and can determine infeasible proposals. Furthermore, genetic operators that can transmit the current proposal to a new one depend upon the representation procedure. For the DSR problem, it is better that each chromosome shows a radial topology. The generation of non-radial solutions is not useful. An illustration of a solution proposal including only open branches and their related independent loops is the best way, because it provides only radial solutions.

A Prim algorithm is adopted here to form an initial population. In this method, the optimal solution can be found because of generation of a minimum spanning tree of the graph. The construction of random radial topologies in each step without load flow calculations can be done using the proposed modified Prim algorithm. If higher weighting values are allocated to more important branches, this algorithm can find radial topologies.

In the adopted algorithm, the minimization process of objective function (19.14) is converted to the maximization of cost function J, where $c_{ij}$and $P_{Loss_{ij}}$ are the value (weight) and power losses of branch ij, respectively:

$$\max J = \sum_{ij \in \Omega^l} \frac{c_{ij}}{P_{Loss_{ij}}} \tag{19.38}$$

In this method, steps mentioned in the flowchart of Figure 19.6 need to be considered.

A typical chromosome (Yi) for the system shown in Figure 19.4 is illustrated in Eq. (19.39). This chromosome consists of two parts. The first part ($NB-1$) includes the branches of the radial topology and the second part ($|\Omega^t|$) consists of open branches:

$$Y_i = [\overbrace{\overbrace{2,4,3,5,8,9}^{NB-1},\overbrace{1,7,6}^{|\Omega^t|}}^{|\Omega^l|}] \tag{19.39}$$

where Yi shows that branches 2, 4, 3, 5, 8, and 9 have to be closed while branches 1, 7, and 6 should be open.

The branch current and nodal voltage magnitude of all chromosomes as well as power losses (PLoss) are calculated using radial power flow (forward–backward sweep load flow). Then, the selection operator chooses more appropriate chromosomes (Yi) for recombination. In the recombination process, two new radial proposals (Eq. (19.61)) are constructed by two existing radial topologies (Eq. (19.60)). First, the recombination position is selected in the first parts of Y_1 and Y_2 between points $(NB-1)/2$ and $(NB-2)$. In the offspring chromosomes (Y_1 and Y_2), genes (elements) that form a loop are allocated to the second parts and those that do not form a loop are incorporated in the first ones, in which repetitive elements are discarded. In this way, only radial offspring chromosomes are constructed.

$$\begin{gathered} Y_1 = [\overbrace{2,4,3,5,\underbrace{8,9}_{\downarrow}}^{\text{Part 1}},\overbrace{1,\underbrace{7,6}_{\downarrow}}^{\text{Part 2}}] \\ \\ Y_2 = [\underbrace{1,3,4,5,\overbrace{7,9}^{\uparrow}}_{\text{Part 1}},\underbrace{\overbrace{2,6}^{\uparrow},8}_{\text{Part 2}}] \end{gathered} \tag{19.60}$$

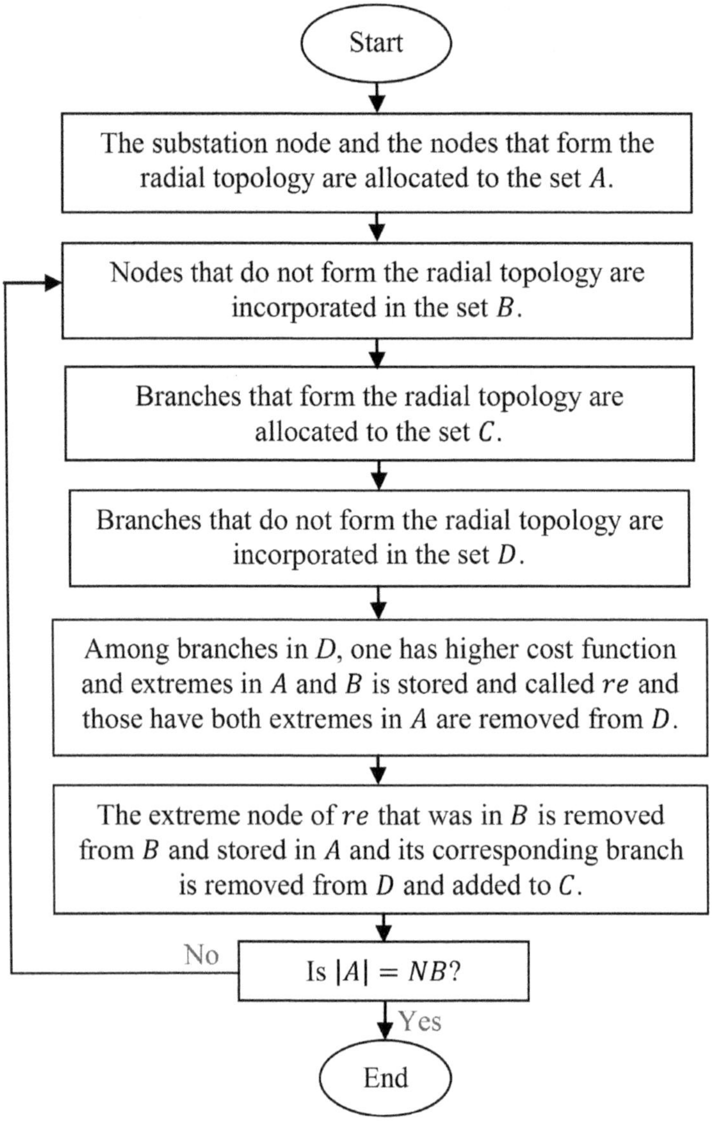

Figure 19.6 Flowchart of the modified Prim algorithm.

$$Y_1' = [2,4,3,5,7,9,1,6,8]$$
$$Y_2' = [1,3,4,5,8,9,2,7,6] \tag{19.61}$$

For example, in Eq. (19.60), the recombination point was chosen between the fourth element and the fifth element. Thus, the branches 2, 4, 3, and 5 are transmitted to Y_1'. Branch 1 is given to the second part of Y_1', branches 3, 4, and 5 are eliminated, branches 7 and 9 are added to the first part, because they do not create a loop, and branches 6 and 8 are added to the second part. This kind of recombination has two main properties: (1) radial solutions are generated and (2) most of the branches of the new radial topologies are in initials chromosomes Y_1 or Y_2.

After that, objective function (19.14) can be evaluated again and this process is repeated until the end condition is satisfied. Figure 19.7 illustrates the EDCGA flowchart.

This technique has a better performance than recent genetic algorithms proposed by other specialized literature. The most important feature of the proposed genetic algorithm is prevention of the complete diversity of the current population. Consequently, only radial solutions exist in the final population. Thus, the size of the population can be bigger for more complex problems.

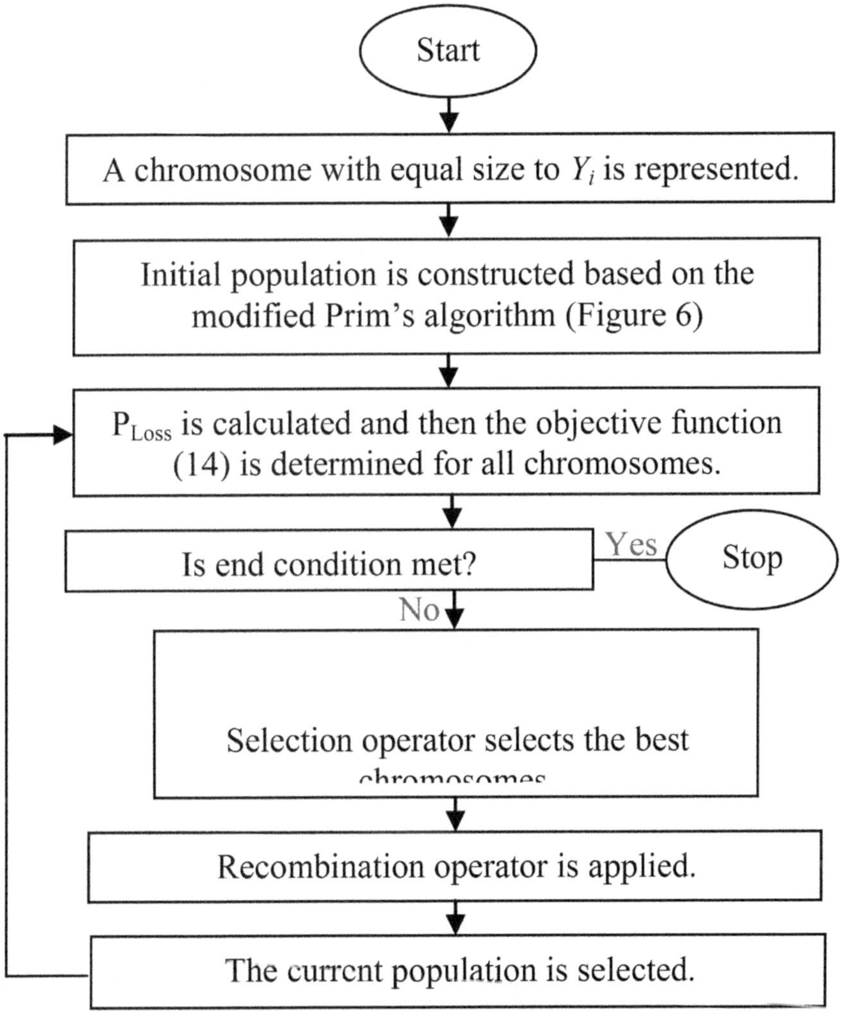

Figure 19.7 Flowchart of the EDCGA method.

19.4 Examples

The proposed GA-based DSR model is applied to several test systems using DCGA, IDCGA, and EDCGA, and the results are presented in comparison with other GA methods. It should be mentioned that V_{min} and V_{max} were considered to be 0.93 and 1 per unit (p.u.), respectively, in which the voltage magnitude of substation nodes has been fixed on 1 per unit.

Example 1: Consider System 1 where all details are given in Section 2 of Chapter 17. Solve the DSR problem for this network, but only where the power loss is to be a minimum. The base power is 1 MVA and the base voltage is 12.66 kV

Solution: The proposed model was applied to System 1 using genetic algorithms and the results are listed in Table 19.1. It should be noted that the best voltage magnitude before and after reconfiguration is 1 p.u. at substation bus 1. Buses 5 and 7 have the worst voltage magnitudes in the original and reconfigured networks, respectively. Furthermore, the solutions were ranked based on the execution time in Table 19.2.

The results of Example 1 show that although DCGA and IDCGA could find optimal solutions as efficient as EDCGA, EDCGA is faster than the DCGA and IDCGA techniques.

Example 2: Consider System 2 described in Section 2 of Chapter 17. Solve the DSR problem for this system

Solution: The model was implemented on System 2 and the results were listed in Table 19.3. It should be noted that the best voltage magnitudes before and after reconfiguration is 1 p.u. at bus 1. Bus 10 has the worst voltage magnitude before and after the network reconfiguration. Table 19.4 ranks the solutions based on the computational time.

According to Table 19.4 of Example 2, IDCGA and EDCGA can find the optimal solution more quickly than DCGA.

Table 19.1 Numerical results for Example 1.

Items		Power loss (kW)		Worst voltage (p.u.)		
Methods	Open switches	Before	After	Before	After	Run time (s)
DCGA	5	1.44	0.82	0.995	0.99	0.06
IDCGA	5	1.44	0.82	0.995	0.99	0.06
EDCGA	5	1.44	0.82	0.995	0.99	0.04

Table 19.2 Ranking of methods based on execution time for Example 1.

Methods	Run time (s)
DCGA and IDCGA	0.06
EDCGA	0.04

Table 19.3 Numerical results for Example 2.

Items		Power loss (kW)		Worst voltage (p.u.)		
Methods	Open switches	Before	After	Before	After	Run time (s)
DCGA	3, 9	3.55	1.31	0.9970	0.998	0.06
IDCGA	3, 9	3.55	1.31	0.9970	0.998	0.05
EDCGA	3, 9	3.55	1.31	0.9970	0.998	0.05

Table 19.4 Ranking of methods based on execution time for Example 2.

Methods	Run time (s)
DCGA	0.06
IDCGA and EDCGA	0.05

Example 3: Solve the DSR problem for System 3 where all of its data, such as resistance and reactance of branches as well as nodal active and reactive demand, have been presented in Chapter 17. The base power and voltage are considered to be 100 MVA and 23 kV

Solution: The proposed GA approaches were applied to this network and the results in comparison with other GA methods are provided in Table 19.5.

It should be noted that the best (maximum) voltage magnitudes before and after reconfiguration are 1 p.u. at substation buses 1, 2, and 3. From a voltage profile aspect, the worst (minimum) voltage magnitude is related to bus 12 in original and reconfigured networks.

Table 19.5 Numerical results for Example 3.

Items		Power loss (kW)		Worst voltage (p.u.)		
Methods	Open switches	Before	After	Before	After	Run time (s)
Refined GA (RGA) [86]	17, 19, 26	511.4	466.1	0.9693	–	–
GA [87, 88]	17, 19, 26	511.4	466.1	0.9693	0.9716	2.32
Efficient GA (EGA) [89]	17, 19, 26	511.44	466.12	0.9693	0.97158	2.0269
GA based on Matroid theory (GAMT) [90]	17, 19, 26	511.5	466.1	0.9693	0.9716	1.35
Non-revisiting GA (NrGA) [91]	17, 19, 26	511.44	466.13	0.969	0.972	0.45
Non-dominated sorting GA (NSGA) [92]	17, 19, 26	511.44	466.13	0.969	0.972	0.27
DCGA	17, 19, 26	511.4	465.2	0.9631	0.9716	1.1
IDCGA	17, 19, 26	511.4	462.2	0.9631	0.9716	0.11
EDCGA	17, 19, 26	511.4	462.2	0.9631	0.9716	0.1

Furthermore, the solution methods presented for solving the DSR problem were ranked based on power loss reduction, worst voltage increase, and computing time in Tables 19.6–19.8, respectively.

Table 19.6 of Example 3 shows that the proposed genetic algorithms (DCGA, IDCGA, and EDCGA) are more precise and accurate than other GA methods, while Table 19.8 indicates a higher convergence speed of IDCGA and EDCGA compared to DCGA and other GA algorithms.

Table 19.6 Ranking of methods based on loss reduction for Example 3.

Methods	Power loss (kW)
[91, 92]	466.13
[89]	466.12
[86–88, 90]	466.1
DCGA, IDCGA and EDCGA	462.2

Table 19.7 Ranking of methods regarding voltage improvement for Example 3.

Methods	Worst voltage (p.u.)
[89]	0.97158
[86–88, 90], DCGA, IDCGA and EDCGA	0.9716
[91, 92]	0.972

Table 19.8 Ranking of methods based on execution time for Example 3.

Methods	Run time (s)
[87, 88]	Long
[89]	6
[90]	4.5
DCGA	2.32
[91]	2.07
[92]	2.027
IDCGA	0.45
EDCGA	0.27

Example 4: Consider System 4 with data described in Chapter 17. Apply the DSR problem to this test system

Solution: Results for this network are illustrated in Tables 19.9 and 19.10.

From results of Example 4, it can be concluded that the EDCGA method can reach the optimal solution faster than DCGA and IDCGA, respectively.

Table 19.9 Numerical results for Example 4.

Items		Power loss (kW)		Worst voltage (p.u.)		
Methods	Open switches	Before	After	Before	After	Run time (s)
DCGA	11, 20, 21, 28	47.05	39.7	0.91	0.94	27
IDCGA	11, 20, 21, 28	47.05	39.7	0.91	0.94	0.9
EDCGA	11, 20, 21, 28	47.05	39.7	0.91	0.94	0.7

Table 19.10 Ranking of methods based on computing time for Example 4.

Methods	Run time (s)
DCGA	27
IDCGA	0.9
EDCGA	0.7

Example 5: Solve the DSR problem for System 5. All data of this system can be found in Chapter 17. The base values are 1 MVA and 18.6 kV

Solution: The proposed genetic algorithms were applied to this test system in comparison with the conventional GA technique and results were provided for the different techniques in Tables 19.11–19.14. The best voltage magnitude before and after reconfiguration is 1 p.u. at substation bus 1, while the worst voltages before and after reconfiguration belong to buses 29 and 14, respectively.

According to tables of Example 5, it can be seen that EDCGA and IDCGA find more optimal solutions compared to DCGA and GA. Furthermore, EDCGA converges to the best solution quicker than other genetic algorithms.

Example 6: Consider System 6 for applying the DSR problem using genetic algorithms. The data of this test system are available in Chapter 17. The MVA and kV base are 100 and 12.66, respectively

Table 19.11 Numerical results for Example 5.

Items		Power loss (kW)		Worst voltage (p.u.)		
Methods	Open switches	Before	After	Before	After	Run time (s)
GA [93]	3, 7, 10, 25, 32, 33	1240	368	0.8251	0.92	–
DCGA	3, 6, 10, 25, 31, 33	1239	373.3	0.8251	0.88	61.24
IDCGA	3, 5, 7, 10, 25, 33	1239	361.04	0.8251	0.94	1.34
EDCGA	3, 5, 7, 10, 25, 33	1239	361.04	0.8251	0.94	1.05

Table 19.12 Ranking of methods based on loss reduction for Example 5.

Methods	Power loss (kW)
DCGA	378.32
[93]	368
IDCGA and EDCGA	361.04

Table 19.13 Ranking of methods regarding voltage improvement for Example 5.

Methods	Worst voltage (p.u.)
DCGA	0.88
[93]	0.92
IDCGA and EDCGA	0.94

Table 19.14 Ranking of methods based on execution time for Example 5.

Methods	Run time (s)
[93]	–
DCGA	61.24
IDCGA	1.34
EDCGA	1.05

Solution: The proposed formulation was applied to this test system and the results of Example 6 are listed in Tables 19.15–19.18. Also, the voltage magnitude of the substation bus (node 0) is adjusted to 1 p.u. The solution methods are arranged in Tables 19.16–19.18 from viewpoints of accuracy, convergence speed, and voltage stability, respectively.

It is obvious that more accurate solutions cannot be found in a very short time because the algorithm needs to search more precisely. It can be seen that the IDCGA and EDCGA solve the DSR problem more precisely than other GA methods in shorter computing times, but EDCGA is much faster than IDCGA.

Example 7: In this example, System 7 is used to solve the DSR problem. The system data and information are available in Chapter 17. There are seven capacitors in buses 9, 13, 24, 32, 36, 42, and 48 with sizes of 100, 75, 100, 200, 200, 350, and 200 kVAr, respectively, in this system

Solution: The results are listed in Tables 19.19–19.22 by implementing the DCGA, IDCGA, and EDCGA on this test system in comparison with the GA proposed in [81].

From Tables 19.20 and 19.22 of Example 7, it can be said that only IDCGA and EDCGA find the optimal solution, while EDCGA is the better method from the viewpoint of convergence speed.

Table 19.15 Numerical results for Example 6.

Items		Power loss (kW)		Worst voltage (p.u.)		
Methods	Open switches	Before	After	Before	After	Run time (s)
GA [94, 95]	7, 9, 14, 32, 37	202.71	141.6	0.9131	0.929	19.1
RGA [86, 95, 96]	7, 9, 14, 32, 37	202.7	139.55	0.9131	0.9378	7.41
Fuzzy GA (FGA) [97]	9, 28, 33, 34, 36	205.81	140.6	0.842	0.95	–
Restricted GA [98]	7, 9, 14, 32, 37	202.68	139.5	–	–	6.3
Binary GA (BGA) [90]	7, 10, 14, 36, 37	202.68	142.68	–	–	160
GAMT [90, 96]	7, 9, 14, 32, 37	202.68	139.55	0.9131	0.9378	7.2
GA + BE [99]	7, 10, 14, 36, 37	202.68	139.551	0.9131	0.9378	10.83
DCGA	7, 11, 14, 32, 37	202.68	140.8	0.9131	0.938	50.7
IDCGA	7, 9, 14, 32, 37	202.68	139.1	0.9131	0.938	2.06
EDCGA	7, 9, 14, 32, 37	202.68	139.1	0.9131	0.938	0.4

Table 19.16 Arrangement of methods according to loss reduction for Example 6.

Methods	Power loss (kW)
BGA [90]	142.68
[94]	141.6
DCGA	140.8
[97]	140.6
[86], GAMT [90, 99]	139.55
[96, 98]	139.5
[95]	139.28
IDCGA	139.1
EDCGA	139.1

Table 19.17 Methods based on computational time in Example 6.

Methods	Run time (s)
BGA [90]	160
DCGA	50.7
[94]	19.1
[99]	10.83
RGA [96]	7.41
GAMT [90], GAMT [96]	7.2
[98]	6.3
IDCGA	2.06
EDCGA	0.4

Table 19.18 Methods according to voltage enhancement in Example 6.

Methods	Minimum voltage (p.u.)
[94]	0.929
GA [95]	0.934
[86, 90], RGA [95, 96]	0.9378
DCGA, IDCGA and EDCGA	0.938
[97]	0.95

Table 19.19 Numerical results compared to other methods for Example 7.

Items		Power loss (kW)		Worst voltage (p.u.)		
Methods	Open switches	Before	After	Before	After	Run time (s)
GA [100]	15, 34, 38, 45, 52	10.593	8.976	0.9957	0.9970	–
DCGA	34, 38, 45, 49, 51	10.48	8.15	0.9957	0.9934	19.97
IDCGA	34, 39, 45, 49, 51	10.48	8	0.9957	0.9970	1.56
EDCGA	34, 39, 45, 49, 51	10.48	8	0.9957	0.9970	0.86

Table 19.20 Ranking of methods based on loss reduction for Example 7.

Methods	Power loss (kW)
[100]	8.976
DCGA	8.15
IDCGA and EDCGA	8

Table 19.21 Ranking of methods regarding voltage improvement for Example 7.

Methods	Worst voltage (p.u.)
DCGA	0.9934
[100], IDCGA and EDCGA	0.9970

Table 19.22 Ranking of methods based on execution time for Example 7.

Methods	Run time (s)
[100]	–
DCGA	19.97
IDCGA	1.56
EDCGA	0.86

Example 8: Consider the DSR problem for System 8. The line and load data for this real distribution network has been given in Chapter 17

Solution: Tables 19.23–19.26 show the simulation results of Example 8 after applying the proposed genetic algorithms to System 8.

It can be seen that only IDCGA and EDCGA converge to the optimal solutions, while EDCGA has a better convergence speed.

Table 19.23 Numerical results compared to other methods for Example 8.

Items		Power loss (kW)		Worst voltage (p.u.)		
Methods	**Open switches**	**Before**	**After**	**Before**	**After**	**Run time (s)**
DCGA	25, 31, 35, 39, 52	178.66	132.82	0.9767	0.9610	69.48
IDCGA	25, 31, 39, 40, 62	178.66	131.69	0.9767	0.9850	6.11
EDCGA	25, 31, 39, 40, 62	178.66	131.69	0.9767	0.9850	3.35

Table 19.24 Ranking of methods based on loss reduction for Example 8.

Methods	Power loss (kW)
DCGA	132.82
IDCGA and EDCGA	131.69

Table 19.25 Ranking of methods regarding voltage improvement for Example 8.

Methods	Worst voltage (p.u.)
DCGA	0.961
IDCGA and EDCGA	0.9850

Table 19.26 Ranking of methods based on execution time for Example 8.

Methods	Run time (s)
DCGA	69.48
IDCGA	6.11
EDCGA	3.35

Example 9: Solve the DSR problem for System 9. Data for this system are available in Chapter 17. The base values for this network are 12.66 kV and 100 MVA

Solution: The proposed GA methods were applied to study the network and the results are compared to other genetic algorithms in Tables 19.27–19.30.

Table 19.27 Numerical results compared to other methods for example 9.

Items		Power loss (kW)		Worst voltage (p.u.)		
Methods	Open switches	Before	After	Before	After	Run time (s)
BGA [101]	12, 20, 58, 64, 69	226.13	113.8	0.9092	0.9328	25
FGA [101]	12, 20, 58, 64, 69	226.13	113.8	0.9092	0.9328	50
NSGA [102]	14, 23, 51, 60, 72	226.92	136.87	0.9092	0.9428	37
GA [103]	14, 53, 61, 69, 70	225	103.29	0.9092	0.9411	–
RGA [103]	13, 17, 55, 61, 69	225	100.28	0.9092	0.9428	–
GA [104]	14, 58, 61, 69, 70	225.05	99.62	0.9092	0.9428	1181
Fast NSGA (FNSGA) [92]	14, 57, 61, 69, 70	224.93	99.62	0.9092	0.9430	20.2
GA [105]	14, 58, 61, 69, 70	225	99.62	0.9092	0.9428	12.5
DCGA	10, 14, 57, 62, 70	225	105.91	0.9092	0.9411	94.15
IDCGA	14, 56, 61, 69, 70	225	99.54	0.9092	0.9431	4.75
EDCGA	14, 56, 61, 69, 70	225	99.54	0.9092	0.9431	3.72

Table 19.28 Ranking of methods based on loss reduction for Example 9.

Methods	Power loss (kW)
[102]	136.87
[101]	113.8
DCGA	105.91
GA [103]	103.29
RGA [103]	100.28
[92, 104, 105]	99.62
IDCGA and EDCGA	99.54

Table 19.29 Ranking of methods based on execution time for Example 9.

Methods	Run time (s)
[104]	1181
DCGA	94.15
FGA [101]	50
[102]	37
BGA [101]	25
[92]	20.2
[105]	12.5
IDCGA	4.75
EDCGA	3.72

Table 19.30 Ranking of methods regarding voltage improvement for Example 9.

Methods	Worst voltage (p.u.)
[101]	0.9328
GA [103] and DCGA	0.9411
[102], RGA [103–105]	0.9428
[92]	0.9430
IDCGA and EDCGA	0.9431

Simulation results of Example 9 show that IDCGA and EDCGA can solve the DSR problem more precisely compared to GA methods. Also, it can be seen that EDCGA is faster than IDCGA and other genetic algorithms.

Example 10: Consider System 10 and solve the DSR problem. Data for this system have been given in Chapter 17

Solution: Tables 19.31–19.34 show the simulation results of Example 10. It can be seen that EDCGA finds the best solution faster than other GA methods.

Example 11: Solve the DSR problem for System 11, in which its data are available in Chapter 17. The substation voltage is 1 p.u. and the power and voltage base values are equal to 100 kVA and 11.4 kV, respectively

Table 19.31 Numerical results compared to other methods for Example 10.

Items		Power loss (kW)		Worst voltage (p.u.)		
Methods	Open switches	Before	After	Before	After	Run time (s)
GAMT [90]	14, 30, 38, 46, 51, 66, 70, 71, 76, 77, 79	227.52	203.17	0.9052	0.9297	160
BGA [90]	13, 30, 45, 51, 66, 70, 75–79	227.52	201.4	0.9052	0.9297	1900
IDCGA	13, 30, 45, 51, 66, 70, 75–79	227.53	201.36	0.9052	0.9312	56.28
EDCGA	13, 30, 45, 51, 66, 70, 75–79	227.53	201.36	0.9052	0.9313	5.28

Table 19.32 Ranking of methods based on loss reduction for Example 10.

Methods	Power loss (kW)
GAMT [90]	203.17
BGA [90]	201.4
IDCGA and EDCGA	201.36

Table 19.33 Ranking of methods regarding voltage improvement for Example 10.

Methods	Worst voltage (p.u.)
[90]	0.9297
IDCGA	0.9312
EDCGA	0.9313

Table 19.34 Ranking of methods based on execution time for Example 10.

Methods	Run time (s)
BGA [90]	1900
GAMT [90]	160
IDCGA	56.28
EDCGA	5.28

Solution: The proposed genetic algorithms were implemented using the study system and the results of Example 11 are listed in Tables 19.35–19.38.

It can be observed that the IDCGA and EDCGA are more efficient than other GA techniques for reconfiguration of medium-size distribution networks, because they can reduce network losses more. Also, EDCGA can solve the DSR problem faster than other genetic algorithms.

Table 19.35 Numerical results of proposed model compared to other methods for Example 11.

Items		Power loss (kW)		Worst voltage (p.u.)		
Methods	Open switches	Before	After	Before	After	Run time (s)
GA [87, 88, 96]	7, 13, 34, 39, 41, 55, 62, 72, 83, 86, 89, 90, 92	531.99	471.08	0.9285	0.9532	303.66
EGA [106]	7, 13, 34, 39, 42, 55, 62, 72, 83, 86, 89, 90, 92	531.99	469.878	0.9285	0.9532	450
GA [107]	7, 33, 55, 61, 72, 83, 86, 88, 89, 90, 92, 93, 95	531.99	471.51	0.9285	0.9532	–
GA [99]	7, 13, 34, 39, 63, 72, 83, 84, 86, 89, 90, 92, 95	531.99	470.214	0.9285	0.9517	30.526
RGA [99]	7, 13, 34, 39, 42, 63, 72, 83, 84, 86, 89, 90, 92	531.99	470.084	0.9285	0.9517	170
GA + BE [99]	7, 13, 34, 39, 42, 55, 62, 72, 83, 86, 89, 90, 92	531.99	469.878	0.9285	0.9532	37.983

Table 19.35 (*continued*)

Items		Power loss (kW)		Worst voltage (p.u.)		
Methods	**Open switches**	**Before**	**After**	**Before**	**After**	**Run time (s)**
IDCGA	7, 13, 34, 39, 42, 55, 62, 72, 83, 86, 89, 90, 92	531.99	468.63	0.9285	0.9554	39.44
EDCGA	7, 13, 34, 39, 42, 55, 62, 72, 83, 86, 89, 90, 92	531.99	468.63	0.9285	0.9554	17.92

Table 19.36 List of methods regarding loss minimization in Example 11.

Methods	Losses (kW)
[107]	471.51
[87, 88]	471.08
GA [99]	470.214
RGA [99]	470.084
[96], GA + BE [99, 106]	469.878
IDCGA and EDCGA	468.63

Table 19.37 Methods based on nodal voltage enhancement of Example 11.

Methods	Minimum voltage (p.u.)
GA [99], RGA [99]	0.9517
[87, 88, 96], GA + BE [99, 106, 107],	0.9532
IDCGA and EDCGA	0.9554

Table 19.38 Classification of solution methods according to running time in Example 11.

Methods	Computing time (s)
[106]	450
[87]	303.66
RGA [99]	170
IDCGA	39.44
GA + BE [99]	37.983
GA [99]	30.526
EDCGA	17.92

Example 12: Consider System 12, where its data have been provided in Chapter 17, and solve the DSR problem

Solution: Tables 19.39–19.42 show relevant results obtained for Example 12.

The results reveal that EDCGA can find the optimal solution in a shorter time than other GA techniques.

Table 19.39 Numerical results compared to other methods for Example 12.

Items		Power loss (kW)		Worst voltage (p.u.)		
Methods	Open switches	Before	After	Before	After	Run time (s)
GA [95]	24, 31, 35, 40, 43, 49, 51, 62, 73, 74, 77, 83,110, 120, 126	1298.09	885.56	0.8783	0.9321	24.45
RGA [95]	23, 27, 33, 40, 43, 49, 52, 62, 73, 74, 77, 83, 110, 126, 131	1298.09	883.13	0.8783	0.9321	17.53
GA [108]	5, 27, 34, 43, 47, 54, 61, 71, 74, 83, 96, 99, 110, 125, 131	1298.09	980.54	0.8687	0.9321	–
GA [105]	24, 27, 35, 40, 43, 52, 59, 72, 75, 96, 98, 110, 123, 130, 131	1298.1	865.86	0.8688	0.9323	42.13
IDCGA	24, 35, 40, 43, 49, 51, 59, 72, 75, 96, 98, 110, 120, 130, 131	1298	874.81	0.869	0.9327	13.6
EDCGA	24, 27, 35, 40, 43, 52, 59, 72, 75, 96, 98, 110, 123, 130, 131	1298	865.86	0.869	0.9328	1.95

Table 19.40 Ranking of methods based on loss reduction for Example 12.

Methods	Power loss (kW)
[108]	980.54
GA [95]	885.56
RGA [95]	883.13
IDCGA	874.81
[105] and EDCGA	865.86

Table 19.41 Ranking of methods regarding voltage improvement for Example 12.

Methods	Worst voltage (p.u.)
[95, 108]	0.9321
[105]	0.9323
IDCGA	0.9327
EDCGA	0.9328

Table 19.42 Ranking of methods based on execution time for Example 12.

Methods	Run time (s)
[105]	42.13
GA [95]	24.45
RGA [95]	17.53
IDCGA	13.6
EDCGA	1.95

Table 19.43 Numerical results of proposed model compared to other methods for Example 13.

Items		Power loss (kW)		Worst voltage (p.u.)		
Methods	Open switches	Before	After	Before	After	Run time (s)
DCGA [96, 109]	7, 51, 53, 84, 90, 96, 106, 118, 126, 128, 137–139, 141, 144, 145, 147, 148,150, 151, 156	320.3	280.22	0.9307	0.9605	3600
Dedicated GA (DGA) [110]	7, 38, 51, 53, 90, 96, 106, 118, 126, 128, 137, 138, 141, 144–148, 150, 151, 156	320.3	280.4	0.9307	0.9605	227
FGA [111]	51, 53, 90, 96, 106, 118, 136–139, 141, 144–148, 150, 151, 154–156	320.3	281.7	0.9307	0.9605	~ 40
IDCGA	7, 38, 51, 53, 90, 96, 106, 118, 126, 137, 138, 141, 144–148, 150 151, 155, 156	320.37	279.52	0.9307	0.9634	159.27
EDCGA	7, 38, 51, 53, 90, 96, 106, 118, 126, 137, 138, 141, 144–148, 150 151, 155, 156	320.37	279.52	0.9307	0.9634	18.55

Example 13: Solve the DSR problem for System 13 with data described in Chapter 17. The nominal voltage and MVA base are 13.8 kV and 1, respectively

Solution: The proposed genetic algorithms were applied to this test system and the results were provided in Tables 19.43–19.45. It should be mentioned that the minimum voltage of all solutions obtained by genetic algorithms after reconfiguration is 0.96 per unit. Therefore, the arrangement of methods based on voltage improvement was ignored.

According to Tables 19.44 and 19.45 of Example 13, the EDCGA approach could find the more accurate solution for this real large-scale distribution network in a reasonable run time.

Example 14: In this example, try to solve the DSR problem for a large realistic distribution network like System 14, in which all data have been provided in Chapter 17

Solution: The proposed genetic algorithms were applied to the system of Example 14 and the results are shown in Tables 19.46–19.49.

Table 19.44 Arrangement of the results based on power loss reduction in Example 13.

Methods	Power loss (kW)
[111]	281.7
[110]	280.4
[109]	280.22
[96]	280.19
IDCGA and EDCGA	279.52

Table 19.45 Computation times for Example 13.

Methods	Time (s)
[109]	3600
[96]	894.2
[110]	227
IDCGA	159.27
[111]	~ 40
EDCGA	18.55

Table 19.46 Numerical results compared to other methods for Example 14.

Items		Power loss (kW)		Worst voltage (p.u.)		
Methods	Open switches	Before	After	Before	After	Run time (s)
IDCGA	27, 52, 104, 112, 120, 125, 134, 156, 169, 180, 191, 197, 198, 202, 208	547.6	510.35	0.9588	0.9638	350
EDCGA	27, 51, 104, 112, 120, 125, 156, 169, 199, 180, 191, 198, 202, 208, 211	547.6	510.57	0.9588	0.9644	38.76

Table 19.47 Ranking of methods based on loss reduction for Example 14.

Methods	Power loss (kW)
EDCGA	510.57
IDCGA	510.35

Table 19.48 Ranking of methods regarding voltage improvement for Example 14.

Methods	Worst voltage (p.u.)
IDCGA	0.9638
EDCGA	0.9644

Although IDCGA could find a little bit better solution, EDCGA can find the optimal solution very quickly.

Example 15: Solve the DSR problem for the large real distribution system described in Subsection 17.2.15 of Chapter 17

Solution: The IDCGA and EDCGA were applied to System 15 and simulation results are shown in Tables 19.50–19.53.

According to Tables 19.52 and 19.53 of Example 15, it can be seen that EDCGA is an efficient method to solve the DSR problem in large distribution networks.

Table 19.49 Ranking of methods based on execution time for Example 14.

Methods	Run time (s)
IDCGA	350
EDCGA	38.76

Table 19.50 Numerical results compared to other methods for Example 15.

Items		Power loss (kW)		Worst voltage (p.u.)		
Methods	Open switches	Before	After	Before	After	Run time (s)
IDCGA	13, 15, 16, 21, 26, 31, 50, 53, 56, 59, 73, 86, 94, 96, 97, 107, 119, 126, 136, 150, 155,156, 158, 165,168, 169, 178, 179, 194,195, 201, 211, 230, 256, 258, 267, 270, 282, 297, 321, 322, 358, 362, 364, 377, 392, 395, 396, 403, 404, 416, 423, 426, 433, 435, 436, 445, 446, 466	708.52	602.82	0.9331	0.9497	61525
EDCGA	1, 2, 13, 15, 16, 30, 31, 40, 41, 50, 59, 73, 75, 82, 94, 96, 97, 111, 115, 136, 142, 150, 155, 156, 158, 163, 168, 169, 178, 179, 191, 195, 209, 214, 230, 254, 256, 270, 294, 314, 317, 325, 358, 362, 385, 389, 392, 395, 403, 404, 423, 424, 426, 436, 437, 439, 446, 449, 466	708.52	582.99	0.9373	0.9523	439.67

Table 19.51 Ranking of methods based on loss reduction for Example 15.

Methods	Power loss (kW)
IDCGA	602.82
EDCGA	582.99

Table 19.52 Ranking of methods regarding voltage improvement for Example 15.

Methods	Worst voltage (p.u.)
IDCGA	0.9497
EDCGA	0.9523

Table 19.53 Ranking of methods based on execution time for Example 15.

Methods	Run time (s)
IDCGA	61525
EDCGA	439.67

19.5 Conclusion

In this chapter, the distribution system reconfiguration problem was solved by meta-heuristic methods such as the decimal codification genetic algorithm (DCGA), the improved DCGA (IDCGA), and the efficient DCGA (EDCGA). The proposed reconfiguration algorithms were applied to several test systems including small-, medium-, and large-sized distribution networks using the proposed methods (DCGA, IDCGA, and EDCGA) and the results are compared to other GA techniques reported in specialized literature. Evaluation of simulation results show that IDCGA and EDCGA solve the DSR problem in small-sized distribution networks more accurately and faster than DCGA and other genetic algorithms, in which EDCGA is the fastest method for solving DSR. In medium-sized distribution networks, it can be said generally that EDCGA and IDCGA find the more optimal solution than DCGA and other GA methods, whereas EDCGA solves the problem in a shorter time than IDCGA. Also, it can be seen that reconfiguration of large-sized distribution systems using EDCGA is more optimal from accuracy and convergence speed points of view. Generally, it is concluded that EDCGA is the best method for studying the reconfiguration of radial distribution systems because of its high accuracy and low computational time.

References

[1] M. Mahdavi and E. Mahdavi, "Transmission expansion planning considering network adequacy and investment cost limitation using genetic algorithm," *International Journal of Electrical and Computer Engineering*, vol. 5, no. 8, pp. 1–5, 2011.

[2] M. Mahdavi and R. Romero, "A new model for dynamic transmission expansion planning," *4th National Conference on Technology in Electrical and Computer Engineering*, Iran, pp. 1–7, 2018.

[3] M. Mahdavi, A.R. Kheirkhah, and R. Romero, "Transmission expansion planning considering line failures and maintenance," *33rd International Power System Conference*, Tehran, Iran, pp. 1–7, 2018.

[4] A. Kazemi, S. Jalilzadeh, H. Haddadian, and M. Mahdavi, "Transmission expansion planning considering network adequacy and losses using DCGA," *IEEE International Conference on Sustainable Energy Technologies*, Singapore, pp. 145–150, 2008.

[5] S. Jalilzadeh, A. Kazemi, H. Shayeghi, and M. Mahdavi, "Technical and economic evaluation of voltage level in transmission network expansion planning using GA," *Energy Conversion and Management*, vol. 49, no. 5, pp. 1119–1125, 2008.

[6] H. Shayeghi, S. Jalilzadeh, M. Mahdavi, and H. Haddadian, "Studying influence of two effective parameters on network losses in transmission expansion planning using DCGA," *Energy Conversion and Management*, vol. 49, no. 11, pp. 3017–3024, 2008.

[7] M. Mahdavi, H. Monsef, and R. Romero, "Reliability and economic effects of maintenance on TNEP considering line loading and repair," *IEEE Transactions on Power Systems*, vol. 31, no. 5, pp. 3381–3393, 2016.

[8] M. Mahdavi, C. Sabillón, M. Ajalli, H. Monsef, and R. Romero, "A real test system for power system planning, operation, and reliability," *Journal of Control, Automation and Electrical Systems*, vol. 29, no. 2, pp. 192–208, 2018.

[9] H. Shayeghi, S. Jalilzadeh, M. Mahdavi, and H. Haddadian, "Role of voltage level and network losses in TNEP Solution," *Journal of Electrical Systems*, vol. 5, no. 3, pp. 1–18, 2009.

[10] S. Jalilzadeh, A. Kazemi, M. Mahdavi, and H. Haddadian, "Transmission expansion planning considering voltage level and network loss using genetic algorithm," *5th International Conference on Electrical and Electronics Engineering*, Bursa, Turkey, pp. 1–5, 2007.

[11] M. Mahdavi and H. Haddadian, "Evaluation of GA performance in TNEP considering voltage level, network losses and number of bundle Lines," *International Journal of Discrete Mathematic*, vol. 3, no. 1, pp. 1–10, 2018.

[12] A. Kimiyaghalam, M. Mahdavi, A. Ashouri, and H. Soheil, "Transmission expansion planning considering uncertainty in fuel price using DABC," *The 9th International Energy Conference*, Tehran, Iran, pp. 1–10, 2013.

[13] A. Kazemi, S. Jalilzadeh, M. Mahdavi, and H. Haddadian, "Evaluation of network losses role in TNEP considering voltage level and number of bundle lines using GA," *3rd IEEE Conference on Industrial Electronics and Applications*, Singapore, pp. 1–6, 2008.

[14] S. Jalilzadeh, A. Kazemi, M. Mahdavi, and H. Haddadian, "Studying of voltage level role in TNEP considering the losses for years after expansion using GA with different trails," *43rd International Universities Power Engineering Conference*, Padova, Italy, pp. 1–6, 2008.

[15] S. Jalilzadeh, A. Kazemi, M. Mahdavi, and H. Haddadian, "Technical and economic evaluation of voltage level effect on TNEP using genetic algorithm," *The 8th International Power Engineering Conference*, Singapore, pp. 1034–1039, 2007.

[16] O. Badran, S. Mekhilef, H. Mokhlis, and W. Dahalan, "Optimal reconfiguration of distribution system connected with distributed generations: A review of different methodologies," *Renewable and Sustainable Energy Reviews*, vol. 73, pp. 854–867, June 2017.

[17] M. Rahmani-Andebili and M. Fotuhi Firuzabad, "An adaptive approach for PEVs charging management and reconfiguration of electrical distribution system penetrated by renewables," *IEEE Transactions on Industrial Informatics*, vol. 14, no. 5, pp. 2001–2010, May 2018.

[18] M. Mahdavi, H. Monsef, A. Bagheri, and A. Kimiyaghalam, "Dynamic transmission network expansion planning considering network losses, DG sources and operational costs-Part 1: Review and problem formulation," *Wulfenia*, vol. 19, no. 10, pp. 242–257, 2012.

[19] M. Mahdavi, H. Monsef, A. Bagheri, and A. Kimiyaghalam, "Dynamic transmission network expansion planning considering network losses, DG sources and operational costs-Part 2: Solution method and numerical results," *Wulfenia*, vol. 19, no. 10, pp. 258–273, 2012.

[20] S. Jalilzadeh, I.D. Kolagar, M. Mahdavi, and V. Nabaei, "Considering the torsional frequencies of an actual wind turbine to prevent of SSR phenomena," *IEEE International Conference on Sustainable Energy Technologies*, Singapore, pp. 222–227, 2008.

[21] M. Mahdavi, "Iran energy network overview: Renewable energies in Iran," Presentation, *Advanced Topics in Bioenergy*, Bioenergy Research Institute, Brazil, 2020. Available at: Available at: https://www.researchgate.net/publication/346719635.

[22] M. Mahdavi, "Temperature effects on the methanogenesis and sulfidogenesis in the wastewater treatment," Presentation, *Advanced Topics in Biorefinery*, Bioenergy Research Institute, Brazil, 2020. Available at: Available at: https://www.researchgate.net/publication/346719672.

[23] M. Mahdavi and L.A. Okamura, "Biofuels production from petrochemical industry wastewater," Presentation, *Advanced Topics in Biorefinery*, Bioenergy Research Institute, Brazil, 2020. Available at: https://www.researchgate.net/publication/346719761.

[24] M. Mahdavi and R.A.V. Ramos, "Optimal allocation of bioenergy distributed generators in electrical energy systems," *IV Bioenergy Workshop*, Bioenergy Research Institute, Brazil, 2020. Available at: https://www.researchgate.net/publication/347294531.

[25] M. Mahdavi and R. Romero, "New models and optimization techniques applied to the problem of optimal reconfiguration of radial distribution systems," Report, São Paulo Research Foundation, no. 16/12190-7, Brazil, 2019. Available at: https://bv.fapesp.br/en/bolsas/168851.

[26] M. Mahdavi, L.H. Macedo, and R. Romero, "A mathematical formulation for distribution network reconfiguration," *7th Regional Conference on Electricity Distribution*, Iran, pp. 1–5, 2019.

[27] A. Azizivahed, H. Narimani, M. Fathi, E. Naderi, H.R. Safarpour, and M.R. Narimani, "Multi-objective dynamic distribution feeder reconfiguration in automated distribution systems," *Energy*, vol. 147, pp. 896–914, March 2018.

[28] A. Lotfipour and H. Afrakhte, "A discrete teaching–learning-based optimization algorithm to solve distribution system reconfiguration in presence of distributed generation," *International Journal of Electrical Power and Energy Systems*, vol. 82, pp. 264–273, November 2016.

[29] A. Kimiyaghalam, M. Mahdavi, A. Ashouri, and M. Bagherivand, "Optimal placement of PMUs for reliable observability of network under probabilistic events using BABC algorithm," *22nd International Conference on Electricity Distribution*, Sweden, pp. 1–4, 2013.

[30] H. Hosseini, S. Jalilzadeh, V. Nabaei, G.R.Z. Govar, and M. Mahdavi, "Enhancing deregulated distribution network reliability for minimizing penalty cost based on

reconfiguration using BPSO," *2nd IEEE International Conference on Power and Energy*, Malaysia, pp. 983–987, 2008.

[31] M. Mahdavi and R. Romero, "Transmission system expansion planning based on LCC criterion considering uncertainty in VOLL using DCGA," *32nd International Power System Conference*, Iran, pp. 1–7, 2017.

[32] M. Khodayari, H. Monsef, and M. Mahdavi, "Customer reliability based energy & SR scheduling at hierarchical level II," *27th International Power System Conference*, Iran, pp. 1–11, 2012.

[33] M. Khodayari, M. Mahdavi, and H. Monsef, "Simultaneous scheduling of energy & spinning reserve considering customer and supplier choice on reliability," *19th Iranian Conference on Electrical Engineering*, Iran, pp. 491–496, 2011.

[34] H. Shayeghi, H.A. Shayanfar, M. Mahdavi, and A. Bagheri, "Application of binary particle swarm optimization for transmission expansion planning considering lines loadings," *The 2009 International Conference on Artificial Intelligence*, Las Vegas, Nevada, USA, pp. 653–659, 2009.

[35] H. Shayeghi, M. Mahdavi, and A. Bagheri, "An improved DPSO with mutation based on similarity algorithm for optimization of transmission lines loading," *Energy Conversion and Management*, vol. 51, no. 12, pp. 2715–2723, 2010.

[36] H. Shayeghi, M. Mahdavi, and A. Bagheri, "Discrete PSO algorithm based optimization of transmission lines loading in TNEP problem," *Energy Conversion and Management*, vol. 51, no. 1, pp. 112–121, 2010.

[37] M. Mahdavi, H. Monsef, and A. Bagheri, "Lines loading optimization in transmission expansion planning based on binary PSO algorithm," *I-manager's Journal on Information Technology*, vol. 1, no. 1, pp. 24–32, 2012.

[38] M. Mahdavi, H. Monsef, and A. Bagheri, "Transmission lines loading enhancement using ADPSO approach," *International Journal of Electrical and Computer Engineering*, vol. 4, no. 3, pp. 556–561, 2010.

[39] H. Shayeghi and M. Mahdavi, "Optimization of transmission lines loading in TNEP using decimal codification based GA," *International Journal of Electrical and Computer Engineering*, vol. 2, no. 5, pp. 942–947, 2008.

[40] M. Mahdavi and A. Bagheri, "BPSO applied to TNEP considering adequacy criterion," *American Journal of Neural Networks and Applications*, vol. 4, no. 1, pp. 1–7, 2018.

[41] M. Mahdavi, A. Bagheri, and E. Mahdavi, "Comparing efficiency of PSO with GA in transmission expansion planning considering network adequacy," *WSEAS Transactions on Power Systems*, vol. 7, no. 1, pp. 34–43, 2012.

[42] H. Shayeghi, M. Mahdavi, and A. Kazemi, "Discrete particle swarm optimization algorithm used for TNEP considering network adequacy restriction," *International Journal of Electrical and Computer Engineering*, vol. 3, no. 3, pp. 521–528, 2009.

[43] H. Shayeghi, M. Mahdavi, and H. Haddadian, "DCGA based-transmission network expansion planning considering network adequacy," *International Journal of Electrical and Computer Engineering*, vol. 2, no. 12, pp. 2875–2880, 2008.

[44] A. Kazemi, H. Haddadian, and M. Mahdavi, "Transmission network adequacy optimization using genetic algorithm," *5th International Conference on Electrical and Electronics Engineering*, Turkey, pp. 1–5, 2007.

[45] A. Bagheri, M. Mahdavi, and H. Monsef, "Suggestion of an economical plan for electrical power transmission," *The 9th International Energy Conference*, Tehran, Iran, pp. 1–10, 2013.

[46] A. Kazemi, S. Jalilzadeh, H. Haddadian, and M. Mahdavi, "Transmission network expansion planning considering network adequacy using genetic algorithm," *The 8th International Power Engineering Conference*, Singapore, pp. 1040–1044, 2007.

[47] S. Jalilzadeh, M. Jabbari, A. Kimiyaghalam, and M. Mahdavi, "Inrush current identification in power transformers using weight functions," *The 2009 IEEE International Conference on Intelligent Computing and Intelligent Systems*, Shanghai, China, pp. 795–797, 2009.

[48] N.G. Paterakis, A. Mazza, S.F. Santos, O. Erdinç, G. Chicco, A.G. Bakirtzis, and J.P.S. Catalão, "Multi-objective reconfiguration of radial distribution systems using reliability indices," *IEEE Transactions on Power Systems*, vol. 31, no. 2, pp. 1048–1062, March 2016.

[49] A. Asrari, S. Lotfifard, and M. Ansari, "Reconfiguration of smart distribution systems with time varying loads using parallel computing," *IEEE Transactions on Smart Grid*, vol. 7, no. 6, pp. 2713–2723, November 2016.

[50] Z. Li, S. Jazebi, and F. de León, "Determination of the optimal switching frequency for distribution system reconfiguration," *IEEE Transactions on Power Delivery*, vol. 32, no. 4, pp. 2060–2069, August 2017.

[51] S.S.F. Souza, R. Romero, J. Pereira, and J.T. Saraiva, "Artificial immune algorithm applied to distribution system reconfiguration with variable demand," *International Journal of Electrical Power and Energy Systems*, vol. 82, pp. 561–568, November 2016.

[52] M.A. Muhammad, H. Mokhlis, K. Naidu, J.F. Franco, H.A. Illias, and L. Wang, "Integrated database approach in multiobjective network reconfiguration for distribution system using discrete optimisation techniques," *IET Generation, Transmission & Distribution*, vol. 12, no. 4, pp. 976–986, 2018.

[53] F. Ding and K.A. Loparo, "Feeder reconfiguration for unbalanced distribution systems with distributed generation: A hierarchical decentralized approach," *IEEE Transactions on Power Systems*, vol. 31, no. 2, pp. 1633–1642, March 2016.

[54] H. Shayeghi and M. Mahdavi, "Application of PSO and GA for transmission network expansion planning," *Analysis, Control and Optimal Operations in Hybrid Power Systems*, ook chapter, Springer, London, pp. 187–226, 2013.

[55] A. Kimiyaghalam, M. Mahdavi, S. Jalilzadeh, and A. Ashouri, "Improved binary particle swarm optimization based TNEP considering network losses, voltage level, and uncertainty in demand," *Journal of Artificial Intelligence in Electrical Engineering*, vol. 1, no. 2, pp. 29–42, 2012.

[56] M. Mahdavi and E. Mahdavi, "Evaluating the effect of load growth on annual network losses in TNEP considering bundle lines using DCGA," *International Journal on Technical and Physical Problems of Engineering*, vol. 3, no. 4, pp. 1–9, 2011.

[57] M. Mahdavi, A. Nazari, V. Hosseinnezhad, and A. Safari "A PSO-based static synchronous compensator controller for power system stability enhancement," *Journal of Artificial Intelligence in Electrical Engineering*, vol. 1, no. 1, pp. 18–25, 2012.

[58] M. Mahdavi, M. Ajalli, and M. Mohammadi, "The Role of Artificial Intelligence Techniques in Induction Motors," *LAMBERT Academic Publishing*, Germany, pp. 1–52, 2017.

[59] S. Jalilzadeh, M. Mahdavi, and A. Kimiyaghalam, "Optimal design of single-phase induction motor with permanent capacitor based on maximum efficiency using PSO algorithm," *6th International Conference on Electrical Engineering Electronics, Computer, Telecommunications and Information Technology*, Pattaya, Thailand, pp. 174–177, 2009.

[60] S. Jalilzadeh, M. Mahdavi, A. Kimiyaghalam, and A. Bagheri, "Using PSO algorithm for parameter estimation of three-phase double-feed induction motor," *6th International Conference on Electrical Engineering Electronics, Computer, Telecommunications and Information Technology*, Pattaya, Thailand, pp. 266–269, 2009.

[61] M. Mahdavi and S. Jalilzadeh, "Advanced particle swarm optimization for parameter identification of three-phase DFIM," *The 2009 IEEE International Conference on Intelligent Computing and Intelligent Systems*, Shanghai, China, pp. 580–584, 2009.

[62] M. Mahdavi and H. Monsef, "Advanced particle swarm optimization used for optimal design of single-phase induction motor," *International Journal on Technical and Physical Problems of Engineering*, vol. 2, no. 1, pp. 55–59, 2010.

[63] A. Jalilvand, A. Kimiyaghalam, A. Ashouri, and M. Mahdavi "Advanced particle swarm optimization-based PID controller parameters tuning," *2008 IEEE International Multitopic Conference*, Karachi, Pakistan, pp. 429–435, 2008.

[64] M. Mahdavi, C.S. Antunez, M. Ajalli, and R. Romero, "Transmission expansion planning: Literature review and classification," *IEEE Systems Journal*, vol. 3, no. 3, pp. 3129–3140, 2019.

[65] L.H. Macedo, J.F. Franco, M. Mahdavi, and R. Romero, "A contribution to the optimization of the reconfiguration problem in radial distribution systems," *Journal of Control, Automation and Electrical Systems*, vol. 29, no. 6, pp. 756–768, 2018.

[66] H. Shayeghi, M. Mahdavi, A. Kazemi, and H.A. Shayanfar, "Studying effect of bundle lines on TNEP considering network losses using decimal codification genetic algorithm," *Energy Conversion and Management*, vol. 51, no. 12, pp. 2685–2691, 2010.

[67] M. Mahdavi, H. Monsef, and R. Romero, "Reliability effects of maintenance on TNEP considering preventive and corrective repairs," *IEEE Transactions on Power Systems*, vol. 32, no. 5, pp. 3768–3781, 2017.

[68] M. Mahdavi, C.S. Antunez, A. Bagheri, and R. Romero, "Line maintenance within transmission expansion planning: A multistage framework," *IET Generation, Transmission and Distribution*, vol. 13, no. 14, pp. 3057–3065, 2019.

[69] M. Mahdavi, H. Shayeghi, and A. Kazemi, "DCGA based evaluating role of bundle lines in TNEP considering expansion of substations from voltage level point of view," *Energy Conversion and Management*, vol. 50, no. 8, pp. 2067–2073, 2009.

[70] H. Shayeghi, H. Hosseini, A. Shabani, and M. Mahdavi, "GEP considering purchase prices, profits of IPPs and reliability criteria using hybrid GA and PSO," *International Journal of Electrical and Computer Engineering*, vol. 2, no. 8, pp. 1619–1625, 2008.

[71] M. Mahdavi, L.H. Macedo, and R. Romero, "Transmission and generation expansion planning considering system reliability and line maintenance," *26th Iranian Conference on Electrical Engineering*, Mashhad, Iran, pp. 1005–1010, 2018.

[72] M. Mahdavi, L.H. Macedo, and R. Romero, "Considering line maintenance and repair in GTEP," *4th Conference on Modern Technologies in Electrical, Telecommunication and Mechatronics Engineering*, Iran, pp. 1–6, 2018.

[73] H. Shayeghi and M. Mahdavi, "Genetic algorithm based studying of bundle lines effect on network losses in transmission network expansion planning," *Journal of Electrical Engineering*, vol. 60, no. 5, pp. 237–245, 2009.

[74] S. Jalilzadeh, H. Shayeghi, M. Mahdavi, and H. Haddadian, "A GA based transmission network expansion planning considering voltage level, network losses and number of bundle lines," *American Journal of Applied Sciences*, vol. 6, no. 5, pp. 987–994, 2009.

[75] S. Jalilzadeh, A. Kimiyaghalam, M. Mahdavi, and A. Ashouri, "STNEP considering voltage level and uncertainty in demand using IABPSO," *International Review of Electrical Engineering*, vol. 7, no. 4, pp. 5186–5195, 2012.

[76] Y.Y. Fu and H.D. Chiang, "Toward optimal multiperiod network reconfiguration for increasing the hosting capacity of distribution networks," *IEEE Transactions on Power Delivery*, vol. 33, no. 5, pp. 2294–2304, October 2018.

[77] N.C. Koutsoukis, D.O. Siagkas, P.S. Georgilakis, and N.D. Hatziargyriou, "Online reconfiguration of active distribution networks for maximum integration of distributed generation," *IEEE Transactions on Automation Science and Engineering*, vol. 14, no. 2, pp. 437–448, April 2017.

[78] H. Haghighat and B. Zeng, "Distribution system reconfiguration under uncertain load and renewable generation," *IEEE Transactions on Power Systems*, vol. 31, no. 4, pp. 2666–2675, July 2016.

[79] M. Mahdavi and R. Romero, "Reconfiguration of radial distribution systems: An efficient mathematical model," *IEEE Latin America Transactions*, vol. 100, no. 1e, pp. 1–10, 2020.

[80] A. Kazemi, S. Jalilzadeh, M. Mahdavi, and H. Haddadian, "Genetic algorithm-based investigation of load growth factor effect on the network loss in TNEP," *3rd IEEE Conference on Industrial Electronics and Applications*, Singapore, vol. 1, pp. 764–769, 2008.

[81] S. Jalilzadeh, A. Kazemi, M. Mahdavi, and H. Haddadian, "DCGA-based evaluation of inflation rate effect on the network loss in transmission expansion planning," *43rd International Universities Power Engineering Conference*, Italy, pp. 1–5, 2008.

[82] M. Mahdavi and H. Monsef, "Review of static transmission expansion planning," *Journal of Electrical and Control Engineering*, vol. 1, no. 1, pp. 11–18, 2011.

[83] H. Shayeghi and M. Mahdavi, "Studying the effect of losses coefficient on transmission expansion planning using decimal codification based GA," *International Journal on Technical and Physical Problems of Engineering*, vol. 1, no. 1, pp. 58–64, 2009.

[84] M. Mahdavi, A.R. Kheirkhah, L.H. Macedo, and R. Romero, "A genetic algorithm for transmission network expansion planning considering line maintenance," *IEEE Congress on Evolutionary Computation (CEC), Glasgow*, United Kingdom, pp. 1–6, 2020.

[85] S. Jalilzadeh, A. Kazemi, M. Mahdavi, and H. Haddadian, "TNEP considering voltage level, network losses and number of bundle lines using GA," *Third International Conference on Electric Utility Deregulation and Restructuring and Power Technologies*, China, pp. 1580–1585, 2008.

[86] J.Z. Zhu, "Optimal reconfiguration of electrical distribution network using the refined genetic algorithm," *Electric Power Systems Research*, vol. 62, no. 1, pp. 37–42, May 2002.

[87] C.T. Su, C.F. Chang, and J.P. Chiou, "Distribution network reconfiguration for loss reduction by ant colony search algorithm," *Electric Power Systems Research*, vol. 75, no. 2–3, pp. 190–199, August 2005.

[88] C.F. Chang, "Reconfiguration and capacitor placement for loss reduction of distribution systems by ant colony search algorithm," *IEEE Transactions on Power Systems*, vol. 23, no. 4, pp. 1747–1755, November 2008.
[89] S. Sivanagaraju, Y. Sreekanth, and E.J. Babu, "An efficient genetic algorithm for loss minimum distribution system reconfiguration," *Electric Power Components and Systems*, vol. 34, no. 3, pp. 249–258, March 2006.
[90] B. Enacheanu, B. Raison, R. Caire, O. Devaux, W. Bienia, and N. HadjSaid, "Radial network reconfiguration using genetic algorithm based on the Matroid theory," *IEEE Transactions on Power Systems*, vol. 23, no. 1, pp. 186–195, August 2008.
[91] C. Wang and Y. Gao, "Determination of power distribution network configuration using non-revisiting genetic algorithm," *IEEE Transactions on Power Systems*, vol. 28, no. 4, pp. 3638–3648, November 2013.
[92] A.M. Eldurssi and R.M. O'Connell, "A fast nondominated sorting guided genetic algorithm for multi-objective power distribution system reconfiguration problem," *IEEE Transactions on Power Systems*, vol. 30, no. 2, pp. 593–601, March 2015.
[93] N. Kagan, "Configuração de redes de distribuição através de algoritmos genéticos e tomada de decisão Fuzzy", PhD Thesis, University of São Paulo, São Paulo, Brazil, pp. 1–187, 1999.
[94] K. Nara, A. Shiose, M. Kitagawa, and T. Ishihara, "Implementation of genetic algorithm for distribution system loss minimum reconfiguration," *IEEE Transactions on Power Systems*, vol. 7, no. 3, pp. 1044–1051, August 1992.
[95] R.S. Rao, S.V.L. Narasimham, M.R. Raju, and A.S. Rao, "Optimal network reconfiguration of large-scale distribution system using harmony search algorithm," *IEEE Transactions on Power Systems*, vol. 26, no. 3, pp. 1080–1088, August 2011.
[96] A. Swarnkar, N. Gupta, and K.R. Niazi, "Adapted ant colony optimization for efficient reconfiguration of balanced and unbalanced distribution systems for loss minimization," *Swarm and Evolutionary Computation*, vol. 1, no. 3, pp. 129–137, September 2011.
[97] Y.Y. Hong and S.Y. Ho, "Determination of network configuration considering multiobjective in distribution systems using genetic algorithms," *IEEE Transactions on Power Systems*, vol. 20, no. 2, pp. 1062–1069, May 2005.
[98] J. Mendoza, R. Lopez, D. Morales, E. Lopez, P. Dessante, and R. Moaaga:, "Minimal loss reconfiguration using genetic algorithms with restricted population and addressed operators: Real application," *IEEE Transaction on Power Systems*, vol. 21, no. 2, pp. 948–954, May 2006.
[99] H.R. Esmaeilian and R. Fadaeinedjad, "Energy loss minimization in distribution systems utilizing an enhanced reconfiguration method integrating distributed generation," *IEEE Systems Journal*, vol. 9, no. 4, pp. 1430–1439, December 2015.
[100] R.A. Abttan, "Electric distribution system reconfiguration for loss reduction using genetic algorithm," *Journal of Baghdad College of Economic Sciences University*, vol. 2015, no. 43, pp. 457–478, 2015.
[101] N.C. Sahoo and K. Prasad, "A fuzzy genetic approach for network reconfiguration to enhance voltage stability in radial distribution systems," *Energy Conversion and Management*, vol. 47, no. 18–19, pp. 3288–3306, November 2006.

[102] S. Chandramohan, N. Atturulu, R.P.K. Devi, and B. Venkatesh, "Operating cost minimization of a radial distribution system in a deregulated electricity market through reconfiguration using NSGA method," *International Journal of Electrical Power and Energy Systems*, vol. 32, no. 2, pp. 126–132, February 2010.

[103] R.S. Rao, K. Ravindra, K. Satish, and S.V.L. Narasimham, "Power loss minimization in distribution system using network reconfiguration in the presence of distributed generation," *IEEE Transactions on Power Systems*, vol. 28, no. 1, pp. 317–325, February 2013.

[104] S.A. Taher and M.H. Karimi, "Optimal reconfiguration and DG allocation in balanced and unbalanced distribution systems," *Ain Shams Engineering Journal*, vol. 5, no. 3, pp. 735–749, September 2014.

[105] B. Khorshid-Ghazani, H. Seyedi, B. Mohammadi-ivatloo, K. Zare, and S. Shargh, "Reconfiguration of distribution networks considering coordination of the protective devices," *IET Generation, Transmission & Distribution*, vol. 11, no. 1, pp. 82–92, January 2017.

[106] E.M. Carreno, R. Romero, and A. Padilha-Feltrin, "An efficient codification to solve distribution network reconfiguration for loss reduction problem," *IEEE Transactions on Power Systems*, vol. 23, no. 4, pp. 1542–1551, November 2008.

[107] F. Mendoza, J.L. Bernal-Agustin, and J.A. Dominguez-Navarro, "NSGA and SPEA applied to multiobjective design of power distribution systems," *IEEE Transactions on Power Systems*, vol. 21, no. 4, pp. 1938–1945, November 2006.

[108] R. Tapia-Juárez and E. Espinosa-Juárez, "Reconfiguration of radial distribution networks by applying a multi-objective technique", *International Conference Artificial Intelligence*, pp. 131–137, 2015.

[109] J.C. Cebrian and N. Kagan, "Reconfiguration of distribution networks to minimize loss and disruption costs using genetic algorithms," *Electric Power Systems Research*, vol. 80, no. 1, pp. 53–62, January 2010.

[110] M.A.N. Guimarães, C.A. Castro, and R. Romero, "Distribution systems operation optimisation through reconfiguration and capacitor allocation by a dedicated genetic algorithm," *IET Generation, Transmission, and Distribution*, vol. 4, no. 11, pp. 1213–1222, April 2010.

[111] N. Gupta, A. Swarnkar, K.R. Niazi, and R.C. Bansal, "Multi-objective reconfiguration of distribution systems using adaptive genetic algorithm in fuzzy framework," *IET Generation, Transmission and Distribution*, vol. 4, no. 12, pp. 1288–1298, December 2010.

Part X

Demand Side Management Mechanisms and a Smart Home Energy Management System

20

Demand Response Techniques and Smart Home Energy Management Systems

Hooman Firoozi, Hosna Khajeh, and Hannu Laaksonen

Flexible Energy Resource (FER), School of Technology and Innovations, University of Vaasa, Vaasa, Finland

20.1 Introduction

Recent advances in information and communication technology (ICT) have introduced new stakeholders in the power system. These advances pave the way for small-scale customers and end-users to monitor their consumption and control their energy resources. They may also decide to reshape their consumption based on some external signals such as energy tariffs or market dynamic prices in order to minimize their costs. In addition to this, residential customers have an enormous potential to provide technical ancillary services, i.e. flexibility services to both transmission and distribution systems. They can change their consumption in real-time to provide the system with downward and upward flexibility services if they are equipped with sufficient ICT technologies and management systems.

Demand response (DR) programs are deployed to incentivize customers to change their consumption according to the different requests. The control requests can come from the system operators, aggregators, retailers, or other entities who need or aggregate the resources that can provide the flexibility services. In the near future, large conventional fuel-based generators are going to be replaced with resources that are economically more profitable and more environmentally friendly [1, 2]. For this purpose, different solutions have been proposed, such as installing large-scale renewable energy sources (RESs) in transmission networks, small-scale distributed energy resources (DERs) in distribution networks and energy storage systems at different voltage levels or at strategic locations (e.g. at substations) between voltage levels. Controllable DER units, energy storages, as well as controllable loads (demand response) can mitigate fluctuations of RES [3] and also provide multiple other flexibility services. In the future, the role of DR at the household level can be important also from the whole power system viewpoint if households can directly or through aggregation provide their flexibility to different needs and marketplaces.

The participation of smart households or homes is beneficial for both power systems and households as participants [4]. First, households can make additional profits and decrease their energy costs by the provision of flexibility services. Second, DR resources can smooth the prices of electricity markets and mitigate price-spikes resulting from the lack of generation power [5, 6]. It is expected that the reliability, security, and resiliency of the power system will be enhanced in the future by the active participation of large numbers of

Active Electrical Distribution Network: A Smart Approach, First Edition. Edited by Baseem Khan, Josep M. Guerrero, Sanjeevikumar Padmanaban, Hassan Haes Alhelou, Om Prakash Mahela, and Sudeep Tanwar.

household customers. However, most of the residential customers prefer not to decrease their comfort level and change their routines when providing DR-based flexibility services [7]. In [8], a meta-heuristic technique is utilized for the optimization of smart home appliances with cost-effective energy management systems. In [9], a comprehensive review of smart home energy management system optimization techniques is presented. In [10], various components of the smart grid system are discussed. In [11], management of the generation and demand inside the smart grid structure is discussed. In [12], smart microgrid energy management using a novel artificial shark optimization is discussed. Therefore, different DR schemes have been deployed in order to motivate customers to participate actively in DR programs.

The main focus of this chapter is to assess how residential household customers can participate in DR programs. For this purpose, first categories of the existing DR schemes associated with residential customers are introduced. After that, the potential DR-related flexibility services of household customers are presented. In this chapter the flexibility services are split into peak-saving services, services provided to distribution networks, and those related to system-wide ancillary services.

Later, this chapter presents the smart homes and their home energy management systems (HEMSs) as well as analyzing the main reasons behind equipping homes with smart technologies. Then the applications of HEMS in smart homes are evaluated thoroughly including, for example, the scheduling appliances based on their characteristics. Finally, the chapter presents briefly different scheduling techniques, such as rule-based, artificial intelligence-based, and optimization-based, which can be used in HEMSs. At the end a summary of the chapter is provided.

20.2 Different Demand Response (DR) Schemes

Typically the main aim of a demand response scheme is to offer incentive programs in order to motivate customers to provide flexibility services for the system operators [7, 13]. In addition, DR programs are expected to increase the awareness of the customers about the financial and environmental benefits of DR [4]. For example, responsibility sensing, monetary benefits (such as cost saving) as well as blackout prevention can be reasons behind the participation of customers in the DR programs [14]. In general, demand response schemes can be categorized into incentive-based, price-based, and hybrid schemes (Figure 20.1) [3, 15]. Figure 20.1 illustrates the different types of DR schemes and provides some examples from incentive-based and price-based ones.

In the incentive-based scheme, a contractual agreement exists based on which consumers are encouraged to change their consumption pattern upon the request sent by the system operator (utility company) [16]. In some programs, the system operator also has a certain degree of authority to directly control the flexible energy resources of the customers. For example, if a household has battery energy storage (BES) for self-sufficiency purposes, it may also negotiate a contract with the distribution system operator to control its BES during major disturbances in order to ensure security of supply in the distribution network.

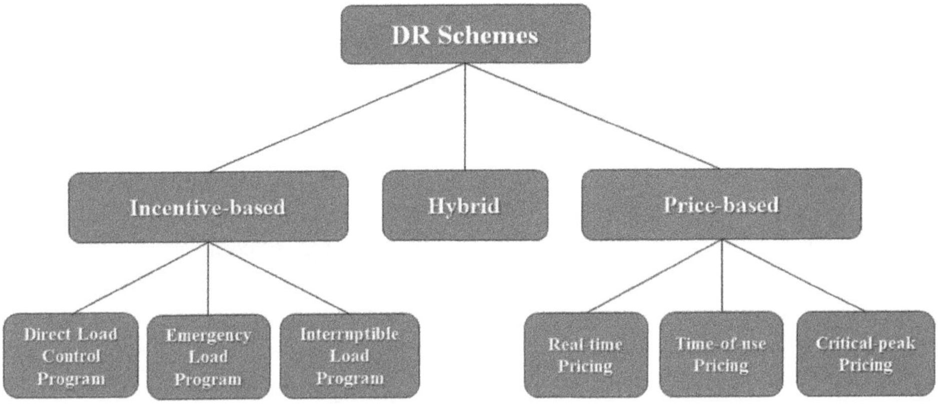

Figure 20.1 Different types of DR schemes.

Previously, industrial and large commercial customers were the only targets of the DR programs. However, recently these programs have been designed for residential end-users as well. In this regard, the real-time consumption as well as the operating state of each device is sent to the centralized controller. Subsequently, the controller makes a decision and takes appropriate control actions. Direct load control, emergency, and interruptible load programs are examples of this type of DR schemes [4, 17]. Although the centralized control of devices increases the response time and is more beneficial for the utility, it is more likely to invade the privacy of customers, which has been considered to be the main shortcoming of incentive-based programs [18]. Moreover, the level of comfortability of the customers may be decreased by the incentive-based DR schemes.

In the price-based scheme, the system operator (utility) can deploy variable-pricing techniques. These techniques can be selected based on the customers' reflection on the value and the costs of electricity in different time slots [19]. In a price-based DR program, customers are able to individually manage their flexible energy resources [20]. Flexible energy resources can be controlled and managed based on their types. For example, for devices whose operating time can be controlled, the customer can shift consumption from peak hours to off-peak hours. In comparison, for those devices that have flexible working power as well as flexible working time, the customer may decide to reduce their consumption power in these time slots [21].

Regarding the price-based program, the system operator can define various tariffs at preset hours or it can set dynamic tariffs based on the time of the day [22]. According to these tariffs, a smart and rational customer reacts to the price changes in order to minimize the electricity costs. Real-time, time-of-use, and critical-peak pricing are popular examples of price-based programs [23–25]. It is worth mentioning that the acceptable response is not guaranteed in these programs. In order to receive an acceptable response from residential customers, the customers should be equipped with suitable energy management systems as well as smart meters in order to receive the data about the prices, which are determined by the system operator. After that, they can schedule their resources accordingly.

The system operator could also deploy a hybrid DR program with the aim to better manage the customers [3]. In this way, the system operator sets dynamic tariffs for different time slots. However, in critical situations, the system operator negotiates contracts with the customers in order to directly control some resources. In return, customers receive financial benefits. However, the privacy of customers and their comfortability level may still be jeopardized by the system operator during the time slots (periods) when their resources are under control.

DR programs are able to incentivize demand-side resources to contribute to the provision of different types of flexibility services for different parties (especially system operators). In the following section, these services are introduced briefly.

20.3 Demand Response-Based Flexibility Services

Demand-side resources have considerable potential to provide different types of flexibility services to the transmission and distribution system operators (TSOs and DSOs) as well as others such as aggregators and balance responsible parties. These flexibility services can range from peak-load shaving and other distribution-network-related services to technical ancillary services for the TSO [26]. Figure 20.2 provides an overview of DR-based flexibility services.

20.3.1 Peak-Load Shaving

Huge capital investments are required to accommodate the demand at peak hours. Moreover, power systems need to be designed to supply the highest load during the highest peak of the year. Hence, since the power system must maintain an instantaneous balance between load and generation, a number of generation units are needed to cover the peak hours. In addition, during peak load hours, losses increase in the power system significantly. Previously, the demand side was not an active participant when it comes to balancing [27]. Hence, the costly infrastructure for a power system has been required to fulfill peak demand.

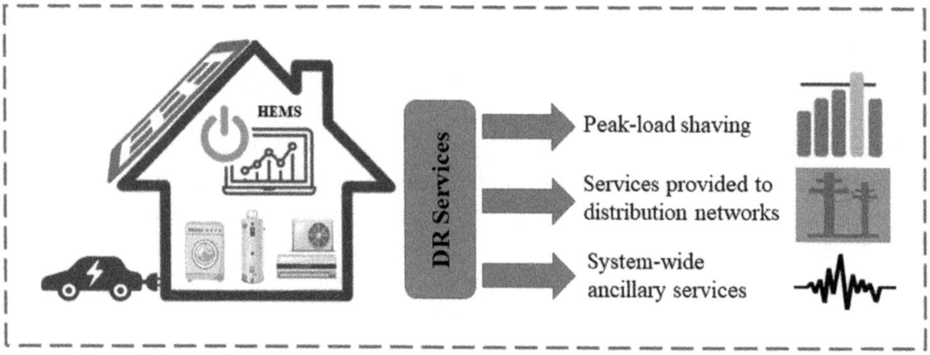

Figure 20.2 An overview of DR-based flexibility services.

One potential solution to avoid the occurrence of peak loads is to reduce consumption (demand response) during peak hours. In this way, customers voluntarily decrease their demand in order to contribute to the peak-load shaving program. To provide this service, the working time of devices should be shifted from peak-load hours to other time slots or the working power of some appliances could be curtailed as well [28]. The customer can also decide to reduce the working power of its controllable devices in order to provide peak-hour reduction services. In return, they are able to save costs or they will receive financial incentives, based on the programs they have chosen.

20.3.2 Other Potential Services for Distribution System Operators (DSOs)

Distribution feeders are often operated close to their capacity (voltage and thermal limits), especially in rural areas as well as in cities with large numbers of customers. In addition, with the high injection of intermittent power from renewable energy resources, the volatility, and the uncertainty of power along with the bi-directional flow of power in distribution networks may result in some unexpected challenges for the DSO. If the flow of the power is not controlled and managed, some distribution feeders can be congested and voltage and thermal capacity may be exceeded. This can be due to the fact that the radial topology of distribution networks accepts no alternative path for alleviating the congestion [4]. Moreover, distribution networks can have an unpredictable behavior due to the diverse customers that are located in this network. Hence, the occurrence of the congestion may not coincide with the whole system's peak hours. As a result, this can lead to an interruption of loads and a decrease in the security of the distribution system. Currently, the alleviation process and congestion management of distribution networks are performed based on two actions taken by the DSO. The first action is the deployment of the transformers with an on-load tap changer (OLTC) in the distribution network. The second action is to re-dispatch the available generation.

DRs can provide an alternative solution instead or in addition to these actions. Demand-side flexible resources like controllable loads and battery energy storage can provide both upward and downward flexibility to the DSO in order to manage the congestion. Regarding upward flexibility services, customers decrease their consumption by shifting the operating times of their devices or reducing their working power. In contrast, to provide downward flexibility services, the customers increase their consumption. For example, they can charge their electric vehicles during hours at which the DSO makes a request for downward flexibility.

The DSO can also decide to coordinate these services by forming local markets in each area. In these markets, the DSO can send signals based on its real-time need and customers can react to them according to their available resources.

20.3.3 System-Wide Flexibility Services for TSOs

Frequency related technical ancillary services are typically flexibility services provided to the TSO in order to control the frequency of the whole system. The specification of these frequency control related flexibility services varies depending on the country or continent and the prevailing flexibility needs of power systems. In Finland, as an example from

Europe, there are three types of frequency control related services. The first type is the frequency containment reserve (FCR), which is procured in both normal conditions (frequency containment reserve for normal condition, FCR-N) and disturbance situations (frequency containment reserve for disturbance condition, FCR-D). In this regard, the frequency containment reserve FCR unit should automatically react to the spontaneous frequency deviations in real-time. In addition, the fast frequency reserve (FFR) market has quite recently been introduced to compensate for rapid fluctuations of power in low-inertia situations. If frequency deviations still exist after activation of primary reserves (FCR, FFR), then the frequency restoration reserve (FRR) is activated to restore the frequency. FRRs can be seen as secondary and tertiary reserves, which are activated automatically or manually. In the automatic case, the reserve is called automatic FRR (aFRR) while the manual FRR (mFRR) is utilized in the case of an outage [29–31]. These services are reviewed in Figure 20.3.

The system-wide ancillary services have been traditionally provided by large-scale synchronous generators due to some market-based and regulatory barriers that have hindered the participation of demand-side resources. In addition, ICT technologies have not been sufficient enough to facilitate the participation of small-scale customers and aggregate them appropriately. Finally, a dedicated control system is required to control the demand-side resources and encounter the uncertain characteristics of demand in the distribution network.

On the other hand, demand-side flexible resources represent the enormous potential to contribute in improving the reliability and security of the power system. In addition, future

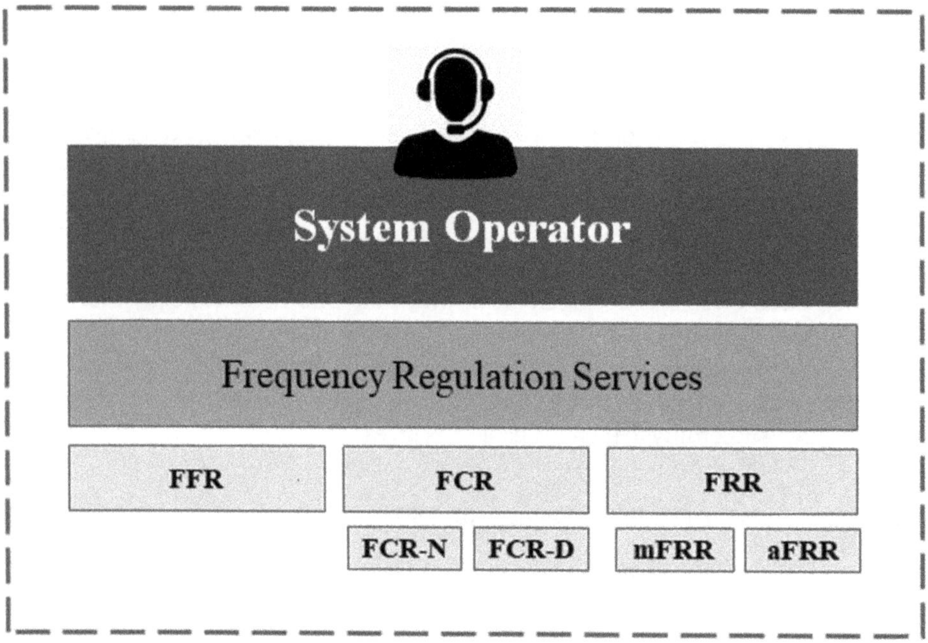

Figure 20.3 System-wide frequency-related ancillary services (in Finland).

power systems try to eliminate fossil fuel-based resources and increase the amount of renewable energy resources. As a result, these systems need more non-fuel-based resources for flexibility services provision. The participation of demand-side resources can create a more competitive environment for frequency-related ancillary service markets. This is due to the fact that the contribution of the demand-side resources increases the number of players competing in the market, which can in turn mitigate market power in these markets.

However, demand-side resources need to satisfy the qualification requirements related to their size, response time, etc., depending on the specific frequency control-related ancillary service market. In the case of tripping large generators, DRs can provide a sufficient amount of contingency reserves (mFRRs) and avoid the system blackout. In some cases, potentially the response time of demand-based reserves could be faster than the actions taken by a generator governor. As a result, they could also potentially improve the reliability of the power system considerably. DRs are also capable of providing balancing (regulating) services. They can compensate for the fluctuations in the demand consumption that was previously performed by generation-side resources. However, these resources should be fast enough and be able to automatically respond to the external signals sent by the system operator. In this regard, the centralized control can be a better option in order to control demand-side resources for providing some ancillary reserves such as FCRs, FFRs, and automatic FRRs.

The capability of DRs to contribute to frequency control-related ancillary services is highly dependent on the type of resources utilized for the DR. Some flexible energy resources are highly flexible, meaning that their working time and working power can be managed by their resources [32]. These resources can respond to external signals fast. However, the response time is naturally also dependent from the used communication network (e.g. optical fiber or wireless 4G/5G) and latencies of it.

20.4 Smart Home and Home Energy Management System (HEMS)

The definition of smart home varies depending on whether the technology is utilized for increasing the owners' comfort or it is deployed for providing services to the DSO [33]. Regarding the first case, the definition focuses on homes equipped with ICT technology through which appliances can be controlled remotely and automatically leading to increase the comfort of the citizens [33]. Regarding the second definition, a smart home can be considered as a flexible energy resource in the electricity distribution network, which is able to react based on the request of the system operator like DSO [34].

An individual household customer (owner) can equip his/her home with state-of-the-art ICT technology in order to remotely control its appliances, which in turn enhances the user's comfort, health, and the security as well as adding value to the owner's property [34]. In this regard, the owner may prefer not to jeopardize comfort by allowing other entities to control the appliances. Besides, the customer does not want to change his/her routines for making profits or saving costs. Typically this kind of smart home does not want to participate in DR programs.

In comparison, the customer may choose to make his/her home smart in order to benefit the smart control of the appliances and have monetary benefits by participating in different

DR programs if such options are available. In this way, the customer can have the opportunity to compensate the investment costs related to equipping the home with smart technologies. However, instead of traditional investments different service-based business models are also possible and increasingly available for customers.

A smart home with flexible energy resources and suitable management systems can provide different services to different parties like system operators. These previously defined services have different prices and effects on the comfort of the customer. Typically appliances (devices) need to be controlled automatically in order to decrease the response time. Accordingly, the customer could choose the services he/she is willing to provide. The customer may choose several services and let the management system solve an optimization problem and decide which services maximize the benefits of the customer.

Home energy management systems (HEMSs) enable scheduling and coordination of home devices according to inputs given by the customer, the network, as well as the data received from different sensors [35, 36]. A HEMS in a smart home plays the role of a brain in the human body. It sends scheduling commands to the appliances based on chosen scheduling techniques. The scheduling techniques of HEMSs are introduced in the final section of this chapter. Figure 20.4 illustrates different tasks of a HEMS.

Figure 20.4 Different tasks of home energy management system (HEMS) in a smart home.

20.5 Demand Response and a HEMS

Active customers with smart home technologies owning electricity production and related management systems are often called prosumers who can provide different flexibility services. Therefore, smart homes will play an increasingly important role when heading towards more environmentally friendly, sustainable, and flexible power systems.

The recent technologies deployed in an HEMS can enable smart homes to follow a certain target while providing various services to the network and utilities. The main target of the HEMS can be minimization of the electricity costs or maximizing the profits by participating in various DR programs or marketplaces. Smart homes may also be equipped with generation resources such as solar panels for photovoltaic (PV) electricity production as well as battery energy storages. HEMSs can be used to manage the production and the consumption of controllable loads of the smart home by scheduling flexible resources based on the main objectives defined by the customer.

In general, home energy management consists of several processes ranging from monitoring, managing, and controlling the controllable devices (loads) to making bidding strategies for the household to participate in the local markets. HEMS has the potential ability to change the consumption pattern of residential household customers while also taking into account the desired comfortability level [37]. Each home can have its own energy management system and individually control its flexible energy resources. On the other hand, in the future it can be increasingly common for neighboring households to form a local energy community with one energy management system. In this way, the flexible resources of households are controlled centrally while following the community's predefined objectives [38].

HEMSs need to control the appliances and flexible energy resources of households based on their type. In this regard, the HEMS must consider the operating constraints of each appliance as well as the customer's needs and preferences regarding each device.

20.5.1 Scheduling Residential Appliances

In terms of a load consumption pattern, residential appliances can be categorized into three different types, such as essential non-shiftable loads (T1), interruptible non-shiftable loads (T2), and shiftable loads (T3) [35, 39]. The HEMS schedules appliances based on their type as well as the inputs associated with the customer and external signals, as illustrated in Figure 20.5.

Essential non-shiftable appliances (T1) consist of loads that need to respond immediately upon the request of their users. There is a wide range of appliances that belong to this category. The appliances, such as building security systems, lighting devices, personal computers and laptops, chargers, entertainment-related appliances like television, game consoles, and DVD players, as well as kitchen appliances like cooking stoves, microwaves, toasters, and coffee makers, fall into the essential non-shiftable category of appliances. The HEMS should hand over the operation of these appliances to the user as the comfort of the user is highly dependent on the working time of the devices. However, the HEMS needs to monitor the consumption of these appliances and forecast these loads on the basis of the customer's routine in order to schedule other type of appliances. HEMS can also send messages to the customer if the total demand of these appliances exceeds the predefined or

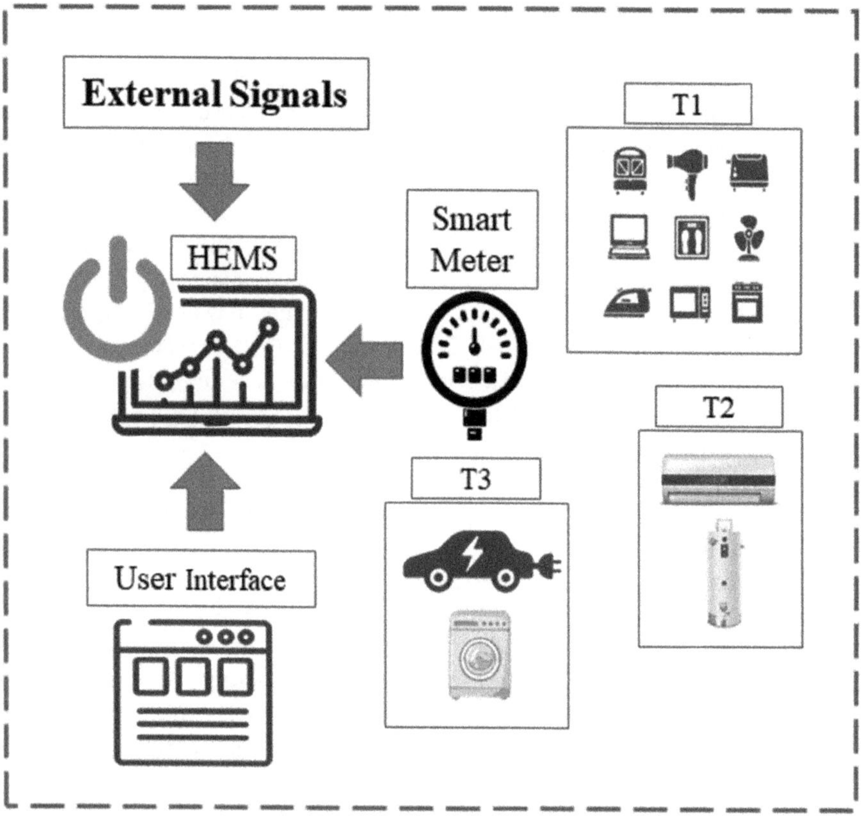

Figure 20.5 The scheduling of HEMS based on the type of appliances and other inputs.

forecasted threshold because it may lead to extra costs for the household. In this way, the customer may decide whether they want to continue using the appliance or not.

The second category is interruptible non-shiftable loads (T2) such as thermostatic loads. The main responsibility of these devices is to keep the operating temperature within a range that was previously set by the customer. Air conditioners, water heaters, and refrigerators can also be included in this category. The owner of these appliances sets a constraint regarding its preferred temperature. After that, the HEMS tries to schedule these resources based on the total objectives and preferences of the household as well as the flexibility services it may provide to the system operators. However, the HEMS applies the constraints set by the customer in the scheduling. The temperature constraints are also dependent on the customer presence and the outdoor temperature. The operational constraints of these appliances are required to be taken into account as well.

In addition to the two mentioned categories of appliances, shiftable loads (T3) comprise those appliances that are able to be flexibly operated in order to achieve the objectives of the household. However, there exists also some constraints related to the operation of these appliances and their nature. In terms of the operating nature, the shiftable loads can be categorized into two types, namely non-preventive shiftable loads and preventive shiftable loads. The HEMS is permitted to schedule preventive shiftable loads regardless of the stage that the appliance is working at. The HEMS may interrupt the operation of these appliances several times in order to reach optimal scheduling. However, the preferred time span set by the owner can be considered by the energy management system (HEMS). The power consumed by charging EVs can be taken into account as preventive shiftable loads. In comparison, the operation of non-preventive shiftable appliances should not be interrupted once it is switched on. Washing machines and dishwashers are two examples of non-preventive shiftable loads.

Regarding the scheduling of these appliances, the power output of T1 appliances can be aggregated for one scheduling time slot. After that, the HEMS may check whether the total power does not exceed the predefined threshold. If it does, the HEMS alerts the customer about the limit violation, which may result in extra costs. In comparison, each appliance of T2 category should be directly controlled by the HEMS. In this regard, the HEMS receives the exact information on the real-time operation mode of the appliance (whether it is on/off or in the stand-by mode), the consumption power, the tolerance limit of the owner, and the operating temperature. In addition, the HEMS needs information from the predefined temperature chosen by the customer and the current indoor and outdoor temperatures. With this information, the HEMS is able to control T2 appliances by sending on/off signals based on the total objectives and preferences of the household. Controlling T3 appliances is similar to that of T2 devices. However, the HEMS should consider the working time span of the non-preventive shiftable appliances and decide on the hour at which the appliance starts to operate. During this time span, the HEMS is not allowed to send control signals to the device. In terms of preventive shiftable appliances, the HEMS should take into account the time range determined by the customer. During this range, it can send on/off signals and control the device according to the objectives of the household customer.

20.5.2 Participation in DR Programs with the HEMS

The scheduling process performed by the HEMS should be in line with the flexibility services that the household has promised to provide. However, the HEMS must also consider the main objectives and preferences of the household, which could be minimization of the total cost or maximization of the benefits of the household from participating in DR programs.

If the household has promised to provide peak shaving and other DSO-related services, the working time of controllable appliances (T2 and T3) should be shifted to other time slots. However, the customer may define a time range during which an appliance should operate. If this time range overlaps with the peak hours of the distribution system, the device can send a message to the customer and warn the customer that high extra costs can be caused if he/she does not stick to his/her promises. Thus, the household customer should then be able to decide whether utilizing the device is worth this cost or not.

If the household promised to provide flexibility services to the system operator, the HEMS should consider the type and flexibility degree of the appliances. For example, inverter-based heating, ventilation, and air conditioning (HVAC) are more suitable for these purposes because the working power and time of the appliances can be controlled.

Moreover, if the household has battery energy storages or electric vehicle (EV) chargers with controllable power, it helps the household to participate in providing frequency reserve services like FCR. It is worth mentioning that there should still be an aggregator who aggregates several households to reach the technical requirements of the power system frequency control-related reserve markets. However, it has been stated for example in [34] that the aggregator of relay-based devices should not be the same as the owner of controllable-power resources. As a result, the HEMS should consider the type of appliances and connect them and their flexibility service potential to the relevant aggregator in order to provide ancillary services for the TSO.

In the future, there might also be increasing amounts of different local flexibility markets, where the household may provide flexibility services to the local DSO needs as well as to other small-scale customers. In this case, the HEMS should build bidding strategies for the household. However, the process of making an optimal bidding strategy is quite complicated and requires information from the behavior of the competitors and different types of forecasts, e.g. on the market prices.

20.6 Scheduling Techniques for HEMS

HEMS can deploy various scheduling approaches in order to control the controllable appliances. In general, there are three scheduling algorithms that can be utilized by the HEMS. These are rule-based, artificial intelligence-based, and optimization-based techniques [40]. Figure 20.6 presents an overview of some of these techniques.

20.6.1 Rule-Based Scheduling

Rule-based algorithms determine conditions for behavioral systems. One rule-based algorithm is a Rete algorithm, which can be deployed in HEMS. In this regard, the HEMS can control energy consumption and production via smart taps [41]. The controllable

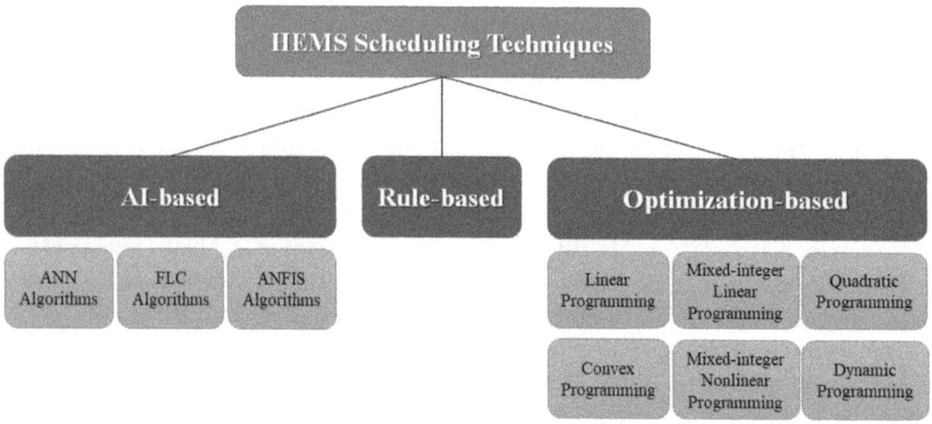

Figure 20.6 Possible scheduling techniques for HEMS.

appliances are distributed to the smart taps so that their data are collected while they are going through some rule processing. The rule-based process creates which rules to take into account the priority of the appliances [42]. For example, a rule-based algorithm can set a rule to shift the working time of high-priority controllable appliances from peak hours to off-peak hours. The algorithm is also able to take into account the uncertainties of the prices and the uncontrollable loads in real-time, like in [43]. In [44], rule-based algorithm has been developed for the HEMS by considering also the DR application. However, the rule-based algorithm cannot be extended to the more advanced applications. This algorithm may also fail to analyze large amounts of data that can restrict the ability of a smart home to provide DR services.

20.6.2 Artificial Intelligence (AI)-Based Scheduling

There are several studies that deploy various artificial intelligence (AI) techniques to schedule controllable appliances within smart homes. In this regard, the controllers of these HEMS are proposed to be designed based on an artificial neural network (ANN), fuzzy logic control (FLC), or adaptive neural fuzzy inference system (ANFIS) in their scheduling [40]. These controllers are designed to imitate human thinking when they are making optimal decisions.

ANN algorithms can be added to the HEMS with the aim to empower the scheduling process. The main applications of ANN algorithms are related to nonlinear decision making. During this process, the input information is processed intelligently in order to optimally schedule the appliances [45]. The obtained solutions can be utilized in scheduling when a prediction of human behavior (e.g. forecasting non-controllable loads) is needed. The HEMS may also be equipped with an FLC algorithm. There are typically four steps in designing FLC. These steps are fuzzification, defuzzification, rule base, and an inference engine. FLC does not require any mathematical model to solve the problem while it is capable of solving nonlinear scheduling models as well as linear ones [46, 47]. With the help of fuzzification, the comfort level of the customers can be modeled and the DR-based scheduling of a smart home can be performed through the use of fuzzy parameters. In comparison, the ANFIS algorithm presents several layers, which also do not require any mathematical models to schedule a controllable appliance [48].

20.6.3 Optimization-Based Scheduling

Different optimization techniques may also be utilized in HEMS in order to find the best possible scheduling for the household appliances. These optimization-based scheduling algorithms model the system mathematically. In these approaches, the HEMS follows predefined objectives. The objective can be minimization of the costs or maximization of the profits as well as maximization of the comfortability level of the customer. The variables of the optimization problem can be the optimal consumption power of the appliances or the binary decision variables that determine the on/off status for those controllable appliances whose working power is not controllable.

Each optimization problem has some constraints in order to restrict the decision variables. With regards to the HEMS, these constraints can be related to the operation of each controllable device such as their permissible working power and the time span specified by the customers. The HEMS needs to consider the type of appliance and define constraints accordingly.

When it comes to the DR application, the HEMS should add constraints related to the promised DR services. In other words, the household appliances should be scheduled in such a way to enable provision of the promised services. Some services may not need to be pre-scheduled.

The scheduling can be solved by defining different types of optimization problems. Below some examples are given [49]:

1) One way is to define a linear programming problem, which is the simplest form of the optimization problem. In this problem, the objective function and constraints are linear functions. Linear programming problems can be defined if the power of all appliances is controllable and there does not exist any nonlinear functions defining the behavior of the households with two or more dependent variables. Most appliances are assumed to be operating with constant working power and therefore this kind of optimization is not very well suited for household scheduling.
2) Defining a quadratic programming problem is also relatively simple. This problem can have a quadratic objective function while it should still have linear constraints. In cases when the objective function is a positive definite function, the optimization problem is able to be solved in polynomial time. However, if not, then the problem is hard to solve. This type of problem is suitable if the objective of the household is to minimize the costs in which the cost is a quadratic function of the power of appliances. However, also in this case the power of the appliances should be defined as a continuous variable.
3) Convex programming optimization problems need to have a convex objective function along with some linear equality constraints as well as concave inequality constraints. This type of problem is guaranteed to converge with the unique solution.
4) A dynamic programming problem is able to solve large complex problems by splitting the main problem into some smaller subproblems. It then recursively solves the subproblems and stores the solution of each subproblem.
5) A mixed-integer linear programming problem should have a linear objective function and constraints. However, it has some integer variables as well as continuous ones, which make the problem nonlinear. There are various algorithms that can solve the problem easily. The branch-and-bound algorithm is an example of such algorithms. Since most of the appliances of a smart home work with constant power, the problem can be defined by multiplying binary variables by the constant power of the appliances. In this way, the HEMS can send on/off signals to the appliances, which are defined by binary decision variables.
6) Defining a mixed-integer nonlinear programming is not an appropriate form of scheduling as it is very difficult to solve. In addition, it does not guarantee a unique solution to the problem. However, in most studies, mixed-integer nonlinear problems are defined and then they utilize linearization techniques to solve the problem. As a result, the nonlinear behavior of households can be covered while the problem can still lead to a unique solution.

20.7 Summary

For the RES-based future resilient power systems and the whole of society, the active participation of distribution network-connected household customers in providing different flexibility services through different DR programs or marketplaces will be increasingly important. This enables the maximization of socioeconomic welfare by keeping the whole power system costs feasible through creating possibilities also for the households to provide increasingly needed flexibility services. In this way household customers can also make profits and their energy costs can be kept at a reasonable level. However, most residential customers prefer not to decrease their comfort level and change their routines in order to provide flexibility services. Therefore, DR programs, for example, have been deployed in order to motivate customers to participate actively in the provision of flexibility services.

DR programs can be categorized into incentive-based, price-based, and hybrid schemes. In the incentive-based scheme, there exists a contractual agreement based on which consumers are encouraged to change their consumption patterns upon the request sent by the system operator. In some programs, the system operator also has a certain degree of authority in order to directly control the flexible energy resources of the customers. In a price-based scheme, the system operator may deploy variable-pricing techniques. These techniques can be selected based on the customers' reflection on the value and cost of electricity at different time slots. On the other hand, price-based DR programs give customers the ability to individually manage their flexible energy resources. A hybrid model adopts both schemes based on the services requested by the operator.

Regarding DR services, they can be related to (i) reducing peak loads in power systems, (ii) different other services provided to the DSOs, e.g. for congestion management (in order to avoid violation of voltage and thermal limits), as well as (iii) system-wide frequency control and balancing related services.

At the household level HEMS is needed to schedule the resources and manage their consumption and possible production as well as enable participation in DR programs or local energy and flexibility markets. In general, HEMS consists of several processes ranging from monitoring, managing, and controlling the controllable devices to making bidding strategies for the household in order to participate in the local markets. HEMS has the ability to change the consumption pattern of residential customers while taking into account the comfort level of the customers. Each home can have its own HEMS and individually control its flexible energy resources depending on their type, controllability, and customer (user) preferences.

In addition, HEMS can deploy various scheduling techniques in order to achieve customer objectives, such as rule-based, artificial intelligence-based, and optimization-based ones. Rule- and artificial intelligence-based techniques do not model the scheduling problem mathematically. In comparison, the optimization-based techniques aim to develop a mathematical model for the scheduling of appliances. In this regard, the operational constraints of each appliance must be mathematically modeled in order to obtain the optimal solution, which can be very different for different customers depending on their controllable resources and preferences.

References

[1] G. Strbac, "Demand side management: Benefits and challenges," *Energy Policy*, vol. 36, no. 12, pp. 4419–4426, 2008.

[2] C. Delmastro, E. Lavagno, and G. Mutani, "Chinese residential energy demand: Scenarios to 2030 and policies implication," *Energy and Buildings*, vol. 89, pp. 49–60, 2015.

[3] H. Khajeh, M. Shafie-khah, and H. Laaksonen, "Blockchain-based demand response using prosumer scheduling." In: Blockchain-Based Smart Grids, Miadreza Shafie-khah (ed.), pp. 131–144. Elsevier, Amsterdam, Netherlands, April 2020.

[4] H.T. Haider, O.H. See, and W. Elmenreich, "A review of residential demand response of smart grid," *Renewable and Sustainable Energy Reviews*, vol. 59, pp. 166–178, 2016.

[5] H. Khajeh, H. Laaksonen, A.S. Gazafroud, and M. Shafie-Khah, "Towards flexibility trading at TSO-DSO-customer levels: A review," *Energies*, vol. 13, p. 165, 2020.

[6] P. Siano, "Demand response and smart grids – A survey," *Renewable and Sustainable Energy Reviews*, vol. 30, pp. 461–478, 2014.

[7] A. Soares, Á. Gomes, and C.H. Antunes, "Categorization of residential electricity consumption as a basis for the assessment of the impacts of demand response actions," *Renewable and Sustainable Energy Reviews*, vol. 30, pp. 490–503, 2014.

[8] T. Molla, B. Khan, B. Moges, H.H. Alhelou, R. Zamani, and P. Siano, "Integrated optimization of smart home appliances with cost-effective energy management system," *CSEE Journal of Power and Energy Systems*, vol. 5, no. 2, pp. 249–258, June 2019.

[9] T. Molla, B. Khan, and P. Singh, "A comprehensive analysis of smart home energy management system optimization techniques," *Journal of Autonomous Intelligence*, vol. 1, no. 1, pp. 15–21, 2018.

[10] B. Khan, H. Getachew, and H.H. Alhelou, "Components of smart grid system." In: Solving Urban Infrastructure Problems Using Smart City Technologies, John Vacca (ed.), pp. 385–395. Elsevier, Amsterdam, Netherlands, September 2020.

[11] B. Khan, E. Gidey, H. Getachew, and H.H. Alhelou, "Managing the generation and demand inside the smart-grid structure." In: Solving Urban Infrastructure Problems Using Smart City Technologies, John Vacca (ed.), pp. 433–445. Elsevier, Amsterdam, Netherlands, September 2020.

[12] P. Singh and B. Khan, "Smart microgrid energy management using a novel artificial Shark optimization," *Complexity*, vol. 2017, Article ID 2158926, 22 pages, 2017.

[13] F. Shariatzadeh, P. Mandal, and A.K. Srivastava, "Demand response for sustainable energy systems: A review, application and implementation strategy," *Renewable and Sustainable Energy Reviews*, vol. 45, pp. 343–350, 2015.

[14] M.A. López, S. De La Torre, S. Martín, and J.A. Aguado, "Demand-side management in smart grid operation considering electric vehicles load shifting and vehicle-to-grid support," *International Journal of Electrical Power & Energy Systems*, vol. 64, pp. 689–698, 2015.

[15] Y. Shimomura, Y. Nemoto, F. Akasaka, R. Chiba, and K. Kimita, "A method for designing customer-oriented demand response aggregation service," *CIRP Annals*, vol. 63, no. 1, pp. 413–416, 2014.

[16] M.H. Albadi and E.F. El-Saadany, "A summary of demand response in electricity markets," *Electric Power Systems Research*, vol. 78, no. 11, pp. 1989–1996, 2008.

[17] A. Conteh, M.E. Lotfy, K.M. Kipngetich, T. Senjyu, P. Mandal, and S. Chakraborty, "An economic analysis of demand side management considering interruptible load and renewable energy integration: A case study of Freetown Sierra Leone," *Sustainability*, vol. 11, no. 10, pp. 2828, 2019.
[18] C.E. Kement, H. Gultekin, and B. Tavli, "A holistic analysis of privacy aware smart grid demand response," *IEEE Transactions on Industrial Electronics*, 2020.
[19] G.H. Philipo, Y.A. Chande Jande, and T. Kivevele, "Demand-side management of solar microgrid operation: Effect of time-of-use pricing and incentives," *Journal of Renewable Energy*, vol. 2020, 2020.
[20] M.A. Mirzaei, A.S. Yazdankhah, and B. Mohammadi-Ivatloo, "Stochastic security-constrained operation of wind and hydrogen energy storage systems integrated with price-based demand response," *International Journal of Hydrogen Energy*, vol. 44, no. 27, pp. 14217–14227, 2019.
[21] H. Khajeh, H. Firoozi, H. Laaksonen, and M. Shafie-khah, "A new local market structure for meeting customer-level flexibility needs," *2020 International Conference on Smart Energy Systems and Technologies (SEST)*, pp. 1–6, 2020.
[22] P. Palensky and D. Dietrich, "Demand side management: Demand response, intelligent energy systems, and smart loads," *IEEE Transactions on Industrial Informatics*, vol. 7, no. 3, pp. 381–388, 2011.
[23] J. Jiang, Y. Kou, Z. Bie, and G. Li, "Optimal real-time pricing of electricity based on demand response," *Energy Procedia*, vol. 159, pp. 304–308, 2019.
[24] B. Celik, G. Rostirolla, S. Caux, P. Renaud-Goud, and P. Stolf, "Analysis of demand response for datacenter energy management using GA and time-of-use prices," *2019 IEEE PES Innovative Smart Grid Technologies Europe (ISGT-Europe)*, 2019, pp. 1–5.
[25] P. Fitzpatrick, F. D'Ettorre, M. De Rosa, M. Yadack, U. Eicker, and D.P. Finn, "Influence of electricity prices on energy flexibility of integrated hybrid heat pump and thermal storage systems in a residential building," *Energy and Buildings*, vol. 223, pp. 110142, 2020.
[26] P. Du, N. Lu, and H. Zhong, "Overview of demand response." In: Demand Response in Smart Grids, Pengwei Du, Ning Lu, and Haiwang Zhong (eds.), pp. 1–27, Springer Nature Switzerland AG, 2019.
[27] M. Džamarija, G. Leclercq, M. Marroquin, and M. Herman, "Modeling of complex systems including transmission, distribution, aggregation, ancillary services markets." In: TSO-DSO Interactions and Ancillary Services in Electricity Transmission and Distribution Networks, Gianluigi Migliavacca (ed.), pp. 25–59. Springer Nature Switzerland AG, 2020.
[28] H. Khajeh, A. Akbari Foroud, and H. Firoozi, "Robust bidding strategies and scheduling of a price-maker microgrid aggregator participating in a pool-based electricity market," *IET Generation, Transmission & Distribution*, vol. 13, no. 4, pp. 468–477, 26 February 2019.
[29] G. De Zotti, S.A. Pourmousavi, H. Madsen, and N.K. Poulsen, "Ancillary services 4.0: A top-to-bottom control-based approach for solving ancillary services problems in smart grids," *IEEE Access*, vol. 6, pp. 11694–11706, 2018.
[30] P. Spodniak, K. Ollikka, and S. Honkapuro, "The relevance of wholesale electricity market places: The Nordic case," 2019.
[31] F. Oyj, "Fingrid open data" [Online]. Available at: https://data.fingrid.fi/en.

[32] W. Tushar, T.K. Saha, C. Yuen, D. Smith, and H.V. Poor, "Peer-to-peer trading in electricity networks: An overview," *IEEE Transactions on Smart Grid*, vol. 11, no. 4, pp. 3185–3200, July 2020.

[33] K. Gram-Hanssen and S.J. Darby, "'Home is where the smart is'? Evaluating smart home research and approaches against the concept of home," *Energy Research & Social Science*, vol. 37, pp. 94–101, 2018.

[34] M. De Groote, J. Volt, and F. Bean, "Smart buildings decoded: The concept beyond the buzzword," *Buildings Performance Institute Europe* Report, 2017, Available at: http://bpie.eu/publication/smart-buildings-decoded-a-concept-beyond-the-buzzword [Accessed on 12 June 2020].

[35] S.L. Arun and M.P. Selvan, "Smart residential energy management system for demand response in buildings with energy storage devices," *Frontiers in Energy*, vol. 13, no. 4, pp. 715–730, 2019.

[36] X. Jin, K. Baker, D. Christensen, and S. Isley, "Foresee: A user-centric home energy management system for energy efficiency and demand response," *Applied Energy*, vol. 205, pp. 1583–1595, 2017.

[37] B.N. Silva, M. Khan, and K. Han, "Futuristic sustainable energy management in smart environments: A review of peak load shaving and demand response strategies, challenges, and opportunities," *Sustainability*, vol. 12, no. 14, pp. 5561, 2020.

[38] T. Sousa, T. Soares, P. Pinson, F. Moret, T. Baroche, and E. Sorin, "Peer-to-peer and community-based markets: A comprehensive review," *Renewable and Sustainable Energy Reviews*, vol. 104, pp. 367–378, 2019.

[39] F. Oyj, "Terms and conditions for providers of manual frequency restoration reserves (mFRR)," 2020.

[40] H. Shareef, M.S. Ahmed, A. Mohamed, and E. Al Hassan, "Review on home energy management system considering demand responses, smart technologies, and intelligent controllers," *IEEE Access*, vol. 6, pp. 24498–24509, 2018.

[41] T. Kawakami, T. Yoshihisa, N. Fujita, and M. Tsukamoto, "A rule-based home energy management system using the Rete algorithm," *2013 IEEE 2nd Global Conference on Consumer Electronics (GCCE)*, 2013, pp. 162–163.

[42] T. Yoshihisa, N. Fujita, and M. Tsukamoto, "A rule generation method for electrical appliances management systems with home EoD," *The 1st IEEE Global Conference on Consumer Electronics 2012*, 2012, pp. 248–250.

[43] S.Z. Althaher and J. Mutale, "Management and control of residential energy through implementation of real time pricing and demand response," *2012 IEEE Power and Energy Society General Meeting*, 2012, pp. 1–7.

[44] M.S. Ahmed, H. Shareef, A. Mohamad, J. Abd Ali, and A.H. Mutlag, "Rule base home energy management system considering residential demand response application," *Applied Mechanics and Materials*, vol. 785, pp. 526–531, 2015.

[45] M.S. Ahmed, A. Mohamed, H. Shareef, R.Z. Homod, and J. Abd Ali, "Artificial neural network based controller for home energy management considering demand response events," *2016 International Conference on Advances in Electrical, Electronic and Systems Engineering (ICAEES)*, 2016, pp. 506–509.

[46] Y. Wu, B. Zhang, J. Lu, and K.L. Du, "Fuzzy logic and neuro-fuzzy systems: A systematic introduction," *International Journal of Artificial Intelligence and Expert Systems*, vol. 2, no. 2, pp. 47–80, 2011.

[47] L. Suganthi, S. Iniyan, and A.A. Samuel, "Applications of fuzzy logic in renewable energy systems – A review," *Renewable and Sustainable Energy Reviews*, vol. 48, pp. 585–607, 2015.
[48] K. Premkumar and B.V. Manikandan, "Fuzzy PID supervised online ANFIS based speed controller for brushless DC motor," *Neurocomputing*, vol. 157, pp. 76–90, 2015.
[49] M. Beaudin and H. Zareipour, "Home energy management systems: A review of modelling and complexity," *Renewable and Sustainable Energy Reviews*, vol. 45, pp. 318–335, 2015.

21

A Sustainable Building Lightning Solution for Energy Conservation in Different Geographical Conditions

Sunil Kumar Goyal[1], Bharat Bhatia[1], Amit Saraswat[1], Bhuwan Pratap Singh[1], and Om Prakash Mahela[2]

[1] *Department of Electrical Engineering, Manipal University, Jaipur, India*
[2] *Power System Planning Division, R.R.V.P.N.L., Jaipur, India*

21.1 Introduction

Energy plays an important role in growth and survival of any country in the world. The increasing energy demand has an adverse effect on the environment and increases the pressure on government to fulfill the requirement. For a developing country like India, the energy criterion decides the growth of the country. Energy demand in developing and developed nations is expanding exponentially. Therefore, the reduction in electricity consumption means saving money, which in return increases the cashflow of the nation. Less pollution and fewer greenhouse gases consuming petroleum derivatives for the creation of power brings about huge measures of contamination into the air. A few instances of ways that power creation can harm the earth include corrosive downpour, residue, carbon dioxide, poisonous gases, and deforestation [1–3]. It is simply from consuming conventional energy sources that creates ozone-depleting substances. Since we do not inherit the Earth from our ancestors, we borrow it from our children, so more emphasis should be laid on sustainable growth and make the Earth a better place to live in.

Burning of fossil fuels for electric power generation releases pollutants in the air and contaminates the environment. It can be reduced either by employing renewable energy resources like solar, wind, tidal, etc., or by reducing the energy generation. Nowadays, electrical energy is an essential commodity and an integral part of day-to-day life. Therefore, less generation is possible only when the demand is less. It can be achieved only when the outcome of electrical devices is the same as the reduced energy consumption. In other words, energy conservation is the effort made to reduce the consumption of energy by using less of an energy service. This can be achieved either by using energy more efficiently or by reducing the amount of service used. Energy conservation can be achieved either by using efficient devices or by reducing the use of electricity. Figure 21.1 shows the possible solutions for energy conservation, i.e. load shading, use of energy efficient devices, and optimum utilization. In load shading, load can be reduced by curtailment or feeding with planning as per the requirement and priority of the utility. Electrical power requirement can also be reduced by employing energy efficient devices, i.e. star

Active Electrical Distribution Network: A Smart Approach, First Edition. Edited by Baseem Khan, Josep M. Guerrero, Sanjeevikumar Padmanaban, Hassan Haes Alhelou, Om Prakash Mahela, and Sudeep Tanwar.

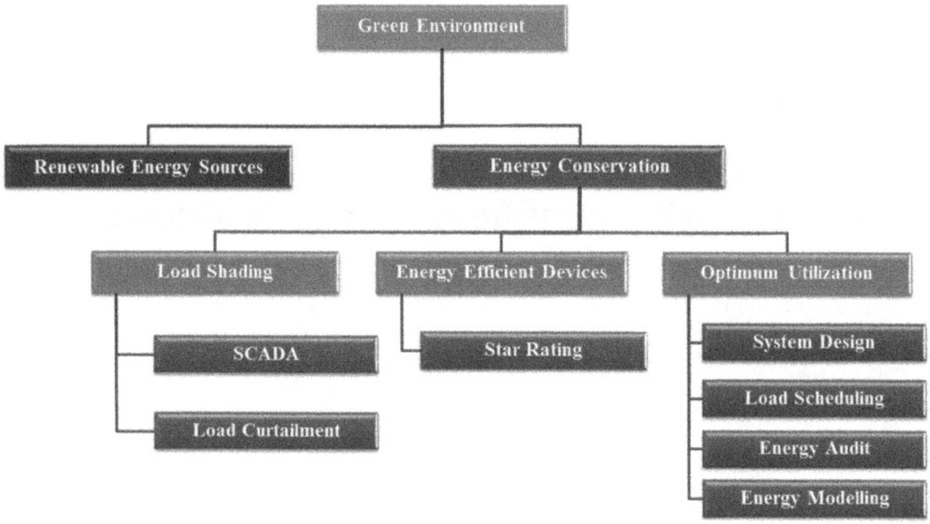

Figure 21.1 Modes of energy conservation.

rating appliances, which provide the same output at less energy consumption. Optimum utilization of available resources can be achieved by load scheduling, energy audit, system design, and energy modeling. System design deals with the architectural, fixtures, controllers, timers, etc., to reduce the energy consumption [4–7]. For example, the lighting and thermal requirement can be satisfied fully or partly by optimizing natural resourced sunlight. It will not only provide the desired level of illumination but help to maintain the desired temperature in the utility space. An energy audit gives the existing energy consumption for installed devices/appliances. It will also provide a probable solution to reduce the electricity consumption. In [8], a meta-heuristic technique is utilized for the optimization of smart home appliances using a cost-effective energy management system. In [9], a comprehensive review of smart home energy management system optimization techniques is presented. In [10], various components of a smart grid system are discussed. In [11], management of the generation and demand inside the smart grid structure is discussed. In [12], smart microgrid energy management using a novel artificial shark optimization is discussed.

Electricity consumption for illumination may be reduced by increasing the use of natural sunlight. The buildings are designed in such a way that adequate natural sunlight will be available for most of the daytime. It not only reduces the electricity bill but also gives pleasure to human eyes and also improves the working efficiency of residents of that building. Energy conservation improves the economy of individuals as well as nations, the environment, and national security. The governments of different countries set certain guidelines under which the higher the savings in terms of electricity consumption the better will be the rating that would be given to the building, i.e. GRIHA (Green Rating for Integrated Habitat Assessment), LEED (Leadership in Energy and Environment Design), and IGBC (Indian Green Building Council) [13–16].

21.2 Energy Modeling

Energy modeling is the development of a simulation model through software in order to analyze the energy consumption and to find the probable energy efficient solution. These solutions reduce the consumption of electrical energy to perform the identical operation. Energy efficiency not only improves the system economy but also has many advantages, i.e. reducing the energy demand and improving fuel consumption, which further improves the quality of the environment as less pollutant particles are released into the air. A reduction in energy imports significantly improves the national economy, sustainability, and reduces the dependency on other countries. An energy efficient building requires less energy, which reduces the operating cost of the building and overall generation constraint of the electrical energy. It further reduces the greenhouse gasses emission, which improves the environmental sustainability of the building.

A computerized or virtual simulation model of a building developed in energy modeling focuses on energy consumption. It also deals with the use of different electrical and thermal devices such as lights, fans, air conditioners, hot water, refrigerators, etc. The pattern of use, energy consumption, and life cycles of these devices has been analyzed for present energy consumption. Different energy efficient solutions are then suggested and employed in the simulation. These solutions include application of renewable energy sources, i.e. solar PV, wind turbines, and employing appliances with high efficiency [17–19]. The process of designing a virtual model of any workplace or building on a computer is called a building simulation. In this, all the physical components of the building are taken into consideration along with variable environmental parameters. All the electrical and thermal appliances installed in the building are put at their actual rating and operating time. It will help to estimate the actual consumption of electrical energy as used in real time. Building simulation provides future quantitative predictions in terms of energy utilization and is usually segregated into two groups: load design and energy analysis. Load design includes estimation of air conditioning loads, volumetric air flow requirements, supply temperature, hydronic plant, equipment capacities, etc. An energy analysis is employed for the prediction of annual energy consumption, electricity bills, CO_2 emission, comparison of various efficiency options, and estimation of the payback life cycle of these options [20].

21.2.1 Green Building

A building that is designed and constructed to either reduce/eliminate negative impacts and/or create positive impacts on the environment is considered to be a 'green' building. The main attributes of a green building include the efficient use of energy and other resources, inclusion of renewable energy equipment, and non-toxic, sustainable, and ethical materials. It takes care of pollution and wastage reduction processes including its reuse and recycling. The design and operation of a green building ensures the air quality and environment to improve the quality of life of occupants, even in a changing environment. Green buildings employ renewable energy sources, which reduce the dependency on energy from conventional sources. Most of the windows that are installed allow more natural light, which further reduces the use of artificial light, electricity bills, and tiredness

[21–23]. The use of rainwater reduced water waste, and recycling of water are fundamental features of green buildings. These efforts reduce the operation and maintenance costs and improve the indoor environment and health of occupants. At the same time they reduce electricity consumption and electricity bills.

Total electricity consumption of a commercial building in the US is shown in Figure 21.2. It can be seen that the maximum amount of electricity is consumed in heating and lighting. More than half of the electricity bill will be paid in lieu of these two applications. The consumption for lighting can be reduced by extracting maximum daylight to maintain the desired illumination level. Similarly, consumption for heating can also be reduced by using the heat of sunlight.

21.2.2 Building Lighting

Building lighting plays a very important role in the overall electricity consumption of any building. Lighting is the purposeful utilization of light to accomplish useful or aesthetic impacts and is defined in "lux." Lighting includes the utilization of both natural and artificial light sources. A natural light source is the incident light through the Sun while artificial light sources are governed by an external supply, such as electrical energy, petroleum, wax, and wood. Examples of light sources using electricity are bulbs, LEDs, and light fixtures. Natural light is captured through windows, doors, and lookout windows. Proper illumination is an essential requirement of any place used in that area, i.e. a living room, office, reception area, theater, library, dining room, etc. Appropriate illumination improves the appearance of the region, upgrades task execution, and has a positive emotional impact

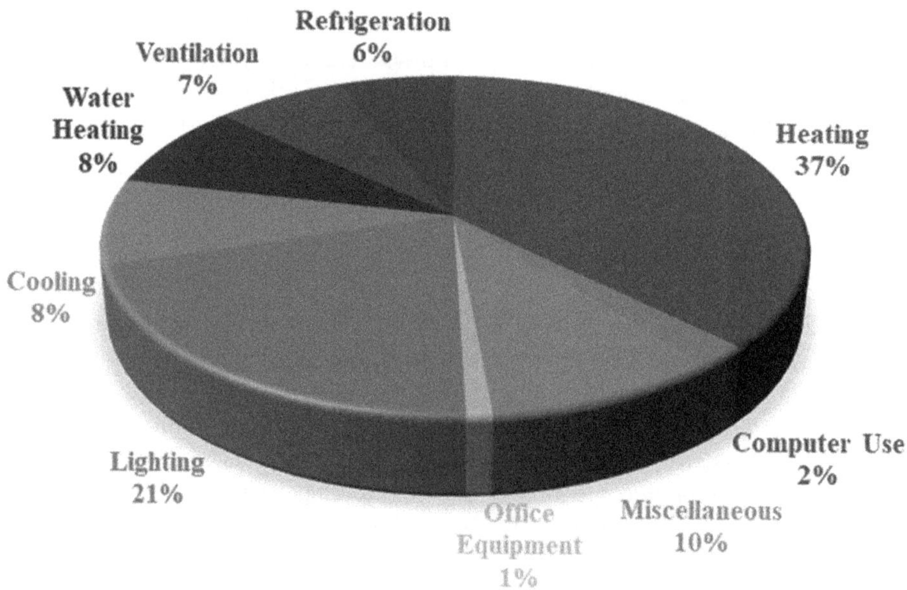

Figure 21.2 Electricity consumption of a commercial building in the US.

on residents of that building. Indoor lighting is generally practiced utilizing light apparatuses and is a key piece of the inside plan. Lighting can likewise be a characteristic part of scene ventures. Daylighting is the most established technique for indoor and outdoor lighting. More daylight in any space increases the illumination of that area, which further reduces the use of artificial light [24]. It also reduces the overall load demand and electricity bill of the consumer. A maximum portion of outdoor lighting is comprised of natural sunlight during daytime. Natural sunlight also has a positive impact on the occupants of the building, such as hospital patients, office employees, school students, etc. Commercial buildings have many utilities, such as a restaurant, theater, shops, activity areas, etc. Illumination levels of these areas are different depending on their use. The architectural design of a building greatly affects the illumination of any space through natural light.

21.2.3 Orientation of a Window

Indoor lighting of an area can be maintained at the required level depending on the utility, through natural sunlight and artificial lighting. Illumination is defined in watts per square meter and depends upon the area of the room/utility. More illumination is required for a big space. Natural daylight generally comes through windows. The illumination may be increased by installing more windows or larger sized windows. These parameters depend on the architectural design and mechanical strength of the walls. The natural sunlight coming through windows also depends on the angle of incident light, which is directly affected by rotation of the Earth around its own axis and around the Sun. The direction/orientation of the window decides the illumination level in the room. It also depends on the working timings of the utility, such as schools/colleges, which work in the morning and noon times and some shops in the noon and evening times. The Sun rises in the east direction and sets in the west direction; hence windows in school buildings will be in the east direction. The angle and amount of sunlight varies with geographical location of any building. It reduces as the location moves from the Equator to the poles. The illumination can be controlled by using different types of glass in the windows.

21.2.4 Glazing

The term "glazing" is basically a translucent material used in a window or a door in any building. Popular types of glazing applied for any architectural application are laminated glass, tempered glass, tinted float glass, coated glass, etc. It may be mounted on the surface of a window/door, generally with the help of PVC/aluminum/wood. The glass may be fitted into a window or door frame using putty. However, a hardened or reinforced and laminated glass may be glazed to a metal frame using bolts that pass through holes made by a drill-machine. The main purpose of glazing is to reduce the temperature inside the room/hall in order to improve the overall thermal efficiency of the building as it is used in thermal collectors with a low temperature by retaining the collected heat. In fact, the energy performance of any window or door may be improved by proper use of an effective glazing. The change in type of glazing, depth of glazing, and the material placed between glazing layers may have a direct impact on the temperature as well as the environment in any residential or commercial building. In general, the glazing may consist of one or two or several layers. An appropriate choice among single-glazed, double-glazed, and multiglazed window applications decides the extent of temperature control inside the building. Some of the distinct features of a single-glazed window and double-glazed window are highlighted in the subsequent paragraphs.

21.2.4.1 Single-Glazed Windows

The single-glazed window, as shown in Figure 21.3(a), is one consisting of a single layer/pane of glazing using a glass material. It is fitted in a wooden or metallic frame of the window using putty. The significant advantages and disadvantages of a single-glazed wind are as follows:

21.2.4.1.1 Advantages of Single Glazing

- If the single glazing is again opted during renovation or routine maintenance, the old window frame does not need to be replaced. New frames are not always required for a new single-glass pane.
- It is very suitable in a hot and tropical climate as there is no need for any kind of heat retention inside the building.

21.2.4.1.2 Disadvantages of single glazing

- It provides very poor insulation because the outside atmospheric temperature impacts the inside temperature. Therefore, a strong climate control system is required to maintain the desired temperature inside the building.
- It does not give protection against outside noise. In fact, any voice pollution existing outside or in the vicinity will come inside the building.
- The strength of single glazing is very poor and it may break easily by an external impact.

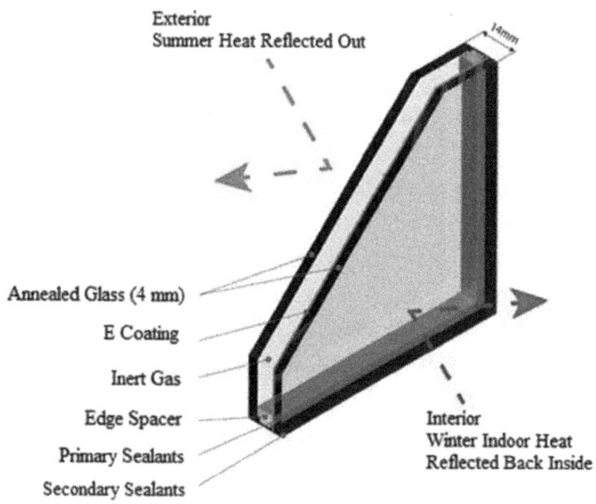

Figure 21.3 (a). Single-glazed window.

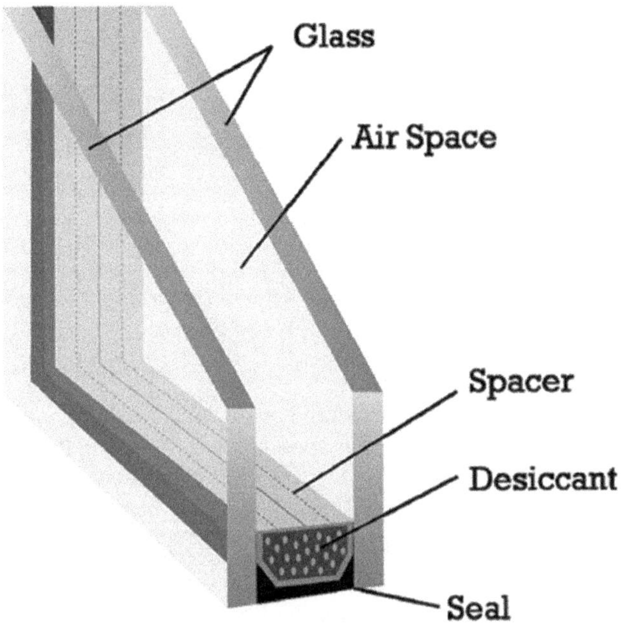

Figure 21.3 (b). Double-glazed window.

21.2.4.2 Double-Glazed Windows

In this glazing structure, there are two distinct layers of glass panes for a double-glazed window, as shown in Figure 21.3 (b). Although the space between the glass layers may be empty, it is generally filled either with a gas or a tight vacuum. The main purpose of this medium (i.e. either a gas or a tight vacuum) between the two glass layers is to provide an additional level of insulation from extreme environmental conditions. Detailed benefits of a double-glazed window over a single-glazed window are highlighted as follows:

- It provides high insulation from the outside environmental or weather conditions because of two glass layers separated by a medium. It helps in maintaining the temperature inside the building irrespective of the outside weather conditions. Moreover, when heating is required inside the building during the winter season, the room temperature does not fall drastically due to the insulation layer of the windows.
- It provides an energy efficient building solution as it has less requirements of a climate control system inside the building to maintain a proper temperature irrespective of the weather outside. This is because the inner glass pane is not at the same as the outside temperature. It ultimately reduces the requirements of electrically driven climate control systems and therefore provides a significant saving in maintenance costs as well as electricity costs.
- Double glazing prevents condensation and therefore makes the entire ambience much more comfortable.

- It reduces noise interference inside the building by cutting out the noise present outside. It ensures the desired reduction in sound pollution.
- It has a higher margin of safety with very high durability because of the extra glass pane. This double-glazed structure is harder to break.

21.2.5 Selection of Glass

The level of illumination and availability of sunlight largely depends on the type of glass used in the window. Different factors are involved while selecting the appropriate glass for the illumination and thermal requirement of the desired space corresponding to the use of that space. Different parameters affecting the performance of window glass are as follows:

- **U-value**: This is the heat flow rate in caused by thermal conduction and the difference between the outdoor and indoor temperature. A lower U-value denotes a lower heat transfer capability of the glass.
- **Solar heat gain coefficient (SHGC)**: This is an indicator of the solar heat blocking capability. SHGC is the ratio of incident sunlight radiation coming through a window to consequently enter another area. A lower SHGC denotes lower heat transmission.
- **Shading coefficient (SC)**: This signifies the limit of a direct solar heat gain and is expressed as

$$Shading\,Coefficient \frac{Solar\,Heat\,Gain\,Coefficient\,(SHGC)}{Heat\ Gain\ through\ 3mm\ clear\ glass}$$

The lower value of the shading coefficient denotes superior glazing in blocking a solar heat gain.

- **Visible transmittance (VT)**: This is an optical characteristic that denotes the amount of visible light coming through a window. A lower value of VT denotes less light passing through the window.
- **Light-to-solar gain ratio (LSGR)**: This gives a measure of relative efficiency of different glass types in transmitting daylight while blocking heat gains. It is defined as the ratio between VT and SHGC.

$$LSGR = \frac{VT}{SHGC}$$

A higher value of LSGR denotes more light transmitted through the window without increasing heat.

- **Air leakage**: It is denoted by air leakage (volume in cubic feet of passing air) through a window area (in square feet). Heat gain and loss by convection happens due to penetration through gaps/cracks in the window assembly. A lower value of air leakage denotes less air passing through the window.
- **Condensation resistance (CR)**: This is a measure of the capability to resist condensation on the inner surface of glass. A lower value of CR denotes less resistive ability against condensation formation.

- **Relative heat gain (RHG):** This is the collective effect of the U-value and SC. The major part is the direct heat gain component (up to 90%) and a minor part is the conductive heat gain (10%).

  $RHG = Direct\ Heat\ Gain + Conductive\ Heat\ Gain$

- **Acoustic performance:** This is an interpretation of the type of sound from outside. It differentiates low-frequency signals (125–800 Hz), such as city traffic, appliance, and music, from high-frequency signals (800–5000 Hz), which includes fast moving trains, factories, airplanes, etc. The cheapest, simplest, and most efficient method to reduce the sound is by using thick glass.

21.3 Methodology

Electrical energy consumption for illumination of any building and its different areas depends upon the daylight available through windows. Orientation, size, pane configuration, and type of glass used in the window decide the total lux in that area. ECOTECT simulation software is employed to calculate the value of lux in different regions of a building to meet the demand for utilization in that region. This software is used by architects and building engineers to evaluate the building performance including energy consumption in heating/cooling equipment, illumination, and airflow for ventilation. It also helps to make the strategic modifications required to minimize the day-to-day electricity consumption for air-conditioning/heater and lighting. The software also takes care of the geometric location of the building and weather conditions on a certain day and time. The process of simulation to make a model of a building is described in Figure 21.4. The base on which models in ECOTECT are built is taken from AUTOCAD software with the help of the architectural plans of the actual site provided by the customer/architect. Then this model is sent to Radiance or additionally DIALux for examination. Radiance and DIALux are both approved programming bundles. These models can be utilized to create several forms of daylight analysis. Energy models are employed that are similar to light models and are specifically designed for Energy Simulation. These modeling and simulations are designed and analyzed by energy modeling software like eQUEST, Energy-Pro, EnergyPlus, etc. By performing the simulations on the above-mentioned software, a contractor is able to compare different design options, verify that a proposed plan will give sufficient sunshine, predict sources of potential glare, look at contrast ratios or balance of daylight within a space, indicate potential energy savings from electric lighting use, and, last but not least, document daylighting code compliance.

The lux of all internal zones can be calculated by repeating the same procedure for each zone as per building drawings. These values of lux are then compared with the minimum value of lux required according to utilization in that zone. The lux of any zone can be adjusted by changing window orientation, size, pane, and glass type. Different Government certifications are issued based on the percentage of area meeting daylight.

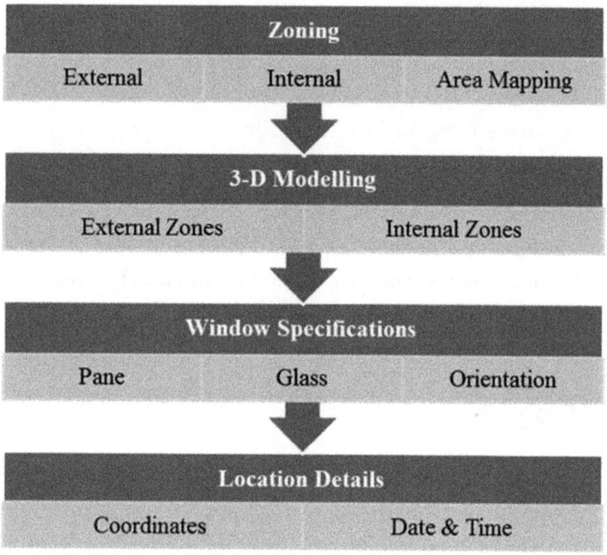

Figure 21.4 Building modeling simulation process.

21.4 Case Study

The Earth revolves around its axis and completes one round in 24 hours (which makes one day and one night). The axis of Earth is not exactly vertical but inclined at an angle of 23.5°; therefore sunlight has a different pattern and intensity in different geographical locations of the northern and southern hemispheres. Sunlight is at a maximum at the Equator and gradually reduces while moving towards the poles. Hence sunlight intensity decreases in buildings considered in distances from the Equator to the poles. The proposed methodology is employed at five places of different geographical locations, which are mentioned in Table 21.1.

These five locations cover all possible combinations of longitude (east or west) and latitude (northern or southern hemisphere). The locations, New Delhi – India, Rome – Italy, and Reykjavik – Iceland, are situated in the northern hemisphere, Nairobi – Kenya is located almost on the Equator, and Sydney – Australia is in the southern hemisphere. Reykjavik is in the western part of prime median, whereas the remaining four locations are in the eastern part of prime median.

In the proposed case study, a room of 5000 × 4000 mm is considered as a test system in each location. A window of 1500 × 1200 mm is assumed in the room. The availability of natural sunlight through the window varies with the geographical location of these places. The impact of the direction of the window, the pane type, and the fitting of different glass is discussed in the next section.

Table 21.1 Geographical coordinates of locations

Place	Geographical coordinates
New Delhi – India	28.6139° N, 77.2090° E
Rome – Italy	41.9028° N, 12.4964° E
Reykjavik – Iceland	64.1466° N, 21.9426° W
Nairobi – Kenya	1.2921° S, 36.8219° E
Sydney – Australia	33.8688° S, 151.2093° E

21.5 Result Analysis

Room lighting design depends upon the availability of natural light from the sun during daytime. It is assumed that the natural sunlight enters the test room only through the window under consideration, which depends upon various factors, i.e. size of room, size of window, direction/orientation of the window, type of pane system, and glass used in the window. In the present analysis, the size of the test room and window is kept constant. Therefore, orientation of the window and the type of pane and glass are the prime factors for analysis. Different factors related to illumination depend on the geographical location, i.e. length of day, intensity of sunlight, sun-path, and angle of incident light. It is a well-known fact that the sunlight is inclined in the morning and evening hours, whereas it is almost perpendicular at noon and in the afternoon hours. Therefore, a single timing in each time domain, i.e. morning (09:00 a.m.), afternoon (02:00 p.m.), and evening (05:00 p.m.), has been considered.

21.5.1 Sun-Path Analysis

The Earth is a satellite of the Sun, so moves around the Sun and completes one round in one year, which makes the Seasons. The Earth also rotates on its own axis and completes one rotation in 24 hours, which creates day and night. Due to these rotations, sunlight is not constant on any location on Earth. The intensity and angle of incident rays from the Sun is not even for all the locations and for a whole day. It is different for both the hemispheres, i.e. northern and southern hemispheres. The duration of the day and night is also different, corresponding to the position of a geographical location with reference to the Sun. The Sun is continuously changing its position with reference to a geographical location on the Earth, which further varies the light intensity at that location. Therefore, tracking of Sun path with reference to a set location is very important when evaluating the availability of light intensity and its locus for a day. The daily Sun path means that the Sun seems to move across the sky as the Earth rotates and orbits the Sun. The relative locus of the Sun affects the illumination and heating of the buildings and the performance of the solar systems. In this section, the Sun path analysis for all five locations under consideration, i.e. New Delhi, Rome, Reykjavik, Nairobi, and Sydney, is done for one day, i.e. 21 March 2020. To observe the impact of variation in the angle of incidence and light intensity during a day, the Sun path is captured in the morning (09:00 a.m.), noon (02:00 p.m.), and evening (05:00 p.m.) for all locations.

21.5.1.1 New Delhi – India

New Delhi is the capital of India and its geographical coordinates are 28.6139° N, 77.2090° E. It is situated in the northern hemisphere near to the Equator. It is around 6835 km from the North Pole, 3166 km from the Equator and 13 168 km from the South Pole [25]. To determine the real time motion and position of the Sun we need to observe the Sun path carefully.

On 21 March 2020, at 9:00 a.m. in the morning the orientation of the Sun is shown in Figure 21.5 [26]. The Sun rises in New Delhi at 05:56 a.m. in an extreme east direction. As the Sun is moving from east to west, its position in the morning at 09:00 a.m. is in the south-east (SE) direction. It can be predicted from Figure 21.5 that the luminance level at the east and south direction will almost be the same, while the shadow of the building lies toward the north and west direction, leading toward lower values of lux in the morning.

The orientation of the Sun at 02:00 p.m. on 21 March 2020 is shown in Figure 21.6. It can been seen that the orientation of the Sun is more towards the south-west (SW) direction, it can be predicted that the illuminance level in the SW direction is at a maximum. The sunlight is perpendicular; hence the length of shadows is small. At this time the illumination is almost at its peak.

The orientation of the Sun at 05:00 p.m. on same day is shown in Figure 21.7. It can be seen that the orientation of the Sun is more towards the west direction, so it can be predicted that the illuminance level in the west direction is at a maximum, while creating a shadow towards the east and north directions, which automatically decreases the lux level in those directions. The sunset was observed at 06:02 p.m. in the extreme west direction.

21.5.1.2 Rome – Italy

Rome is the capital of the Lazio region and is situated in south-central Europe. Its geographical coordinates are 41.9028° N, 12.4964° E. It is also located in the northern

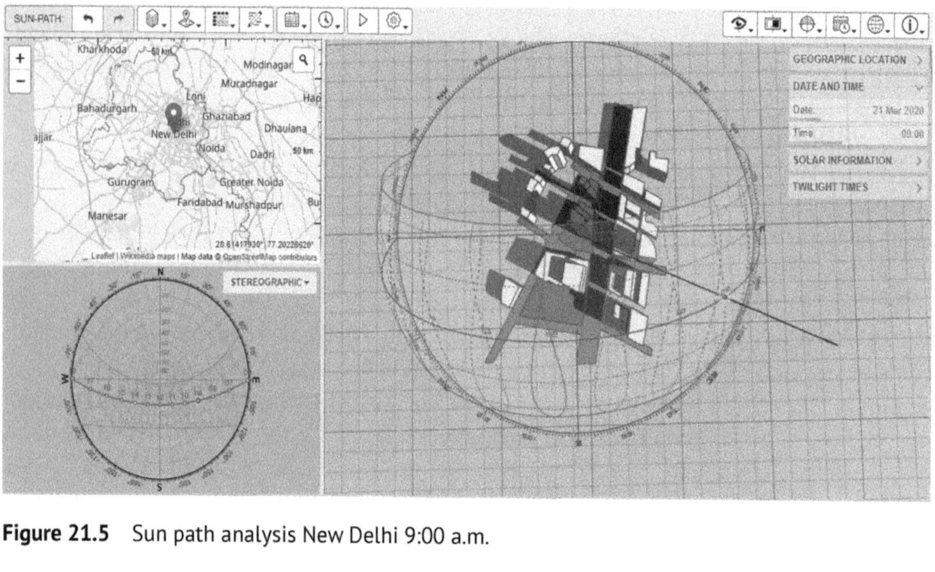

Figure 21.5 Sun path analysis New Delhi 9:00 a.m.

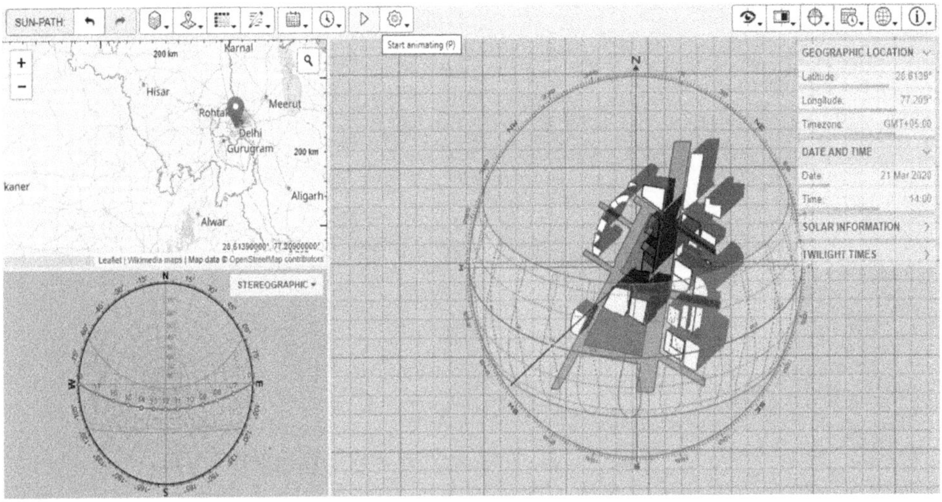

Figure 21.6 Sun path analysis New Delhi 2:00 p.m.

hemisphere between the North Pole and the Equator. It is around 5362 km from the North Pole, 4640 km from the Equator and 14 642 km from South Pole [27]. To determine the real time motion and position of the Sun we need to observe the Sun's path carefully.

On the test date, i.e. 21 March 2020, at 9:00 a.m. in the morning the orientation of the Sun is shown in Figure 21.8. The Sun rose in Rome at 06:13 a.m. in the extreme east direction. As the Sun moves from east to west, its position in the morning at 09:00 a.m. is around 30° in the south-east (SE) direction. It can be predicted from Figure 21.5 that the illuminance level in the east and south directions will almost be the same, while the shadow of the

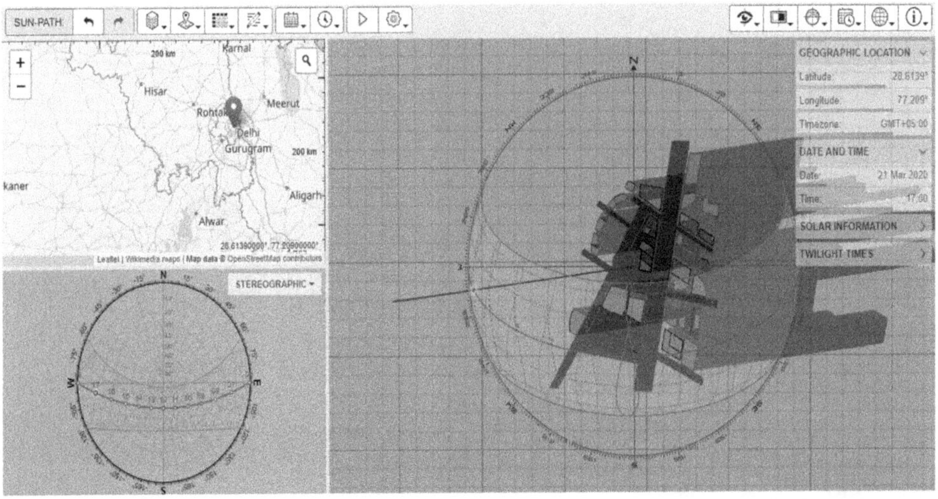

Figure 21.7 Sun path analysis New Delhi 5:00 p.m.

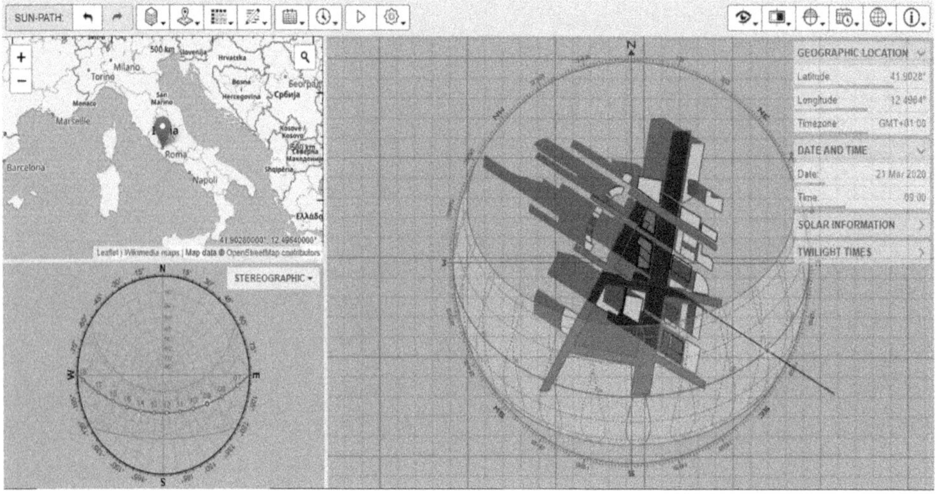

Figure 21.8 Sun path analysis Rome 09:00 a.m.

building lies towards the north and west direction, leading toward lower values of lux in the morning.

The orientation of the Sun at 02:00 p.m. is more towards the south-west (SW) direction, as shown in Figure 21.9. The sunlight is perpendicular at 02:00 p.m. and hence the illumination is at its peak and lengths of shadows are small.

The orientation of the Sun at 05:00 p.m. on the test day is shown in Figure 21.10. The orientation of the Sun was more towards the west direction, so it can be predicted that the illuminance level in the west direction is maximum, while creating a shadow toward the east and north directions, which automatically decreases the lux level in those directions. The sunset was observed at 06:22 p.m. in the extreme west direction.

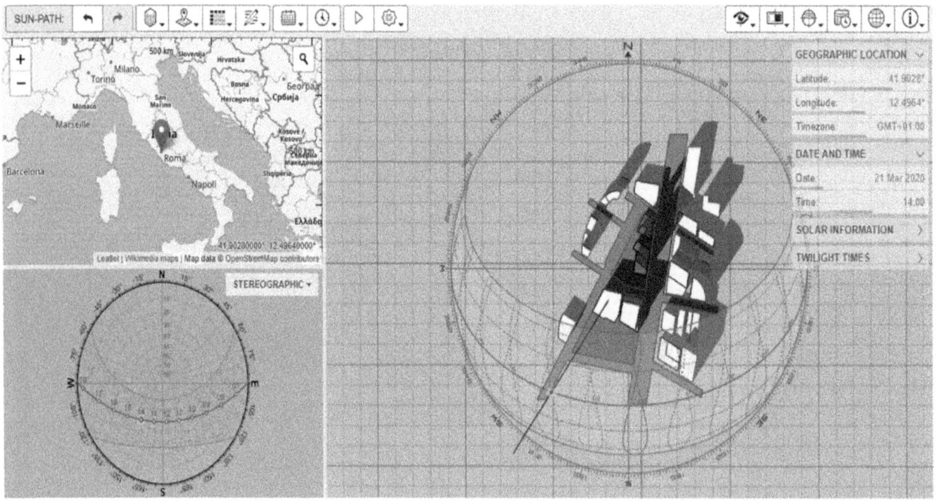

Figure 21.9 Sun path analysis Rome 02:00 p.m.

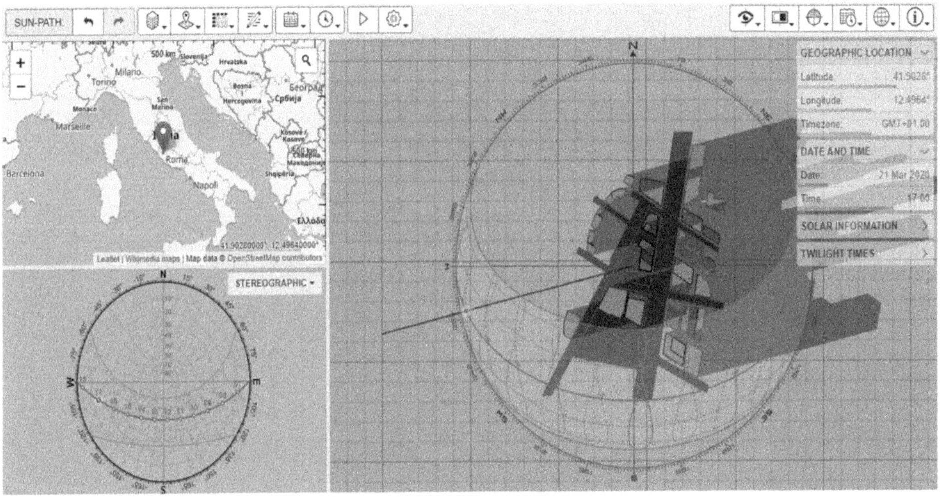

Figure 21.10 Sun path analysis Rome 05:00 p.m.

21.5.1.3 Reykjavik – Iceland

Reykjavik is the largest city and capital of Iceland and its geographical coordinates are 64.1466° N, 21.9426° W. It is the world's northernmost capital of a sovereign state situated on the southern shore of Faxaflói Bay in the northern hemisphere. Its distances from the North Pole, Equator, and South Pole are 2886 km, 7116 km, and 17 118 km, respectively [28]. The Sun's path needs to be observed precisely to determine the real time motion and position of the Sun.

The orientation of the Sun at 09:00 a.m. on 21 March 2020 is shown in Figure 21.11. The Sun rose in Reykjavik at 06:28 a.m. in the east direction. The position of the Sun was in the south-east (SE) direction. It has been observed that the incident sunlight is more inclined and hence the shadows are very long. The time of sunrise is delayed as the location moves towards the North Pole.

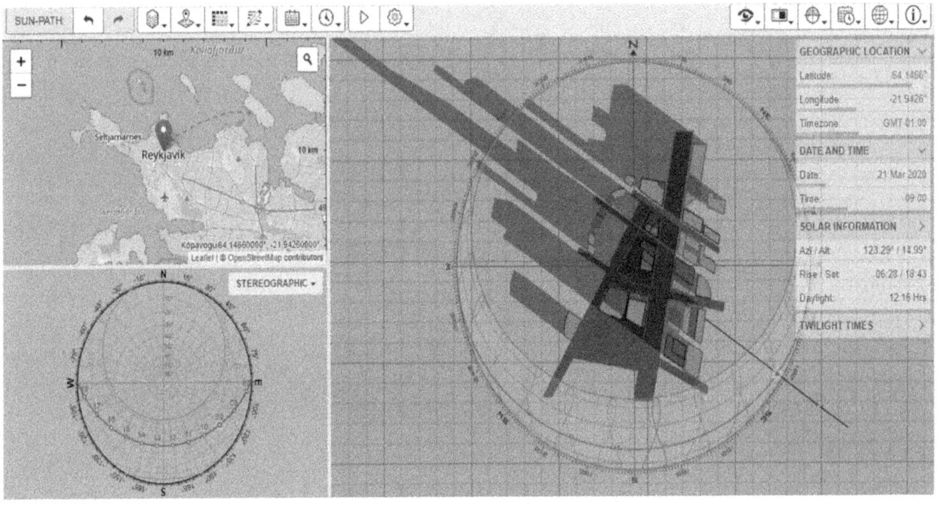

Figure 21.11 Sun path analysis Reyjavik 9:00 a.m.

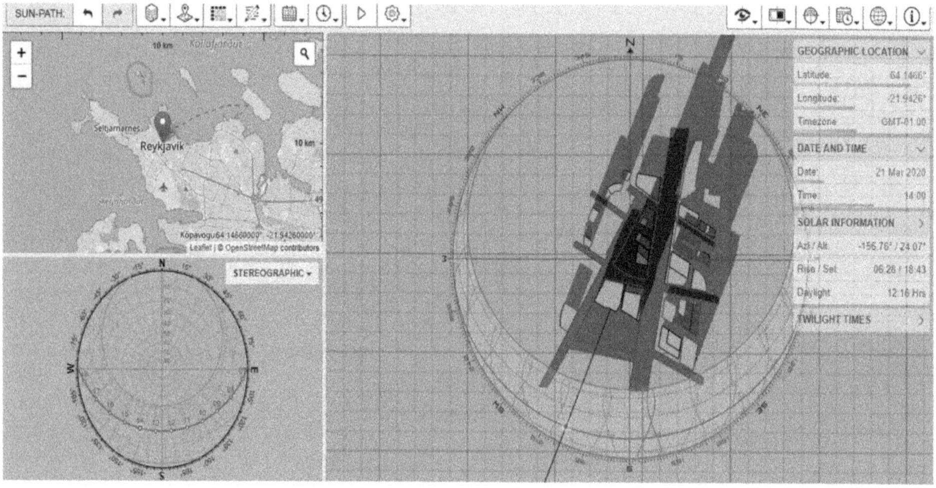

Figure 21.12 Sun path analysis Reyjavik 2:00 p.m.

The orientation of the Sun at 02:00 p.m. is shown in Figure 21.12. The orientation of the Sun was more toward the south-west (SW) direction. The inclination of incident rays is more if the geographical locations shift away from the Equator (towards the poles). Therefore, the shadows are in ascending order from New Delhi, Italy, and Reykjavik.

The orientation of the Sun was very near to the west direction at 05:00 p.m., as shown in Figure 21.13. The sunset was observed at 06:43 p.m. in the west direction. The time of the sunset is also delayed as the location moves from the Equator to the poles. The duration of the day or daylight does not vary much as both timings for sunrise and sunset are delayed.

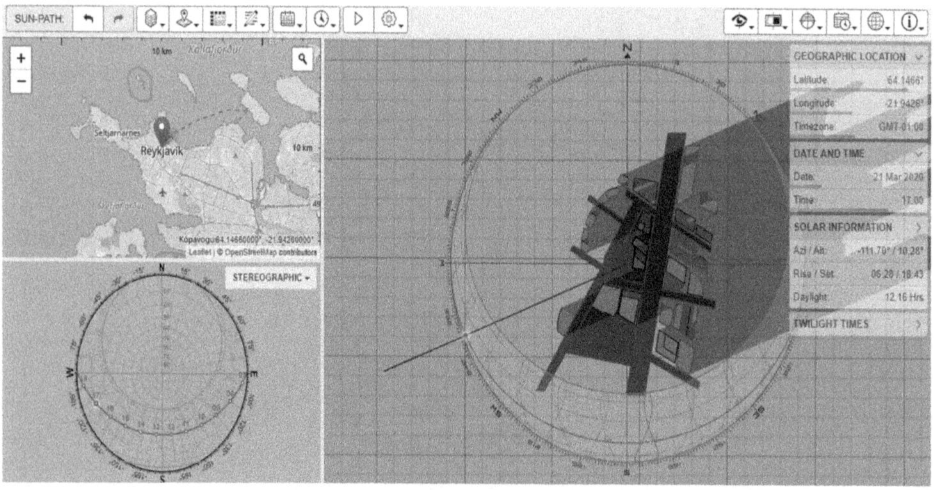

Figure 21.13 Sun path analysis Reyjavik 5:00 p.m.

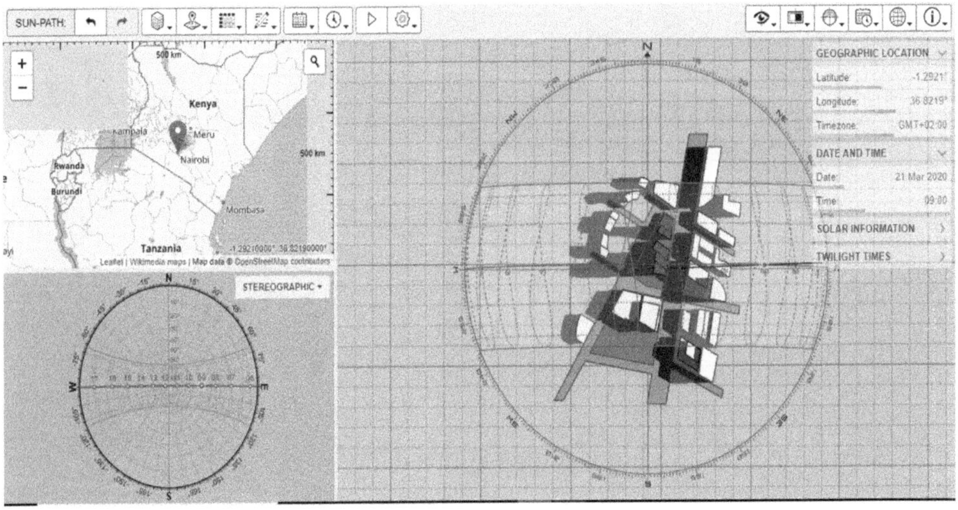

Figure 21.14 Sun path analysis Nairobi 09:00 a.m.

21.5.1.4 Nairobi – Kenya

Nairobi is the largest city and capital of Kenya and its geographical coordinates are 1.2921° S, 36.8219° E. It is located almost on the Equator. Its distances from the North Pole, Equator, and South Pole are 10144 km, 142 km, and 9860 km, respectively [29]. The Sun's path needs to be observed precisely to determine the real time motion and position of the Sun.

The orientation of the Sun at 09:00 a.m. on 21 March 2020 is shown in Figure 21.14. The Sun rose in Nairobi at 05:37 a.m. in the east direction. As Nairobi is located almost at the Equator, the locus of the Sun is from east to west. It does not go into the south-east or south-west directions, as in the case of the previous three locations. It can be seen from

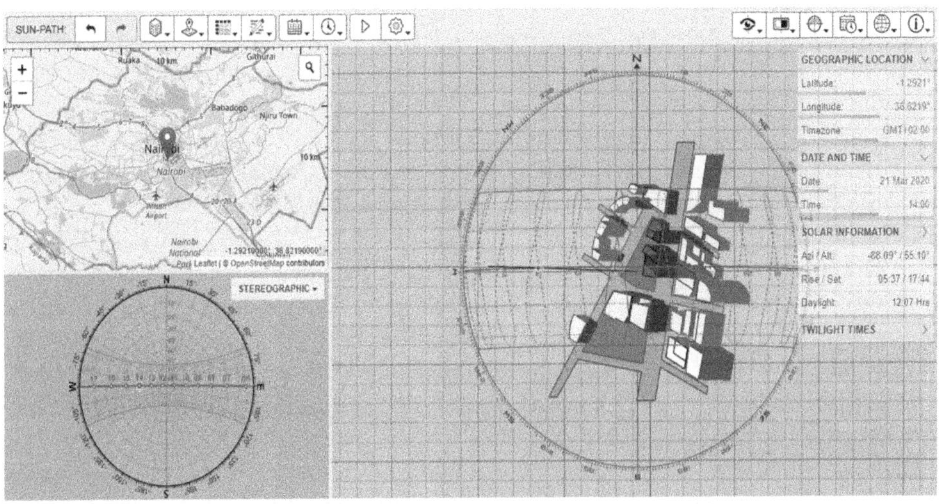

Figure 21.15 Sun path analysis Nairobi 02:00 p.m.

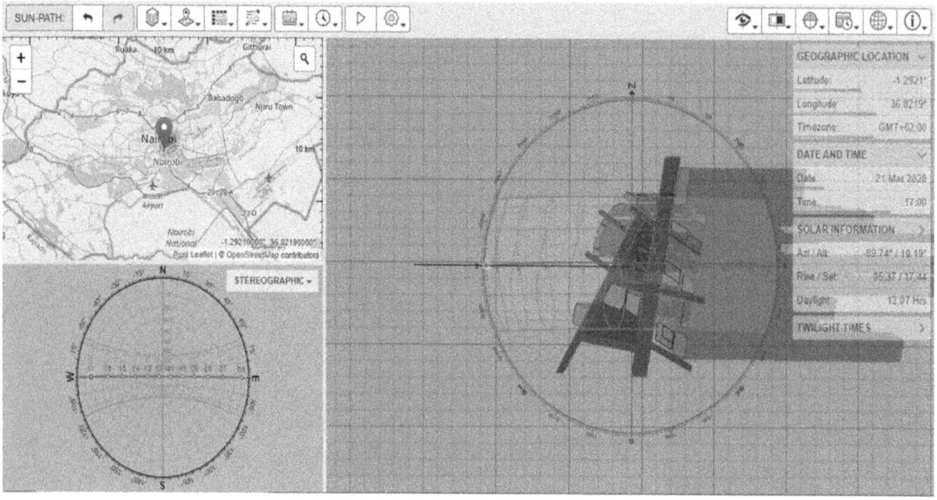

Figure 21.16 Sun path analysis Nairobi 05:00 p.m.

Figure 21.14 that the incident sunlight is perpendicular; hence the shadows are very short, even at 09:00 a.m. in the morning.

The orientation of the Sun at 02:00 p.m. is shown in Figure 21.15. The sunlight is perpendicular throughout the day. Therefore, the shadows are very small. The orientation of the Sun is more towards the west, while the shadow of the buildings lies toward the east.

The orientation of the Sun was almost in the extreme west direction at 05:00 p.m., as shown in Figure 21.16. Therefore, the shadows are very long. The sunset was observed at 05:44 p.m. Sunrise and sunset both were early as compared to the other three locations under consideration. The duration of the day was almost the same, equal to approximately 12 hours.

21.5.1.6 Sydney – Australia

Sydney is the state capital of New South Wales and its geographical coordinates are 33.8688° S, 151.2093° E. It is situated on Australia's east coast and is surrounded by the Blue Mountains (west), Macarthur (south-west), Royal National Park (south), and Hawkesbury (north). It is in the southern hemisphere and closer to the Equator than to the South Pole. It is around 13 750 km from the North Pole, 3749 km from the Equator, and 6253 km from the South Pole [30]. To determine the real time motion and position of the Sun we need to observe the Sun path carefully.

On 21 March 2020 at 9:00 a.m. in the morning the orientation of the Sun was as shown in Figure 21.17. The Sun rose in Sydney at 05:59 a.m. in the east direction. As the Sun is moving from east to west, its position in the morning at 09:00 a.m. was in the north-east (NE) direction. It can be predicted that the illuminance level at the east and north directions will almost be the same. As the shadow of the building lies towards the south and west directions this leads toward lower values of lux in the morning.

The orientation of the Sun at 02:00 p.m. was in the north-west (NW) direction, as shown in Figure 21.18. The shadows were in the south-east direction. The sunlight is less inclined and hence the length of the shadows is moderate.

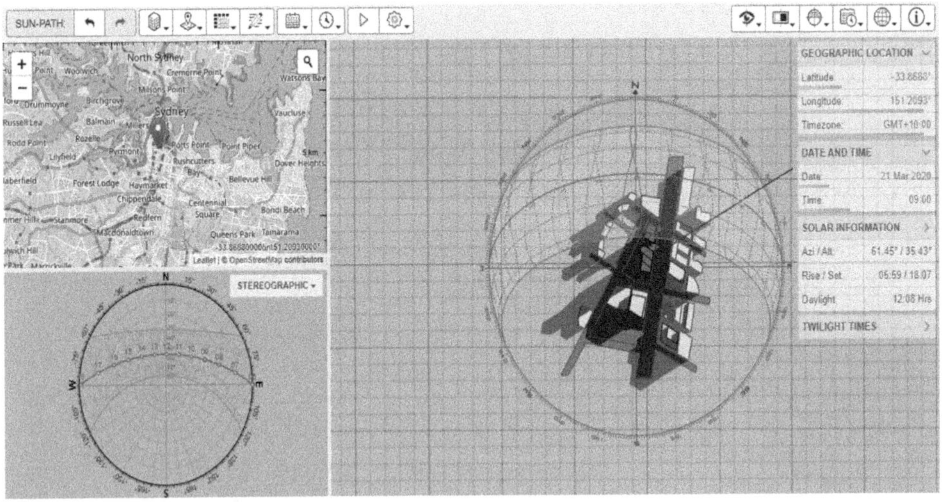

Figure 21.17 Sun path analysis Sydney 9:00 a.m.

The orientation of the Sun at 05:00 p.m. was in the west direction, as shown in Figure 21.19. The shadow was created towards the east direction, thus automatically decreasing the lux level in this direction. The sunset was observed at 06:07 p.m. and the daylight or duration of sunlight was 12:00 hours, which was almost the same in all five cases under consideration.

21.5.1.7 Summary

It has been seen from Figures 21.5 to 21.19 that while moving from the Equator to the poles (either north or south), the time of sunrise as well as sunset becomes delayed. The duration

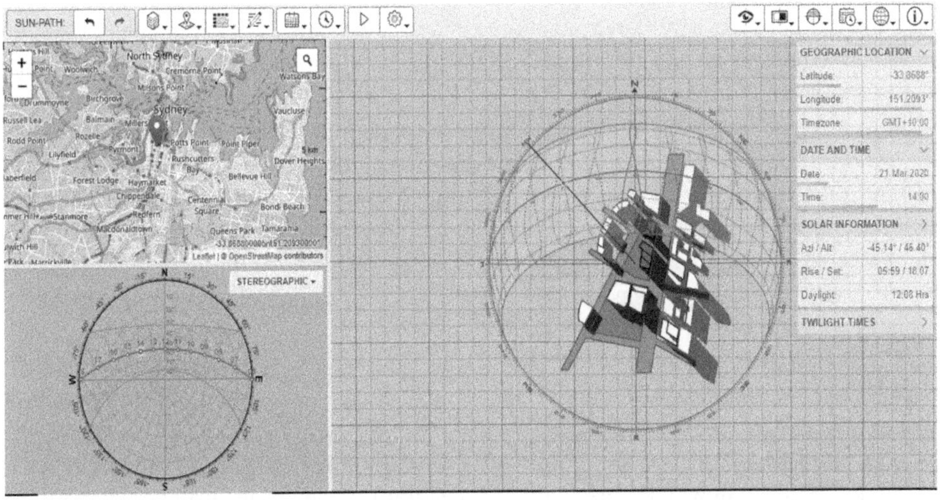

Figure 21.18 Sun path analysis Sydney 2:00 p.m.

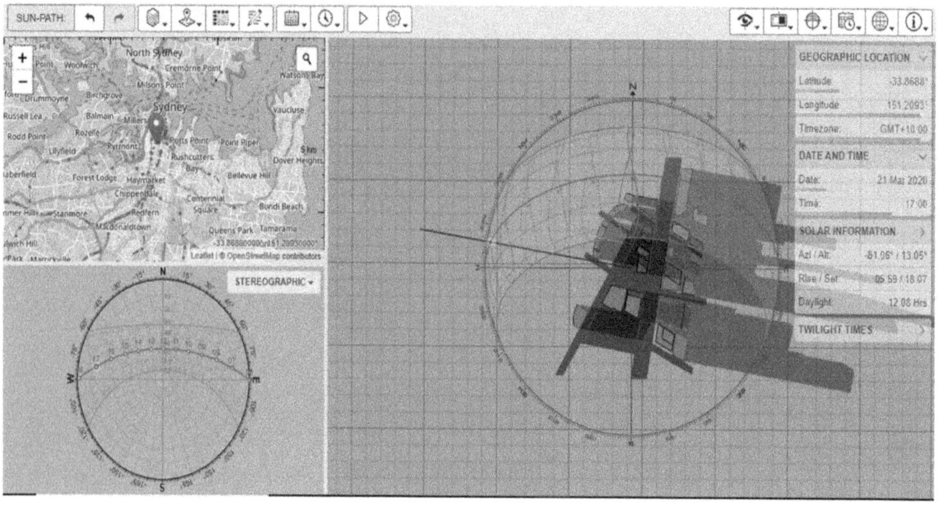

Figure 21.19 Sun path analysis Sydney 5:00 p.m.

of availability of sunlight or daylight is almost constant, i.e. 12 hours. The duration of the day rapidly increased just near the poles. The incident sunlight is perpendicular near the Equator. The inclination of sunlight is increased while moving from the Equator to the poles. The illumination is greatly affected by the sunlight and hence the geographical locations.

21.5.2 Impact of Orientation of a Window

A Sun-path analysis for five different locations has been discussed in the previous subsection. The time of sunrise, angle of incident sunlight, and direction of light affect the illumination of any building and its part. The drawings of any building provide the information regarding dimensions of different amenities and their relative measurements and directions. In the present study, a test room is considered for illumination calculation. The same size of test room is assumed in all of the five locations considered in Subsection 5.1. One window of 1500 mm × 1200 mm is assumed to be fitted in the test room. The orientation of the window is considered in eight different directions one by one, i.e. north, south, west, east, north-east, north-west, south-east, and south-west. The illumination level of the test room is evaluated in all window orientations for all five locations. To make the study more comprehensive, illumination is evaluated at three different timings, in the morning (09:00 a.m.), afternoon (02:00 p.m.), and evening (05:00 p.m.).

21.5.2.1 New Delhi – India

New Delhi is located near the Equator; hence an average illumination of the test room is higher. Figure 21.5 (09:00 a.m.), Figure 21.6 (02:00 p.m.), and Figure 21.7 (05:00 p.m.) are referred to for evaluation of the illumination. The illumination level of the test room is calculated for all eight window orientations. The illumination level in lux at proposed three timings for all orientations are given in Table 21.2.

Table 21.2 Illumination level for window orientation at New Delhi

	Lux value		
Window orientation	**09:00 a.m.**	**02:00 p.m.**	**05:00 p.m.**
North	720	843.5	651
South	1100	1354	928
West	650	1080	1049.2
East	1051.2	1102	662
North-east	773	950	510.3
North-west	457	678.5	873
South-east	1350	1230	820
South-west	928	1600	989

It has been observed from Table 21.2 that at 09:00 a.m. the Sun is in the south-east (SE) direction. Therefore, maximum illumination is obtained in the SE direction. While the shadow of the building lies in the north and west directions, lower values of lux are observed in the morning. The lowest illumination level is found in the NW direction. It is also clear from Table 21.2 that moving from the SE to NW through any path (clockwise/anticlockwise), the illumination level decreases. At the time of sunrise, the incident rays are inclined. As time passes, the sun moves from east to west and the angle of inclination is increasing till midday and then is decreasing till sunset in the evening. Therefore, the illumination level is the minimum at the time of sunrise, which increases till midday and again decreases till sunset. This implies that the illumination in the test room is at a maximum if the window is in the south-east wall, whereas illumination is at a minimum in the north-west direction.

At 02:00 p.m., the maximum and minimum illumination levels increase by 1600 and 678.8 in comparison with morning, i.e. 1350 and 457. Maximum illumination in the south-west direction infers that window orientation in the south-west will give maximum illumination during the afternoon. Similarly, in the evening at 05:00 p.m., the incident rays are inclined and maximum and minimum illumination are in the west and east directions, respectively.

21.5.2.2 Rome – Italy

Geographically, Rome – Italy is almost equidistant from the Equator and the North Pole. Therefore, inclination of incident sunlight is more compared to New Delhi but its intensity is less, which is shown in Table 21.3.

It has been observed by Sun path analysis of New Delhi and Rome that the length of shadow is longer in Rome at all three times under consideration. This proves the fact that the incident light is more inclined. Table 21.3 gives a clear picture of the illumination level in different directions at different timings. Maximum illumination in the test room can be achieved if the window orientation is in the south-east, south, and west directions at

Table 21.3 Illumination level for window orientation at Rome

	Lux value		
Window orientation	**09:00 a.m.**	**02:00 p.m.**	**05:00 p.m.**
North	430	546.7	650
South	780	1047	629
West	395	690	850
East	739	800	558
North-east	550	666	490
North-west	305	575	710
South-east	966	998	620
South-west	545	850	751

09:00 a.m., 02:00 p.m., and 05:00 p.m., respectively. Similarly, minimum illumination in the test room is observed if the window orientation is in the north-west, north, and north-east directions at 09:00 a.m., 02:00 p.m., and 05:00 p.m., respectively.

21.5.2.3 Reykjavik – Iceland

Iceland is located away from the Equator and hence the inclination of incident sunlight differs in comparison to New Delhi and Rome. This can be seen from the Sun path analysis of these locations. The length of shadows increases when moving from the Equator to the poles.

The Sun rises in the east direction and the time of sunrise is delayed when moving towards the poles. Table 21.4 indicates that Reykjavik shows almost similar behavior to Rome when considering the direction of maximum and minimum illumination.

Table 21.4 Illumination level for window orientation at Reykjavik

	Lux value		
Window orientation	**09:00 a.m.**	**02:00 p.m.**	**05:00 p.m.**
North	445	540	606
South	1050	1229	851
West	406	775	1030
East	914.9	702	387
North-east	602.3	689.7	493
North-west	289	595	790
South-east	1197	918	585
South-west	670	1060	982

Table 21.5 Illumination level for window orientation at Nairobi

	Lux value		
Window orientation	**09:00 a.m.**	**02:00 p.m.**	**05:00 p.m.**
North	626	800.1	705
South	819	989.7	699
West	416.5	619	952.8
East	924	632	401.2
North-east	552.8	690	352
North-west	385.5	445	608.5
South-east	1070	899.5	680.2
South-west	420	580.5	490.9

21.5.2.4 Nairobi – Kenya

Nairobi is at the Equator and hence the incident sunlight is perpendicular most of the time. Table 21.5 shows the illumination level of the test room with different window orientations in the morning, afternoon, and evening timings.

It is clear from Table 21.5 that the difference between the maximum and minimum illumination levels is in narrow band for Nairobi. The sunrise is at 05:37 AM in the morning; hence, at morning 09:00 a.m. the illumination level is high. The illumination is at the maximum at around 12:30 p.m. and afterwards it starts to decrease. The window in the south-east wall provides maximum illumination at 09:00 a.m. in the morning. This maximum illumination transfers to the south at 02:00 p.m., due to rotation of the Earth on its own axis. The Sun sets at 05:54 p.m. in the evening; hence, at 05:00 p.m. the illumination level is very low. The Sun is about to set at this time. Maximum illumination in the test room can be achieved if the window orientation is in the south-east, south, and west directions at 09:00 a.m., 02:00 p.m., and 05:00 p.m., respectively. Similarly, minimum illumination in the test room is observed if the window orientation is in the north-west, north-west, and north-east directions at 09:00 a.m., 02:00 p.m., and 05:00 p.m., respectively.

21.5.2.5 Sydney – Australia

In the present analysis, Sydney is the only location considered in the southern hemisphere. Geographically, it is nearer to the Equator than to the South Pole. Table 21.6 shows the illumination level in the test room corresponding to different window orientations at morning, afternoon, and evening.

21.5.2.6 Summary

The Sun is rises in the east and sets in the west, as in the case of the locations of the northern hemisphere. However, the direction of inclination varies in Sydney as its geographical location is in the southern hemisphere. This was observed in the Sun path analysis given in Figures 21.17 to 21.19. It can easily be seen that the Sun is closer to the north-east direction

Table 21.6 Illumination level for window orientation at Sydney

	Lux value		
Window orientation	**09:00 a.m.**	**02:00 p.m.**	**05:00 p.m.**
North	750	1090	791.2
South	406.1	983	599.3
West	402.3	910.5	690
East	890	729.8	580
North-east	1190	883	620
North-west	318	738	651
South-east	410	350	292.4
South-west	250	399	403

at 09:00 a.m., resulting in more lux in this direction. The inclination of incident sunlight is in this direction and hence in the test room the window in the north-east wall gets maximum illumination. Maximum illumination in the test room can be achieved if the window orientation is in the north-east, north-west, and west directions at 09:00 a.m., 02:00 p.m., and 05:00 p.m., respectively. Similarly, minimum illumination in the test room is observed if the window orientation is in the south-west, south-east, and east directions at 09:00 a.m., 02:00 p.m., and 05:00 p.m., respectively.

21.5.3 Impact of Pane

Natural illumination through sunlight varies with time, i.e. less in the morning and evening and more in the afternoon. The requirement for an illumination level is different depending on the utility of the test room. Increasing the number of panes in the window increases the thermal insulation level and reduces the illumination level. This is due to an additional layer of air/vacuum or any gas between the glass layers. Better thermal insulation keeps the area cool and hence less amount of electricity is required for cooling the area. The illumination level can also be controlled by employing different numbers of panes in the window. A single-pane system of windows allows more sunlight to enter the room; hence more illumination is available in the area under consideration. However, a double-pane or multiple-pane system reduces the amount of sunlight entering the area, and hence illumination is decreased.

Table 21.7 shows the illumination level for single-pane and double-pane systems at 02:00 p.m. for all five locations and window orientations under consideration. Only one timing, i.e. 02:00 p.m., is enough to demonstrate the significance of increasing panes as a similar kind of impact is observed for all the timings. It has been noticed from Table 21.7 that by replacing the pane from single to double, the illumination level is reduced. However, by adopting different pane systems, illumination may be controlled to achieve the desired level.

Table 21.7 Illumination level for single- and double-pane windows for all orientations

	Illumination level at 02:00 p.m.									
	New Delhi – India		Rome – Italy		Reykjavik – Iceland		Nairobi – Kenya		Sydney – Australia	
Window orientation	Single pane	Double pane	Single pane	Double pane	Single pane	Double pane	Single pane	Double pane	Single pane	Doublepane
North	843.5	620.5	546.7	315.5	540	398	800.1	682	1090	870
South	1354	1170	1047	739	1229	1103	989.7	880	983	332
West	1080	840	690	471	775	645	619	517.5	910.5	505
East	1102	900.5	800	598	702	597	632	539	729.8	524
North-east	950	725	666	450	689.7	579	690	580	883	684
North-west	678.5	462	575	361	595	468	445	340	738	523
South-east	1230	1050	998	768	918	768	899.5	790	350	225
South-west	1600	1295	850	629	1060	924	580.5	487.5	399	284

Table 21.8 Illumination level for single- and double-glazed window glass for all orientations

	Illumination level at 02:00 p.m. (lux)							
	Single-glazed glass				Double-glazed glass			
Window orientation	Dew Drop	Sparkling Ice	Clear Cosmos	Graphite	Moon Shine	Icy Menthol	Pristine White	Iris
North	1240	818	1003	1090	990	817	840	870
South	618	233	409	483	391	233	249	332
West	814	433	632	690	593	410	265	505
East	842.8	483	669	729	630	460	490	524
North-East	1002	603	804	883	792	607	671	684
North-West	545	475	676	738	633	458	497	523
South-East	497	120	305	350	256	105	148	225
South-West	858	165	331	399	301	159	199	284

Table 21.9 Glass specification sheet (Blues)

Glass details		Light factors			Energy factors		
			Reflection				
Glazing type	Color shade	Transmission	External	Internal	Solar factor (SF)	Shading coefficient (SC)	U-value (W/m^2K)
Single Glazed Unit	Sapphire Blue	20	22	53	0.37	0.43	5.7
	Blue Ray	43	11	18	0.47	0.54	5.6
	Royale Blue	33	10	16	0.4	0.46	5.6
	Titanium Blue	22	21	29	0.31	0.36	5.2
	Tranquil Blue	13	16	27	0.26	0.3	5.2
	Blue Breeze	24	12	18	0.34	0.39	5.5
	Blue Isle	5	21	38	0.15	0.18	3.6
	Orion Blue	18	14	9	0.23	0.27	3.85
	Blue	32	8	10	0.37	0.42	5 (0.88)

(*Continued*)

Table 21.9 (*continued*)

Glass details		Light factors			Energy factors		
			Reflection				
Glazing type	Color shade	Transmission	External	Internal	Solar factor (SF)	Shading coefficient (SC)	U-value (W/m^2K)
Double-glazed unit	Sapphire Blue	19	22	52	0.27	0.32	2.8
	Blue Ray	38	12	22	0.36	0.42	2.8
	Royale Blue	30	11	21	0.30	0.34	2.8
	Titanium Blue	20	21	31	0.23	0.27	2.7
	Tranquil Blue	12	16	29	0.17	0.19	2.6
	Blue Breeze	22	12	22	0.24	0.27	2.7
	Blue Isle	5	21	39	0.09	0.11	1.9
	Satin Blue	49	8	10	0.35	0.40	1.8
	Misty Blue	30	10	10	0.24	0.28	1.9
	Twilight Blue	24	12	11	0.2	0.23	1.8
	Orion Blue	16	15	15	0.17	0.19	2.0
	Lumosa	26	11	11	0.18	0.2	1.6
	Quasar	32	10	25	0.2	0.23	1.5
	SKN	39	9	16	0.24	0.27	1.5
	Blue	29	9	16	0.27	0.32	2.6 (0.46)
	Ultramarine	23	19	19	0.18	0.21	1.6
	Lagoon	27	17	19	0.2	0.23	1.6

21.5.4 Impact of Glass Type

The selection of glass for a window depends on various factors, i.e. U-value, solar heat gain coefficient (SHGC), visible transmittance (VT), light-to-solar gain ratio (LTSGR), air leakage, and condensation resistance, etc. The amount of sunlight entering in the room varies with the type of glass used in the window. Glazing also affects the illumination level of any space. Four different types of glass, i.e. Dew Drop, Sparkling Ice, Clear Cosmos, and Graphite, are used in the single-glazed category of glass and four types of double-glazed glass, i.e. Moon Shine, Icy Menthol, Pristine White, and Iris, are used for illumination calculations. Different specifications of these glasses are given in the appendices from Tables 21.8–21.11 [31]. The amount of sunlight entering is significantly reduced by increasing the pane. Therefore, double-glazed windows are preferred as they provides thermal comfort and energy savings as well.

Table 21.10 Glass specification sheet (metals)

Glass details		Light factors			Energy Factors		
			Reflection				
Glazing type	Color shade	Transmission	External	Internal	Solar factor (SF)	Shading coefficient (SC)	U-value (W/m^2K)
Single-glazed unit	Bronze	18	17	53	0.43	0.49	5.7
	Dark Grey	5	5	53	0.35	0.40	5.7
	Platinum	8	43	38	0.14	0.17	3.6
	Sterling Silver	20	32	27	0.29	0.33	5.2
Double-glazed unit	Bronze	16	18	52	0.33	0.38	2.8
	Dark Grey	4	5	51	0.24	0.27	2.8
	Platinum	7	43	39	0.10	0.11	1.9
	Sterling Silver	17	32	30	0.21	0.24	2.6

Table 21.11 Glass specification sheet (greens)

Glass details		Light factors			Energy factors		
			Reflection				
Glazing type	Color shade	Transmission	External	Internal	Solar factor (SF)	Shading coefficient (SC)	U-value (W/m^2K)
Single-glazed unit	Blue Green	43	14	17	0.43	0.49	5.6
	Turquoise	31	17	18	0.37	0.43	5.5
	Green	27	34	53	0.39	0.44	5.7
	Emerald Glaze	56	15	18	0.50	0.58	5.6
	Deep Green	7	33	38	0.15	0.17	3.6
	Aquamarine	17	24	27	0.27	0.31	5.2
	Green Aura	24	21	9	0.26	0.28	3.85
	Green	41	10	11	0.39	0.45	5 (0.88)

(*Continued*)

Table 21.11 (*continued*)

Glass details		Light factors			Energy factors		
			Reflection				
Glazing type	Color shade	Transmission	External	Internal	Solar factor (SF)	Shading coefficient (SC)	U-value (W/m^{2}K)
Double-glazed unit	Blue Green	39	16	21	0.33	0.38	2.8
	Turquoise	28	18	22	0.26	0.30	2.8
	Green	24	35	52	0.29	0.34	2.8
	Emerald Glaze	50	17	23	0.40	0.46	2.8
	Deep Green	6	33	39	0.10	0.11	1.9
	Aquamarine	15	24	30	0.18	0.21	2.6
	Mint Green	63	10	11	0.40	0.46	1.8
	Olive	39	13	10	0.28	0.32	1.9
	Tropca Green	31	18	12	0.22	0.25	1.8
	Green Aura	21	22	15	0.18	0.21	2.0
	Futura	34	16	12	0.20	0.23	1.6
	Nebula	42	14	25	0.23	0.27	1.5
	SKN	50	13	16	0.27	0.32	1.5
	Green	37	12	17	0.30	0.34	2.6 (0.46)
	Echo	30	29	19	0.20	0.23	1.6
	Celeste	34	25	20	0.22	0.26	1.6

Table 21.12 Glass specification sheet (Neutrals)

Glass details		Light factors			Energy Factors		
			Reflection				
Glazing type	Color shade	Transmission	External	Internal	Solar factor (SF)	Shading coefficient (SC)	U-value (W/m^{2}K)
Single-glazed unit	Light Gold	32	45	54	0.5	0.58	5.7
	Dew Drop	51	18	17	0.54	0.62	5.6
	Sparkling Ice	65	19	19	0.66	0.75	5.6
	Graphite	37	22	18	0.42	0.48	5.5
	Clear Cosmos	28	27	9	0.29	0.33	3.8
	Neutral	50	13	11	0.5	0.58	5 (0.88)

(*Continued*)

Table 21.12 (*continued*)

Glass details		Light factors			Energy Factors		
			Reflection		Solar factor (SF)	Shading coefficient (SC)	U-value (W/m^{2}K)
Glazing type	Color shade	Transmission	External	Internal			
Double-glazed unit	Light Gold	29	46	52	0.43	0.49	2.8
	Dew Drop	45	20	21	0.44	0.51	2.8
	Sparkling Ice	59	22	23	0.56	0.64	2.8
	Graphite	33	23	22	0.33	0.38	2.7
	Pristine White	74	12	12	0.54	0.62	1.8
	Moonshine	46	17	10	0.36	0.41	1.9
	Icy Menthol	36	23	12	0.27	0.31	1.8
	Clear Cosmos	25	28	15	0.23	0.26	2.0
	Iris	39	20	12	0.23	0.26	1.6
	Magma	49	18	26	0.27	0.31	1.5
	SKN	59	16	17	0.32	0.37	1.5
	Neutral	45	15	17	0.42	0.48	2.6 (0.46)
	Shine	36	40	19	0.26	0.29	1.6
	Chroma	42	33	19	0.29	0.34	1.6

Table 21.12 presents the illumination level at 02:00 p.m. with all window orientations taken into consideration. A total of eight glass types (four single and four double glazed) were used for the calculations. Each and every glass has its own significant properties, which causes the variance of the lux, which is clearly visible from Table 21.12. The selection of glass type and glazing depends on the type of utility, window orientation, and occupancy level. For example, a library requires more amounts of lux than a drawing room. It has been observed that the illumination level of Dew Drop provides maximum illumination whereas Sparkling Ice gives minimum illumination in the single-glazed category. Sparkling Ice reduces the illumination to a level of 100–200 lux and Dew Drop increases the illumination level up to 1240 lux. The double-glazed glasses reduce the overall illumination of the test room for all types of glasses. The maximum illumination level varies in the range of 850–1000 lux whereas the minimum illumination level varies in the range of 100–200 lux. Moon Shine and Iris provide better illumination in comparison to Icy Menthol and Pristine White. Out of the above-mentioned types, Iris is the most suitable glass because for both DG and SG glass it shows nearly the same properties and it is cheaper as well, reducing the cost involved in the purchase.

21.6 Conclusion

Energy conservation is very essential for reducing the consumption of fuel and improves utilization of available resources and reliability of the power system. Existing consumption of any building can be estimated by energy modeling. Energy consumption in lighting various parts of the building is significant in the total electricity bill for that building. It can be reduced by fixing the windows in an appropriate direction for use of the space. The Sun path has been analyzed for five different geographical locations, i.e. New Delhi – India, Rome – Italy, Reykjavik – Iceland, Nairobi – Kenya, and Sydney – Australia. The impact of orientation of the window on the amount of illumination has been estimated. Every space needs different illumination corresponding to the use of that area, such as the library, restaurant, theater, shop, etc. The natural sunlight will increase the total illumination level of the room if the window is fixed in an appropriate direction. The illumination level is decreased while moving from the Equator to the poles, as the angle of incident sunlight is decreased. Distinct illumination is recorded at different timings for any location, which can be utilized to assign space utility. The impact of pane type and glazing type has also been discussed. It can be concluded that the use of artificial lights can be reduced by employing proper window orientation, pane type, and glazing of glass to maintain the desired illumination level. The consumer will benefit as electricity bills are reduced. Electric utilities will also benefit as loading of the transmission and distribution line is reduced and at the same time generation requirements go down. The reduction in generation yields less consumption of fuel and less emissions to the environment. Therefore, overall energy demand, loading of feeders, generation, fuel consumption, and environment pollution can be reduced by making appropriate architectural designs that fix proper orientation, pane, and glazing of windows.

References

[1] J. van de Weijer, T. Gevers, and A. Gijsenij, "Edge-based color constancy," *IEEE Transactions on Image Processing*, vol. 16, no. 9, pp. 2207–2214, September 2007.

[2] A. Gijsenij, T. Gevers, and J. van de Weijer, "Improving color constancy by photometric edge weighting," *IEEE Transactions on Pattern Analysis and Machine Intelligence*, vol. 34, no. 5, pp. 918–929, May 2012.

[3] D. Cheng, D.K. Prasad, and M.S. Brown, "Illuminant estimation for color constancy: Why spatial-domain methods work and the role of the color distribution," *Journal of the Optical Society of America. A, Optics, Image Science*, vol. 31, no. 5, pp. 1049–1058, May 2014.

[4] N. Baniá and S. Lonäari, "Green stability assumption: Unsupervised learning for statistics-based illumination estimation," *Journal of Imaging*, vol. 4, no. 11, pp. 127, 2018.

[5] V.C. Cardei, B. Funt, and K. Barnard, "Estimating the scene illumination chromaticity by using a neural network," *Journal of the Optical Society of America. A, Optics, Image Science*, vol. 19, no. 12, pp. 2374–2386, December 2002.

[6] S. Ram Ola, A. Saraswat, S.K. Goyal, V. Sharma, B. Khan, O.P. Mahela, H. Haes Alhelou, and P. Siano, "Alienation coefficient and wigner distribution function based protection scheme for hybrid power system network with renewable energy penetration," *Energies*, vol. 13, no. 5, pp. 1120, March 2020.

[7] A. Chakrabarti, K. Hirakawa, and T. Zickler, "Color constancy with spatiospectral statistics," *IEEE Transactions on Pattern Analysis and Machine Intelligence*, vol. 34, no. 8, pp. 1509–1519, August 2012.

[8] T. Molla, B. Khan, B. Moges, H.H. Alhelou, R. Zamani, and P. Siano, "Integrated optimization of smart home appliances with cost-effective energy management system," *CSEE Journal of Power and Energy Systems*, vol. 5, no. 2, pp. 249–258, June 2019.

[9] T. Molla, B. Khan, and P. Singh, "A comprehensive analysis of smart home energy management system optimization techniques," *Journal of Autonomous Intelligence*, vol. 1, no. 1, pp. 15–21, 2018.

[10] B. Khan, H. Getachew, and H.H. Alhelou, "Components of smart grid system." In: *Solving Urban Infrastructure Problems Using Smart City Technologies*, John Vacca (ed.), pp. 385–395. Elsevier, Amsterdam, Netherlands, September 2020.

[11] B. Khan, E. Gidey, H. Getachew, and H.H. Alhelou, "Managing the generation and demand inside the smart-grid structure." In: *Solving Urban Infrastructure Problems Using Smart City Technologies* John Vacca (ed.), pp. 433–445. Elsevier, Amsterdam, Netherlands, September 2020.

[12] P. Singh and B. Khan, "Smart microgrid energy management using a novel artificial Shark optimization," *Complexity*, vol. 2017, Article ID 2158926, 22 pages, 2017.

[13] N. Banic and S. Loncaric, "Color cat: Remembering colors for illumination estimation," *IEEE Signal Processing Letters*, vol. 22, no. 6, pp. 651–655, June 2015.

[14] S. Ram Ola, A. Saraswat, S.K. Goyal, S.K. Jhajharia, B. Khan, O.P. Mahela, H. Haes Alhelou, and P. Siano, "A protection scheme for a power system with solar energy penetration," *Applied Sciences*, vol. 10, no. 4, pp. 1516, February 2020.

[15] D. Cheng, B. Price, S. Cohen, and M.S. Brown, "Effective learning-based illuminant estimation using simple features," *Proceedings of the IEEE Conference on Computer Vision and Pattern Recognition (CVPR)*, China, pp. 1000–1008, June 2015.

[16] A. Akbarinia and C.A. Parraga, "Colour constancy beyond the classical receptive field," *IEEE Transactions on Pattern Analysis and Machine Intelligence*, vol. 40, no. 9, pp. 2081–2094, September 2018.

[17] B.P. Singh, S.K. Goyal, and S.A. Siddiqui, "Analysis and classification of maximum power point tracking (MPPT) techniques: A review." In: *Intelligent Computing Techniques for Smart Energy Systems*, Lecture Notes in Electrical Engineering (LNEE), vol. 607, Akhtar Kalam, Khaleequr Rehman Niazi, Amit Soni, Shahbaz Ahmed Siddiqui, and Ankit Mundra (eds.), Springer, Singapore.

[18] S.-M. Woo, S.-H. Lee, J.-S. Yoo, and J.-O. Kim, "Improving color constancy in an ambient light environment using the Phong reflection model," *IEEE Transactions on Image Processing*, vol. 27, no. 4, pp. 1862–1877, April 2018.

[19] S.K. Goyal, B.P. Sungh, A. Kumar, P. Kumar, and A. Saraswat, "Modelling and simulation of a solar PV system: A comprehensive study," *2020 International Conference on Computation, Automation and Knowledge Management (ICCAKM)*, Dubai, United Arab Emirates, pp. 367–372, 2020

[20] D.A. Forsyth, "A novel algorithm for color constancy," *International Journal of Computer Vision*, vol. 5, no. 1, pp. 5–35, August 1990.

[21] G.D. Finlayson, S.D. Hordley, and I. Tastl, "Gamut constrained illuminant estimation," *International Journal of Computer Vision*, vol. 67, no. 1, pp. 93–109, April 2006.

[22] J. Long, E. Shelhamer, and T. Darrell, "Fully convolutional networks for semantic segmentation," *Proceedings of the IEEE Computer Society Conference on Computer Vision and Pattern Recognition*, China, pp. 3431–3440, October 2015.

[23] N. Banić and S. Lonärić, "A perceptual measure of illumination estimation error," *Proceedings of the 10th International Conference on Computer Vision Theory Applications*, pp. 136–143, 2015.

[24] W.J. Hee, M.A. Alghoul, B. Bakhtyar, O. Elayeb, M.A. Shameri, M.S. Alrubaih, and K. Sopian, "The role of window glazing on daylighting and energy saving in buildings," *Renewable and Sustainable Energy Reviews*, vol. 42, pp. 323–343, February 2015.

[25] https://www.timeanddate.com/worldclock/distances.html?n=176.

[26] http://andrewmarsh.com/apps/staging/sunpath3d.html.

[27] https://www.timeanddate.com/worldclock/distances.html?n=215.

[28] https://www.timeanddate.com/worldclock/distances.html?n=211.

[29] https://www.timeanddate.com/worldclock/distances.html?n=170.

[30] https://www.timeanddate.com/worldclock/distances.html?n=240.

[31] https://in.saint-gobain-glass.com/en-in/sgg-truvision-neutral.

Part XI

Smart Meter Technology

22

Smart Metering

Transforming from One-Way to Two-Way Communication

Siddharth Jain, Sanjana Babu, Ashwini Ramachandran Nair, and Yashwnat Sawle

School of Electrical Engineering, Vellore Institute of Technology, Vellore, India

22.1 Introduction

Electrical energy is important for recent civilization. The request for electrical energy is growing at a rapid rate, especially in urban areas. Electricity has been one of the greatest gifts that science has provided to mankind [1]. Failures throughout the electric field fall into two categories: technology and management. Based on the practical uses of machinery in a power plant of a distribution station, technical failures of electric power arise. Non-technical failures sometimes occur, leading to a shortage of maintenance professionals, and hence an increase in outage duration. In emerging countries such as India, these losses are much greater. The modern world is heavily dependent on electricity, and disruptions of the power supply, as well as shutdowns, have a massive negative effect on living conditions [2]. According to the US National Energy Technology Laboratory, approximately 100 million dollars per annum of losses occur on account of power outage. There is a need for well-organized supervision of electric power and a supply of reliable electric power to consumers. Smart meters, moreover, generate data at a rate or volume that surpasses the capability of several traditional systems.

As a consequence, the data are gathered but not analyzed. Therefore, electric utilities can be utilized by using smart meter data to calculate the utilization duration for consumer billing. This will open the gate for greater opportunities, which would enable electrical companies to manage the electrical grid and allow consumers to track their utilization better.

22.2 Existing/Traditional System

Traditional electric meters exhibit the information in kWh by constantly computing the instant potential drop (volts) and electric current (amperes) that deliver used power in joules (J). The fundamental kinds of electrical meter are the electromechanical induction meter and the electronic meter. In the electromechanical induction meter, the total number of rotations of the aluminum disc is directly proportional to the power consumed.

Active Electrical Distribution Network: A Smart Approach, First Edition. Edited by Baseem Khan, Josep M. Guerrero, Sanjeevikumar Padmanaban, Hassan Haes Alhelou, Om Prakash Mahela, and Sudeep Tanwar.

Electronic meters illustrate energy used by a consumer, by use of liquid crystal display (LCD) or light emitting diode (LED) [3]. By utilizing various communication technologies, power consumption readings can be delivered to isolated regions. This represents a one-way connection. To evaluate the power consumed, electronic meters register other variables that are utilized for load and feed, such as an estimate of the utilization needs.

22.2.1 Obstacles of a Traditional Meter

- There is a lot of scope for human error that cannot be evaded while taking readings.
- There is no re-verification of the data taken by human readers for power usage.
- There are major chances for dishonesty.
- There is scope to change the readings of the meter when taking pictures of the electronic meter by using various software.
- To take meter readings a company should hire more employees, which is an additional expense to the company.

22.3 Smart Grid

The smart grid (SG) is a smart network that comprises various communication technologies, including an energy flow network. It is an automatic way to gain information [4]. In addition, it addresses energy flow and even deals with the data flow. Various transmission assistance programs and utilization of real-time computation methods enhance resiliency and forecasting of the SG. It also offers protection against cyber threats [5]. The SG uses advanced metering infrastructure (AMI) technology to gather and work data from the SMs. There is a lot of complexity in the communication architecture of an SG. For incorporated communication design and stability, an appropriate foundation is needed [6–15].

22.4 Smart Meter

The smart meter (SM) is a very innovative meter that observes the consumer's electricity usage and provides information to utility companies with extra details in comparison to the standard energy meter.

SMEs are influential instruments that essentially modify the process of energy grids. In addition to implementing roles for traditional meter, SMs acts like sensors throughout the total network.

When AMI technology is used, SM undertakes the measure and registers true power consumption for a particular period [16]. This gathered information is transmitted to the main data control organization over a safe channel through wireless transmission. Numerous utility companies use a variety of sensors to discover various faults. By using this data, it permits the utility companies to deliver reliable power. This helps in enhancing planning, functioning, and a quicker response for the grid.

Utility companies use SM for various applications, such as:

- It has very fast outage detection, response, and restoration.
- It keeps customers briefed about the condition of the electrical network.
- Companies can communicate required data such as the reason for failure, field estimation, and duration for the restoration of power and notices for protection of communities.

22.4.1 Development of a Meter System

Electric meters are utilized in measuring the amount of electric current delivered to the user and calculating the charges for the electrical vendors and system participants. There are various types of electricity meters. Traditionally, meter reading was done manually. That involves physically visiting the meter to take readings and also to carry out maintenance. The most commonly used meter is an accumulation meter. It registers the power consumed over a certain period. This type of meter is commonly called a flat rate electric meter. These electric meters need to be viewed to note down the collected meter readings. Interval meters register the exact usage of electricity consumed in 30 minutes. The usage of electricity differs from day to day, which is dependent on the tariff signed with the electricity retailer. Some advantages associated with this kind of meter are that detailed information on energy used is available and there is an opportunity for tariff plans, which helps to reduce the demand for electricity during peak times. Interval electric meters are commonly called time use electric meters [17]. These are mainly electronic meters. There is an inbuilt display that is scheduled to mention the date, time, and total usage of electricity. The electric meter reader is comprised of an optical probe to the electric meter, which helps in reading the data every 30 minutes and transfers it to a machine. This data is then delivered to the retailer to calculate energy bills. The electric meter registering technology has recently been enhanced with the help of automated technologies. It uses its local grouping of usage of data from the customer side by utilizing telephonic lines, satellite communications, radiofrequency, and other technologies to process the data for bill generation [18]. Water and gas meters are also connected so that we are able to send our consumption status to the electricity meter for billing. Electric meters share data using wireless protocols like ZigBee [19]. Figure 22.1 shows the various types of meter systems and their uses.

22.4.2 Comparison Between a Traditional Meter and an SM

In Table 22.1, we find a comparison of functions between the traditional meter and SMs for your reference.

22.4.3 Smart Meter Architecture

Compared to a conventional energy meter, a smart meter is an energy efficiency meter that facilitates two-way communication. It can, therefore, evaluate a customer's energy consumption information and afterwards transfers information to the utility to enable reasonably good centralized distributed generation and energy storage devices and charge the

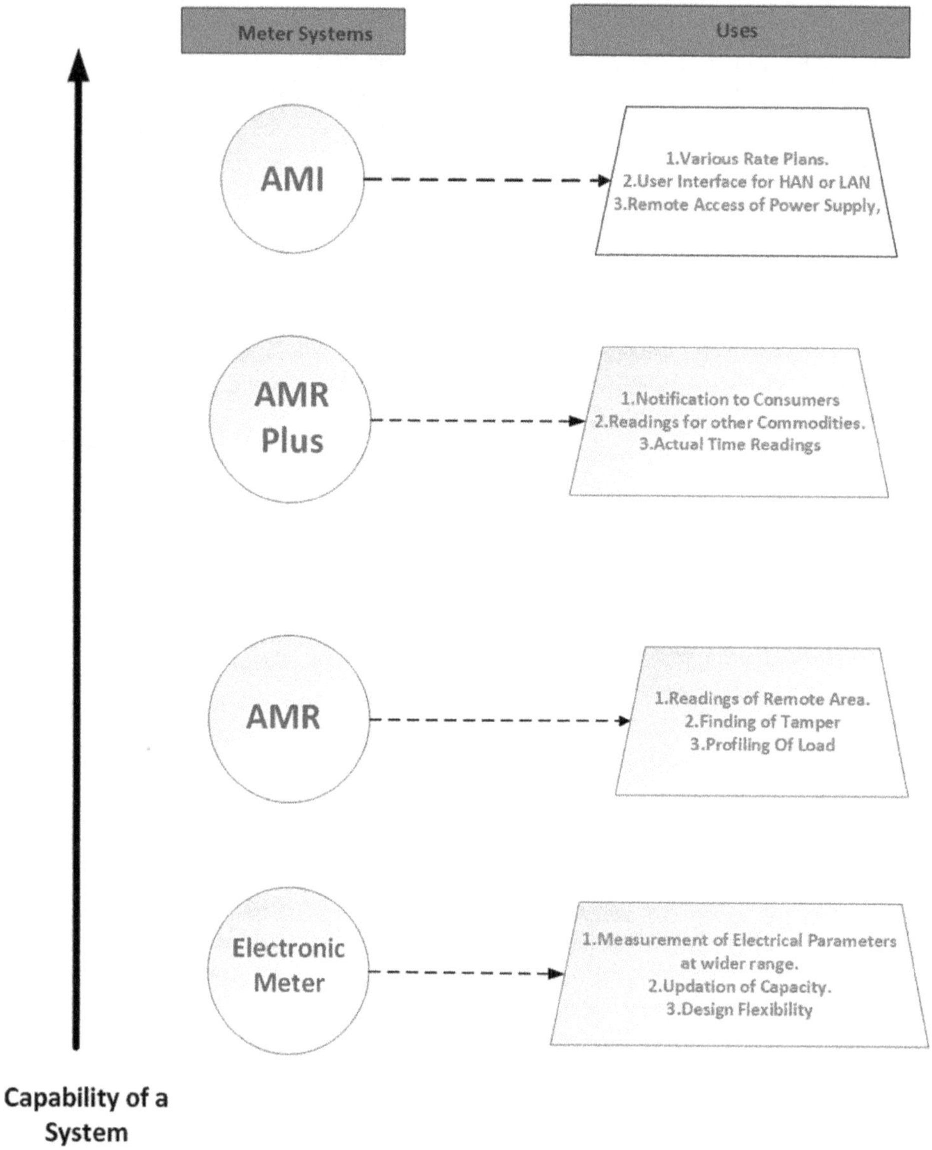

Figure 22.1 Development of a meter system.

client directly. In action, smart meters can monitor user information on energy consumption in real-time, including voltage values, periodicities, etc., and can transmit data to control systems [20]. By using two-way data transmission, an SM will gather data about the values of consumer premises for electricity consumption. Information gathered through smart meters is often a mix of variables, including a unique meter code, data timestamp, and the value of electricity usage. Based on these details, the SM can track and perform

Table 22.1 Comparison between traditional and SM functions.

Function	Traditional meter	SM
Data storage	No data storage	Data is stored for processing bills
Meter reading	A person has to physically come to read the meter	No manual reading
Outage detection	No outage detection as distribution companies cannot react to interruption in the power supply quickly	Automated outage detection so that distribution companies restore power quickly
Connections and disconnections	Must be done manually	Faster as they are managed remotely
Errors	Since meter reading is manual, an error can occur	Error is less compared to a traditional meter
Effects due to signal weakening	Due to manual reading, no issues arise due to signal weakening	A poor signal causes SMs to work ineffectively
Health effects	No ill effects on the health of human beings and the environment	Can cause adverse ill effects on human beings

orders across entire equipment and systems, including remotely as well as locally at the customer's premises. An SM system comprises numerous pieces of equipment, such as scanners that can recognize variables and conditions for SG. It then transfers the information to the command station or sends control instructions to the equipment. Collecting electricity usage information from all consumer equipment on a standard basis helps utilities to monitor electrical needs more effectively [21]. It provides the consumers with valuable data on cost-effective techniques of using their equipment. In Figure 22.2, we can see the architecture of a traditional meter vs an SM system.

22.4.4 SM Functionalities

Smart meters primarily incorporate two essential functions, which are communication and measurement. Therefore each meter is configured accordingly as communication and metrology with two components. The transmission section includes authentication as well as protection that specify an appropriate approach to information transmission [22]. The measurements vary according to various characters, such as neighborhood, precision, and functions, as well as data security. The SM should have six basic functionalities, as stated below:

- **Numerical evaluation**: SMs must be very accurate in measuring the quantity of the medium.
- **Command and standardization**: SMs should be provided only if they can balance minor differences according to respective classifications.
- **Protection of transmission**: Electric meters have the capability of getting effective orders and advancements for its firmware.

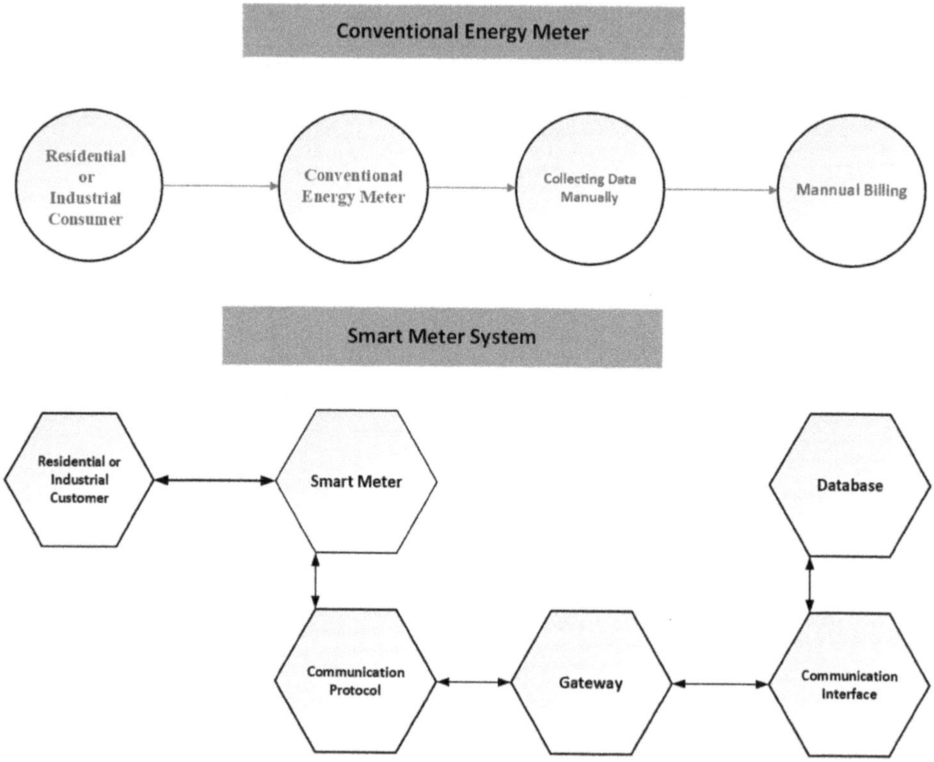

Figure 22.2 Architecture of a traditional meter and SM.

- **Supervision of power**: SMs should support the network to precisely preserve its functions when the main source of power is absent.
- **Presentation**: SMs will refer and exhibit data on the consumption of electricity to users for generating an invoice. It also presents data on real-time usage, which it displays on the SM.

22.4.5 SM Operating Principles

The SM provides a varied range for control devices, hardware as well as software. The core elements of smart meters are metrology, security, and communication. An SM system consists of a real-time clock, data transmission unit, system-on-chip (SOC), metering unit, and protection unit [23]. An SM hardware module is based on the SOC processor, which includes the infrastructure to help the measurements. A meter's analog front-end comprises analog-to-digital converters (ADCs), which support diverse inputs. A built-in gain offers minimum out-turn sensors. The principal parameters measured during power measurements are a root mean square (RMS) electric current and potential drop, active and reactive power, a power factor, and frequency.

22.4.6 SM Components

The power grid, microcontroller, and communication interface are the primary three major components of SMs. To guarantee the working of metering electronics, even after the main is shut off, it is fitted with a switched-mode power supply and a battery backup, and, to obtain the information, a microcontroller unit (MCU) usually contains an analog to digital converter (ADC) and a digital to analog converter (DAC). A wired or wireless device helps in communication with the rest of the system. Power grid and microcontroller components are introduced here. The communication interface is discussed in detail in the next subsections.

- **Power system**: A switched-mode power supply mainly powers the electronics, which transforms alternating current (AC) to direct current (DC). When there is no supply from the utility grid during regular operation, the battery is disconnected from the primary grid.
- **Microcontroller**:
  Though the design for a microcontroller is very important, integration with components of the system is also possible. Analog front end (AFE), ADC, digital signal processor (DSP), or MCU are three feasible architectures of SG. The two-chip approach provides leverage in updating the system, as a single-chip approach makes it less able to modify.

22.4.7 Benefits of SMs

The functioning of SM provides a beneficial impact to users, to companies, and also our community.

Advantages to the user:

- Consumers are provided with more information about their energy consumption.
- Increase in awareness of the standard of power distribution and comprehensive response on the usage of power.
- Invoices are generated based on genuine utilization.
- The need for invoice calculation is reduced.

Advantages to companies:

- Local supervisory bodies help in improved administration of invoices and extra issues linked to consumers.
- The electrical network is supervised more rapidly.
- Empowering more well-organized utilization of power resources.
- Dynamic pricing is enabled.
- Enhancing revenue with current assets.
- Vital share in the expansion of SGs.
- Operational costs will be minimized.

Advantages to nature:

- SMs interact directly with the companies and eliminate the necessity to put utility automobiles on streets.
- SMs remove the necessity to construct new power plants as an outcome, which minimizes pollution.

22.4.8 Barriers for SM

Several barriers exist while deploying SM.

Difficulty to the utilities:

- The transformation from traditional technology to innovative technology.
- Management of consumers for approval of innovative technology.
- Complex, innovative, and time-consuming projects.
- An additional charge for electric meters and organization.
- Management of protection of electric meter information.
- Enhancing distribution of SMs.

Difficulties to users:

- Authentication of exactness of the information gathered and administered by the innovative electric meter.
- Security of their private information.
- Supplementary fees for a new innovative meter.

Difficulties of nature:

- Discarding old electric meters.

22.4.9 Summary of Smart Meter Section

In this section, we have discussed how existing meters function and their drawbacks. We have also made a comparison between the traditional meter and the SM.

22.5 Requirements for SM Applications

The following main application can be considered for SMs:

- Readings from an isolated meter
- The option of flexible time to time use of tariffs
- Management of a remote meter
- Reduction in remote demand
- Disconnection/connection in remote areas
- Reliability of supply
- Signal value to user
- Finding a reason for the power outage

22.5.1 Vigorous Costing

Vigorous costing is now a means for production for a peak flat. Users are recommended to minimize their usage of power intake during peak periods. In short, it decreases the average market price and users also minimize their electric current. There are three types of costing dynamics:

- Time-of-use pricing
- Critical peak pricing
- Real-time pricing

22.5.2 Duplex Transmission

Power information is collected as well as processed at regular time intervals by smart meters (generally daily for housing, every 15 minutes for manufacturing organizations). On request, this information is forwarded to the utility's meter data management system (MDMS).

22.5.3 Isolated Service

The SM is digitally innovative equipment that permits isolated control by the utility company for energy associated applications, such as:

- Isolated disconnection/connection of energy to the consumer
- Stealing of power recognition and rectification
- Detection of power outage
- Sending warnings to customers for power used over a specified demand

22.5.4 Error Finding

The SM can support companies by identifying wherever an error has happened. It notifies the utility when the current will be re-established. To have an error finding skill, the advanced metering infrastructure (AMI), the SMs should have the following structures:

- The capability to handle previous gaps
- Verification capability of the power
- Notification sending options for the restoration of power

22.5.5 Peak Demand Decreases

Smart meters notify users regarding timing as well as payment terms for power consumption. Consumers would also be expected to change their power usage patterns to off prime times for reduced pricing. That could lower the peak load. By handling peak loads power loss through high stress can be prevented from occurring on the grid. The services should manage electricity flows during peak times and utilize synchronized and networked smart meters to function effectively through grid incidents.

22.5.6 Analyzing the Efficiency of Delivery

SM is also used to calculate the effectiveness of the supply control system (recloses, switches, and faulted circuit indicators (FCIs)) throughout emergencies, such as storms and blackouts. The SM also offers technology to mechanize the verification of many devices with proper working.

22.5.7 Software Program Upgradation

The electric meter is capable of updating meter firmware from a district or remote location, without disrupting previously approved meterological functions. Redeveloping the software meter program will also allow new features to be introduced with fixed bugs in the earlier version of the system [24]. This will prevent smart meters from becoming obsolescent early.

22.5.8 Reclosure

Smart meters have always been resistant to reclosure of such a delivery system. A reclosure arises when an energy failure is produced by the journey of shielding equipment between the meters as well as the service network. At every single power system activity, four various reclosures can happen in sequence over a period of 10 to 30 seconds.

22.6 Introduction to the AMI System

AMI is the enabling technology for the SG on the distribution side. It is a collective term used to define the entire architecture, starting from the SM to the bi-directional transmission system. It acts as a controller for the leading equipment as well as for all the functions that permit the transfer of power consumption data in real-time [25]. It is considered as a support to the SG and makes two-way communication possible. The purposes of AMI are as follows:

- Local meter observation for fault-free data
- Recognition of network issues
- Auditing of power
- Limitation of a partial load instead of load shedding

AMI consists of several hardware as well as software components that perform various functions, such as calculating power usage, transfer of data about power, usage of water, and gas to retailers and users. The practical mechanisms of AMI are:

- **SMs**: As already discussed, these are the advanced meters that can collect information about energy at several intermissions and transmit it to retailers over a secure transmission network. It also receives information like price from retailers and conveys it to customers.

- **Communication network**: This communication network helps in two-way communication between consumers and utility companies. Network systems like Broadband over Power Line (BPL), Power Line Communications, Fiber Optic Communications, Fixed Radio Frequency, or other public network systems, such as landline, cellular lines, etc., are utilized for communication.
- **Meter data acquisition systems**: These are software programs used on Control Centre hardware and the Data Concentrator Units that are used to acquire data from meters via a transmission network and deliver it to the Meter Data Management System.
- **Meter data management system**: This is the host system that accepts, stocks, and examines the meter data.

22.6.1 Meter Data Management System (MDMS)

The MDMS represents a vital component for AMI. It includes gathering data from concentrators that join the operation center, and then properly storing and processing it. In fact, the identical kind of information is stored at a similar location and therefore the extraction of the required data will become simpler. The MDMS includes analytical capabilities that help various operating and administration network parts, which help to communicate with it and collect whatever data is required [26]. The key components of the operating and management system are the Failure Management System (FMS), Customer Information System (CIS), Geographic Information System (GIS), and Distribution Management System (DMS). Figure 22.3 shows the basic blocks of the MDMS.

22.6.2 Data Concentrator

This plays a vital part, which is linked to numerous utility servers and SMs within the AMI. It allows data exchange between both the smart meters as well as the MDMS. Data concentrators collect information securely from a variety of meters and transmit it to the MDMS. SM information is collected by a neighborhood area network (NAN) that is then communicated to the MDMS via a wide-area network (WAN).

22.6.3 Advantages of AMI

The advantages of AMI are as follows:

- **Operational benefits**: AMI improves accuracy of electric meter readings, power theft detection, and comeback from power blackouts. It also eradicates the necessity for an electric meter reading.
- **Benefits of funding**: AMI provides funding assistance to organizations, such as water and gas authorities, by minimizing the cost of equipment and care, enabling quicker electric service reinstatement when there is a blackout, and restructuring the invoice process.

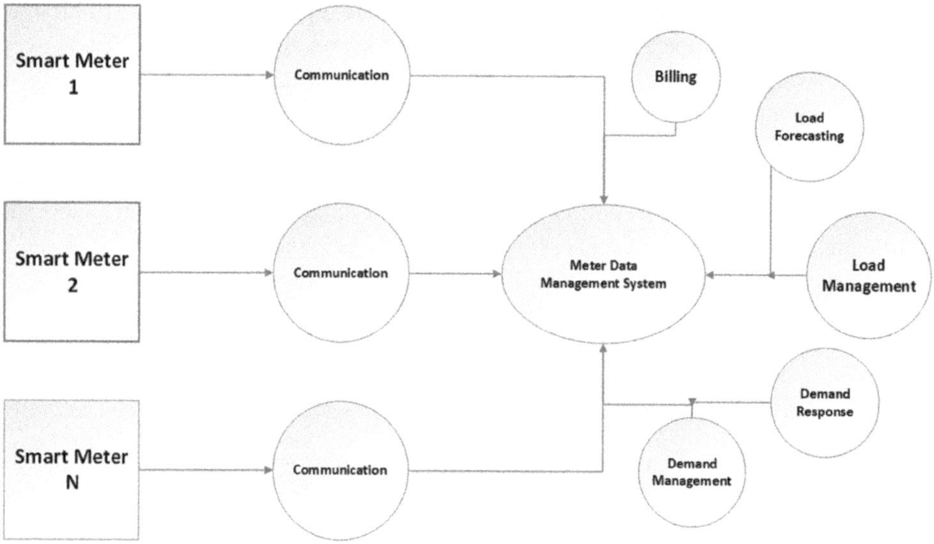

Figure 22.3 Blocks of MDMS.

- **Advantages to users**: AMI assists users through early detection of electric meter failures, assisting with quicker service recovery, and enhancing billing accuracy and flexibility. Also, AMI provides a time-based rate option, which benefits users when they reduce their amount as well as supervise their power usage.
- **Benefits of security**: AMI innovation allows improved system resource observation, which lowers potential grid attacks from cyber-terror attacks.

22.6.4 Protection in AMI

AMI protection is required to secure the transmission networks as well as the energy grid. The two systems need to guarantee their accessibility and durability in diverse situations. The protection of the transmission networks and energy grid varies in numerous ways. In the transmission system, the response time needs to be reduced as well, as the band needs to be ensured, whereas the protection of the energy network requires guaranteeing authenticity as well as durability [27]. AMI is a vital element in a SG. AMI protection is given unique consideration in the SG.

22.6.5 Protection of Confidential Information at the Consumer End

In AMI, SMs are accomplished by gathering data of every user every 15 minutes. If invaders access the information, they can hack "user profiles". The access of profile permits intruders to copy the manner of energy usage of users without the absence of software or algorithms, which is very dangerous for the security of the smart grid. Therefore, protection of the SM data is a very critical issue.

22.6.6 Protection Against Cyber Threats or Manual Attacks

The AMI-Sec Task community is developed by the security department authority. The commercial people and standards bodies form what is necessity for AMI protection. It ensures advice and stability protocols to departments for enforcement of AMI solutions. It should meet the following requirement stated as follows:

- **Privacy**: In AMI, measurement, as well as usage data, has to be secret to safeguard the user information. The information is only accessible to permitted people.
- **Reliability**: The AMI system should guarantee the reliability of communicated information; the functioning of AMI is highly subjected to the reliability of transmitted data. The reliability of data must be secured through encrypting algorithms to protect it from invaders. In AMI, SMs should identify cyber threats and neglect all statements from invaders to protect the SG.
- **Availability**: Any system resources, such as information, periodicity, and devices, will always be accessible to an official body. In addition, the primary motive for the absence of information is element error, like transmission error (due to broken lines, path change, network issues, etc.), software issues, etc.

22.6.7 Issues and Solutions in AMI

As AMI is a complicated system, we know that there would be many issues. In this part, the issues, as well as the solutions in AMI, are identified in two modules, which are protection and transmission between networks.

22.6.7.1 Trouble in Finding Failures

In AMI protection, the main issue is the high level of reliance among grid modules, such that self-sufficient casual incidents can provide scope for failures in the network [28]. A significant complication in AMI increases the chances of the possibility of shortcomings and unexpected entry issues raise the chances of threats. Innovations, such as electric automobiles, etc., are waiting to be included in the network. Studies on security brought by the teams have received some recognition. Hence, it is tough to find and tackle upcoming issues in such a system.

22.6.7.2 Dependence Among Grids and Transmission Modules of AMI

We realize the risk to transmission modules of AMI a grid, and we need to ensure that to a bit level how the dangers related to SG transmission architectural influence the grid [29]. Moreover, it is uncertain how dangers in energy grids disturb the transmission modules of AMI.

22.6.7.3 Issues to Find Network Constructed Attacks

The most significant issues derive from the global link in the device software in AMI. Network-constructed attacks may circulate swiftly to overcome the entire system of AMI. In sum, the general link and numerous admission ideas make AMI more susceptible to outbreaks (such as denial-of-service (DoS)). It is essential to depend on automatic identification plans to reply to network-constructed attacks.

22.6.7.4 Interruption Recognition, Protection, and Regaining for AMI

Usually, DoS is mainly the cause of dangerous threats to AMI. If a similar threat cannot be noticed and isolated quickly enough, it will let down the functions in critical ways and endanger AMI. It requires innovative solutions intended for threat valuation grounded on preceding information. In case an outbreak cannot be recognized and prohibited, suitable interruption retrieval methods must be executed to cure the significance of the outbreak.

22.6.7.5 Vital Organization Practices for AMI

Nowadays, key administration projects were mainly intended only for safe transmission within the SG to deal with the difficulties of essential creation for the transmitting bodies within the supervisory control and data acquisition systems (SCADA) network to secure important content and response information about the power usage of users. Very little knowledge is available at present on practices for AMI. Hence, studies should be concentrated on suggestions for different fundamental administration methods specifically planned for AMI [30].

22.6.8 Safety Analysis

It is important to create a danger system that can independently detect errors and to confine the damage caused to the transmission of AMI. The result should trigger the impact of diverse dangers on the electric grid. It is a need to create an extensive background that includes various effects of different attacks together. The risk is related to the interplay among cyber security and physical systems. It is practically not possible to view all the combinations of danger. According to automation tools, various breakdowns (attacks) in both cyberspace and physical systems will be an essential sum of all of the threats. Further, their behavior is also learned. An emergency assessment has already been performed for the assessment of the balance of AMI. However, that will need to be extended to integrate the hazard due to risk arriving from different transmission networks in AMI. A more accurate discovery of techniques used in various factors for accurately forecasting danger will need to be the development to reduce misreporting probability [31]. Based on the earlier risk results, the computation can independently find the errors in AMI to reduce the risks brought by corrupted safety performance.

22.6.8.1 Safety Standards

On the other hand, worldwide security principles are also needed for transmission in AMI. Currently, various independent attempts have been made to enlarge protection standards. Safety standards are being implemented to meet the need to be future-proof. Standard experiment schemes need to be established for the studies by developing computation techniques. Furthermore, we must launch uniform testing demands for the protection of all functions in AMI. It is also vital to generate auditing needs to guarantee fulfilment of safety legislation for companies, device producers, and generators for the neighborhood, nationwide, and local legislation authorities.

22.6.8.2 Quantum Key Distribution (QKD) in AMI

The use of QKD assists in enhancing the protection of the transmission in AMI. Quantum transmission is a developing skill with possible applications to energy grids. QKD has been projected as a method to enhance the protection of transmission among the grids. It could be applied to current fiber-optic networks and free-space optical links, within generation networks and energy circulation systems. Scholars are discovering its uses in more demanding and as well as stimulating situations. One possible usage of AMI is in quantum position confirmation. As present-day power system organization modules tend to be motionless, quantum transmission methods could be used to enhance safety with respect to identifying the position of an SM. This enhances an additional level of protection by safeguarding the condition that an SM is located at a static position in the grid. There are possibly countless additional claims of quantum transmission approaches that will be beneficial to guarantee the safety of AMI [32].

22.6.9 Technologies Required for SM

SM uses a variety of technologies and layouts but functions through a basic method. SMs collect the data from consumers and transmit the data to the utilities through the communication network. This transmission process is carried out as frequently as required. The process involved in data collection transmission and processing in the SM is as follows.

22.6.9.1 Signal Acquirement

SM is utilized for collecting network variables precisely as well as uninterruptedly for the calculations and transmissions purposes. Essential variables are size and rate of occurrences of voltage and size. Power and potential drop converters evaluate the electric current that registers the load and the potential drop at that point. Some of the electric meters are linked straight to the power cable by usage of current-detection shunt resistance. The detection of the current shunt is an extraordinary firmness resistor where the voltage drop is directly proportional to the current flowing through it. This shunt resistor's current rating is restricted by its self-preservation such that it is generally only used for domestic purposes [33]. To match the voltage across this resistor with an ADC, a programmable gain amplifier (PGA) is used at this stage before the ADC.

22.6.9.2 Signal Conditioning

This stage of signal conditioning involves preparing the input signals for the next stage of the process, the ADC. The signal conditioning stage involves addition, subtraction, attenuation, amplification, filtering, etc. In numerous situations, the input communication will need an offset to be attenuated, amplified, or added/subtracted in such a way that its extreme size rests under the ADC stage input. To avoid inaccuracy occurring due to aliasing, the removal of elements of the input signal more than the NY Quist regularity is essential. Therefore, before passing the input to the ADC phase, a short-pass strainer is required.

22.6.9.3 Analog-to-Digital Conversion

The current and voltage signals obtained as output from sensors are sampled and digitized by metering software. Since two signals (current and voltage) are in a single-phase meter, a multiplexer is required to send the signals to the ADC in turn if a single ADC is used. Since the quantity of stages presented for the ADC is low, ADC transformation constantly performs in a diverse way.

There are many methods involved in the transformation from an incoming signal to a digitalized output. Popular techniques involve the planning of comparators and registers with a synchronizing clock impulse [34]. Successive approximation and the sigma-delta method is the most common ADC used for metering.

22.6.9.4 Estimation

These are divided into mathematical processes on incoming frequency and preparation of data transmission. A DSP processor is utilized for a relatively large number of arithmetic operators. For computation, volatile memory and non-volatile memory are used. Volatile memory is used as a temporary storage of data to support the processor. The quantity of unstable storage utilized relies on the capacity, value, and complications in the calculation and the amount of transmission to and from ports. Permanent storage is utilized for the collection of data, such as the unit sequential digit. A present watch is also utilized as an interval citation that is attached to every acquired data.

22.6.9.5 Input/Output

The SM has a screen that displays data as text or graphs for users. LCD or LED is preferred for this. SMs offer a mini keypad or else monitor touch for interaction. They also offers isolated adjustment and control ability with the help of transmission links. Power usage and invoices are presented on different user platforms. This helps customers to reduce their energy usage.

22.6.10 Summary of the AMI Section

In this chapter, we have discussed the various types of functionalities that an SM requires. We have also discussed what an AMI system is and what are the advantages and security challenges in AMI.

22.7 Transmission Architecture

Advancement of innovative energy network organizations, alternatively recognized as SG, is technologically advancing with the aim to enable bi-directional transmission of electric meter data. An SM is incorporated into the SG architecture system, the core objective of which is to computerize the observation of consumer electricity usage for invoices. SM architecture, or AMI, is a vital element of the SG infrastructure network. The electric meter network observes the power intake of consumers by gathering data, which includes usage and sharing this data with companies for load observation and generation of invoices [35]. SM architecture recommends users with enhanced observation of energy usage and a transparent invoice scheme. Utilities can apply various invoice schemes for electricity usage created using the time of day and seasonal energy usage.

22.7.1 Various Transmission Architectures

AMI is composed of two systems, i.e. SM and MDMS. The transmission architecture is divided into three sections [36]. These are defined as follows:

- Home area network (HAN)
- Neighborhood area network (NAN)
- Wide area network (WAN)

A model of the transmission architecture for the SM is shown in Figure 22.4.

22.7.1.1 Han

HAN acts as a means of transmission for the gadgets interface. It is a single unidirectional or bi-directional transmission medium, or grouping of several technologies such as power line carrier (PLC), Ethernet, or wireless communication tools.

22.7.1.2 Nan

NAN acts as the intermediate level that connects numerous HANs mutually in the SG. Its motive is to fetch knowledge on power usage from residence HANs in a district. It dispatches the statistics to companies for the generation of invoices and for observation. According to several studies, there are numerous ways to transmit the usage to organizations for illustration with the help of a power line carrier (PLC), the global positioning system (GPS), etc.

22.7.1.3 Wan

This system is a public network that can be used to transmit AMI data to utility organizations at a local distance. It should be noted that these transmission networks are application reliant.

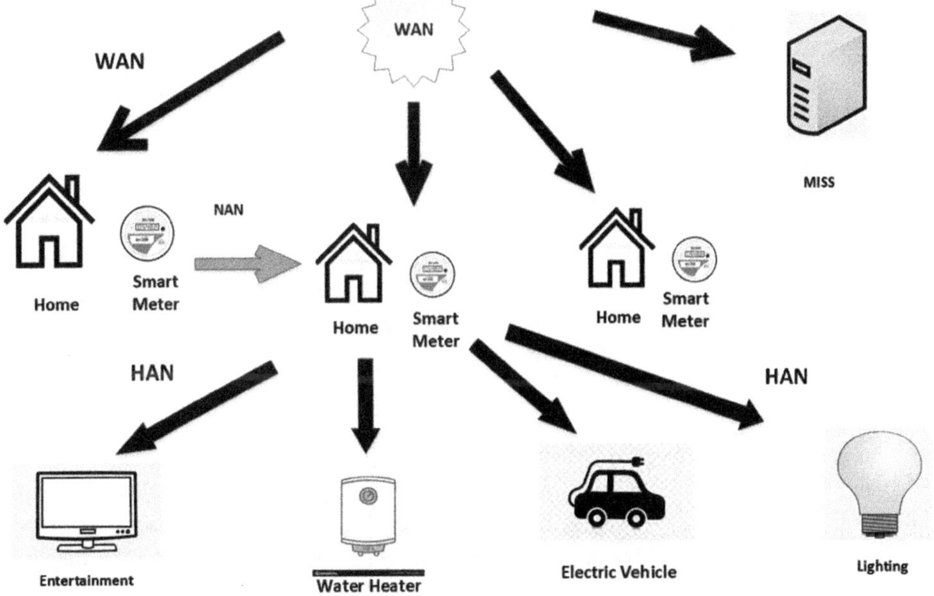

Figure 22.4 Transmission architecture.

22.7.2 Transmission Technologies for AMI

Used together, wired and wireless transmission technologies can be employed to support AMI distribution [37, 38]. Standard wired technologies are PLC as well as fiber optic. General wireless transmission technologies comprise cellular, wireless local area network (WLAN), and ZigBee.

22.7.2.1 Wired Transmission Technologies

A. PLC

PLC uses traditional power cables to understand the communication system. PLC is a vital module in smart grid applications because of the accessibility of the prevailing architecture. PLC is a useful technology for numerous applications, such as SM, home-based automation, etc. Its design is specifically suitable for local zones that are accessible to energy, but other transmission architectures are not accessible. PLC faces numerous technical issues. The second drawback of PLC is security issues.

B. Fiber optic communication

From the benefits of a maximum data rate and resistance to noise, the fiber optic option has developed as a common transmission equipment to offer backbone to transmission and to support numerous SG functions. The high installation price is its key drawback.

22.7.2.2 Wireless Transmission Technologies

A. Cellular network

The current cellular network is a decent selection for setting up AMI. It helps in data transmission among concentrators and control stations. As this cellular architecture prevails, there is no additional period, and the price for a utility to set up as a system for SG protection of a cellular network that is very substantial. The major disadvantage comprises sharing cellular networks with other users, which results in system crowding in many backup circumstances. Cellular networks may not always provide a guaranteed amenity during rare circumstances.

B. WLAN

WLAN, also recognized as Wi-Fi, is based on the IEEE 802.11 series of standards and functions on 2.4 GHz, 3.6 GHz, and 5 GHz. WLAN can offer consistent, safe, and maximum-speed data. Its application cost and energy usage are moderately more advanced than additional short-range technologies, like ZigBee. Mesh topology is used in WLAN.

C. ZigBee

This is a private area network protocol grounded on the IEEE 802.15.4 standard. ZigBee can cover a distance of up to 100 meters, while that of ZigBee pro can cover a distance of up to 1600 meters. Its data tariff ranges from 20 kbps to 250 kbps. ZigBee is a low-cost, low-energy usage technology and is safe. It has a severe interference challenge and low dispensation capabilities.

22.7.3 Summary of Transmission Architecture

This chapter deals with transmission architecture. Here we have discussed HAN, WAN, and MAN. It includes details about wired transmission technologies and wireless transmission technologies in AMI.

References

[1] R.R. Mohassel, A. Fung, F. Mohammadi, and K. Raahemifar, "A survey on advanced metering infrastructure," *International Journal of Electrical Power & Energy Systems*, vol. 63, pp. 473–484, 2014.

[2] S. McLaughlin, D. Podkuiko, and P. McDaniel, "Energy theft in the advanced metering infrastructure," *International Workshop on Critical Information Infrastructures Security*, Springer, Berlin, Heidelberg, 2009.

[3] S.C. Yip, C. Tan, W.N. Tan, M.T. Gan, K. Wong, and R.C.W. Phan, "Detection of energy theft and metering defects in advanced metering infrastructure using analytics," *2018 International Conference on Smart Grid and Clean Energy Technologies (ICSGCE)*, Kajang, pp. 15–22, 2018.

[4] J. Zhou, R.Q. Hu, and Y. Qian, "Scalable distributed communication architectures to support advanced metering infrastructure in smart grid," *IEEE Transactions on Parallel and Distributed Systems*, vol. 23, no. 9, pp. 1632–1642, 2012.

[5] S. McLaughlin, D. Podkuiko, S. Miadzvezhanka, A. Delozier, and P. McDaniel, "Multi-vendor penetration testing in the advanced metering infrastructure," *Proceedings of the 26th Annual Computer Security Applications Conference (ACSAC'10)*, Association for Computing Machinery, New York, USA, pp. 107–116, 2010.

[6] T.F. Agajie, B. Khan, H.H. Alhelou, and O.P. Mahela, "Optimal expansion planning of distribution system using grid-based multi-objective harmony search algorithm," *Computers & Electrical Engineering*, vol. 87, p. 106823, 2020.

[7] Y. Kifle, B. Khan, and J. Singh, "Designing and modeling grid connected photovoltaic system (Case study: EEU Building at Hawassa city)," *International Journal of Convergence Computing*, vol. 3, no. 1, pp. 20–34, 2018.

[8] O.P. Mahela, B. Khan, H.H. Alhelou, and P. Siano, "Power quality assessment and event detection in distribution network with wind energy penetration using stockwell transform and Fuzzy Clustering," *IEEE Transactions on Industrial Informatics*, vol. 16, no. 11, pp. 6922–6932, November 2020. doi: 10.1109/TII.2020.2971709.

[9] O.P. Mahela, B. Khan, H. Haes Alhelou, and S. Tanwar, "Assessment of power quality in the utility grid integrated with wind energy generation," *IET Power Electronics*, vol. 13, no. 13, pp. 2917–2925, 14 October 2020. doi: 10.1049/iet-pel.2019.1351.

[10] S. Ram Ola, A. Saraswat, S.K. Goyal, V. Sharma, B. Khan, O.P. Mahela, H. Haes Alhelou, and P. Siano, "Alienation coefficient and Wigner distribution function based protection scheme for hybrid power system network with renewable energy penetration," *Energies*, vol. 13, p. 1120, 2020.

[11] K. Negash, B. Khan, and E. Yohannes, "Artificial intelligence versus conventional mathematical techniques: A review for optimal placement of phasor measurement units," *Journal of Technology and Economics of Smart Grids and Sustainable Energy*, vol. 1, p. 10, 2016.

[12] K. Negash, B. Khan, D. Tesfaye, and M. Gayathri, "Optimal placement of phasor measurement unit for system observability," *Proceedings of IEEE International Conference on Computing Methodologies and Communication, (ICCMC 2017)*, Erode, India, pp. 73–77, 18–19 July 2017.
[13] Y. Kifle, B. Khan, and P. Singh, "Assessment and enhancement of distribution system reliability by renewable energy sources and energy storage," *Journal of Green Engineering*, vol. 8, no. 3 (2), pp. 219–262, July 2018.
[14] D. Anteneh and B. Khan, "Reliability enhancement of distribution substation by using network reconfiguration a case study at debre berhan distribution substation," *International Journal of Economy Energy and Environment*, vol. 4, no. 2, pp. 33–40, April 2019.
[15] B. Khan, H.H. Alhelou, and F. Mebrahtu, "A holistic analysis of distribution system reliability assessment methods with conventional and renewable energy sources," *AIMS Energy*, vol. 7, no. 4, pp. 413–429, 2019.
[16] H. Li, S. Gong, L. Lai, Z. Han, R.C. Qiu and D. Yang, "Efficient and secure wireless communications for advanced metering infrastructure in smart grids," *IEEE Transactions on Smart Grid*, vol. 3, no. 3, pp. 1540–1551, 2012.
[17] C. Selvam, K. Srinivas, G.S. Ayyappan, and M. Venkatachala Sarma, "Advanced metering infrastructure for smart grid applications," *2012 International Conference on Recent Trends in Information Technology*, Chennai, Tamil Nadu, pp. 145–150, 2012.
[18] M.P. McHenry, "Technical and governance considerations for advanced metering infrastructure/smart meters: Technology, security, uncertainty, costs, benefits, and risks," *Energy Policy*, vol. 59, pp. 834–842, 2013.
[19] M.A. Faisal, Z. Aung, J.R. Williams and A. Sanchez, "Data-stream-based intrusion detection system for advanced metering infrastructure in smart grid: A feasibility study," *IEEE Systems Journal*, vol. 9, no. 1, pp. 31–44, 2014.
[20] R.R. Mohassel, A.S. Fung, F. Mohammadi, and K. Raahemifar, "A survey on advanced metering infrastructure and its application in Smart Grids," *2014 IEEE 27th Canadian Conference on Electrical and Computer Engineering (CCECE)*, Toronto, ON, Canada, pp. 1–8, 2014.
[21] Y. Yan, Y. Qian, and H. Sharif, "A secure and reliable in-network collaborative communication scheme for advanced metering infrastructure in Smart Grids," *2011 IEEE Wireless Communications and Networking Conference*, Cancun, Quintana Roo, pp. 909–914, 2011.
[22] H. Sui, Y. Sun, and W. Lee, "A demand side management model based on advanced metering infrastructure," *2011 4th International Conference on Electric Utility Deregulation and Restructuring and Power Technologies (DRPT)*, Weihai, Shandong, pp. 1586–1589, 2011.
[23] A. Mohammadali, M. Sayad Haghighi, M.H. Tadayon, and A. Mohammadi-Nodooshan, "A novel identity-based key establishment method for advanced metering infrastructure in Smart Grids," *IEEE Transactions on Smart Grid*, vol. 9, no. 4, pp. 2834–2842, July 2018.
[24] Z. Wan, G. Wang, Y. Yang, and S. Shi, "SKM: Scalable key management for advanced metering infrastructure in Smart Grids," *IEEE Transactions on Industrial Electronics*, vol. 61, no. 12, pp. 7055–7066, December 2014.

[25] P. Deng and L. Yang, "A secure and privacy-preserving communication scheme for Advanced Metering Infrastructure," *2012 IEEE PES Innovative Smart Grid Technologies (ISGT)*, Washington, DC, USA, pp. 1–5, 2012.

[26] A. Ikpehai, B. Adebisi, and K.M. Rabie, "Broadband PLC for clustered advanced metering infrastructure (AMI) architecture," *Energies*, vol. 9, no. 7, p. 569, 2016.

[27] J. Matanza, S. Alexandres, and C. Rodríguez-Morcillo, "Advanced metering infrastructure performance using European low-voltage power line communication networks," *IET Communications*, vol. 8, no. 7, pp. 1041–1047, 2014.

[28] G.R. Barai, S. Krishnan, and B. Venkatesh, "Smart metering and functionalities of smart meters in smart grid – A review," *2015 IEEE Electrical Power and Energy Conference (EPEC)*, London, ON, Canada, pp. 138–145, 2015.

[29] J.-H. Huh and K. Seo, "Hybrid advanced metering infrastructure design for micro grid using the game theory model," *International Journal of Software Engineering and Its Applications*, vol. 9, no. 9, pp. 257–268, 2015.

[30] M. Alizadeh, A. Scaglione, and Z. Wang, "On the impact of SmartGrid metering infrastructure on load forecasting," *2010 48th Annual Allerton Conference on Communication, Control, and Computing (Allerton)*, Allerton, IL, USA, pp. 1628–1636, 2010.

[31] R. Shein, "Security measures for advanced metering infrastructure components," *2010 Asia-Pacific Power and Energy Engineering Conference*, Chengdu, pp. 1–3, 2010.

[32] F. Ye, Y. Qian, and R.Q. Hu, "A security protocol for advanced metering infrastructure in Smart Grids," *2014 IEEE Global Communications Conference*, Austin, TX, USA, pp. 649–654, 2014.

[33] R.R. Mohassel, A.S. Fung, F. Mohammadi, and K. Raahemifar, "Application of advanced metering infrastructure in Smart Grids," *22nd Mediterranean Conference on Control and Automation* (2014), Palermo, Italy, pp. 822–828, 16–19 June 2014.

[34] Trong Nghia Le, Wen-Long Chin, Dang Khoa Truong, and Tran Hiep Nguyen, "Advanced metering infrastructure based on smart meters in smart grid," In: Smart Metering Technology and Services – Inspirations for Energy Utilities, Moustafa Eissa (ed.), IntechOpen, London, UK, 2016. doi: 10.5772/63631.

[35] Ž. Popović and V. Čačković, "Advanced metering infrastructure in the context of Smart Grids," *2014 IEEE International Energy Conference (ENERGYCON)*, Cavtat, pp. 1509–1514, 2014.

[36] E. Bou-Harb, C. Fachkha, M. Pourzandi, M. Debbabi, and C. Assi, "Communication security for smart grid distribution networks," *IEEE Communications Magazine*, vol. 51, no. 1, pp. 42–49, January 2013.

[37] P. Yi, T. Zhu, Q. Zhang, Y. Wu, and J. Li, "A denial of service attack in advanced metering infrastructure network," *2014 IEEE International Conference on Communications (ICC)*, Sydney, NSW, Australia, pp. 1029–1034, 2014.

[38] W. Luan, D. Sharp, and S. Lancashire, "Smart grid communication network capacity planning for power utilities," *IEEE PES T&D 2010*, New Orleans, LA, USA, pp. 1–4, 2010.

Index